Conversion table (metric to English units)

	To convert metric/SI unit	Multiply by	To obtain
Area	cm²	0.155	in²
	m²	10.76	ft²
	km²	0.3861	mi²
	hectare (ha) = 10^4 m²	2.471	acre
Energy (work or quantity of heat)	joule (J) = N · m	0.0009482	Btu (60°F)
	J	0.2389	cal (15°C)
	W/m²	0.001433	langley/min
Flowrate	m³/s = 10^3 l/s	35.31	cfs
	m³/s = 10^3 l/s	22.82	mgd
	l/s = 10^{-3} m³/s	15.85	gal/min
Force	newton (N)	0.2248	lb
Length	mm	0.03937	in
	m	3.281	ft
	km	0.6214	mi
	km	0.5400	nmi
Power	W = J/s = N · m/s	0.7376	ft · lb/s
	W	0.001341	hp
Pressure	N/m² = Pa	0.0001450	lb/in²
	N/m²	0.02089	lb/ft²
	kPa = 10 mb	0.2953	in Hg (32°F)
Specific heat	J/kg · K	5.981	ft · lb/slug · °R
Specific weight	N/m³	0.006364	lb/ft³
Velocity	m/s	3.281	ft/s
	km/hr	0.6214	mi/hr
Viscosity, dynamic	N · s/m² = Pa · s	0.02089	lb · s/ft²
Viscosity, kinematic	m²/s = 10^4 stokes	10.76	ft²/s
Volume	m³	35.31	ft³
	l = 10^{-3} m³	0.2642	U.S. gal
Weight (see Force)			

Important equivalents

1 cfs-day/mi² = 0.03719 in of runoff

1 in of runoff/mi² = 26.89 cfs-day = 2,323,200 ft³ = 53.33 acre · ft

1 acre = 43,560 ft² = 4047 m² = 0.4047 ha

1 mi² = 640 acres = 2.59 km²

1 mm Hg (0°C) = 1.333 mb = 0.03937 in Hg = 133.3 Pa

1 Btu (60°F) = 0.99529 Btu (39°F) = 777.9 ft · lb = 252.0 cal (15°C)

N = 100,000 dynes

mgd = 1.547 cfs

gal/min = 0.002228 cfs

U.S. gal = 0.1337 ft³

hp = 550 ft · lb/s

slug = lb · s²/ft

HYDROLOGY FOR ENGINEERS

McGraw-Hill Series in Water Resources and Environmental Engineering

Ven Te Chow, Rolf Eliassen, Paul H. King, and Ray K. Linsley
Consulting Editors

Bailey and Ollis: *Biochemical Engineering Fundamentals*
Biswas: *Models for Water Quality Management*
Bockrath: *Environmental Law for Engineers, Scientists, and Managers*
Bouwer: *Groundwater Hydrology*
Canter: *Environmental Impact Assessment*
Chanlett: *Environmental Protection*
Gaudy and Gaudy: *Microbiology for Environmental Scientists and Engineers*
Graf: *Hydraulics of Sediment Transport*
Haimes: *Hierarchical Analysis of Water Resources Systems: Modelling and Optimization of Large-Scale Systems*
Hall and Dracup: *Water Resources Systems Engineering*
James and Lee: *Economics of Water Resources Planning*
Linsley and Franzini: *Water Resources Engineering*
Linsley, Kohler, and Paulhus: *Hydrology for Engineers*
Metcalf & Eddy, Inc.: *Wastewater Engineering: Collection and Pumping of Wastewater*
Metcalf & Eddy, Inc.: *Wastewater Engineering: Treatment, Disposal, Reuse*
Nemerow: *Scientific Stream Pollution Analysis*
Rich: *Environmental Systems Engineering*
Rich: *Low-Maintenance, Mechanically-Simple Wastewater Treatment Systems*
Sawyer and McCarty: *Chemistry for Environmental Engineering*
Schroeder: *Water and Wastewater Treatment*
Steel and McGhee: *Water Supply and Sewerage*
Tchobanoglous, Theisen, and Eliassen: *Solid Wastes: Engineering Principles and Management Issues*
Walton: *Groundwater Resources Evaluation*

HYDROLOGY FOR ENGINEERS

Third Edition

Ray K. Linsley, Jr.

Professor Emeritus of Hydraulic Engineering
Stanford University
President, Linsley Kraeger, Assoc. Ltd.

Max A. Kohler

Consulting Hydrologist
Formerly Associate Director, Hydrology
U.S. National Weather Service

Joseph L. H. Paulhus

Consulting Hydrometeorologist
Formerly Chief, Water Management Information Division
U.S. National Weather Service

McGraw-Hill Book Company

New York   St. Louis   San Francisco   Auckland   Bogotá   Hamburg
Johannesburg   London   Madrid   Mexico   Montreal   New Delhi
Panama   Paris   São Paulo   Singapore   Sydney   Tokyo   Toronto

This book was set in Times Roman by Bi-Comp, Incorporated.
The editors were Julienne V. Brown and Susan Hazlett;
the production supervisor was Rosann E. Raspini.
The drawings were done by VIP Graphics.
R. R. Donnelley & Sons Company was printer and binder.

HYDROLOGY FOR ENGINEERS

1234567890 DODO 8987654321

Library of Congress Cataloging in Publication Data

Linsley, Ray K.
    Hydrology for engineers.

    (McGraw-Hill series in water resources and
environmental engineering)
    Includes bibliographies and indexes.
    1. Hydrology.   I. Kohler, Max Adam, date
II. Paulhus, Joseph L. H.   III. Title.   IV. Series.
GB661.2.L56 1982   551.48   81-3689
ISBN 0-07-037956-4   AACR2

CONTENTS

Chapter 3   Precipitation   47

PREFACE

The first edition of "Hydrology for Engineers" was published in 1958 and the second in 1975. The book has been increasingly used as a text in both senior and graduate courses. There have been many important developments in the science of hydrology which are reflected in improved techniques of today. This third edition represents a further extensive revision. Chapters on hydraulic routing, water quality, and applications have been added, and many changes have been made throughout the balance of the book. The importance of the digital computer as a tool for hydrologic analysis continues to be stressed, but recognizing that not everyone has access to a computer, we still discuss the more traditional methods. The basic processes of hydrology are also stressed in the belief that an understanding of these processes is essential to the application of all modern tools of the science.

Since most nations of the world now use metric units of measurement while the United States has only begun a conversion from the English system, both systems of units are given in the text, tables, and figures. The metric units (SI in most cases) are usually given first with English equivalents in parentheses. It is hoped that this arrangement will serve students of those countries where the metric system is in use and at the same time ease the transition for those of countries where the change is yet to occur. Problems are included in both systems of units.

The student should find hydrology an interesting subject but one very different from most engineering courses. The natural phenomena with which hydrology is concerned do not lend themselves to such rigorous analyses as are possible in engineering mechanics. There is therefore a greater variety of methods, more latitude for judgment, and a seeming lack of accuracy in problem solutions. Actually, the accuracy of sound hydrologic solutions compares favorably with other types of engineering computations. Uncertainties in engineering are frequently hidden by use of factors of safety, rigidly standardized working procedures, and conservative assumptions regarding properties of materials.

The authors gratefully acknowledge the helpful suggestions, data, and other assistance from their colleagues in the National Weather Service, Stanford University, Hydrocomp, Inc., and elsewhere. Special mention should be made of the major contributions of Delbert Franz, Stephen Burges, and John Imhoff to Chapters 10, 14, and 15, respectively.

Ray K. Linsley, Jr.
Max A. Kohler
Joseph L. H. Paulhus

LIST OF SYMBOLS AND ABBREVIATIONS

Symbols

A = area

$\mathbf{A}$ = boldface symbol indicates a matrix

$\mathbf{A}^T$ = transpose of $\mathbf{A}$

a = coefficient

B = width; coefficient

b = coefficient

C = Chézy coefficient; coefficient

C_p = synthetic-unit-hydrograph coefficient of peak

C_t = synthetic-unit-hydrograph coefficient of lag

c = coefficient; concentration; wave celerity

D = depth; drainage density; molecular diffusion rate

d = diameter; coefficient

E = evaporation; sediment washed from impervious areas

E_a = reference evaporation rate

E_p = pan evaporation

E_T = evapotranspiration

$E[\ \ ]$ = expected value of [ ]

e = vapor pressure; base of napierian logarithms

e_s = saturation vapor pressure

F = fall; force; infiltration volume; flux

f = relative humidity

$f(\ \ )$ = function of

f_c = final infiltration capacity

f_i = infiltration rate

f_0 = initial infiltration capacity

f_p = infiltration capacity

G = safe yield of a groundwater basin; gully-erosion rate; coefficient of skew; heat flux

G_i = bed-load transport

g = gage height; acceleration of gravity

H_v = latent heat of vaporization

h = height; head; Hurst coefficient

I = inflow; antecedent-precipitation index

I_r = retention index

I_s = season index

i = rainfall intensity; index number

i_e = rainfall excess (effective rainfall)

i_s = supply rate (rainfall less retention)

J = probability in N years; first moment of area about water surface

j = probability; number

K = Muskingum storage constant; frequency factor; compaction coefficient; permeability coefficient (hydraulic conductivity); conveyance

K_r = recession constant

k = coefficient; number; exponent; intrinsic permeability

L = length

L_c = distance from outlet to center of basin

M = snowmelt rate; heat storage rate in layer above surface

m = coefficient or exponent; molecular weight; rank of event

N = normal precipitation; number

n = Manning roughness coefficient; coefficient or exponent; number

O = outflow

O_g = subsurface seepage

P = precipitation; pollutant; probability of event in any year

P_e = precipitation in excess of interception and infiltration

P_r = radar returned power

p = pressure; porosity; computed probability

pH = $-\log_{10}$ of hydrogen-ion concentration

Q = volume or rate of discharge or runoff

Q_a = net longwave radiation

Q_{ar} = reflected longwave radiation

Q_e = energy used for evaporation

Q_g = groundwater-flow volume

Q_h = sensible-heat transfer

Q_{ir} = incident minus reflected radiation
Q_n = net radiant energy
Q_0 = emitted longwave radiation
Q_r = reflected shortwave radiation
Q_S = volume of surface streamflow
Q_s = shortwave radiation; suspended sediment load
Q_v = advected energy
Q_θ = change in energy storage
q = discharge rate; volatilization rate; light intensity
q_b = base-flow discharge
q_d = direct-runoff discharge
q_e = equilibrium flowrate
q_h = specific humidity
q_o = overland-flow rate
q_p = peak discharge
R = hydraulic radius; Bowen's ratio; soil splash; rainfall rate
R_g = gas constant
R_n = range of cumulative departures
R_s = sediment residue on ground surface
r = radius; range; adjusted correlation coefficient; ratio
S = storage; volume of surface retention; transport of sediment; standard error
S_c = storage constant of an aquifer
S_d = depression-storage capacity
S_g = groundwater storage
S_i = interception storage
S_s = surface storage
s = slope
T = temperature; transmissibility; time base of unit hydrograph
T_a = air temperature
T_d = dewpoint temperature
T_L = lag time
T_r = return period or recurrence interval
T_w = wet-bulb temperature
t = time; random variable; routing period
t_e = time to equilibrium
t_p = basin lag
t_R = duration of rain
t_r = unit duration for synthetic unit hydrograph
u = wave velocity; factor in well hydraulics
V = volume of surface detention
V_i = volume of interception storage

V_s = volume of depression storage

v = velocity

v_s = settling velocity

v_* = friction velocity

W = infiltration index; total sediment wash load

W_p = precipitable water

$W(u)$ = well function of u

w = specific weight

w_r = mixing ratio

X = variable; supply rate (precipitation minus interception)

$\bar{X}$ = mean of X

X_m = mode of X

x = distance; constant or exponent; weighting factor

Y = variable; aquifer thickness

$\bar{Y}$ = mean of Y

y = vertical distance; reduced variate in frequency analysis

Z = drawdown in a well; drop-size function; a variable

z = vertical distance

α = ratio; evaporation portion of advected energy; solubility; function of recession factor

β = constant

Γ = gamma function

γ = Bowen's ratio coefficient

Δ = slope of vapor pressure-temperature curve; an increment

ϵ = mixing coefficient; emissivity; random variable; error term

η = extinction coefficient

θ = angle; reaction constant; weighting factor

Λ = total potential

λ = saturation vapor concentration

μ = absolute viscosity; mean

μ_2 = variance

μ_3 = third moment about mean

ν = kinematic viscosity

π = 3.1416 . . .

ρ = density; correlation coefficient

Σ = summation

σ = standard deviation; Stefan-Boltzmann constant

τ = shear; time in past

Υ = du Boy's coefficient

Φ = infiltration index; bed-load function

ϕ = number or ratio; weighting factor

ψ = capillary potential; bed-load function

Abbreviations

Å	=	angstrom (10^{-10} m)
acre · ft	=	acre-foot
AR(p)	=	autoregressive process of order p
ARMA(p,q)	=	autoregressive, moving average process of orders p and q
atm	=	atmosphere
Btu	=	British thermal unit
°C	=	degrees Celsius
cal	=	calorie
cfs	=	cubic feet per second
cm	=	centimeter (10^{-2} m)
csm	=	cubic feet per second-square mile
CV	=	coefficient of variation
D	=	darcy
deg	=	degree
exp	=	exponential
°F	=	degrees Fahrenheit
FGN	=	fractional gaussian noise
FFGN	=	fast fractional gaussian noise
ft/s	=	feet per second
ft	=	foot
g	=	gram
gal	=	gallon
gal/min	=	gallons per minute
hr	=	hour
Hg	=	mercury
IUH	=	instantaneous unit hydrograph
hp	=	horsepower
in	=	inch
J	=	joule
K	=	kelvins
kn	=	knot
l	=	liter
lat	=	latitude
lb	=	pound
lb/ft²	=	pounds per square foot
lb/in²	=	pounds per square inch
ln	=	logarithm to the base e
log	=	logarithm to the base 10
LN	=	log normal
LZS	=	lower-zone storage

LZSN = nominal capacity of lower zone
Ly = langley
m = meter
MA(q) = moving average process of order q
mi = mile
mb = millibar
min = minute
mm = millimeter (10^{-3} m)
mgd = million gallons per day
mi/hr = miles per hour
mps = meters per second
msl = mean sea level
N = newton
nmi = nautical mile
Pa = pascal
ppm = parts per million
°R = degrees Rankine
SI = International System of Units
s = second
UZS = upper-zone storage
UZSN = nominal capacity of upper zone
W = watt

SI Prefixes

Factor	Prefix	Symbol
10^9	giga	G
10^6	mega	M
10^3	kilo	k
10^2	hecto	h
10^{-2}	centi	c
10^{-3}	milli	m
10^{-6}	micro	μ

INTRODUCTION

"*Hydrology* treats of the waters of the Earth, their occurrence, circulation, and distribution, their chemical and physical properties, and their reaction with their environment, including their relation to living things. The domain of hydrology embraces the full life history of water on the Earth" [1].† Engineering hydrology includes those segments of the field pertinent to planning, design, and operation of engineering projects for the control and use of water. The boundaries between hydrology and other earth sciences such as meteorology, oceanography, and geology are indistinct, and no good purpose is served by attempting to define them rigidly. Likewise, the distinctions between engineering hydrology and other branches of applied hydrology are vague. Indeed, engineers owe much of their present knowledge of hydrology to agriculturists, foresters, meteorologists, geologists, and others in a variety of fields.

1-1 The Hydrologic Cycle

The concept of the *hydrologic cycle* is a useful, if academic, point from which to begin the study of hydrology. This cycle (Fig. 1-1) is visualized as beginning with the evaporation of water from the oceans. The resulting vapor is transported by moving air masses. Under the proper conditions, the vapor is condensed to form clouds, which in turn may result in precipitation. The precipitation which falls upon land is dispersed in several ways. The greater part is temporarily retained in the soil near where it falls and is ultimately returned to

† Numbered references will be found at the end of each chapter.

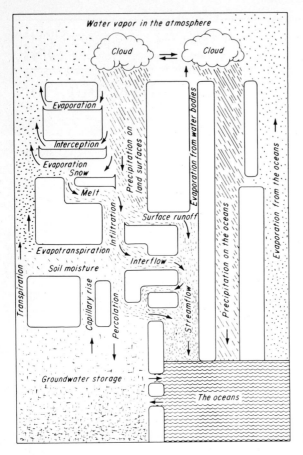

Figure 1-1 The hydrologic cycle.

the atmosphere by evaporation and transpiration by plants. A portion of the water finds its way over and through the surface soil to stream channels, while other water penetrates farther into the ground to become part of the groundwater. Under the influence of gravity, both surface streamflow and groundwater move toward lower elevations and may eventually discharge into the ocean. However, substantial quantities of surface and underground water are returned to the atmosphere by evaporation and transpiration before reaching the oceans.

This description of the hydrologic cycle and the schematic diagram of Fig. 1-1 are enormously oversimplified. For example, water from some surface streams may percolate to the groundwater, while in other cases groundwater is a source of streamflow. Some precipitation may remain on the ground as snow for months before melting releases the water to streams or groundwater. The hydrologic cycle is a convenient means for delineating the scope of hydrology as that portion between precipitation on the land and the return of this water to the atmosphere or ocean. The cycle also emphasizes the four phases of interest to the hydrologist: precipitation, evaporation and transpiration, surface streamflow, and groundwater. These topics are the subject of more detailed discussion later in this text.

The discussion of the hydrologic cycle should not give an impression of a continuous mechanism through which water moves steadily at a constant rate. The movement of water through the cycle is erratic, both in time and over area. On occasion, nature provides torrential rains which tax surface-channel capacities to the utmost. At other times it seems that the machinery of the cycle has stopped completely and, with it, precipitation and streamflow. In adjacent areas the variations in the cycle may be quite different. It is precisely these extremes of flood and drought that are of most interest to the engineering hydrologist, for hydraulic engineering projects are designed to protect against the ill effect of extremes. The reasons for these climatic extremes are found in the science of meteorology and should be understood, in broad detail at least, by the hydrologist (Chaps. 2 and 3).

Hydrologists are interested in more than obtaining a qualitative understanding of the hydrologic cycle and measuring the quantities of water in transit in this cycle. They must be able to deal quantitatively with the interrelations between factors so that they can predict the influence of human activities on these relationships. They must concern themselves with the frequency with which extremes of the cycle may occur, for this is the basis of economic analysis, an important determinant for all hydraulic projects. The final chapters of this text deal with these quantitative problems.

1-2 History

We do not know when the first hydraulic project was constructed. There is evidence that irrigation canals existed in Egypt during the reign of King Scorpion about 3200 B.C. King Menes (ca. 3000 B.C.) is said to have constructed a diversion dam to redirect the flow of the Nile so that Memphis, his capital, could be constructed in the fertile riverbed. About 2850 B.C., Sadd el-Kafara Dam was built across a wadi a few miles south of Cairo. There was apparently no spillway, and the evidence suggests that the dam failed during the first rainy season after its construction. Biswas [2] summarizes the archeological evidence for hydraulic structures in the Middle East, such as extensive irrigation, a water code by Hammurabi (Babylon, ca. 1700 B.C.), construction of kanats in Persia, and many other water-control activities. It is not known when quantitative estimates of flow were first made. Neither is it known with certainty who might have first perceived the hydrologic cycle as we know it today.

A brief reference in an essay, the Critias, by Plato (427–347 B.C.) seems to advocate the pluvial origin of rivers and springs, but Plato is better known for the idea of the great underground sea, Tartarus, as the source of all surface waters. Aristotle (ca. 350 B.C.) suggested that cold converted air into water, and since the mountains were cold, they converted much air into water as the source of springs and streams. Theophrastus (ca. 320 B.C.) wrote a number of treatises on meteorological subjects, but only a fragmentary abstract remains.

However, the writings of Marcus Vitruvius Pollio (ca. 100 B.C.) attribute to Theophrastus views much like our modern concepts. In any case, however, there is little evidence of any meaningful application of hydrology until the Renaissance (1450–1600). Gages to measure the stage of the Nile (*nilometers*) were installed as early as 3000 B.C., and an extensive network of rain gages appears to have existed in India by 300 B.C. In both cases the target was the same—to provide a base for tax collection. The variations in climate were known from experience, but the underlying processes were not understood.

It remained for Pierre Perrault [3] (1608–1680) to compare measured rainfall with the estimated flow of the Seine River to show that the streamflow was about one-sixth of the precipitation. The English astronomer Halley [4] (1656–1742) measured evaporation from a small pan and estimated evaporation from the Mediterranean Sea from these data.

Satisfactory methods for measuring streamflow are a recent development. Frontinus, water commissioner of Rome in A.D. 97, based estimates of flow on cross-sectional area alone without regard to velocity. In the United States, organized measurement of precipitation started under the Surgeon General of the Army in 1819, was transferred to the Signal Corps in 1870, and finally, in 1891, to a newly organized U.S. Weather Bureau, renamed the National Weather Service in 1970. Scattered streamflow measurements were made on the Mississippi River as early as 1848, but a systematic program was not started until 1888, when the U.S. Geological Survey undertook this work. Generally, these activities began slightly earlier in Europe, but for most of the world systematic observations of precipitation and streamflow have been available for relatively short periods of time. It is not surprising, therefore, that little quantitative work in hydrology was done before the early years of the twentieth century, when men such as Horton, Mead, and Sherman began to explore the field. The great expansion of activity in flood control, irrigation, soil conservation, and related fields which began about 1930 gave the first real impetus to organized research in hydrology, as need for more precise design data became evident. Most of today's concepts of hydrology date from 1930.

1-3 Hydrology in Engineering

Hydrology is used in engineering mainly in connection with the design and operation of hydraulic structures. What flood flows can be expected over a spillway, at a highway culvert, or in an urban storm drainage system? What reservoir capacity is required to assure adequate water for irrigation or municipal water supply during droughts? What effect will reservoirs, levees, and other control works exert on flood flows in a stream? What are reasonable boundaries for the floodplain? These are typical questions that the hydrologist is expected to answer. The need to answer these questions has been the principal incentive for the development of the techniques of quantitative hydrology discussed in this text.

Large organizations such as federal and state water agencies can maintain staffs of hydrologic specialists to analyze their problems, but smaller offices often have insufficient hydrologic work for full-time specialists. Hence, many civil engineers are called upon for occasional hydrologic studies. It is probable that these civil engineers deal with a larger number of projects and greater annual dollar volume than the specialists do. In any event, it seems that knowledge of the fundamentals of hydrology is an essential part of the civil engineer's training.

1-4 Subject Matter of Hydrology

Hydrology deals with many topics. The subject matter as presented in this book can be broadly classified into two phases: data collection and methods of analysis. Chapters 2 to 6 deal with the basic data of hydrology. Adequate basic data are essential to any science, and hydrology is no exception. In fact, the complexity of hydrologic phenomena makes it difficult to apply rigorous deductive reasoning. It is necessary to interpret observed data and from this analysis to establish the systematic pattern that governs these events. Without adequate historical data for the particular problem area, the hydrologist is in a difficult position. Most countries have one or more agencies with responsibility for data collection [5]. It is important to know how these data are collected and published, the limitations on their accuracy, and the proper methods of interpretation and adjustment.

Typical hydrologic problems involve estimates of extremes rarely observed in a small data sample, hydrologic characteristics at locations where no data have been collected (such locations are much more numerous than sites with data), or estimates of the effects of human actions on the hydrologic characteristics of an area. Generally, each hydrologic problem is unique in that it deals with a distinct set of physical conditions within a specific river basin. Hence, quantitative conclusions of one analysis are often not directly transferable to another problem. However, the general solution for most problems can be developed from application of a few relatively basic concepts. Chapters 6 to 16 describe these concepts and explain how they are utilized to solve hydrologic problems.

PROBLEMS

1-1 List the agencies in your state which have responsibilities of a hydrologic nature. What is the special responsibility of each agency?

1-2 Repeat Prob. 1-1 for federal agencies.

1-3 List the major hydraulic projects in your area. What specific hydrologic problems did each project involve?

REFERENCES

1. Federal Council for Science and Technology, "Scientific Hydrology," Washington, June 1962.
2. Biswas, A. K.: "History of Hydrology," 2d ed., North-Holland, Amsterdam, 1972.

3. Perrault, P.: "De l'Origine des fontaines," Paris, 1674, trans. by A. LaRocque, Hafner, New York, 1967.
4. Halley, E.: An Account of the Evaporation of Water, *Phil. Trans. R. Soc. London,* vol. 18, pp. 183–190, 1694.
5. World Meteorological Organization, Organization of Hydrometeorological and Hydrological Services, *Rep. WMO/IHD Proj.* 10, Geneva, 1969.

BIBLIOGRAPHY

Bruce, J. P., and R. H. Clark: "Introduction to Hydrometeorology," Pergamon, New York, 1966, revised 1980.
Chow, V. T. (ed.): "Handbook of Hydrology," McGraw-Hill, New York, 1964.
Eagleson, P.: "Dynamic Hydrology," McGraw-Hill, New York, 1970.
Kalinin, G. P.: "Problemi Globolnoy Gidrologii" ("Problems of Global Hydrology"), in Russian, Gidrometeoizdat, Leningrad, 1963. English translation available from U.S. Dept of Commerce National Technical Information Service, Springfield, Va.
Kazmann, R. G.: "Modern Hydrology," 2d ed., Harper & Row, New York, 1972.
Kirkby, M. J. (ed.): "Hillslope Hydrology," Wiley, New York, 1978.
Linsley, R. K., and J. B. Franzini: "Water-Resources Engineering," 3d ed., McGraw-Hill, New York, 1978.
——, M. A. Kohler, and J. L. H. Paulhus: "Applied Hydrology," McGraw-Hill, New York, 1949.
Mead, D. W.: "Hydrology," 1st ed., McGraw-Hill, New York, 1919; 2d ed., rev. by H. W. Mead, 1950.
Meinzer, O. E. (ed.): "Hydrology," vol. 9, Physics of the Earth Series, McGraw-Hill, New York, 1942; reprinted by Dover, New York, 1949.
Meyer, A. F.: "Elements of Hydrology," 2d ed., Wiley, New York, 1928.
Roche, M.: "Hydrologie de surface," Gauthier-Villars, Paris, 1963.
Rodda, J. C. (ed.): "Facets of Hydrology," Wiley, New York, 1976.
UNESCO, World Meteorological Organization, and International Association of Hydrological Sciences: "Three Centuries of Scientific Hydrology," Tercentary Celebration, Paris, September 1974.
Viessman, W., Jr., J. W. Knapp, G. L. Lewis, and T. E. Harbaugh: "Introduction to Hydrology," 2d ed., Dun-Donnelley, New York, 1977.
Ward, R. C.: "Principles of Hydrology," McGraw-Hill, London, 1967.
Wisler, C. O., and E. F. Brater: "Hydrology," 2d ed., Wiley, New York, 1959.

WEATHER AND HYDROLOGY

The hydrologic characteristics of a region are determined largely by its geology and geography, climate playing a dominant part. Among climatic factors that establish the hydrologic features of a region are the amount and distribution of precipitation; the occurrence of snow and ice; and the effects of wind, temperature, and humidity on evapotranspiration and snowmelt. Hydrologic problems in which meteorology plays an important role include determination of probable maximum precipitation and critical snowmelt conditions for spillway design, forecasts of precipitation and snowmelt for reservoir operation, and determination of probable maximum winds over water surfaces for evaluating resulting waves in connection with the design of dams and levees. Obviously, the hydrologist should have some understanding of the meteorological processes determining a regional climate. The general features of climatology are discussed in this chapter.

SOLAR AND EARTH RADIATION

2-1 Solar and Earth Radiation

Solar radiation, the earth's chief source of energy, determines weather and climate. Both earth and sun radiate essentially as *blackbodies*; i.e., they emit for every wavelength almost the theoretical maximum amount of radiation for their temperatures.

Radiation wavelengths are usually given in micrometers (μm)† (10^{-6} m) or in angstroms (Å) (10^{-10} m). Maximum energy of solar radiation is shortwave in the visible range of 0.4 to 0.8 μm. Earth radiation is about 10 μm (longwave).

---

† Formerly called the micron (μ).

The rate at which solar radiation reaches the upper limits of earth's atmosphere on a surface normal to the incident radiation and at earth's mean distance from the sun is called the *solar constant*. Measurements have ranged from 1.89 to 2.05 Ly/min,[†] with most of the uncertainty resulting from corrections for atmospheric effects rather than from fluctuations in solar activity, which are considered relatively small. High-altitude observations with airborne instruments, which minimize atmospheric effects, indicate a narrow range of values centering about 1.97 Ly/min, and that value is often used as the solar constant.[‡]

2-2 Solar Radiation at Earth's Surface

A large part of the solar radiation reaching the outer limits of the atmosphere is scattered and absorbed in the atmosphere or reflected from clouds and the earth's surface. Scattering of radiation by air molecules is most effective for the shortest wavelengths. With sun overhead and clear sky, over half the radiation in the blue range (short wavelengths about 0.45 μm) is scattered, thus accounting for the blue sky. Very little radiation in the red range (about 0.65 μm) is scattered. Estimates of radiation scattered to space average about 8 percent of the incident solar radiation (*insolation*).

Clouds reflect much incident solar radiation to space. The amount reflected depends on the amount of clouds and their albedo.[§] The albedo (and absorption) of clouds varies greatly with thickness and liquid-water content and inversely with solar elevation. A high, thin overcast may reflect less than 20 percent of incident radiation, while a 2000-ft (600-m) layer of stratus or stratocumulus clouds may reflect over 80 percent.

About half the incident radiation at the outer limits of the atmosphere eventually reaches earth's surface. Much of it is absorbed, but some is reflected to the atmosphere and to space. The albedo of earth's surface varies widely, depending on solar altitude and type of surface. It is less for moist soil surfaces than for dry and tends to be less for high solar altitudes than for low. The albedo (in percent) ranges from 10 to 20 for green forests; 15 to 30 for grass-covered plains; 15 to 20 for marshy lands; 15 to 25 for crop-covered, cultivated fields; 10 to 25 for dark, bare soils when dry, and 5 to 20 when moist; 20 to 45 for dry,

---

[†] The abbreviation Ly is for langley; 1 Ly = 1 cal/cm^2.

[‡] The units of solar radiation in the International System of Units (SI) are the watt per square meter (W/m^2) and the kilojoule per square meter (kJ/m^2): 1 cal (15°C)/cm^2 · min = 697.6 W/m^2, and 1 cal (15°C)/cm^2 = 41.86 kJ/m^2. The solar constant equivalents of 1.97 cal/cm^2 · min in SI units are 1374 W/m^2 and 82.46 kJ/m^2 · min.

[§] *Reflectivity* is the ratio of the amount of electromagnetic radiation (within any specified range of wavelength) reflected by a body to the amount incident upon it, commonly expressed as a percentage. *Albedo* is the ratio of the amount of solar radiation (or sometimes visible radiation) reflected by a surface to the amount incident upon it, also expressed as a percentage. For example, the reflectivity of fresh snow for infrared wavelengths (terrestrial radiation) is nearly zero, but the albedo is about 85 percent. The albedo is the reflectivity for the solar or visible range of radiation.

light, sandy soils; 40 to 50 for old, dirty snow; and 80 to 95 for pure, white snow, the highest albedo being for fresh, clean, dry snow and low solar altitude.

The albedo of ocean surfaces depends on surface roughness as well as on solar altitude. The albedo (in percent) of a calm sea is 2 to 3 with solar altitude between 90 and 50°, increases to 12 with sun at 20°, and is about 40 with sun at 5°. The albedo of a rough sea is greater than that of a calm sea for solar altitudes greater than about 45° and is less for lower altitudes. Estimates of average albedo for all ocean surfaces range from 6 to 8 percent. For lakes and rivers reflectivities are approximately 5 percent for solar and 3 percent for atmospheric longwave radiation. The mean weighted albedo of the earth's surface has been estimated at about 14 percent. The albedo for the planet as a whole, i.e., including atmosphere, clouds, etc., is estimated to be about 30 percent.

The above discussion generally treats radiation scatter, reflection, and absorption in the mean. Very little of earth's surface is normal to incident solar radiation; and the greater the angle of the surface from normal, the less the radiation intensity. Thus, less solar radiation reaches high latitudes than reaches low latitudes. Differences in insolation are one of the primary factors in determining the general circulation of the earth's atmosphere.

2-3 Heat Balance of Earth's Surface and Atmosphere

Since the surface area of a sphere is 4 times that of a great circle, solar radiation intercepted by the planet earth averages, for the globe as a whole, one-fourth the solar constant, or about 0.5 Ly/min. This amount is arbitrarily set at 100 units in Fig. 2-1, which presents estimates of the various components of earth's heat balance. On the scale of Fig. 2-1, 26 units of the incoming solar radiation at the top of the atmosphere are reflected back to space by the atmosphere (including clouds, dust, water vapor, etc.) and 4 units by the earth's surface. Of the remaining 70 units, 19 are absorbed by the atmosphere, and 51 by the earth's surface.

The earth's surface radiates at a mean temperature of about 15°C (59°F).† This emission has been estimated to be about 2½ times the solar radiation absorbed. Net loss of heat is prevented and a heat balance maintained because the atmosphere reflects back to the surface about 85 percent of the emitted radiation. Were it not for this *greenhouse effect*, the mean temperature of the earth would be about −40°C (−40°F).

Figure 2-1 shows the upward emission from earth's surface to be 21 units more than the downward emission from the atmosphere. Fifteen of these 21

---

† There are two temperature scales in common use. These are the *Celsius scale* (formerly called centigrade) with freezing point of water at 0° and boiling point at 100°, and the *Fahrenheit scale*, with freezing point at 32° and boiling point at 212°. Temperatures on one scale are converted to the other by the relationships $F = \frac{9}{5}C + 32$ and $C = \frac{5}{9}(F - 32)$. Hence, −40° on both scales represents the same degree of heat. The Celsius scale is the recommended standard scale for international use and is generally used for meteorological and hydrologic purposes.

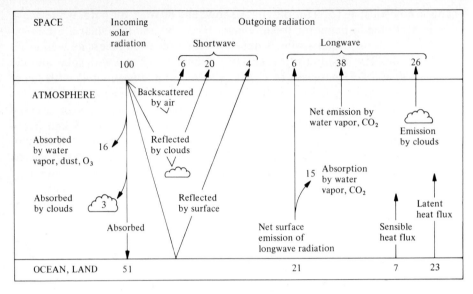

Figure 2-1 Components of earth's average annual heat balance in percentage units. *(Reproduced from "Understanding Climatic Change," p. 18, 1975, with permission of the National Academy of Sciences, Washington, D.C.)*

units are absorbed by the atmosphere, and 6 units escape to space. These 6 units, together with the 64 units emitted to space by the atmosphere, make a total of 70 units of infrared radiation lost to space. The 30-unit difference between the 51 units of solar radiation absorbed and the 21 units of infrared radiation emitted by the surface is balanced by 7 units of turbulent heat transfer and 23 units of latent heat transfer (evaporation).

Figure 2-2 shows the average annual values of the radiation balance of the earth's surface, i.e., the difference between the absorbed shortwave radiation and the effective (or net) longwave radiation. Figure 2-3 shows the average amount of heat utilized by evaporation.

2-4 Measurement of Radiation

Actinometer and *radiometer* are general names for instruments used to measure intensity of radiant energy. There are five types:

Pyrheliometer. For measuring intensity of direct solar radiation
Pyranometer. For measuring hemispherical shortwave radiation, i.e., the combined intensity of direct solar radiation and diffuse sky radiation (radiation reaching earth's surface after being scattered from direct solar beam by molecules and suspensoids in the atmosphere)

Figure 2-2 Mean annual heat balance at earth's surface, in kilocalories per square centimeter. *(From M. I. Budyko, N. A. Yefimova, L. I. Aubenok, and L. A. Strokina, The Heat Balance of the Surface of the Earth, Soviet Geog.: Rev. and Translation, vol. 3, pp. 3–16, May 1962.)*

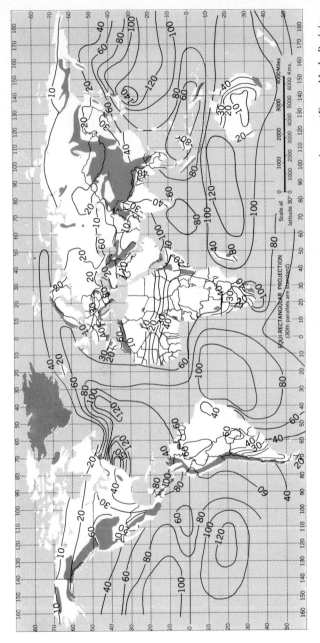

Figure 2-3 Mean annual amount of heat utilized by evaporation, in kilocalories per square centimeter. (*From M. I. Budyko, N. A. Yefimova, L. I. Aubenok, and L. A. Strokina, The Heat Balance of the Surface of the Earth, Soviet Geog.: Rev. and Translation, vol. 3, pp. 3–16, May 1962.*)

Pyrgeometer. For measuring hemispherical longwave radiation, used face up to measure atmospheric radiation or inverted to measure terrestrial and reflected atmospheric radiation

Pyrradiometer, or *total hemispherical radiometer.* For measuring all-wave radiation flux, used face up to measure hemispherical longwave radiation plus global radiation or inverted to measure terrestrial and reflected atmospheric radiation plus reflected solar radiation

Net Pyrradiometer, or *net radiometer.* For measuring net all-wave radiation flux

Details of radiation-measuring instruments can be found in texts listed in the Bibliography at the end of this chapter.

Net radiometers (net pyrradiometers) are not generally used for network observations because their measurements are applicable only to the type of ground surface at the installation site and their area of representativity is thus limited. Greatest use of radiometers in hydrology is in studies of evaporation and snowmelt. For most studies of evaporation, incident all-wave radiation data are adequate, since the reflectivity of water is nearly constant (average daily values of 5 and 3 percent, shortwave and longwave, respectively). The reflectivity of snow, however, is highly dependent upon wavelength, and its albedo may range from 40 to 95 percent. Hence, both incident shortwave and longwave radiation data are required. Because of the sparsity of all-wave radiation measurements, application of the energy-budget technique in snowmelt computations frequently requires that incident longwave radiation be computed. Several procedures, some empirical and some based on theoretical considerations, have been developed [1–4] for computing longwave radiation from regularly observed surface and upper-air data, such as temperature, vapor pressure, cloudiness, and incident solar radiation.

THE GENERAL CIRCULATION

2-5 Thermal Circulation

If the earth were a nonrotating sphere, a purely thermal circulation (Fig. 2-4) would result. The equator receives more solar radiation than the higher latitudes. Equatorial air, being warmer, is lighter and tends to rise. As it rises, it is replaced by cooler air from higher latitudes. The only way the air from the higher latitudes can be replaced is from above by the poleward flow of air rising from the equator. The true circulation differs from that of Fig. 2-4 because of the earth's rotation and the effects of land and sea distribution and landforms.

2-6 Effects of Earth's Rotation

The earth rotates from west to east, and a point at the equator moves at about 1670 km/hr (1040 mi/hr) while one at 60° lat moves at one-half this speed. From

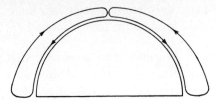

Figure 2-4 Simple thermal circulation on nonrotating earth (Northern Hemisphere).

the principle of conservation of angular momentum, it follows that a parcel of air at rest relative to earth's surface at the equator would attain a theoretical eastward velocity of 2505 km/hr (1560 mi/hr), relative to earth's surface, if moved northward to 60° lat. Conversely, if a parcel of air at the North Pole were moved southward to 60° lat, it would reach a theoretical westward velocity of 835 km/hr (520 mi/hr). However, wind speeds of this magnitude are never observed in nature because of friction.

The direction of movement of a parcel of air is affected by the *Coriolis force,* which is an apparent force arising from the earth's rotation. This apparent force acts as a deflecting force normal to the velocity of the parcel, and has no effect on its speed. The deflection is to the right in the Northern Hemisphere and to the left in the Southern Hemisphere. Were it not for this deflecting force the horizontal flow of air would be in the direction of the pressure gradient (perpendicular to the isobars). The deflecting force and frictional effects lead to flow at an angle across the isobars, thus causing, in the Northern Hemisphere, coun-

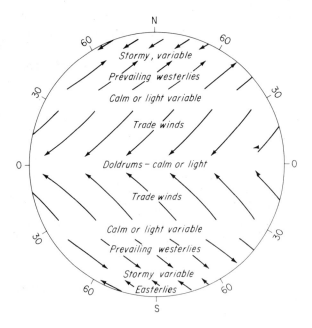

Figure 2-5 Idealized circulation at earth's surface assuming a smooth surface of uniform composition.

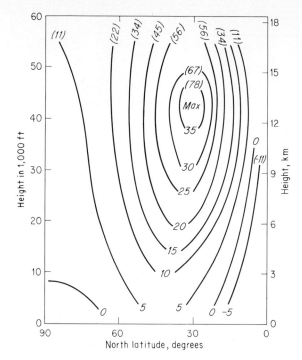

Figure 2-6 Vertical north-south cross section of average speed, in meters per second and (miles per hour), of eastward component of wind in Northern Hemisphere in winter. (Negative value indicates westward component.) Maximum wind center marks mean location of jet-stream core.

terclockwise flow around low-pressure centers and clockwise flow around high-pressure centers. The opposite applies to the Southern Hemisphere.

An oversimplified pattern of the general circulation at the surface is shown in Fig. 2-5. The wind and temperature distributions are shown for the Northern Hemisphere winter season in Figs. 2-6 and 2-7. The physical reasons for these patterns are only partly known. Early attempts to determine a natural mechanism for the general circulation were based on the idea of a convectionally driven meridional circulation. Eddy transport of angular momentum is now considered most important.

2-7 Jet Streams

Jet streams are a prominent feature of the general circulation. They are caused by air masses being brought into motion by strong pressure-gradient forces resulting from steep meridional temperature gradients and by angular momentum imparted by rotation of the earth's surface. *Jet streams* are quasi-horizontal, sinuous, undulating currents of air traveling near the tropopause at speeds ranging from about 30 m/s (70 mi/hr) to over 135 m/s (300 mi/hr). The *tropopause,* the boundary between troposphere and stratosphere, ranges in height from about 8 km (5 mi) at the poles to about 16 km (10 mi) at the equator (Fig. 2-7). The *troposphere,* which extends from earth's surface to the

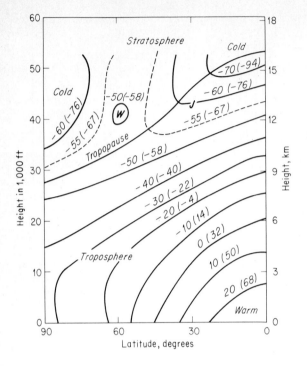

Figure 2-7 Vertical north-south cross section of mean temperature, in degrees Celsius (and Fahrenheit), over the Northern Hemisphere in winter. Tropopause marks boundary between troposphere, where temperature generally decreases with height, and stratosphere, which is relatively isothermal. *J* marks the mean position of the jet-stream core. (*Adapted from H. Riehl, "Introduction to the Atmosphere," 3d ed., Copyright 1978, McGraw-Hill Book Company. Used with permission of McGraw-Hill Book Company.*)

tropopause, is characterized by generally decreasing temperature with height, considerable vertical wind motion, most of the water vapor in the atmosphere, and weather. The *stratosphere* is the relatively isothermal layer extending from the tropopause to about 20 to 25 km (12 to 16 mi), where temperature begins to increase with height.

Being located near the tropopause, which slopes downward toward the poles, jet streams are positioned where the horizontal temperature gradient reverses. In westerly jets of the Northern Hemisphere, warmer air is found to the south below the jet-core level and to the north at heights above the jet core. The reverse applies to easterly jets. Jet streams often appear to produce a break, or discontinuity, in the tropopause, and a transfer of cold, stratospheric air to the troposphere is believed to take place through the gap.

Jet streams apparently provide the mechanism for generating low-level pressure systems that determine the weather. Convergence and divergence of air entering or leaving the maximum-wind centers of the jet result in the addition or removal of air in the jet layer which must be compensated by an opposite effect. This is provided by the layers below jet level, where the thermal structure is more favorable for vertical motion. Thus, convergence of air at jet level results in a piling of air reflected as high pressure and accompanying divergence at the earth's surface. Similarly, divergence at jet level results in removal of air that is compensated by low pressure and convergence at the surface. Paradoxically, the low-pressure systems of the lower levels appear to provide the energy for jet streams. Research has not yet disclosed which is the originator; although it can be said that there is always a jet stream associated with any major general storm [5], there are jet streams not so related.

There are several types of jet streams. A detailed description of the meteorological factors associated with the different jet streams is outside the scope of this text but can be obtained from books listed in the Bibliography. The mean location of the Northern Hemisphere tropospheric jet streams in January is shown in Fig. 2-8.

2-8 Effect of Land and Water Distribution

The horizontal flow of air in any layer of the atmosphere always has a component directed toward low pressure. Thus, the converging flows of air in the surface layer at the equator and at about 60° lat, as indicated by the idealized

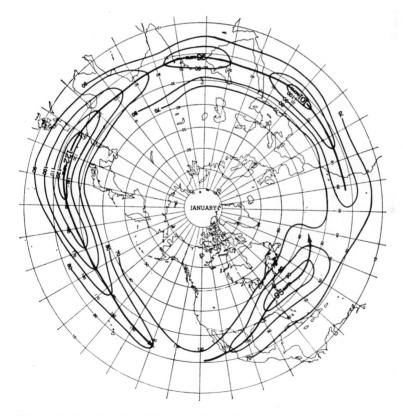

Figure 2-8 Mean location of Northern Hemisphere tropospheric jet streams and average speeds, in miles per hour, in January. *(From J. Namias, and P. F. Clapp, Confluence Theory of High Tropospheric Jet Stream, J. Meteorol., vol. 6, pp. 330–336, October 1949. Used with permission of American Meteorological Society.)*

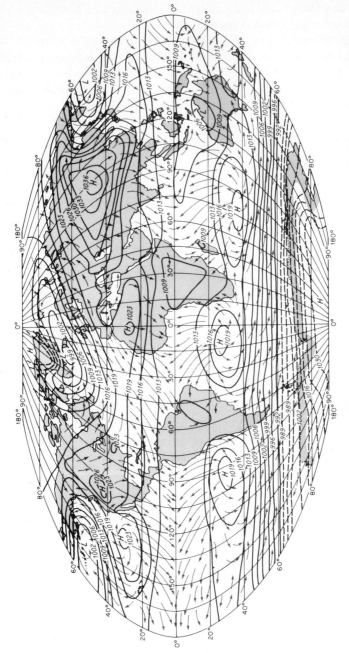

Figure 2-9 January mean sea-level pressure, in millibars, and prevailing winds. (*From V. C. Finch, G. T. Trewartha, A. H. Robinson, and E. H. Hammond, "Physical Elements of Geography," 4th ed., Copyright 1957, McGraw-Hill Book Company. Used with permission of McGraw-Hill Book Company.*)

circulation of Fig. 2-5, imply belts of low pressure at these latitudes. Similarly, high pressure would be expected at about 30° lat and at the poles.

The idealized circulation and implied simple pressure distribution are greatly distorted (Figs. 2-9 and 2-10) by differences in the specific heats, reflectivities, and mixing properties of water and land and by the existence of barriers to airflow. Heat gains and losses are distributed through relatively great depths in large bodies of water by mixing, while land is affected only near the surface. Consequently, land-surface temperatures are more variable than those of the surface of large bodies of water. This condition is further emphasized by the lower specific heat of the soil and its higher albedo, especially in winter, when snow cover reflects most of the incident radiation back to space. In winter there is a tendency for the accumulation of cold dense air over land masses and warm air over oceans. In summer, the situation is reversed.

The pressure and wind patterns of Figs. 2-9 and 2-10 show the effects of the earth's rotation, land and water distribution, and seasonal changes. They also show that winds blow clockwise around high pressure and counterclockwise around low pressure in the Northern Hemisphere and vice versa in the Southern Hemisphere.

2-9 Migratory Systems

The semipermanent features of the general, or mean, circulation (Figs. 2-9 and 2-10) are statistical and at any time may be distorted or displaced by transitory, or migratory, systems. Both semipermanent and transitory features are classified as cyclones or anticyclones. A *cyclone* is a more or less circular area of low atmospheric pressure in which the winds blow counterclockwise in the Northern Hemisphere. *Tropical cyclones* form at low latitudes and may develop into *hurricanes,* or *typhoons,* with winds exceeding 75 mi/hr (33 m/s) over areas as large as 200 mi (300 km) in diameter. *Extratropical cyclones* usually form along the boundaries between warm and cold air masses. Such cyclones are usually larger than tropical cyclones and may produce precipitation over thousands of square miles. Figure 2-11 shows average tracks of cyclones, or storm tracks, in January and July in the Northern Hemisphere. An *anticyclone* is an area of relatively high pressure in which the winds tend to blow spirally outward in a clockwise direction in the Northern Hemisphere. Details on the general circulation and on the structure of cyclones and anticyclones can be found in meteorological textbooks.

2-10 Fronts

A *frontal surface* is the boundary between two adjacent air masses of different temperature and moisture content. Frontal "surfaces" are actually layers or zones of transition. Their thickness, however, is small with respect to dimensions of air masses. The line of intersection of a frontal surface with the earth is called a *surface front.* An *upper-air front* is formed by the intersection of two

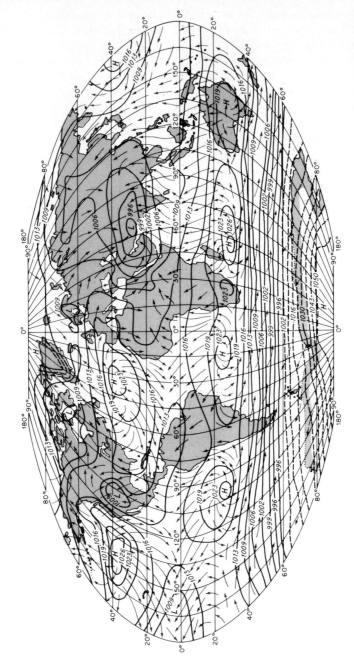

Figure 2-10. July mean sea-level pressure, in millibars, and prevailing winds. (*From V. C. Finch, G. T. Trewartha, A. H. Robinson, and E. H. Hammond, "Physical Elements of Geography," 4th ed., Copyright 1957, McGraw-Hill Book Company. Used with permission of McGraw-Hill Book Company.*)

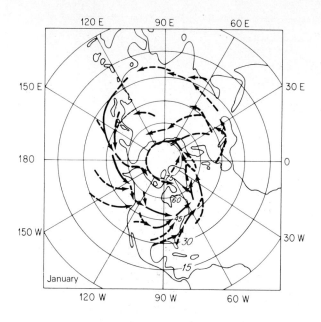

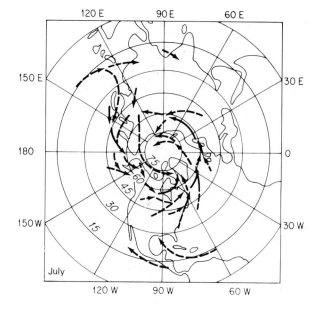

Figure 2-11 Principal tracks of Northern Hemisphere sea-level cyclones in January and July. Solid lines denote most frequent, well-defined tracks; dashed lines, less frequent and less well-defined tracks. Locally preferred regions of genesis are indicated where tracks begin. Arrowheads end where cyclone frequency is a local minimum. *(From W. H. Klein, Principal Tracks and Mean Frequencies of Cyclones and Anticyclones in the Northern Hemisphere. U.S. Weather Bur. Res. Pap. 40, 1957.)*

frontal surfaces aloft and hence marks the boundary between three air masses. If the air masses are moving so that warm air displaces colder air, the front is called a *warm front;* conversely, it is a *cold front* if cold air is displacing warmer air. If the front is not moving, it is called a *stationary front.*

The life history of a typical extratropical cyclone is shown in Fig. 2-12. For

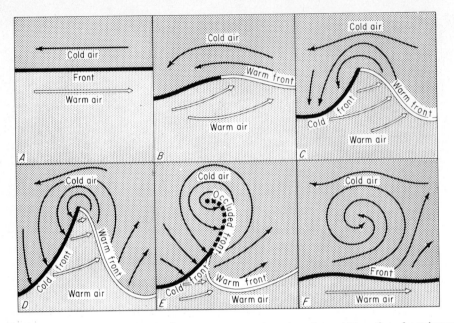

Figure 2-12 Life cycle of a Northern Hemisphere frontal cyclone: (*A*) surface front between cold and warm air; (*B*) wave beginning to form; (*C*) cyclonic circulation and wave have developed; (*D*) faster-moving cold front is overtaking retreating warm front and reducing warm sector; (*E*) warm sector has been eliminated, and (*F*) cyclone is dissipating. (*U.S. National Weather Service.*)

reasons not yet completely understood but with the jet stream apparently often a factor, a wave is generated on the boundary between two air masses (Fig. 2-12*B*). Under conditions of dynamic stability, this wave moves along the front with little change in form and with little or no precipitation. If the wave is unstable, and particularly with a jet stream aloft, it progresses through the successive stages of Fig. 2-12. In stage *C* the cyclone is deepening and has a well-formed warm sector. Warm air from the warm sector is forced upward along the warm-front surface, causing widespread precipitation ahead of the surface front. At the same time, the advancing wedge of cold air behind the cold front is lifting the warm air and causing convective showers behind the cold front. Showers also frequently occur in the warm sector.

Cold fronts move faster than warm fronts and usually overtake them (Fig. 2-12*D*, *E*). This process is called *occlusion*, and the resulting surface front is called an *occluded front*. Cloudiness associated with an occluding system is illustrated in Fig. 2-13, which also shows the position of the jet stream in relation to the cloudiness and surface position of the occluding system. As occlusion progresses, the warm sector is displaced farther from the cyclone center (Fig. 2-12*E*), which is eventually deprived of the warm air necessary to maintain its energy. Cold air replaces the warm, the center fills, and the occluded front is destroyed. A new cyclone center may form, however, at the

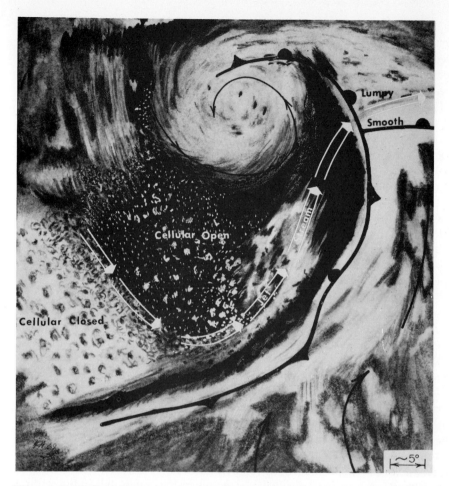

Figure 2-13 Idealized illustration showing relation of jet stream to occluding surface system. Note that the jet stream is over the boundary between open and closed cellular clouds and crosses the occluded front just above the junction of cold and warm fronts. *(U.S. National Environmental Satellite Center.)*

peak of the remaining warm sector. The time from initial wave development to complete occlusion is usually of the order of 3 to 4 days.

TEMPERATURE

2-11 Terminology

A knowledge of terminology and methods of computation is required in order to avoid misuse of published temperature data. The terms *average, mean,* and *normal* are arithmetic means. The first two are used interchangeably, but the *normal* [6], generally used as a standard of comparison, is the average value for a particular date, month, season, or for the year as a whole over a specific 30-year period (1951–1980 as of 1981). Plans call for recomputing the 30-year normals every decade, dropping off the first 10 years and adding the most recent 10 years.

The *mean daily temperature* may be computed by several methods [7]. The most accurate practical method is to average hourly temperatures. Acceptably accurate results can be obtained by averaging 3- or 6-hr observations. In some countries, climatological observations (usually three) are made at hours selected to permit computation of acceptable daily means by a special formula.

In the United States the *mean daily temperature* is the average of the daily maximum and minimum temperatures. This yields a value usually less than a degree above the true daily average. Once-daily temperature observations are usually made about 7 a.m. or 5 p.m. Temperatures are published as of the date of the reading even though the maximum or minimum may have occurred on the preceding day. Mean temperatures computed from evening readings tend to be slightly higher than those from midnight readings. Morning readings yield mean temperatures with a negative bias, but the difference is less than that for evening readings. The maximum effect [8] on mean temperature of arbitrary changes in observation time varies with place and season, and may exceed 3 Fahrenheit (1.6 Celsius) degrees.

The *normal daily temperature* is the average daily mean temperature for a given date computed for a specific 30-year period. The *daily range* in temperature is the difference between the highest and lowest temperatures recorded on a particular day. The *mean monthly temperature* is the average of the mean monthly maximum and minimum temperatures. The *mean annual temperature* is the average of the monthly means for the year.

The *degree-day* is a departure of one degree for one day in the mean daily temperature from a specified base temperature. For snowmelt computations, the number of degree-days for a day is equal to the mean daily temperature minus the base temperature, all negative differences being taken as zero. The number of degree-days in a month or other time interval is the total of the daily values. Published degree-day values are for heating and cooling purposes and are based on departures below and above 65°F (18°C).

2-12 Measurement of Temperature

In order to measure air temperature properly, thermometers must be placed where air circulation is relatively unobstructed, and yet they must be protected from the direct rays of the sun and from precipitation. In the United States thermometers are placed in white, louvered, wooden *instrument shelters* (Fig. 2-14) through which the air can move readily. The shelter location must be typical of the area for which the measured temperatures are to be representative. Because of marked vertical temperature gradients just above the soil surface, all shelters should be about the same height above the ground for

Figure 2-14 Instrument shelter with maximum and minimum thermometers. *(U.S. National Weather Service.)*

recorded temperatures to be comparable. In the United States shelters are set at about 4½ ft (1.4 m) above the ground.

There are about 8000 stations in the United States for which official temperature records are compiled. Except for a few hundred stations equipped or staffed to obtain continuous or hourly temperatures, most make a daily observation consisting of the current, maximum, and minimum temperatures. The *minimum thermometer,* of the alcohol-in-glass type, has an index which remains at the lowest temperature occurring since its last setting. The *maximum thermometer* has a constriction near the bulb which prevents the mercury from returning to the bulb as the temperature falls and thus registers the highest temperature since its last setting. The *thermograph,* with either a bimetallic strip or a metal tube filled with alcohol or mercury for its thermometric element, makes an autographic record on a chart. Electrical-resistance thermometers, thermocouples, gas-bulb thermometers, and other types of instruments are used for special purposes. Electrical-resistance thermometers, for example, are widely used for measuring upper-air temperatures and in dew-cell hygrometers, which are described briefly in the section on humidity.

2-13 Lapse Rates

The *lapse rate,* or vertical temperature gradient, is the rate of change of temperature with height in the free atmosphere. The mean lapse rate in the lower troposphere is a decrease of about 0.7 Celsius degrees per 100 m (3.8 Fahrenheit degrees per 1000 ft) increase in height. The greatest variations in lapse rate are found in the layer of air just above the land surface. The earth radiates heat energy to space at a relatively constant rate which is a function of its absolute temperature, in kelvins.† Incoming radiation at night is less than the outgoing, and the temperature of the earth's surface and of the air immediately above it decreases. The surface cooling sometimes leads to an increase of temperature with altitude, or *temperature inversion,* in the surface layer. This condition usually occurs on still, clear nights because there is little turbulent mixing of air and because outgoing radiation is unhampered by clouds. Temperature inversions are also observed at higher levels when warm-air currents overrun colder air.

In the daytime there is a tendency for steep lapse rates because of the relatively high temperatures of the air near the ground. This daytime heating usually destroys a surface radiation inversion by early forenoon. As the heating continues, the lapse rate in the lower layers of the air steepens until it may reach the *dry-adiabatic lapse rate* (1 Celsius degree per 100 m, or 5.4 Fahrenheit degrees per 1000 ft), which is the rate of temperature change of unsaturated air resulting from expansion or compression as the air rises (lowering pressure) or descends (increasing pressure) without heat being added or removed.

Air having a dry-adiabatic lapse rate mixes readily, whereas a temperature inversion indicates a stable condition in which warm lighter air overlies cold denser air. Under optimum surface heating conditions, the air near the ground may be heated sufficiently for the lapse rate in the lowest layers to become *superadiabatic,* i.e., exceeding 1 Celsius degree per 100 m (5.4 Fahrenheit degrees per 1000 ft). This is an unstable condition since any parcel of air lifted dry-adiabatically remains warmer and lighter than the surrounding air and thus has a tendency to continue rising.

If a parcel of saturated air is lifted adiabatically, its temperature will decrease and its water vapor will condense, releasing latent heat of vaporization. This heat reduces the cooling rate of the ascending air. Hence, the *saturated-adiabatic lapse rate* is less than the dry-adiabatic. It varies inversely with the water-vapor content, and hence temperature, of the air. The average value for the lower layers at temperatures above freezing is roughly half the dry-adiabatic. At very low temperatures or high altitudes there is little difference between the two lapse rates because of the very small amounts of water vapor available.

If the moisture in the rising air is precipitated as it is condensed, the temperature of the air will decrease at the *pseudo-adiabatic lapse rate,* which differs very little from the saturated-adiabatic. Actually, the process is not strictly adiabatic, as heat is carried away by the falling precipitation. A layer of saturated air having a saturated- or pseudo-adiabatic lapse rate is said to be in *neutral equilibrium.* If its lapse rate is less than the saturated- or pseudo-adiabatic, the air is stable; if greater, unstable.

---

† In meteorology, absolute temperature is usually given in degrees Kelvin, called kelvins.

2-14 Geographic Distribution of Temperature

In general, surface air temperature tends to be highest at low latitudes and to decrease poleward (Fig. 2-15). However, this trend is greatly distorted by the influence of land and water masses, topography, and vegetation. In the interior of large islands and continents, temperatures are higher in summer and lower in winter than on coasts at corresponding latitudes. Temperatures at high elevations are colder than at low levels, and southern slopes have warmer temperatures than northern slopes. The average rate of decrease of surface air temperature with elevation is usually between $\frac{1}{2}$ and 1 Celsius degree per 100 m (3 and 5 Fahrenheit degrees per 1000 ft). Forested areas have higher daily minimum and lower daily maximum temperatures than barren areas. The mean temperature in a forested area may be 1 to 2 Celsius (2 to 4 Fahrenheit) degrees lower than in comparable open country, the difference being greater in summer.

The heat from a large city, which may roughly equal one-third of the solar radiation reaching it, produces local distortions in the temperature pattern so that temperatures recorded in cities may not represent the surrounding region. The mean annual temperature of cities averages about 1 Celsius degree (2 Fahrenheit degrees) higher than that of the surrounding region, most of the difference resulting from higher daily minima in the cities. Any comparison of city and country temperatures must allow for differences in exposure of thermometers. In cities the instrument shelters are sometimes located on roofs. On still, clear nights, when radiational cooling is particularly effective, the temperature on the ground may be as much as 8 Celsius (15 Fahrenheit) degrees colder than that at an elevation of 30 m (100 ft). A slight gradient in the opposite direction is observed on windy or cloudy nights. Daytime maxima tend to be lower at rooftop level than at the ground. In general, the average temperature from roof exposures is slightly lower than that on the ground.

2-15 Time Variations of Temperature

In continental regions the warmest and coldest points of the annual temperature cycle lag behind the solstices by about 1 month. In the United States, January is usually the coldest month and July the warmest. At oceanic stations the lag is nearer 2 months, and the temperature difference between the warmest and coldest months is much less.

The daily variation of temperature lags slightly behind the daily variation of solar radiation. The temperature begins to rise shortly after sunrise, reaches a peak 1 to 3 hr (about $\frac{1}{2}$ hr at oceanic stations) after the sun has reached its highest altitude, and falls through the night to a minimum about sunrise. The daily range of temperature is affected by the state of the sky. On cloudy days the maximum temperature is lower because of reduced insolation, and the minimum is higher because of reduced outgoing radiation. The daily range is also smaller over oceans.

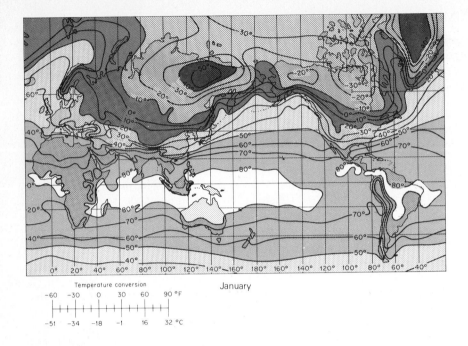

Temperature conversion

January

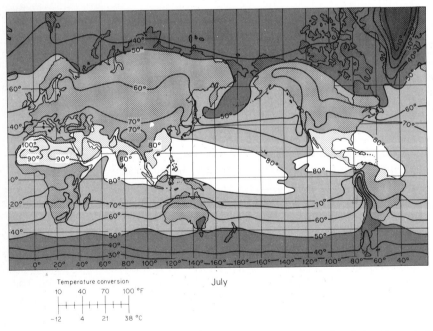

Temperature conversion

July

Figure 2-15 Average January and July temperatures, in degrees Fahrenheit. *(From Arthur N. Strahler, "Introduction to Physical Geography," 3d ed., pp. 68, 69. Copyright © 1965, 1970, 1973, by John Wiley & Sons, Inc. Reprint by permission.)*

HUMIDITY

2-16 Properties of Water Vapor

The process by which liquid water is converted into vapor is called *vaporization* or *evaporation*. Molecules of water having sufficient kinetic energy to overcome the attractive forces tending to hold them within the body of liquid water are projected through the water surface. Since kinetic energy increases and surface tension decreases as temperature rises, the rate of evaporation increases with temperature. Most atmospheric vapor is the product of evaporation from water surfaces. Molecules may leave a snow or ice surface in the same manner as they leave a liquid. The process whereby a solid is transformed directly to the vapor state is called *vaporization* or *evaporation* also, but in meteorology the direct transformation from ice to vapor, and vice versa, is usually called *sublimation*. The process by which vapor changes to the liquid or solid state is called *condensation,* which thus includes sublimation unless otherwise indicated.

In any mixture of gases, each gas exerts a *partial pressure* independent of the other gases. The partial pressure exerted by water vapor is called *vapor pressure.* If all the water vapor in a closed container of moist air with an initial total pressure p were removed, the final pressure p' of the dry air alone would be less than p. The vapor pressure e would be the difference between the pressure of the moist air and that of the dry air, or $p - p'$.

The maximum amount of water vapor that can exist in any space is a function of temperature and is practically independent of the coexistence of other gases. When the maximum amount of water vapor for a given temperature is contained in a given space, the space is said to be *saturated.* The more common expression "the air is saturated" is not strictly correct. The pressure exerted by the vapor in a saturated space is called the *saturation vapor pressure,* which, for all practical purposes, is the maximum vapor pressure possible at a given temperature.

At any temperature below the freezing point the saturation vapor pressure over ice is less than that over water.† The difference is at a maximum at about $-12°C$ ($10°F$), but the ratio of the vapor pressures increases with decreasing temperature. Hence, introduction of ice into a space saturated with respect to liquid water at the same or higher temperature will result in condensation of vapor on the ice.

In a space in contact with a water surface, condensation and vaporization always go on simultaneously. If the space is not saturated, the rate of vaporization will exceed the rate of condensation, resulting in a net evaporation.‡ If the

---

† Vapor pressure over liquid water is generally used for most meteorological purposes regardless of temperature, and the values of saturation vapor pressure in Appendix Tables A-16 and A-17 are for liquid water.

‡ In hydrology, net evaporation is termed simply *evaporation.*

space is saturated, the rates of vaporization and condensation balance, provided that the water and air temperatures are the same.

Vaporization removes heat from the liquid being vaporized, while condensation adds heat. The *latent heat of vaporization* is the amount of heat absorbed by a unit mass of a substance, without change in temperature, while passing from the liquid to the vapor state. The change from vapor to the liquid state releases an equivalent amount of heat.

The heat of vaporization of water H_v in calories per gram varies with temperature but can be determined accurately up to 40°C (104°F) by

$$H_v = 597.3 - 0.564T \tag{2-1}$$

where T is the temperature in degrees Celsius.

The *latent heat of fusion* for water is the amount of heat required to convert one gram of ice to liquid water at the same temperature. When one gram of liquid water at 0°C (32°F) freezes into ice at the same temperature, the latent heat of fusion [79.7 cal (15°C)/g, or 0.337 kJ/g] is liberated.

The *latent heat of sublimation* for water is the amount of heat required to convert one gram of ice into vapor at the same temperature without passing through the intermediate liquid state. It is equal to the sum of the latent heat of vaporization and the latent heat of fusion. At 0°C (32°F) it is about 677 cal (15°C)/g or 2.83 kJ/g. Direct condensation of vapor into ice at the same temperature liberates an equivalent amount of heat.

The *specific gravity* of water vapor is 0.622 times that of dry air at the same temperature and pressure. The density of water vapor ρ_v in grams per cubic centimeter is

$$\rho_v = 0.622 \frac{e}{R_g T} \tag{2-2}$$

where T is the absolute temperature in kelvins (K) and R_g, the specific gas constant for dry air, equals 2.87×10^3 cm^2 s^{-2} K^{-1} when the vapor pressure e is in millibars.†

The density of dry air ρ_d in grams per cubic centimeter is

$$\rho_d = \frac{p_d}{R_g T} \tag{2-3}$$

where p_d is the pressure in millibars.

The density of moist air is equal to the mass of water vapor plus the mass of dry air in a unit volume of the mixture. If p_a is the total pressure of the moist air, $p_a - e$ will be the partial pressure of the dry air alone. Adding Eqs. (2-2) and

---

† The millibar (mb) is the unit of pressure commonly used in meteorology. It is equivalent to a force of 1000 dynes/cm², 0.0143 lb/in², or 0.0295 inch of mercury (abbreviated in Hg). Mean sea-level air pressure is 1013.2 mb. The SI unit of pressure is the kilopascal (kPa), which is equal to 1000 newtons/m², 10^8 dynes/m², 10^4 dynes/cm², 10 mb, or 0.295 in Hg.

(2-3) and substituting $p_a - e$ for p_d gives

$$\rho_a = \frac{p_a}{R_g T}\left(1 - 0.378\,\frac{e}{p_a}\right) \tag{2-4}$$

This equation shows that moist air is lighter than dry air.

2-17 Terminology

There are many expressions used for indicating the moisture content of the atmosphere. Each serves special purposes, and only those expressions common to hydrologic uses are discussed here. Vapor pressure by itself provides no direct measure of atmospheric moisture content, but it is the basic element in practically all expressions of air density and vapor content.

The computation of saturation vapor pressure is somewhat complicated, and its values are obtained generally from psychrometric tables such as those in the Appendix and in Smithsonian Meteorological Tables [9], which are based on the Goff-Gratch formula [10]. This formula yields values of saturation vapor pressure over water that are approximated to within 1 percent in the range of -50 to $+55°C$ (-58 to $+131°F$) by the much simpler equation [11],

$$e_s \approx 33.8639\,[(0.00738T + 0.8072)^8 - 0.000019|1.8T + 48| + 0.001316] \tag{2-5}$$

where e_s is in millibars and T is in degrees Celsius.

It is sometimes necessary to convert vapor pressure over water to that over ice, or vice versa. Ratios for effecting these conversions are found in Appendix Table A-12. Equation (2-6), derived for making the conversions on a computer [12], yields ratios to within 0.1 percent for the range 0 to $-50°C$ (32 to $-58°F$):

$$\frac{e_{s,\text{ice}}}{e_{s,\text{water}}} \approx 1 + 0.00972T + 0.000042T^2 \tag{2-6}$$

where T is in degrees Celsius.

The *dewpoint* T_d is the temperature at which space becomes saturated when air is cooled under constant pressure and with constant water-vapor content. It is the temperature having a saturation vapor pressure e_s equal to the existing vapor pressure e.

When the relative humidity is known, the dewpoint can be approximated to within 0.3 Celsius (0.5 Fahrenheit) degrees in the temperature range of -40 to $50°C$ (-40 to $122°F$) by Eq. (2-7), which yields the dewpoint depression to be subtracted from the temperature to obtain the dewpoint [13],

$$T - T_d \approx (14.55 + 0.114T)X + [(2.5 + 0.007T)X]^3$$
$$+ (15.9 + 0.117T)X^{14} \tag{2-7}$$

where T is in degrees Celsius and X is the complement of the relative humidity expressed as a decimal fraction, or $X = 1.00 - f/100$.

The *relative humidity* f is the percentage ratio of the actual to the saturation vapor pressure and is therefore a ratio of the amount of moisture in a given space to the amount the space could contain if saturated,

$$f = 100\ \frac{e}{e_s} \tag{2-8}$$

Relative humidity can also be computed directly from air temperature T and dewpoint T_d by an approximate formula [14] in convenient form for computer use:

$$f \approx 100\ \left(\frac{112 - 0.1T + T_d}{112 + 0.9T}\right)^8 \tag{2-9}$$

where temperatures are in degrees Celsius. The formula approximates relative humidity to within 0.6 percent in the range of -25 to $45°C$ (-13 to $113°F$).

Psychrometric tables like those in the Appendix usually give dewpoint and relative humidity as a function of air temperature and *wet-bulb depression*, i.e., the difference between air and wet-bulb temperatures (Sec. 2-18).

The *specific humidity* q_h, usually expressed in grams per kilogram, is the mass of water vapor per unit mass of moist air:

$$q_h = 622\ \frac{e}{p_a - 0.378e} \approx 622\ \frac{e}{p_a} \tag{2-10}$$

where p_a is the total pressure of the air in millibars.

The *mixing ratio* w_r is the mass of water vapor per unit mass of perfectly dry air in a humid mixture. Usually expressed in grams per kilogram of dry air, it is given by

$$w_r = 622\ \frac{e}{p_a - e} \tag{2-11}$$

The total amount of water vapor in a layer of air is often expressed as the depth of *precipitable water* W_p, in millimeters or inches, even though there is no natural process capable of precipitating the entire moisture content of the layer. The amount of precipitable water in any air column of considerable height is computed [15] by increments of pressure or height from surface and upper-air observations of temperature, humidity, and pressure. A convenient formula [16] for computing W_p in millimeters is

$$W_p = \sum 0.01\bar{q}_h\ \Delta p_a \tag{2-12}$$

where the pressure p_a is in millibars and $\bar{q}_h$, in grams per kilogram, is the average of the specific humidities at the top and bottom of each layer. Precipitable water may be approximated conveniently by nomograms [17] using dewpoints at specific pressure levels. Less accurate approximations, which may be considerably in error under certain conditions, can be made from surface dewpoints [18] or vapor pressure and assumed temperature and humidity lapse rates or relations based on observed precipitable water. Figure 2-16 gives the depth of precipitable water in a column of saturated air with its base at the 1000-millibar level and its top anywhere up to 200 millibars. Tables for comput-

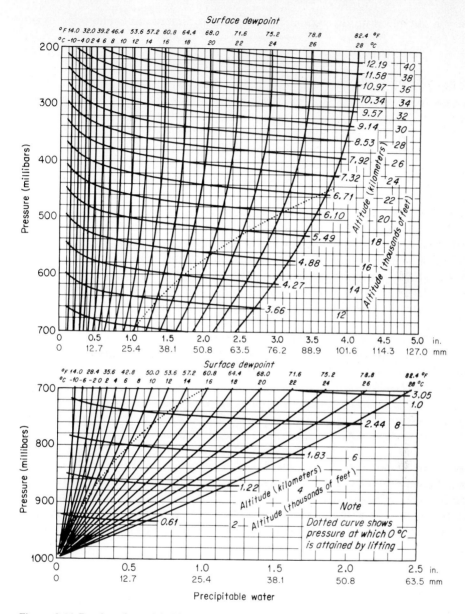

Figure 2-16 Depths of precipitable water in a column of air of any height above the 1000-millibar level as a function of the 1000-millibar dewpoint, assuming saturation and pseudo-adiabatic lapse rate. *(U.S. National Weather Service.)*

ing precipitable water in various layers of the saturated atmosphere are available [19].

2-18 Measurement of Humidity

In general, measurements of humidity in the surface layers of the atmosphere are made with a psychrometer, which consists of two thermometers, one with its bulb covered by a jacket of clean muslin saturated with water. The thermometers are ventilated by whirling or by use of a fan. Because of the cooling effect of evaporation, the moistened, or *wet-bulb*, thermometer reads lower than the dry, the difference in degrees being known as the *wet-bulb depression*. The air and wet-bulb temperatures are used to obtain various expressions of humidity by reference to psychrometric tables (Appendix).

The *hair hygrometer* makes use of the fact that the length of a hair varies with relative humidity. The changes are transmitted to a pointer indicating the relative humidity on a graduated scale. The *hair hygrograph* is a hair hygrometer operating a pen marking a trace on a chart. The *hygrothermograph,* combining the features of both the hair hygrograph and thermograph, records both relative humidity and temperature on one chart. A *dewpoint hygrometer,* which measures dewpoint directly and is used mostly for laboratory purposes, consists of a highly polished metal vessel containing a suitable liquid which is cooled by any of several methods. The temperature of the liquid at the time condensation begins to occur on the exterior of the metal vessel is the dewpoint. The *dew-cell hygrometer* measures the dewpoint by regulating the temperature of a saturated aqueous lithium chloride solution so that the water-vapor pressure of the solution is equal to that of the surrounding atmosphere. The *spectral hygrometer* measures the selective absorption of light in certain bands of the spectrum by water vapor. With the sun as a light source, it has been used to measure total atmospheric moisture [20]. Other humidity instruments have been developed for special purposes but are not generally used in routine operational activities.

Measurement of humidity is one of the least accurate instrumental procedures in meteorology. The standard psychrometer invites many observational errors. The two thermometers double the chance of misreading. At low temperatures, misreading by a few tenths of a degree can lead to absurd results. There is always the chance that the readings are not made when the wet-bulb thermometer is at its lowest temperature. In addition, there are errors with a positive bias resulting from insufficient ventilation, dirty or too thick muslin, and impure water.

Any instrument using a hair element is subject to appreciable error. The hair expands with increasing temperature, and its response to changes in humidity is very slow, the lag increasing with decreasing temperature until it becomes almost infinite at about $-40°C$ ($-40°F$). This can lead to significant error in upper-air soundings, where large ranges in temperature and sharp variations of humidity with altitude are observed. Consequently, sounding instruments are now equipped with electrical hygrometers using various humidity elements. A carbon element [21] has been found most satisfactory of those tested so far.

At freezing temperatures there is some uncertainty whether the dewpoint or frost point is being measured by dewpoint hygrometers. The difference may lead to appreciable errors in computing relative humidity [22] and vapor pressure.

2-19 Geographic Distribution of Humidity

Atmospheric moisture tends to decrease with increasing latitude, but relative humidity, being an inverse function of temperature, tends to increase. Atmospheric moisture is greatest over oceans and decreases with distance inland. It

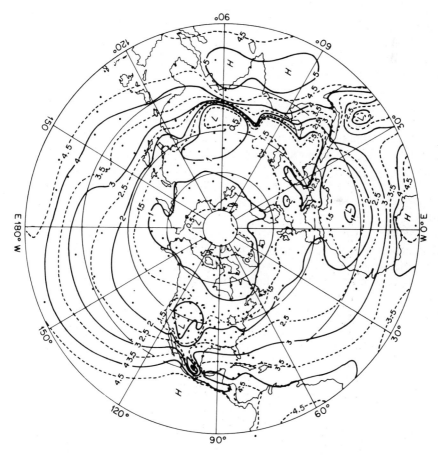

Figure 2-17 Average precipitable water, in centimeters, over the Northern Hemisphere for 1958. *(From V. P. Starr, J. P. Peixoto, and A. R. Crisi, Hemispheric Water Balance for the IGY, Tellus, vol. 17, no. 14, pp. 463–472, 1965. Used with permission of Svenska Geofysiska Foreningen.)*

also decreases with elevation and is greater over vegetation than over barren soil. The distribution of the average precipitable water over the Northern Hemisphere for 1958 is shown in Fig. 2-17. The distribution depicted is fairly representative of the mean annual pattern, since there is generally relatively little year-to-year variation. Distributions of monthly means and semimonthly maxima over the United States are available [23].

2-20 Time Variations in Humidity

Like temperature, atmospheric water vapor is at a minimum in winter and at a maximum in summer. In the Northern Hemisphere the driest months are January and February, and the most humid are July and August. In the middle and high latitudes average monthly precipitable water over continental interiors in the driest months is about half the mean annual; in the most humid months, about twice the mean annual. The seasonal variation is much less over oceans and coastal areas and is at a minimum over tropical seas. Unlike actual water vapor content, relative humidity is at a minimum in summer and at a maximum in winter.

The diurnal variation of atmospheric moisture is normally small, except where land and sea breezes bring air of differing characteristics. Near the ground surface, condensation of dew at night and reevaporation during the day may result in a minimum moisture content near sunrise and a maximum by noon. Relative humidity, of course, behaves in a manner opposite to that of temperature, being at a maximum in the early morning and at a minimum in the afternoon.

WIND

Wind, which is air in motion, is a very influential factor in several hydrometeorological processes. Moisture and heat are readily transferred to and from air, which tends to adopt the thermal and moisture conditions of the surfaces with which it is in contact. Stagnant air in contact with a water surface eventually assumes the vapor pressure of the surface, so that no evaporation takes place. Similarly, stagnant air over a snow or ice surface eventually assumes the temperature and vapor pressure of the surface, so that melting by convection and condensation ceases. Consequently, wind exerts considerable influence in evaporative and snowmelt processes. It is also important in the production of precipitation, since it is only through sustained inflow of moist air into a storm that precipitation can be maintained.

2-21 Measurement of Wind

Wind has both speed and direction. The *wind direction* is the direction *from* which it is blowing. Direction is usually expressed in terms of 16 compass points (N, NNE, NE, ENE, etc.) for surface winds, and for winds aloft in degrees or tens of degrees from north, measured clockwise. Wind speed is given in kilometers per hour, miles per hour, meters per second, or knots (1 km/hr = 0.621 mi/hr = 0.278 m/s = 0.540 kn, and 1 kn = 1.852 km/hr = 1.151 mi/hr = 0.514 m/s).

Wind speed is measured by *anemometers,* of which there are several types. The *three-* or *four-cup anemometer* with a vertical axis of rotation is most commonly used for official observations. It tends to register too high a mean speed in a variable wind because the cups accelerate faster than they lose speed. Vertical currents (turbulence) tend to rotate the cups and cause overregistration of horizontal speeds. Most cup anemometers will not record speeds below 1 or 2 mi/hr because of starting friction. The *propeller anemometer* has a horizontal axis of rotation. *Pressure-tube anemometers,* of which the Dines is the best known, operate on the pitot-tube principle.

While wind speed varies greatly with height above the ground, no standard anemometer level has been adopted. Differences in wind speed resulting from differences in anemometer height, which may range anywhere from 30 to several hundred feet above the ground, usually exceed the errors from instrumental deficiencies. However, approximate adjustment can be made for differences in height [Eq. (2-14), (2-15), or (2-16)].

2-22 Geographic Variation of Wind

In winter there is a tendency for surface winds to blow from the colder interior of land masses toward the warmer oceans (Sec. 2-8). Conversely, in summer the winds tend to blow from the cooler bodies of water toward the warmer land (*monsoons*). Similarly, diurnal land and sea breezes may result from temperature contrasts between land and water.

On mountain ridges and summits wind speeds at 10 m (30 ft) or more above the ground are higher than in the free air at corresponding elevations because of the convergence of the air forced by the orographic barriers. On lee slopes and in sheltered valleys wind speeds are light. Wind direction is greatly influenced by orientation of orographic barriers. With a weak pressure system, diurnal variation of wind direction may occur in mountain regions, the winds blowing upslope in the daytime and downslope at night.

Wind speeds are reduced and directions deflected in the lower layers of the atmosphere because of friction produced by trees, buildings, and other obstacles. These effects become negligible above about 600 m (2000 ft), and this lower layer is referred to as the *friction layer.* Over land the surface wind speed averages about 40 percent of that just above the friction layer, and at sea about 70 percent.

Because of its location in the general circulation, most of the conterminous United States has prevailing westerly winds. However, the winds are generally variable since most of the country is affected by migratory pressure systems, with winds circulating clockwise in the high-pressure areas and counterclockwise in low-pressure systems as they move across the country.

The variation of wind speed with height, or the *wind profile,* in the friction layer is usually expressed by one of two general relationships, namely, the logarithmic velocity profile or the power-law profile. In hydrology, the relationships are most often used to estimate the wind speed in the *surface boundary layer*, i.e., the thin layer of air between the ground surface and the anemometer level, usually about 10 m (30 ft) but often lower at special test sites or experimental stations. The common requirement is the wind speed above a snow or water surface for computations of snowmelt and evaporation.

One of the more common forms [24] of the logarithmic velocity profile for meteorological purposes is

$$\frac{\bar{v}}{v_*} = \frac{1}{k} \ln \frac{z}{z_0} \qquad z \geq z_0 \qquad (2\text{-}13)$$

where $\bar{v}$ is the mean wind speed for at least a few minutes at height z above the ground; k is the von Kármán constant, generally taken equal to 0.4; z_0 is the *roughness length*, which is a measure of surface roughness and presumably the height at which wind speed becomes zero and must therefore be less than z; and v_* is the *friction velocity*, which is equal to $|\tau/\rho|^{1/2}$, τ being the shearing stress, or *Reynolds stress*, and ρ the air density.

In meteorological investigations of the surface boundary layer, τ is generally considered independent of height, with the surface value assumed to apply throughout the layer, that is, $\tau = \tau_0$. Hence, the friction velocity v_* depends on the nature of the surface and the mean wind speed $\bar{v}$. It usually ranges [25] from about 3 to 12 percent of the mean wind speed $\bar{v}$, the lower values being associated with smooth surfaces. A rough assumption sometimes made in meteorological studies is that v_* is approximately equal to $\bar{v}/10$. Some measurements of roughness length z_0 and friction velocity v_* for a mean wind speed $\bar{v}$ of 5 m/s (11 mi/hr) at 2 m (6½ ft) above the ground are presented in Table 2-1.

Table 2-1 Representative values of roughness length z_0 and friction velocity v_* for natural surfaces

Neutral stability; values of v_* corresponding to mean velocity $\bar{v}$ of 5.0 m/s (11 mi/hr) at 2 m (6½ ft)

Type of surface	z_0		v_*	
	cm	in	cm/s	ft/s
Very smooth (mud flats, ice)	0.001	0.0004	16	0.5
Lawn, grass up to 1 cm (0.4 in) high	0.1	0.4	26	0.9
Downland, thin grass up to 10 cm (4 in) high	0.7	0.28	36	1.2
Thick grass, up to 10 cm (4 in) high	2.3	0.91	45	1.5
Thin grass, up to 50 cm (20 in) high	5	2.0	55	1.8
Thick grass, up to 50 cm (20 in) high	9	3.5	63	2.1

Source: O. G. Sutton, "Micrometeorology," copyright 1953, McGraw-Hill Book Company. Used with permission of McGraw-Hill Book Company.

A more detailed listing of z_0 for a wide range of surface roughness is given in Table 2-2.

A value of one-thirtieth the average height of surface irregularities is often taken as a fair estimate of z_0, but the data of Table 2-2 suggest that this simple approximation could be greatly in error under certain conditions, especially in

Table 2-2 Roughness length z_0

Surface	Wind speed at $z = 2$ m ($6\frac{1}{2}$ ft)		Roughness length z_0	
	m/s	mi/hr	cm	in
Open water	2.1	4.7	0.001	0.0004
Smooth mud flats	. . .	. . .	0.001	0.0004
Smooth snow on short grass	. . .	. . .	0.005	0.002
Wet soil	1.8	4.0	0.02	0.008
Desert	. . .	. . .	0.03	0.012
Snow on prairie	. . .	. . .	0.10	0.04
Mown grass:				
1.5 cm (0.6 in)	. . .	. . .	0.2	0.08
3.0 cm (1.2 in)	. . .	. . .	0.7	0.28
4.5 cm (1.8 in)	2	4.5	2.4	0.94
4.5 cm (1.8 in)	6–8	13–18	1.7	0.67
Alfalfa:				
20–30 cm (8–12 in)	1.9	4.3	1.4	0.55
30–40 cm (12–16 in)	1.9	4.3	1.3	0.51
Long grass:				
60–70 cm (24–28 in)	1.5	3.4	9.0	3.54
60–70 cm (24–28 in)	3.5	7.8	6.1	2.40
60–70 cm (24–28 in)	6.2	13.9	3.7	1.46
Maize:				
90 cm (35 in)	. . .	. . .	2.0	0.79
170 cm (67 in)	. . .	. . .	9.5	3.74
300 cm (118 in)	. . .	. . .	22.0	8.66
Sugarcane:				
100 cm (39 in)	. . .	. . .	4.0	1.57
200 cm (79 in)	. . .	. . .	5.0	1.97
300 cm (118 in)	. . .	. . .	7.0	2.76
400 cm (157 in)	. . .	. . .	9.0	3.54
Brush, 135 cm (53 in)	. . .	. . .	14.0	5.51
Orange orchard, 350 cm (138 in)	. . .	. . .	50.0	19.7
Pine forest:				
5 m (16 ft)	. . .	. . .	65.0	25.6
27 m (89 ft)	. . .	. . .	300.0	118.1
Deciduous forest, 17 m (56 ft)	. . .	. . .	270.0	106.3

Source: Adapted from P. S. Eagleson, "Dynamic Hydrology," copyright 1970, McGraw-Hill Book Company. Used with permission of McGraw-Hill Book Company.

the case of brush and trees. Table 2-2 shows that z_0 varies inversely with wind speed in the case of tall grasses, which tend to flatten out as speed increases [26]. In the case of a water surface, which roughens as wind speed increases, z_0 tends to be greater for higher wind speeds. However, it appears that there are discontinuities in this relationship, and, at least in the range of 2 to 10 m/s (4 to 22 mi/hr) at 10 m (30 ft) above the surface, z_0 alternately increases and decreases as wind speed increases [27].

A convenient form [28] of the logarithmic velocity profile for relating mean wind speed $\bar{v}$ at some height z to the measured mean wind speed $\bar{v}_1$ at some standard height z_1 is

$$\frac{\bar{v}}{\bar{v}_1} = \frac{\ln(z/z_0 + 1)}{\ln(z_1/z_0 + 1)} \tag{2-14}$$

Another convenient form [29] of the logarithmic velocity profile for computing mean wind speed $\bar{v}_2$ at some intermediate height z_2 when mean wind speeds $\bar{v}_1$ and $\bar{v}_3$ at heights z_1 and z_3 are known is

$$\bar{v}_2 = \bar{v}_3 - (\bar{v}_3 - \bar{v}_1)\frac{\ln(z_3/z_2)}{\ln(z_3/z_1)} \tag{2-15}$$

For most meteorological purposes the power-law profile is usually expressed as

$$\frac{\bar{v}}{\bar{v}_1} = \left(\frac{z}{z_1}\right)^k \tag{2-16}$$

with exponent k varying with surface roughness and atmospheric stability and usually ranging from 0.1 to 0.6 (Table 2-3) in the surface boundary layer [30, 31].

Table 2-3 Representative values of k for various temperature lapse rates

Surface	Height range		Super-adiabatic	Neutral	Stable	Inversion
	m	ft				
Meadows	10–70	33–230	0.25	0.27	...	0.61
Flat field	11–49	36–161	0.16	0.20	0.25	0.36
Grass field†	8–120	26–394	0.14	0.17	0.27	0.32–0.77‡
Airfield	9–27	30–89	0.09	0.08	0.18	...
Desert	6–61	20–200	0.15	0.18	0.22	...
Near wooded area	11–124	36–407	0.19	0.29	0.35	...‡

Source: Adapted by permission from [29].
† Captive-balloon observations [34]; all other measurements from towers.

‡ $T_{400 ft} - T_{5 ft}$, °F	0–2	2–4	4–6	6–8	8–10	10–12
k	0.32	0.44	0.59	0.62	0.63	0.77

Comparative tests of the two profiles have yielded inconclusive results. The literature suggests that the logarithmic law should be the more representative of the two relationships, but this is not supported by tests except under certain conditions. The logarithmic law has been found generally more representative of the wind profile in the lowest 5 to 8 m (15 to 25 ft) above the ground when the atmospheric temperature lapse rate was adiabatic or near adiabatic [32, 33]. However, Johnson [30] found greater wind-speed increases with height over prairie grass and a snow surface than was indicated by the logarithmic law even under adiabatic conditions.

The power-law profile is considered by some investigators [30, 34–36] to be more representative of the wind profile in the layer from several meters to about 100 m (300 ft) above the ground. Analyzing results of his own investigations, which were based on 1-hr averaging periods, and those of other investigators, DeMarrais [36] concluded that for this range of elevation exponent k increases with surface roughness and atmospheric stability except in the case of large superadiabatic lapse rates, when it increases with instability. Exponent k also varies with height of layer considered. Higher layers have lesser values of k when the lapse rate is superadiabatic, and they have greater values of k when the lapse rate is adiabatic or less. For adiabatic and superadiabatic lapse rates, k tends to range from 0.1 to 0.3, the variation being mainly in proportion to surface roughness. This is shown in Table 2-3, which shows also that for stable conditions k tends to range from 0.2 to 0.8, the higher values being associated with temperature inversions. A value of $\frac{1}{7}$ has been found applicable for a wide range of conditions in the layer 0 to 10 m (0 to 30 ft). Over a snow surface, under conditions favoring melting, a value of $\frac{1}{8}$ may be more appropriate.

Despite differences between the two types of relationships, when z_1/z_0 and k in the logarithmic and power-law profiles, respectively, are determined from observed winds at two levels in the surface boundary layer, say at 1 and 10 m (3 and 30 ft), differences between wind speeds indicated by the two profiles are usually within the limits of accuracy of wind-measuring devices.

2-23 Time Variation of Winds

Wind speeds are highest and most variable in winter, whereas middle and late summer is the calmest period of the year. In winter westerly winds prevail over the United States up to at least 6 km (20,000 ft), except near the Gulf of Mexico, where there is a tendency for southeasterly winds up to about $1\frac{1}{2}$ km (5000 ft). In summer, while there is still a tendency for westerly winds, there is generally more variation of direction with altitude. In the plains west of the Mississippi River there is a tendency for southerly winds up to $1\frac{1}{2}$ km (5000 ft), and on the Pacific Coast the winds at the lower altitudes are frequently from the northwest.

The diurnal variation of wind is significant only near the ground and is most pronounced during the summer. Surface-wind speed is usually at a minimum about sunrise and increases to a maximum in early afternoon. At about 300 m

(1000 ft) above the ground, the maximum occurs at night and the minimum in the daytime.

PROBLEMS

2-1 What would be the theoretical eastward velocity (constant angular momentum), in kilometers per hour, of a particle of air initially at rest relative to the earth's surface at the equator if displaced to 45°N lat?

2-2 A parcel of moist air at 15°C initially at 500 m, mean sea level, is forced to pass over a mountain ridge at 3000 m, mean sea level, then descends to its original elevation. Assuming that a rise of 1500 m produces saturation and precipitation and that the average pseudo-adiabatic lapse rate is one-half the dry-adiabatic, what is the final temperature of the parcel?

2-3 What is the heat of vaporization, in calories per gram, for water at (a) 8°C and (b) 68°F?

2-4 What is the density, in kilograms per cubic meter, of (a) dry air at 20°C and a pressure of 1000 millibars and (b) moist air with relative humidity of 70 percent at the same temperature and pressure?

2-5 Assuming dry- and wet-bulb temperatures of 60 and 52°F, respectively, at a pressure of 30.00 in, and using the psychometric tables in the Appendix determine (a) dewpoint temperature, (b) relative humidity, (c) saturation vapor pressure, and (d) actual vapor pressure, in millibars.

2-6 A radio sounding in saturated atmosphere shows temperatures of 14.0, 9.2, and 3.5°C at the 900-, 800-, and 700-millibar levels, respectively. Compute the precipitable water, in millimeters, in the layer between 900 and 700 millibars, and compare the result with that obtained from Fig. 2-16. (The temperature of 14°C at the 900-millibar level reduces pseudo-adiabatically to 18°C at 1000 millibars.)

2-7 Compute the weight, in kilograms, of 1 m³ of dry air at a temperature of (a) 10°C and a pressure of 900 millibars and (b) 30°C at the same pressure.

2-8 If air temperature is 40°F and dewpoint is 30°F, what is the relative humidity as determined by (a) Eq. (2-9) and (b) psychrometric tables in the Appendix?

2-9 What is the dewpoint as determined by Eq. (2-7) when air temperature is 20°C and relative humidity is 44 percent?

2-10 The saturation vapor pressure over water at −10°C is 2.86 millibars. What is it over ice at the same temperature? Compute by means of Eq. (2-6) and by Table A-12, and compare results.

2-11 Convert Eq. (2-9) so it can be used for Fahrenheit temperatures.

2-12 How many calories are required to evaporate 1 U.S. gallon of water at 15°C? How many pounds of ice at −10°C would the same amount of heat melt? (Specific heat of ice = 0.5.)

2-13 How many calories per square meter are required (a) to melt a 20-cm layer of ice with specific gravity of 0.90 at −10°C and (b) to evaporate the resulting meltwater without raising its temperature above 0°C? (Specific heat of ice = 0.5.)

2-14 A formula for estimating potential evapotranspiration from an orange grove requires wind speed at 5 m above the surface. What is the estimated speed at this level if the roughness length is 50 cm and an anemometer 10 m above the ground indicates a wind speed of 4.0 m/s?

2-15 A tower anemometer shows a wind speed of 40 kn at 100 m above the ground. What is the estimated speed, in miles per hour, 10 m above the ground indicated by the power-law profile with values for exponent k of (a) $\frac{1}{7}$ and (b) $\frac{1}{5}$?

2-16 Anemometers at 5 and 60 m on a tower record wind speeds of 5.0 and 10.0 m/s, respectively. Compute wind speeds at 10 and 30 m using (a) Eq. (2-15) and (b) Eq. (2-16).

2-17 Given a wind speed of 5.0 m/s at 2 m above the ground and a roughness length of 0.8 cm, compute (a) the friction velocity, in centimeters per second, and (b) the wind speed, in meters per second, at 0.5 m.

REFERENCES

1. Anderson, E. R.: Water-Loss Investigations: Lake Hefner Studies, Technical Report, *U.S. Geol. Surv. Prof. Pap.* 269, pp. 71–119, 1954.
2. Myers, V. A.: Infrared Radiation from Air to Underlying Surface, *ESSA Weather Bur. Tech. Note* 44-HYDRO-1, 1966.
3. Anderson, E. A., and D. R. Baker: Estimating Incident Terrestrial Radiation under All Atmospheric Conditions, *Water Resour. Res.,* vol. 3, no. 4, pp. 975–988, 1967.
4. Cooley, K. R., and S. B. Idso: A Comparison of Energy Balance Methods for Estimating Atmospheric Thermal Radiation, *Water Resour. Res.,* vol. 7, no. 1, pp. 39–45, February 1971.
5. Smith, W., and R. J. Younkin: An Operationally Useful Relationship between the Polar Jet Stream and Heavy Precipitation, *Mon. Weather Rev.,* vol. 100, no. 6, pp. 434–440, June 1972.
6. World Meteorological Organization: A Note on Climatological Normals, *Tech. Note* 84, Geneva, 1967.
7. World Meteorological Organization: Guide to Climatological Practices, WMO no. 100, *Tech. Pap.* 44, pp. V.9–V.11, Geneva, 1960.
8. Mitchell, J. M., Jr.: Effect of Changing Observation Time on Mean Temperature, *Bull. Am. Meteorol. Soc.,* vol. 39, no. 2, pp. 83–89, February 1958.
9. List, R. J. (ed.): "Smithsonian Meteorological Tables," 6th ed., pp. 350–373, Smithsonian Institution, Washington, 1966.
10. Goff, J. A., and S. Gratch: Low-Pressure Properties of Water from -160 to 212°F, *Trans. Am. Soc. Heat. Vent. Eng.,* vol. 52, pap. 1286, pp. 95–122, 1946. The formula presented in this paper was adopted as standard in 1947 by Resolution 965 of the Twelfth Conference of Directors of the International Meteorological Organization.
11. Bosen, J. F.: A Formula for Approximation of the Saturation Vapor Pressure over Water, *Mon. Weather Rev.,* vol. 88, no. 8, p. 275, August 1960.
12. Bosen, J. F.: Formula for Approximation of the Ratio of the Saturation Vapor Pressure over Ice to That over Water at the Same Temperature, *Mon. Weather Rev.,* vol. 92, no. 11, p. 508, November 1961.
13. Developed by J. F. Bosen but never published.
14. Bosen, J. F.: An Approximation Formula to Compute Relative Humidity from Dry Bulb and Dew Point Temperatures, *Mon. Weather Rev.,* vol. 86, no. 12, p. 486, December 1958.
15. Harrison, L. P.: Calculation of Precipitable Water, *ESSA Tech. Mem. WBTM TDL* 33, June 1970.
16. Solot, S.: Computation of Depth of Precipitable Water in a Column of Air, *Mon. Weather Rev.,* vol. 67, no. 4, pp. 100–103, April 1939.
17. Peterson, K. R.: A Precipitable Water Nomogram, *Bull. Am. Meteorol. Soc.,* vol. 42, no. 2, pp. 119–121, February 1961.
18. Smith, W. L.: Note on the Relationship between Total Precipitable Water and Surface Dew Point, *J. Appl. Meteorol.,* vol. 5, no. 5, pp. 726–727, October 1966. Comments by L. Berkofsky, *J. Appl. Meteorol.,* vol. 6, no. 5, pp. 959–961, October 1967, and F. K. Schwarz, *J. Appl. Meteorol.,* vol. 7, no. 3, pp. 509–510, June 1968.
19. For English units see Tables of Precipitable Water, *U.S. Weather Bur. Tech. Pap.* 14, 1951. For metric units see W. O. Eihle, R. J. Powers, and R. A. Clark, Meteorological Tables for Determination of Precipitable Water, Temperatures, and Pressures Aloft for a Saturated Pseudoadiabatic Atmosphere—in the Metric System, *Tex. A & M Univ. Water Resour. Inst. Rep.* 16, December 1968.
20. Foster, N. B., D. T. Volz, and L. W. Foskett: A Spectral Hygrometer for Measuring Total Precipitable Water, in A. Wexler (ed.): "Humidity and Moisture: Measurement and Control in Science and Industry," vol. 1, pp. 455–464, Reinhold, New York, 1965.
21. Stine, S. L.: Carbon Humidity Elements: Manufacture, Performance and Theory, in A. Wexler (ed.): "Humidity and Moisture: Measurement and Control of Science and Industry," vol. 1, pp. 316–330, Reinhold, New York, 1965.

22. Appleman, H. S.: Relative Humidity Errors Resulting from Ambiguous Dew-Point Hygrometer Readings, *J. Appl. Meteorol.,* vol. 3, no. 1, pp. 113–115, February 1964.
23. Lott, G. A., J. T. Riedel, and F. P. Ho: Precipitable Water over the United States, Volume 1: Monthly Means; Volume 2: Semi-Monthly Maxima, *NOAA Tech. Rep. NWS* 20, vol. 1, 1976; vol. 2, 1979.
24. Sutton, O. G.: "Micrometeorology," pp. 78–85, 229–241, McGraw-Hill, New York, 1953.
25. See [24], p. 233.
26. Szeicz, G., G. Endrodi, and S. Tajchman: Aerodynamic and Surface Factors in Evaporation, *Water Resour. Res.,* vol. 5, no. 2, pp. 380–394, April 1969.
27. Ruggles, K. W.: The Vertical Mean Wind Profile over the Ocean for Light to Moderate Winds, *J. Appl. Meteorol.,* vol. 9, no. 3, pp. 389–395, June 1970.
28. See [24], p. 84.
29. DeMarrais, G. A.: Wind-Speed Profiles at Brookhaven National Laboratory, *J. Meteorol.,* vol. 16, no. 2, pp. 181–190, April 1959.
30. Johnson, O.: An Examination of the Vertical Wind Profile in the Lowest Layers of the Atmosphere, *J. Meteorol.,* vol. 16, no. 2, pp. 144–148, April 1959.
31. Geiger, R.: "The Climate near the Ground," pp. 116–117, Harvard University Press, Cambridge, Mass., 1965.
32. Best, A. C.: Transfer of Heat and Momentum in the Lowest Layers of the Atmosphere, *Meteorol. Off. Geophys. Mem.,* vol. 7, no. 65, 1935.
33. Thornthwaite, C. W., and P. Kaser: Wind-Gradient Observations, *Trans. Am. Geophys. Union,* vol. 24, pt. 1, pp. 166–182, 1943.
34. Frost, R.: The Velocity Profile in the Lowest 400 Feet, *Meteorol. Mag.,* vol. 76, no. 1, pp. 14–18, 1947.
35. Frost, R.: Atmospheric Turbulence, *Q. J. R. Meteorol. Soc.,* vol. 74, nos. 321–322, pp. 316–338, July–October 1948.
36. See [29], p. 189.

BIBLIOGRAPHY

General circulation

Manabe, S.: The Atmospheric Circulation and the Hydrology of the Earth's Surface, *Mon. Weather Rev.,* no. 11, pp. 739–774, November 1969.
Peixoto, J. P.: Atmospheric Vapor Flux Computations for Hydrological Purposes, *Rep. WMO/IHD Proj. 20, Publ. 357,* World Meteorological Organization, Geneva, 1974.
Rasmusson, E.: Atmospheric Water Vapor Transport and the Water Balance of North America, *Mon. Weather Rev.,* vol. 95, no. 7, pp. 403–426, July 1967.
——: A Study of the Hydrology of Eastern North America Using Atmospheric Vapor Flux Data, *Mon. Weather Rev.,* vol. 99, no. 2, pp. 119–135, February 1971.
Rasmusson, E. M.: Hydrological Application of Atmospheric Vapour-Flux Analysis, *Opera. Hydrol. Rep.* 11, WMO no. 476, World Meteorological Organization, Geneva, 1977.

General climatology

Critchfield, H. J.: "General Climatology," 3d ed., Prentice-Hall, Englewood Cliffs, N.J., 1974.
Sellers, W. D.: "Physical Climatology," University of Chicago Press, Chicago, 1965.
Trewartha, G. T., and L. H. Horn: "An Introduction to Climate," 5th ed., McGraw-Hill, New York, 1980.

General and jet-stream meteorology

Battan, L. J.: "Fundamentals of Meteorology," Prentice-Hall, Englewood Cliffs, N.J., 1979.
Byers, H. R.: "General Meteorology," 4th ed., McGraw-Hill, New York, 1974.

Cole, F. W.: "Introduction to Meteorology," 2d ed., Wiley, New York, 1975.

Donn, W. L.: "Meteorology," 4th ed., McGraw-Hill, New York, 1975.

Reiter, E. R.: "Jet-Stream Meteorology," University of Chicago Press, Chicago, 1963.

Riehl, H.: "Introduction to the Atmosphere," 3d ed., McGraw-Hill, New York, 1978.

Meteorological instruments and observations

Great Britain Meteorological Office: "Handbook of Meteorological Instruments," 2 vols., London, 1961.

Middleton, W. E. K., and A. F. Spilhaus: "Meteorological Instruments," 3d ed., University of Toronto Press, Toronto, 1953.

Sterzat, M. S.: "Instruments and Methods for Meteorological Observations," *Proc. All-Am. Meteorol. Conf.,* vol. 9, 1966; available from U.S. Dept. of Commerce National Technical Information Center, Springfield, Va.

U.S. National Weather Service: Substation Observations, *Obs. Handb.,* no. 2, 1970; rev. December 1972.

World Meteorological Organization: Guide to Meteorological Instrument and Observing Practices, WMO no. 8, *Tech. Pap.* 3, Geneva, 1969.

Solar and terrestrial radiation

Coulson, K. L.: "Solar and Terrestrial Radiation, Methods and Measurements," Academic Press, New York, 1975.

Drummond, A. J. (ed.): "Precision Radiometry," Advances in Geophysics, vol. 14, Academic Press, New York, 1969.

Kondratyev, K. Ya.: "Radiation Processes in the Atmosphere," World Meteorological Organization, Geneva, 1972.

Paltridge, G. W., and C. M. R. Platt: "Radiative Processes in Meteorology and Climatology," Elsevier, New York, 1976.

Thompson, E. S.: Computation of Solar Radiation from Sky Cover, *Water Resour. Res.,* vol. 12, no. 5, pp. 859–865, October 1976.

U.S. National Weather Service: Solar Radiation Observations, *Obs. Handb.,* no. 3, March 1977.

Wallace, J. M., and P. V. Hobbs: "Atmospheric Science, An Introductory Survey," Academic Press, New York, 1977.

World Meteorological Organization: Preciseness of Pyrheliometric Measurements, WMO no. 209, *Tech. Pap.* 109, Geneva, 1966.

Urban effects on temperature

Bornstein, R. D.: Observations of the Heat Island Effect of New York City, *J. Appl. Meteorol.,* vol. 7, no. 4, pp. 575–582, August 1968.

Griffiths, J. F.: Bibliography of Urban Modification of the Atmospheric and Hydrologic Environment, *NOAA Tech. Memo. EDS* 21, February 1974.

Moffitt, B. J.: The Effects of Urbanization on Mean Temperatures at Kew Observatory, *Weather,* vol. 27, pp. 121–129, March 1972.

Munn, R. E., M. S. Hirt, and B. F. Findlay: A Climatological Study of the Urban Temperature Anomaly in the Lakeshore Environment at Toronto, *J. Appl. Meteorol.,* vol. 8, no. 3, pp. 411–422, June 1969.

Wind profiles

Bernstein, A. B.: An Examination of Three Wind Profile Hypotheses, *J. Appl. Meteorol.,* vol. 5, no. 2, pp. 217–219, April 1966.

Bradley, E. F.: A Micrometeorological Study of Velocity Profiles and Surface Drag in the Region Modified by a Change in Surface Roughness, *Q. J. R. Meteorol. Soc.,* vol. 94, no. 401, pp. 361–379, July 1968.

McVehil, G. E.: Wind and Temperature Profiles near the Ground in Stable Stratification, *Q. J. R. Meteorol. Soc.*, vol. 90, no. 384, pp. 136–146, April 1964.

Swinbank, W. C.: The Exponential Wind Profile, *Q. J. R. Meteorol. Soc.*, vol. 90, no. 384, pp. 119–135, April 1964.

Taylor, P. A.: On Wind and Shear Stress Profiles above a Change in Surface Roughness, *Q. J. R. Meteorol. Soc.*, vol. 95, no. 403, pp. 77–91, January 1969.

Webb, E. K.: Profile Relationships: The Log Linear Range and Extension to Strong Stability, *Q. J. R. Meteorol. Soc.*, vol. 96, no. 407, pp. 67–90, January 1970.

UNITED STATES DATA SOURCES

The main sources of data on temperature, humidity, and wind are the monthly bulletins entitled *Climatological Data* published by the Environmental Data Service of the National Oceanic and Atmospheric Administration (NOAA). Hourly and daily totals and monthly means of solar radiation are published for 38 stations by the National Climatic Center in its *Monthly Summary Solar Radiation Data*. The *Climatic Summary of the United States* summarizes monthly and annual data from the beginning of record through 1960. *Monthly Weather Review* contains maps summarizing the weather of the previous month for the country. Summaries of normals may be found in the following:

Daily Normals of Temperature and Heating and Cooling Degree Days, *Climatography of the United States,* no. 84, NOAA Environmental Data Service, September 1973.

Monthly Normals of Temperature, Precipitation, and Heating and Cooling Degree Days, *Climatography of the United States,* no. 81 (by states), NOAA Environmental Data Service, August 1973.

"Climatic Atlas of the United States," ESSA Environmental Data Service, 1968. (This publication presents summaries of wind, temperature, precipitation, humidity, evaporation, and solar radiation on maps.)

Other summaries may be found in:

Climates of the States, *Climatography of the United States,* no. 60-1–60-52 (by states and other areas), U.S. Weather Bureau.

Decennial Census of the United States Climate: Summary of Hourly Observations 1951–1960, *Climatography of the United States,* no. 82-1–82-48 (by cities), U.S. Weather Bureau. (This publication gives monthly and annual frequencies of occurrence of various values of temperature, precipitation, relative humidity, wind speed and direction, and cloudiness for over 250 cities.)

In cooperation with the World Meteorological Organization, the Environmental Data Service also publishes *Monthly Climatic Data for the World,* which contains monthly precipitation amounts and averages of pressure, temperature, and relative humidity for about 1000 stations distributed throughout the world. It also includes monthly averages of upper-air temperature and dewpoint for about 250 stations.

Information on all climatological data published by the Environmental Data Service is contained in its Selective Guide to Climatic Data Sources, *Key Meteorol. Rec. Docum.,* no. 4.11, 1979.

THREE

PRECIPITATION

Hydrologists have long known that only about one-fourth the total precipitation that falls on continental areas is returned to the seas by direct runoff and underground flow. Hence, it was generally believed that continental evaporation constituted the principal source of moisture for continental precipitation. Many ideas for increasing precipitation were based on the premise (now known to be erroneous) that more precipitation would result from increased atmospheric moisture through local evaporation. Impounding streamflow in lakes and ponds and selecting farm crops with high transpiration rates were among the methods suggested. That such methods are ineffective is demonstrated by the Caspian Sea. Although this sea has an area of about 438,000 km² (169,000 mi²), which is larger than California, and its annual evaporation is roughly estimated [1] to be about 500 to 600 billion cubic meters (400 to 500 million acre-feet), the annual precipitation along its shores is generally less than 250 mm (10 in).

It is now recognized that evaporation from ocean surfaces is the chief source of moisture for precipitation, and probably no more than about 10 percent of continental precipitation can be attributed to continental evaporation. Nearness to oceans, however, does not necessarily lead to adequate precipitation, as evidenced by many desert islands. The location of a region with respect to the general circulation, latitude, and distance to a moisture source is primarily responsible for its climate. Orographic barriers often exert more influence on the climate of a region than nearness to a moisture source does. These climatic and geographic factors determine the amount of atmospheric moisture over a region, the frequency and kind of precipitation-producing storms passing over it, and thus its precipitation.

3-1 Formation of Precipitation

Moisture is always present in the atmosphere, even on cloudless days. For precipitation to occur, some mechanism is required to cool the air sufficiently to bring it to or near saturation. The large-scale cooling needed for significant amounts of precipitation is achieved by lifting the air. This is accomplished by convective systems resulting from unequal radiative heating or cooling of the earth's surface and atmosphere or by convergence caused by orographic barriers. Saturation, however, does not necessarily lead to precipitation.

Condensation and freezing nuclei. Assuming that the air is close to saturation, formation of fog or cloud droplets or ice crystals generally requires the presence of *condensation* or *freezing nuclei* on which the droplets or crystals form. These nuclei are small particles of various substances, not necessarily hygroscopic, usually ranging in size from about 0.1 to 10 μm in diameter. Those with diameters less than 3 μm are within the size range of aerosols and might remain airborne indefinitely were it not for precipitation fallout. Condensation nuclei usually consist of products of combustion, oxides of nitrogen, and salt particles. The last are most effective and may result in condensation with relative humidity as low as 75 percent.

Freezing nuclei usually consist of clay minerals, kaolin being the most common [2]. Freezing nuclei serve only to nucleate the liquid phase and thus initiate the growth of ice crystals. Pure water droplets may remain in the liquid state to temperatures as low as $-40°C$ ($-40°F$), and it is only in the presence of such supercooled droplets that natural freezing nuclei are active. Carbon dioxide, CO_2, can produce ice crystals in a supercooled cloud at any temperature up to about 0°C (32°F); silver iodide, AgI, has a nucleating threshold of about $-4°C$ (25°F).

Growth of water droplets and ice crystals. Upon nucleation the droplet or ice crystal grows to visible size in a fraction of a second through diffusion of water vapor to it, but growth thereafter is slow. Diffusion by itself leads only to fog or cloud elements generally smaller than 10 μm in diameter, with some reaching 50 μm. Since condensation tends to enlarge droplets or ice crystals at about the same rate, differences in size result mainly from differences in size of the nuclei on which they are formed. While cloud elements tend to settle, the average element weighs so little that only a slight upward motion of the air is needed to support it. Most droplets in nonprecipitating stratus have diameters under 10 μm, and an upward current of less than 0.5 cm/s (0.2 in/s) is sufficient to keep them from falling. Ice crystals of the same weight, because of their shape and larger size, can be supported by even lower velocities.

Upward velocities under and in clouds often greatly exceed those required to support cloud elements. For precipitation to occur, cloud elements must increase in size until their falling speeds exceed the ascensional rate of the air. The cloud element must also be large enough to penetrate the unsaturated air below the cloud base without evaporating completely before reaching the ground. A drop falling out of a cloud base at 1 km in air of 90 percent relative humidity rising at 10 cm/s (0.3 ft/s) would require [3] a diameter of about 440 μm in order to reach the ground with a diameter of 200 μm, which is often considered to mark the boundary between cloud droplet size and precipitation size.

Some growth of cloud elements through diffusion is to be expected because of differences in vapor pressure resulting from differences in size and temperature. Electric charges appear to have some effect on the growth of cloud elements also. Diffusion is most effective when ice crystals and liquid-water droplets are both present in a cloud. Saturation vapor pressure over ice is less than that over liquid water at the same temperature, and this results in a net transport of moisture from liquid droplets to ice crystals. Growth is particularly rapid when liquid droplets greatly outnumber ice crystals, and this is usually the case. This process is considered important in causing heavy precipitation, but since heavy rainfall can occur from *warm clouds,* with above-freezing temperatures throughout, other factors must sometimes be involved.

Collision and coalescence of cloud and precipitation elements (*accretion*) are considered the most important factors leading to significant precipitation. Collisions between cloud and precipitation particles arise mostly from differences in falling speeds as a result of differences in size. Heavier particles fall faster (or ascend more slowly in rising currents) than smaller particles do. Particles that collide usually coalesce to form larger particles, and the process may be repeated a number of times. It has been estimated [4] that in a typical heavy rain seven collisions occur for every kilometer of fall.

Raindrops may grow as large as 6 mm (0.25 in) in diameter. Maximum falling speeds, or *terminal velocities* (Table 3-1), tend to level off as the drops approach maximum size because of increasing air resistance due to flattening. For large diameters, the deformation may be sufficient to break up the drops before they can attain terminal velocity.

Single ice crystals may reach the ground, but usually a number of them collide and coalesce to form a cluster and fall as *snowflakes.* Coalescence is most effective when temperatures are near freezing. The fall velocity of snow and ice crystals is usually no more than about 1 m/s (3 ft/s), thus providing considerable time for growth by diffusion. At temperatures near freezing, rime (Sec. 3-2) is formed. Since riming enhances both coalescence and inequalities in falling speeds, it increases the probability of collision; it is then that the largest snowflakes are observed.

Maximum liquid-water content of clouds. Speeds of ascending air in vigorous convective systems usually far exceed terminal velocities (Table 3-1) of the largest raindrops. Radar observations have indicated ascensional rates as high as 40 to 50 m/s (90 to 110 mi/hr) in cumulonimbus. Aircraft observations indicate a horizontal diameter of about 1.5 km (5000 ft) for updrafts in a thunderstorm cell, which usually ranges from about 6 to 10 km (4 to 6 mi) in diameter. Downdrafts are about the same size as updrafts. Ascensional rates in

Table 3-1 Terminal velocity of water drops in still air
Pressure 1013.3 mbars, temperature 20°C, relative humidity 50 percent

Drop diameter		Terminal velocity	
mm	in	cm/s	ft/s
0.5	0.02	206	6.8
1.0	0.04	403	13.2
1.5	0.06	541	17.7
2.0	0.08	649	21.3
3.0	0.12	806	26.4
4.0	0.16	883	29.0
5.0	0.20	909	29.8
5.5	0.22	915	30.0
5.8	0.23	917	30.1

Source: From R. Gunn and G. D. Kinzer, The Terminal Velocity of Fall for Water Droplets in Stagnant Air, *J. Meteorol.,* vol. 6, pp. 243–248, August 1949. Used with permission of American Meteorological Society.

updrafts are not uniform, however, and high-speed vertical jets in updrafts may have diameters of no more than a few hundred meters.

The stronger updrafts prevent even the largest raindrops from falling and carry all precipitation elements to the upper portions of the clouds to produce an accumulation of liquid water far exceeding that of the ordinary cloud elements. Theory and radar observations suggest [5] that the height of the accumulation depends on updraft speed. Eventually, the accumulated water is precipitated as a result of the weakening of the updraft or by horizontal displacement from the supporting updraft to a weaker one or, as often happens, to a downdraft, which possibly may be initiated [6] by the mass of accumulated water. When suddenly precipitated in a downdraft, the resulting torrential downpour lasts but a few minutes, and point rainfall is usually less than 100 mm (4 in). In a thunderstorm there may be several such downpours, or bursts, from a number of cells, and total point rainfall in 1 hr may exceed 200 mm (8 in).

In thick nonprecipitating cumuli the maximum concentration of liquid water may be near 4 g/m^3, but the mean for the cloud might be only half this value. Mean values for the cloud as a whole apparently have little significance in relation to natural precipitation. The maximum liquid-water content of nonprecipitating clouds usually ranges from about 0.5 g/m^3 in thin stratus to about 4 g/m^3 in thick cumulus, but values exceeding 30 g/m^3 have been measured [7]. Clouds having concentrations of 4 g/m^3 or more usually produce precipitation that reaches the ground. Rainfall rates tend to be correlated with maximum liquid-water content. For heavy rains it appears that the rainfall rate increases about 25 mm/hr (1 in/hr) for each gram per cubic meter.

3-2 Forms of Precipitation

Any product of condensation of atmospheric water vapor formed in the free air or at earth's surface is a *hydrometeor*. Since the hydrologist is primarily interested in precipitation, only those hydrometeors referring to falling moisture are here defined. Among those hydrometeors not included are damp haze, fog, ice fog, drifting snow, blowing snow, and frost.

Drizzle, sometimes called *mist,* consists of tiny liquid water droplets, usually with diameters between 0.1 and 0.5 mm (0.004 and 0.02 in), with such slow settling rates that they occasionally appear to float. Drizzle usually falls from low stratus and rarely exceeds 1 mm/hr (0.04 in/hr).

Rain consists of liquid water drops mostly larger than 0.5 mm (0.02 in) in diameter. *Rainfall* usually refers to amounts of liquid precipitation. In the United States rain is reported in three intensities:

Light.   For rates of fall up to 0.10 in/hr (2.5 mm/hr) inclusive
Moderate.   From 0.11 to 0.30 in/hr (2.8 to 7.6 mm/hr)
Heavy.   Over 0.30 in/hr (7.6 mm/hr)

In the United States, rainfall is also labeled *excessive precipitation* when amounts R, in hundredths of an inch, for durations t from 5 to 180 min equal or exceed those indicated by $R = t + 20$. Tabulations of excessive-precipitation data have been compiled and published since 1896, but caution should be exercised in their use because of changes in the threshold formulas and methods of compilation during the period of record.

Glaze is the ice coating, generally clear and smooth, formed on exposed surfaces by the freezing of supercooled water deposited by rain or drizzle. Its specific gravity may be as high as 0.8 to 0.9.

Rime is a white, opaque deposit of ice granules more or less separated by trapped air and formed by rapid freezing of supercooled water drops impinging on exposed objects. Its specific gravity may be as low as 0.2 to 0.3.

Snow is composed of ice crystals, chiefly in complex, branched hexagonal form, and often agglomerated into *snowflakes,* which may reach several inches in diameter. The density of freshly fallen snow varies greatly; 125 to 500 mm (5 to 20 in) of snow is generally required to equal 25 mm (1 in) of liquid water. The average density (specific gravity) is often assumed to be 0.1.

Hail is precipitation in the form of balls of ice, produced in convective clouds, mostly cumulonimbus. *Hailstones* may be spheroidal, conical, or irregular in shape, and range from about 5 to over 125 mm (0.2 to over 5 in) in diameter. They are generally composed of alternating layers of glaze and rime, and their specific gravity is about 0.8. The largest hailstone of record in the United States fell at Coffeyville, Kan., Sept. 3, 1970. It measured 44 cm (17.5 in) in circumference, and weighed 766 g (1.67 lb).

Sleet consists of transparent, globular, solid grains of ice formed by the freezing of raindrops or refreezing of largely melted ice crystals falling through a layer of subfreezing air near the earth's surface.

3-3 Types of Precipitation

Precipitation is often typed according to the factor mainly responsible for the lifting which causes it. *Cyclonic precipitation* results from the lifting of air converging into a low-pressure area, or cyclone. Cyclonic precipitation may be either frontal or nonfrontal. *Frontal precipitation* results from the lifting of warm air on one side of a frontal surface over colder, denser air on the other side. *Warm-front precipitation* is formed in the warm air advancing upward over a colder air mass. The rate of ascent is relatively slow, since the average slope of the frontal surface is usually between 1/100 and 1/300. Precipitation may extend 300 to 500 km (200 to 300 mi) ahead of the surface front and is generally light to moderate and nearly continuous until passage of the front. *Cold-front precipitation,* on the other hand, is of showery nature and is formed in the warm air forced upward by an advancing mass of cold air, the leading edge of which is the surface cold front. Cold fronts move faster than warm fronts, and the frontal surfaces, with slopes usually averaging 1/50 to 1/150, are steeper. Consequently, the warm air is forced upward more rapidly than by the warm front, and precipitation rates are generally much higher. Heaviest amounts and intensities occur near the surface front.

Convective precipitation is caused by the rising of warmer, lighter air in colder, denser surroundings. The difference in temperature may result from unequal heating at the surface, unequal cooling at the top of the air layer, or mechanical lifting when the air is forced to pass over a denser, colder air mass or over a mountain barrier. Convective precipitation is spotty, and its intensity may range from light showers to cloudbursts. *Orographic precipitation* results from mechanical lifting over mountain barriers. In rugged terrain the orographic influence is so marked that storm precipitation patterns tend to resemble that of mean annual precipitation. In nature, the effects of these various types of cooling are often interrelated, and the resulting precipitation cannot be identified as being of any one type.

3-4 Artificially Induced Precipitation

Weather modification, sometimes referred to as *weather control,* is the general term for efforts to alter artificially the natural meteorological phenomena of the atmosphere. Attempts to increase or decrease precipitation, suppress hail and lightning, mitigate hurricanes, dissipate fog, prevent frost, alter radiation balance, etc., are all included under weather modification. *Cloud modification,* or *cloud seeding,* is one type of weather modification, and usually has as its goal either dissipation of the cloud or stimulation of precipitation. Urban air pollution is an effective modifier of weather.

It was demonstrated in 1946 that dry ice can cause precipitation in a cloud† containing supercooled water droplets. This discovery soon led to further discoveries that certain salts, notably silver iodide, can also induce precipitation. Both dry ice and silver iodide, the two most commonly used seeding agents, act as freezing nuclei in supercooled clouds. Seeding clouds with dry ice requires delivery into the cloud by aircraft, balloons, or rockets. Silver iodide, which is most effective when heated to vaporization, may be delivered into the cloud by either airborne or ground-based generators but has the disadvantage that its effectiveness is reduced by exposure to sunlight, the number of effective particles decreasing by a factor of about 10 for every hour of exposure. Nevertheless, the low operation cost of ground-based generators has made this method the most commonly used for augmentation of precipitation.

The nucleating threshold temperature of silver iodide, however, is $-4°C$ (25°F), and for cloud-base temperatures warmer than freezing, silver iodide released from a ground-based generator may become embedded in liquid droplets. In this case, crystallization will not occur until the temperature has been lowered by convective lifting to between -10 and $-15°C$ (14 and 5°F). Thus, unless there is enough convective lifting to provide the necessary cooling, seeding from a ground-based generator is ineffective. On the other hand, silver iodide introduced into the cloud at levels where temperatures are $-5°C$ (23°F) or colder will result in immediate crystallization of pure liquid droplets upon contact.

The effectiveness of cloud seeding depends on many factors such as height of cloud base and top, cloud temperature, difference between density inside the cloud and that outside (*buoyancy*), updraft velocity distribution, amount and concentration of liquid water in the cloud, number and distribution of natural freezing or condensation nuclei, the number of artificial nuclei added, and where they are introduced into the cloud. Tests of the effectiveness of cloud seeding are now made where results can be observed over relatively dense rain-gage networks and by radar and satellites, both of which are also used to identify favorable seeding conditions [8]. Precipitation over the target area is usually compared with that over nearby unseeded areas, or control areas.

There are two general approaches to cloud seeding for augmentation of precipitation. The *static approach* consists of introducing about one artificial nucleus per liter (61 in³) of cloud air, to produce ice crystals which, through diffusion and accretion, eventually grow into precipitation particles. The *dynamic approach* involves massive seeding, say 100 to 1000 nuclei per liter of cloud air, of cumulus clouds to stimulate, through release of heat of fusion, the buoyancy forces and circulations that sustain them. With favorable tempera-

---

† Limited inconclusive experiments with solid CO_2 were made in Holland in 1930 by August W. Veraart.

ture and moisture distributions, massive seeding of supercooled cumulus clouds with tops at heights below the $-40°C$ ($-40°F$) level may raise the cloud tops 3 to 5 km (10,000 to 16,000 ft) within a few minutes so that they extend well above the level where spontaneous natural crystallization can occur, thus increasing the probability of natural precipitation.

The dynamic approach is particularly useful in lower latitudes, where tops of cumuli are often below the $-25°C$ ($-13°F$) level. Tests over Florida indicated that seeded clouds are generally larger and longer-lasting than unseeded clouds and produce about twice as much rainfall.

Stratus clouds offer poor prospects for effective seeding with the purpose of increasing rainfall. The moisture content is relatively low, and atmospheric conditions attending such clouds are too stable. Seeding may result in weak turbulence which distributes the ice nuclei so that the cloud may be dissipated with little or no precipitation if the cloud temperature is below the nucleating threshold temperature of the seeding agent used.

Orographic clouds generally have a much lower moisture content than large summer cumuli, but their formation is continuous so long as there is a moist air flow directed upslope. Seeding of orographic clouds under favorable conditions has been fairly effective. A study [9] of seven randomized winter orographic cloud-seeding projects in the Rocky Mountains and on the Pacific Coast of the United States indicated that seeding effectiveness depends on the stability and water content of the clouds and on cloud-top temperature.

Seeding orographic clouds over Kings River Basin in the Sierra Nevada of California over a 20-year period (1955–1974) is reported [10] to have produced significant effects in 14 years. Positive effects could not be identified statistically in the other 6 years. While the annual increases in streamflow over the 20-year period averaged only 4 percent, they were significant at the 1 percent level. In terms of water volume, the 14 positive years produced a total of 1.6×10^9 m³ (1.3×10^6 acre·ft) of additional water, an average of about 116×10^6 m³ (94,000 acre·ft). The average annual increase for the entire 20-year period was 81×10^6 m³ (66,000 acre·ft) at a cost of about one million dollars, or about \$0.60/1000 m³ (\$0.77/acre·ft).

Seeding with freezing nuclei is ineffective in warm clouds because of the absence of supercooled droplets. In such clouds, the only way raindrops can be formed is through collision and coalescence, which cannot occur until some cloud droplets have reached a radius of about 20 μm. This growth of cloud droplets can be initiated or stimulated by introducing into the cloud either large condensation nuclei or droplets of the required size or larger. Experiments have been conducted using sodium chloride and water drops sprayed from aircraft, apparently with some success [11, 12]. The better results of the relatively few experiments were obtained when the seeding was done in the cloud base.

There is little question that seeding can increase precipitation from a cloud under favorable conditions. The effectiveness of localized seeding in increasing precipitation over large areas is difficult to evaluate. While massive seeding of cumulus clouds may produce appreciable rainfall per cloud, it is important to know whether this artificially induced rainfall represents a net increase, a decrease, or merely a redistribution of the rainfall over the area encompassing the cloud. The effect of individual seedings of many supercooled cumuli over hundreds of square kilometers has yet to be determined.

Seeding of precipitating cloud systems of extratropical cyclones over flat or hilly terrain has failed to produce any statistically significant increases. The processes producing steady precipitation over large areas may be so efficient in utilizing the total cloud condensate that any possible increases from seeding would be small [13].

Cities, power plants, steel mills, and agricultural activities such as irrigation and deforestation can modify the weather on local and regional scales. MET-ROMEX (Metropolitan Meteorological Experiment), a 5-year project initiated in 1971 to determine the effects of large cities on clouds and precipitation, reported [14] the following local increases for St. Louis, Mo.: 30 percent in total precipitation, 40 percent in heavy rainfall rates, 45 percent in thunderstorms, and 100 percent in hailfall intensities.

Evaluation of cloud-seeding effectiveness is difficult [15] because increases of 10 to 20 percent are hard to detect against natural variability. In reviewing cloud-seeding results, one finds that until about 1960 reports on seeding experiments were highly conflicting. Some differences undoubtedly arose from indiscriminate seeding without regard to cloud characteristics. Some resulted from inadequate controls for proper evaluation. Also, some reports were probably biased in favor of the rainmakers, some of whom may well have been more interested in the monetary than in the scientific aspects of cloud seeding. Much confusion has been eliminated by (1) increased knowledge of cloud physics, (2) more stringent, statistically designed controls, and (3) more and better radar and rain-gage network observations. The effects of prolonged cloud seeding on the climate, hydrology, ecology, and economy of a region are largely unknown. A discussion of these effects [16–18] and the legal problems [19] arising from cloud seeding is outside the scope of this text.

In 1972, cloud-seeding projects covered 221,000 km^2 (85,300 mi^2). By 1977, 88 projects in 23 states covered 676,000 km^2 (261,000 mi^2), or about 7 percent of the area of the United States. A 1978 survey of the benefits of cloud seeding for augmenting precipitation led to the conclusion that there was a need for better evidence on which users could base decisions about seeding [20].

MEASUREMENT OF PRECIPITATION

A variety of instruments and techniques have been developed for gathering information on precipitation. Instruments for measuring amount and intensity of precipitation are the most important. Other instruments include devices for measuring raindrop-size distribution and for determining the time of beginning and ending of precipitation. All forms of precipitation are measured on the basis of the vertical depth of water that would accumulate on a level surface if the precipitation remained where it fell. In the United States precipitation is measured in inches and hundredths. In the metric system precipitation is measured in millimeters and tenths.

3-5 Precipitation Gages

Any open receptacle with vertical sides is a convenient rain gage, but because of varying wind and splash effects the measurements are not comparable unless the receptacles are of the same size [21] and shape [22] and similarly exposed. The standard gage [23] (Fig. 3-1) of the U.S. National Weather Service has a collector (receiver) of 8-in (20.3-cm) diameter. Rain passes from the collector into a cylindrical measuring tube inside the overflow can. The measuring tube has a cross-sectional area one-tenth that of the collector so that a 0.1-in rainfall will fill the tube to a depth of 1 in. With a measuring stick marked in tenths of an inch, rainfall can be measured to the nearest 0.01 in (0.25 mm). The collector and tube are removed when snow is expected. The snow caught in the outer container, or overflow can, is melted, poured into the measuring tube, and measured.

Three types of recording gages in common use are the tipping-bucket gage, the weighing gage, and the float gage. In the *tipping-bucket gage* (Fig. 3-2) the

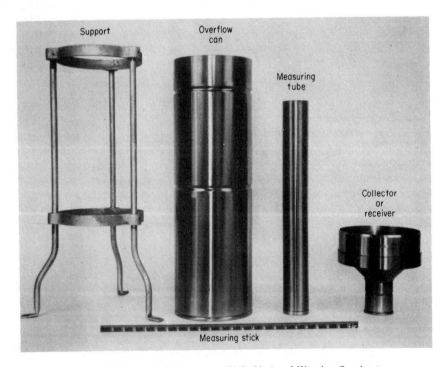

Figure 3-1 Standard 8-in precipitation gage. *(U.S. National Weather Service.)*

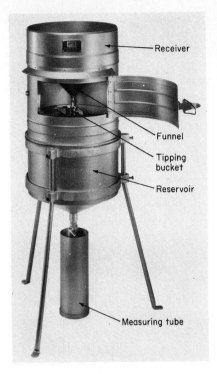

Receiver

Funnel

Tipping
bucket

Reservoir

Measuring tube

Figure 3-2 Tipping-bucket rain gage. (*Belfort Instrument Co.*)

water caught in the collector is funneled into a two-compartment bucket; 0.01 in, 0.1 mm, or some other designed quantity of rain will fill one compartment and overbalance the bucket so that it tips, emptying into a reservoir and moving the second compartment into place beneath the funnel. As the bucket is tipped, it actuates an electric circuit. This type of gage is not suitable for measuring snow without heating the collector. Heating of any gage, however, results in deficient catch because of convective currents and increased evaporation.

The *weighing-type gage* (Fig. 3-3) weighs the rain or snow which falls into a bucket set on the platform of a spring or lever balance. The increasing weight of the bucket and its contents is recorded on a chart. The record thus shows the accumulation of precipitation.

In *float recording gages* the rise of the float with increasing catch of rainfall is recorded. Some gages must be emptied manually, while others are emptied automatically by self-starting siphons. In most gages the float is placed in the receiver, but in some the receiver rests in a bath of oil or mercury and the float measures the rise of the oil or mercury displaced by the increasing weight of the receiver as the rainfall catch accumulates. Floats may be damaged if the rainfall catch freezes.

Most gages record by a pen trace on a chart. The *punched-tape recorder* punches the amount of precipitation accumulated in the collector on a tape in digital code, which later can be run through a translator for adapting to compu-

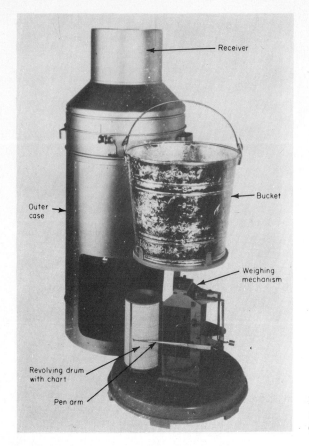

Figure 3-3 Weighing-type precipitation gage, 12-in dual traverse. (*Belfort Instrument Co.*)

ter evaluation of the record. Gages recording on magnetic tape or solid-state memory device are now coming into use. Their advantage is that they can be read directly into a computer.

Storage gages are used in remote regions where frequent servicing is impracticable. Weighing-type storage gages operate from 1 to 2 months without servicing, and some nonrecording storage gages are designed to operate for an entire season without attention. Storage gages located in heavy-snowfall areas should have collectors in the form of a frustrum of a cone to prevent wet snow from clinging to the inside walls and clogging the orifice. The orifice should be above the maximum snow depth expected (Fig. 3-4). Storage gages are customarily charged with calcium chloride or other antifreeze solution to liquefy the snow and prevent damage to the gage. Interim measurements of the gage catch are made by stick or tape, while the initial charge and final measurement of the seasonal catch are usually made by weighing. Losses due to evaporation of the gage contents are practically eliminated by a thin film of oil [24].

Figure 3-4 Tower-mounted storage gage with Alter shield. (*U.S. National Weather Service.*)

Precipitation measurements are subject to various errors, most being individually small but with a general tendency to yield measurements that are too low. Except for mistakes in reading the scale of the gage, observational errors are usually small but cumulative; e.g., light amounts may be neglected, and extended immersion of the measuring stick may result in water creeping up the stick. Errors in scale reading, although large, are usually random and compensating. Instrumental errors may be quite large and are cumulative. The water displaced by the measuring stick increases the reading about 1 percent. Dents in the collector rim may change its receiving area. It is estimated that 0.01 in (0.25 mm) of each rain measured with a gage initially dry is required to moisten the funnel and inside surfaces. This loss could easily amount to 1 in/year (25 mm/year) in some areas. Another loss results from raindrop splash from the collector.

In rainfall of 5 to 6 in/hr (125 to 150 mm/hr) the bucket of a tipping-bucket gage tips every 6 to 7 s. About 0.3 s is required to complete the tip, during which some water is still pouring into the already filled compartment. The recorded rate may be 5 percent too low [25]. However, the water, which is all caught in the gage reservoir, is measured independently of the recorder count, and the difference is prorated through the period of excessive rainfall.

Frictional effects in the weighing mechanism of weighing-type gages, in float guides in float-type gages, or in the recording-pen linkages of both types can result in inaccurate indications of rainfall rates. In self-emptying float-type gages the siphoning takes at least a few seconds, and rain falling into the receiver during the siphoning period is recorded inaccurately. Another source of error is that the rainfall amounts siphoned out are not always the same for all emptying cycles.

Of all the errors the most serious is the deficiency of measurement due to wind. The vertical acceleration of air forced upward over a gage imparts an upward acceleration to precipitation about to enter and results in a deficient catch. The deficiency is greater for small raindrops than for large and is thus greater for light than for heavy rain. The deficiency is greater for snow than for rain and larger for "dry" than for "wet" snow; hence it is inversely related to air temperature [26]. Reliable evaluation of wind errors is difficult because of problems involved in determining the true, or actual, precipitation reaching the ground. Attempts at assessing wind errors usually consist of comparing gage measurements with weight changes in nearby lysimeters or by comparing measurements of shielded and unshielded gages. Figure 3-5 shows average deficiencies in rain and snow measurements. The curve shown for rain is actually an average of separate curves for shielded and unshielded gages which showed a spread of only 3 percent at 10 mi/hr (4.5 m/s) and about 5 percent at 20 mi/hr (8.9 m/s), the curve for unshielded gages showing the greater deficiencies.

Various types of shields [27] have been used, but the Alter shield (Fig. 3-4) is customarily used in the United States. Its open construction provides less opportunity than solid shields for snow buildup, and the flexible design allows wind movement to help keep the shield free from accumulated snow. Artificial windshields, including snow fences, cannot overcome the effects of inherently poor gage exposure. Also, the higher the gage, the greater the wind error. Roof installations and windswept slopes should be avoided. The best site is on level ground with bushes or trees serving as a windbreak, provided that they are not so close that they reduce the gage catch [28, 29]. Trees or other obstacles serving as a windbreak should reach no higher than twice their distance from

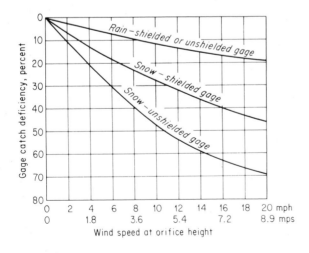

Figure 3-5 Effect of wind speed on the catch of precipitation gages. (*From L. W. Larson and E. L. Peck, Accuracy of Precipitation Measurements for Hydrologic Modeling, Water Resour. Res., vol. 10, no. 4, pp. 857–863, August 1974.*)

the gage and should fairly well surround the gage to provide protection from all directions. An ideal site would be a clearing in a coniferous forest.

Various attempts have been made to estimate true, or actual, precipitation from gage measurements. One is based on the premise that there is a relationship between the ratio of the unshielded-gage catch P_{UG} to the actual precipitation P_A and the ratio of the unshielded-gage catch P_{UG} to the shielded-gage catch P_{SG}, or

$$\ln \frac{P_{UG}}{P_A} = b \ln \frac{P_{UG}}{P_{SG}} \qquad (3\text{-}1)$$

where b is a calibration coefficient, which depends on the type of gage. For weighing gages (U.S. National Weather Service standard) with an Alter shield, coefficient b was found to be about 1.7. The model [30] is presumably independent of wind and form and type of precipitation. Equally accurate measurements are claimed [31] for the Wyoming shield gage system, which consists of a single National Weather Service standard weighing gage with two concentric shields 3 and 6 m (10 and 20 ft) in diameter. The advantage of this system over the dual-gage system is that the cost of gage maintenance and data reduction is only about half.

When rain is falling vertically, a gage inclined 10° from the vertical will catch 1.5 percent less than it should. If a gage on level ground is inclined slightly toward the wind, it will catch more precipitation. Some investigators [32–34] feel that gages should be perpendicular to land slopes. However, the area of a basin is its projection on a horizontal plane, and measurements from tilted gages must be reduced by multiplying by the cosine of the angle of inclination. Considering the variability of slope, aspect, and wind direction, it is virtually impossible to install a network of tilted gages for general purposes. Practically speaking, no gage has been designed which will give reliable measurements on steep slopes experiencing high winds, and such sites should be avoided. All official precipitation records in the United States are obtained from vertical gages.

3-6 The Precipitation-Gage Network

The spatial variability of precipitation and the intended uses of the data should determine network density. A relatively sparse network of stations would suffice for studies of large general storms or for determining annual averages over large areas of level terrain. A very dense network is required to determine the rainfall pattern in thunderstorms. The probability that a storm center will be recorded by a gage varies with network density (Fig. 3-6). A network should be planned to yield a representative picture of the areal distribution of precipitation. On the other hand, the cost of installing and maintaining a network and accessibility of the gage site to an observer are always important considerations. The U.S. National Weather Service publishes precipitation data for about

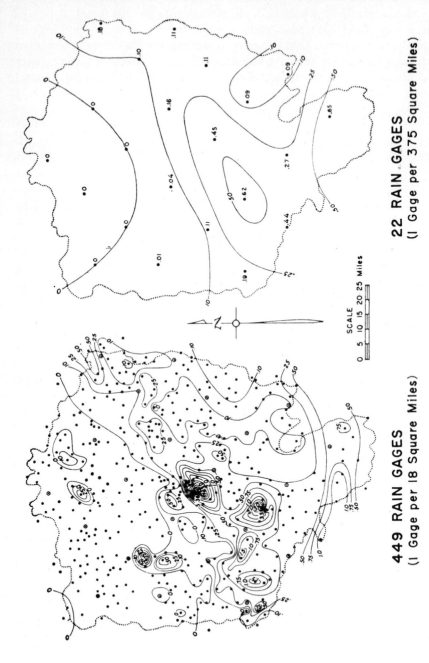

449 RAIN GAGES
(1 Gage per 18 Square Miles)

22 RAIN GAGES
(1 Gage per 375 Square Miles)

SCALE
0 5 10 15 20 25 Miles

Figure 3-6 Isohyetal maps of the storm of Aug. 3, 1939, in the Muskingum Basin, Ohio, showing the effect of network density on the apparent storm pattern. (*U.S. National Weather Service.*)

11,000 gages in the United States. A large number of additional gages are operated by other agencies or individuals. Anyone needing additional data might find a search for these records to be rewarding.

The error of rainfall averages computed from networks of various densities has been investigated [35]. Figure 3-7, based on an analysis [36] of storm rainfall in the Muskingum Basin, Ohio, shows the standard error of rainfall averages as a function of network density and area. In general, sampling errors, in terms of depth, tend to increase with increasing areal mean precipitation and to decrease with increasing network density, duration of precipitation, and size of area. Thus, a particular network would yield greater average errors for storm precipitation than for monthly or seasonal precipitation. Average errors also tend to be greater for summer than for winter precipitation because of the generally greater spatial variability of summer rainfall. Summer-storm rainfall may require a network density 2 to 3 times that required for winter storms in order to maintain equivalent degrees of accuracy.

The following minimum densities of precipitation networks have been recommended [37] for general hydrometeorological purposes:

1. For flat regions of temperate, mediterranean, and tropical zones, 600 to 900 km² (230 to 350 mi²) per station
2. For mountainous regions of temperate, mediterranean, and tropical zones, 100 to 250 km² (40 to 100 mi²) per station
3. For small mountainous islands with irregular precipitation, 25 km² (10 mi²) per station
4. For arid and polar zones, 1500 to 10,000 km² (600 to 4000 mi²) per station

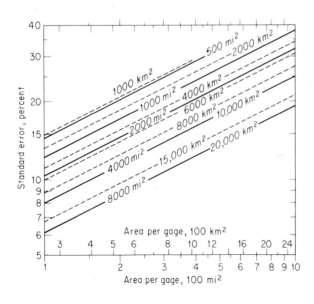

Figure 3-7 Standard error of storm precipitation averages as a function of network density and area for the Muskingum Basin. *(U.S. National Weather Service.)*

Information on the difference between calculated and true catchment precipitation is of climatological interest but does not answer the hydrologic question: What error results in estimation of streamflow because of imperfect precipitation gaging? The answer depends on the climatic characteristics of precipitation in the area (including the effects of catchment topography), hydrologic characteristics of the catchment, the streamflow characteristics being estimated, and possibly the method used to estimate streamflow.

If only one gage is used as an index to catchment precipitation and if storm patterns are randomly distributed over the catchment, the observed precipitation will have a greater variance than the true average precipitation; i.e., the gage may on occasion record amounts much greater or much less than the basin average. Streamflow estimated from this index will display a greater variance than observed streamflow. On the other hand, over a sufficient period of time the gage should indicate average precipitation close to the catchment mean, and thus the mean estimated streamflow should be reasonably accurate.

While not enough work has been done to yield general results, the problem of minimum precipitation-network density for various hydrologic purposes has been investigated. Eagleson [38], for example, found that two properly located gages might be adequate for the determination of long-term basin average precipitation. Johanson [39] explored the errors involved in simulating streamflow from a dense gage network in central Illinois and found that the number of gages is more important than the density. Figure 3-8 indicates the calibration and dispersion errors to be expected when simulation methods (Chap. 12) are used to estimate annual streamflow. In the calibration phase one attempts to match

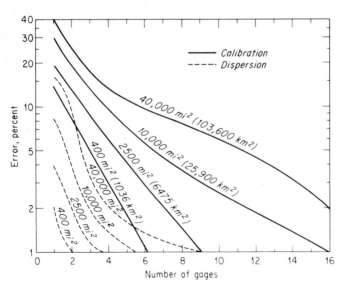

Figure 3-8 Error of simulation of annual volume of streamflow because of imperfect precipitation gaging. (*Adapted from* [39].)

each event in the historic record with an event in the simulated record. The calibration error is the ratio (percent) of the standard error of estimate to the standard deviation of the observed flows. When simulation is used to estimate flows for planning or design, the main concern is in accurately reproducing the mean and dispersion of the flow series. The dispersion error is expressed as a percentage of the coefficient of variation of the historic record. Figure 3-8 shows that the calibration error is much the larger.

Figure 3-9 presents similar data for simulation of storm runoff under summer thunderstorm conditions. Errors are larger but show a similar trend. Johanson's data precluded extending this case to larger areas. His sample period was too short to yield conclusive results for flood peaks, but the evidence indicates that errors are somewhat larger than for direct runoff, as might be expected.

3-7 Radar Measurement of Precipitation

A radar transmits a pulse of electromagnetic energy as a beam in a direction determined by a movable antenna. The beam width and shape are determined by the antenna size and configuration. The radiated wave, which travels at the speed of light, is partially reflected by cloud or precipitation particles and returns to the radar, where it is received by the same antenna.

Energy returned to the radar is called the *target signal,* the amount is termed *returned power,* and its display on the radarscope is called an *echo.* The brightness of an echo, or *echo intensity,* is an indication of the magnitude of returned power, which in turn is a measure of the *radar reflectivity* of the hydrometeors. The reflectivity of a group of hydrometeors depends on (1) drop-size distribution, (2) number of particles per unit volume, (3) physical state, i.e., solid or liquid, (4) shape of the individual elements, and (5) if asym-

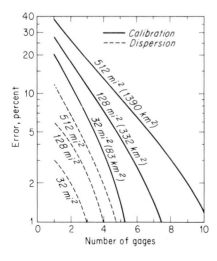

Figure 3-9 Error of simulation of storm direct runoff because of imperfect precipitation gaging. (*Adapted from* [39].)

metrical, their aspect with respect to the radar. Generally speaking, the more intense the precipitation, the greater the reflectivity.

The time interval between emission of the pulse and appearance of the echo on the radarscope is a measure of the distance, or *range,* of the target from the radar. Direction of the target from the radar is determined by the orientation of the antenna at the time the target signal is received. Both range and direction are displayed in their proper perspectives by the locations of the echoes on the radarscope. The areal extent of an entire storm can be obtained by rotating the antenna.

Loss of radar energy due to passage through precipitation is called *attenuation.* Part of the loss results from scattering and part from absorption. The larger the rain and snow diameter-to-wavelength ratios, the greater the attenuation. For a given particle diameter, the shorter the wavelength, the greater the attenuation. Thus, at short wavelengths the total energy may be greatly diminished or completely dissipated by a relatively short penetration into a storm. Even the lightest precipitation intensities will seriously attenuate radar energy of wavelengths less than 1 cm. Wavelengths less than 5 cm are considered unsuitable for measuring precipitation [40]. These short wavelengths, however, are useful for delineating very light rainfall within short distances, drizzle, and cloud forms. Wavelengths of 5 to 10 cm with beam width less than 2° are generally recommended for precipitation measurement.

The *average returned power* $\bar{P}_r$ is a measure of the radar reflectivity of all particles at range r intercepting the radiated beam, or

$$\bar{P}_r = \frac{C}{r^2} \sum d^6 \tag{3-2}$$

where C is dependent on wavelength, beam shape and width, pulse length, transmitted power, antenna gain, and refractive index of the target, and d is the diameter of the individual particles.

Many studies have been made of the relationship of raindrop-size distribution to rainfall intensity, and radar measurement of precipitation is based on empirical relationships between $\sum d^6$, usually represented by Z, and rainfall rate R in the form

$$Z = aR^b \tag{3-3}$$

Values of a and b can be obtained by direct measurement of raindrop-size distribution or comparison of radar and rain-gage measurements.

The main obstacle to the accurate determination of the Z-R relationship arises from the fact that radar measures precipitation in the atmosphere while gages measure it at the ground. In order to avoid interference from hills, trees, buildings, etc., i.e., *ground clutter,* the radar beam is directed upward at an angle of $\frac{1}{2}$ to 1° above the horizontal, and the height of the beam thus increases with distance. The magnitude of the discrepancy between radar and gage measurements varies with the angle of the beam elevation, beam width, and range. Another factor leading to error is evaporation of precipitation before reaching the ground, which happens frequently in arid areas. Also, winds may carry precipitation away from beneath the producing cloud. Solid forms of precipitation, especially if covered with liquid water, and discontinuities in the vertical distribution of precipitation in the cloud affect radar reflectivity and thus are also sources of error.

With Z in mm^6/m^3 and R in mm/hr, values of a and b have been found to vary from about 15 to 1100 and from 1.2 to 3.2, respectively, for rainfall. While errors in the determination of the relationship are undoubtedly responsible for part of the large variation of a and b values, differences in climate and storm types are apparently important factors [41]. It is generally agreed that values of 200 and 1.6 for a and b, respectively, yield the most reliable results on the average, and the Z-R relationship using these values is sometimes referred to as the standard. Also well known is the so-called *Miami relationship,* which uses a and b values of 300 and 1.4, respectively. Few studies of the Z-R relationship have been made for solid precipitation, but values of about 2000 and 2.0 for a and b, respectively, appear applicable for snowflake aggregates [42]. Differences in radar precipitation estimates resulting from different Z-R relationships may be significant for the smallest time and space scales. However, when the biases in the estimates have been removed, based on comparisons of rainfall totals from the radar and rain gages, the differences in estimates would seldom be significant for durations over 3 hr and areas over 4000 km^2 (1500 mi^2) [43].

The hydrologist is usually more interested in precipitation amounts for specified periods (1 hr, 6 hr, 24 hr, etc.) than in instantaneous rates, and much attention has been given to the development of procedures for integrating radar-indicated intensities with respect to time and area. The returned power for incremental areas within the beam can be measured electronically and converted to equivalent rainfall rates, which are then integrated with respect to time. The totals for any duration are displayed on a computer printout in grid form on which *isohyets* (contours of equal precipitation) can be drawn on the basis of both radar and gage measurements. Tests of the method have led to improvements in the configuration of the system and the output products [44, 45].

Because of factors that affect returned power, the accuracy of radar measurements of precipitation varies with duration, area, storm type, and range. Numerous comparisons suggest that radar measurements of rainfall are within one-half to twice the gage measurements within a 60-nmi (110-km) range, with larger deviations for longer distances. The areal extent of the rainfall may be depicted reliably by radar for ranges up to about 125 nmi (230 km). Since measurements by ordinary gage networks may be appreciably in error (Fig. 3-7) as a result of inadequate sampling, and since radar can detect and estimate precipitation between gages in a network of ordinary density, conjunctive use of radar and gage network should yield more accurate areal averages than can be obtained from either one alone [46, 47]. It is expected that a method will be developed eventually for introducing telemetered rain-gage measurements into the system so that the radar estimates would be automatically adjusted. The use of digital radar data alone for developing storm depth-area relations has been investigated [48] and determined to be feasible but requires operational systems designed to record more intensity levels than the 10 available at the time of the study.

3-8 Satellite Estimates of Precipitation

Studies of water balance on a global scale require information on precipitation where gage or radar networks are inadequate or nonexistent, as over oceans. Satellites cannot measure precipitation directly, and their use for estimating it is based on relating brightness of cloud photographs to rainfall intensities. The degree of brightness is an indication of the temperature, or height, of the cloud tops—the brighter the image, the higher the cloud top. The tallest and densest clouds produce the heaviest intensities. The relation between brightness and rainfall intensity is usually determined by calibration with gage measurements [49] and radar estimates [50, 51]. An indirect procedure is to determine the percentage of an area covered by various cloud types and to multiply by empirically established coefficients to obtain the rainfall contribution from each cloud type [52, 53].

Except for precipitation estimates from a few shipboard weather radars, the above calibration procedures are necessarily based on data over land surfaces, and the reliability of satellite precipitation estimates over seas depends on the precipitation processes being similar. The calibrations can also be applied to satellite cloud photographs for remote regions to estimate precipitation intensity, frequency, and extent [54].

A major problem in estimating precipitation by satellites is that the photographs often do not reveal precipitation-producing clouds because of overlying cloud layers. Future developments in instrumentation and techniques may lead eventually to reasonably accurate satellite estimates of precipitation. Greene et al. [55] describe a precipitation data-management and -analysis system under development to place data from satellites, radars, and rain gages into files available for hydrologic forecasting and other operational purposes.

In 1979, the conterminous United States was being scanned thrice daily by polar-orbiting satellites (once with video imagery and twice with infrared) and half-hourly by geostationary satellites at two fixed locations above the equator. This frequency of observations makes possible the estimation of short-term precipitation and the monitoring of river ice and extent of snow cover and flood areas.

INTERPRETATION OF PRECIPITATION DATA

In order to avoid erroneous conclusions it is important to give the proper interpretation to precipitation data, which often cannot be accepted at face value. For example, a mean annual precipitation value for a station may have little meaning if the gage site has been changed significantly during the period for which the average is computed. Also, there are several ways of computing average precipitation over an area, each of which may give a different answer.

3-9 Estimating Missing Precipitation Data

Many precipitation stations have short breaks in their records because of absences of the observer or because of instrumental failures. It is often necessary to estimate this missing record. In the procedure [56] used by the U.S. Environmental Data Service, precipitation amounts are estimated from observations at three stations as close to and as evenly spaced around the station with the missing record as possible. If the normal annual precipitation at each of the index stations is within 10 percent of that for the station with the missing record, a simple arithmetic average of the precipitation at the index stations provides the estimated amount.

If the normal annual precipitation at any of the index stations differs from that at the station in question by more than 10 percent, the *normal-ratio method* is used. In this method, the amounts at the index stations are weighted by the ratios of the normal-annual-precipitation values. That is, precipitation P_X at station X is

$$P_X = \frac{1}{3}\left(\frac{N_X}{N_A} P_A + \frac{N_X}{N_B} P_B + \frac{N_X}{N_C} P_C\right) \tag{3-4}$$

in which N is the normal annual precipitation.

Multiple linear regression will yield an equation of the form

$$P_X = a + b_A P_A + b_B P_B + b_C P_C \tag{3-5}$$

where a should be near zero and the b's will approximate the three coefficients of Eq. (3-4) divided by 3. The advantage of the regression approach is that it allows for some weighting of the stations and adjusts, to some extent, for departures from the normal ratio assumption of Eq. (3-4). If a large amount of data must be estimated, a random error term tS_y should be added to the equation. In this term t is a normal random number with mean of zero and standard deviation of 1, and S_y is the standard error of estimate of P_X. Inclusion of the term recognizes the departures from the regression, and maintains the standard deviation of the estimated values of P_X near the observed standard deviation.

Another method, used by the U.S. National Weather Service [57] for river forecasting, estimates precipitation at a point as the weighted average of that at four stations, one in each of the quadrants delineated by north-south and east-west lines through the point. Each station is the nearest in its quadrant to the point for which precipitation is being estimated. The weight applicable to each station is equal to the reciprocal of the square of the distance between the point and the station. Multiplying the precipitation for the storm (or other period) at each station by its weighting factor, adding the four weighted amounts, and dividing by the sum of the weights yields the estimated precipitation for the point. If one or more quadrants contain no precipitation stations, as might be the case for a point in a coastal area, then the estimation involves only the remaining quadrants. A shortcoming of the method is that it can never yield a point estimate greater than the largest amount observed or less than the smallest. In mountainous regions the normal ratio and multiple linear regression methods should therefore yield more reliable estimates.

Estimates of missing precipitation data are generally most reliable for general-type storms over flat terrain or over relatively smooth windward mountain slopes. Severe and spotty convective activity and rugged terrain lessen the reliability. Estimates for long intervals (month or year) are more reliable than those for short intervals such as a day.

3-10 Double-Mass Analysis

Changes in gage location, exposure, instrumentation, or observational procedure may cause a relative change in the precipitation catch. Frequently these changes are not disclosed in the published records. Current U.S. Environmental Data Service practice calls for a new station identification whenever the gage location is changed by as much as 5 mi (8 km) and/or 100 ft (30 m) in elevation.

Double-mass analysis [58] tests the consistency of the record at a station by comparing its accumulated annual or seasonal precipitation with the concurrent accumulated values of mean precipitation for a group of surrounding stations. In Fig. 3-10, for example, a change in slope about 1961 indicates a change in the precipitation regime at Dillon, Colo. A change due to meteorological

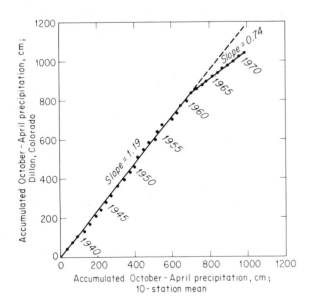

Figure 3-10 Adjustment of precipitation data for Dillon, Colo., by double-mass curve.

causes would not cause a change in slope, as all base stations would be similarly affected. The station history for Dillon discloses a change in gage location in June 1961. To make the record prior to 1961 comparable with that for the more recent location, it should be adjusted by the ratio of the slopes of the two segments of the double-mass curve (0.74/1.19). The consistency of the record for each of the base stations should be tested, and those showing inconsistent records should be dropped before other stations are tested or adjusted.

Considerable caution should be exercised in applying the double-mass technique. The plotted points always deviate about a mean line, and changes in slope should be accepted only when marked or substantiated by other evidence. The double-mass analysis can be made on a computer [59].

3-11 Average Precipitation over Area

The average depth of precipitation over a specific area, on a storm, seasonal, or annual basis, is required in many types of hydrologic problems. The simplest method of obtaining the average depth is to average arithmetically the gaged amounts in the area (Fig. 3-11a). This method yields good estimates in flat country if the gages are uniformly distributed and the individual gage catches do not vary widely from the mean. These limitations can be partially overcome if topographic influences and areal representativity are considered in the selection of gage sites [60].

The *Thiessen method* attempts to allow for nonuniform distribution of gages by providing a weighting factor for each gage. The stations are plotted on a map, and connecting lines are drawn (Fig. 3-11b). Perpendicular bisectors of these connecting lines form polygons around each station. The sides of each polygon are the boundaries of the effective area assumed for the station. The area of each polygon is determined by planimetry and is expressed as a percentage of the total area. Weighted average rainfall for the total area is computed by multiplying the precipitation at each station by its assigned percentage of area and totaling. The results are usually more accurate than those obtained by simple arithmetical averaging. The greatest limitation of the Thiessen method is its inflexibility, a new Thiessen diagram being required every time there is a change in the gage network. Also, the method does not allow for orographic influences. It simply assumes linear variation of precipitation between stations and assigns each segment of area to the nearest station.

The *grid-point method* averages the estimated precipitation at all points of a superimposed grid. This approach has certain advantages over the Thiessen method but is practical only with the aid of a computer. The reliability of the approach depends on the method used to estimate precipitation at the grid points [57].

The *isohyetal method,* when used by an experienced analyst, is the most accurate method of averaging precipitation over an area. Station locations and amounts are plotted on a suitable map, and contours of equal precipitation (*isohyets*) are then drawn (Fig. 3-11c). The average precipitation for an area is

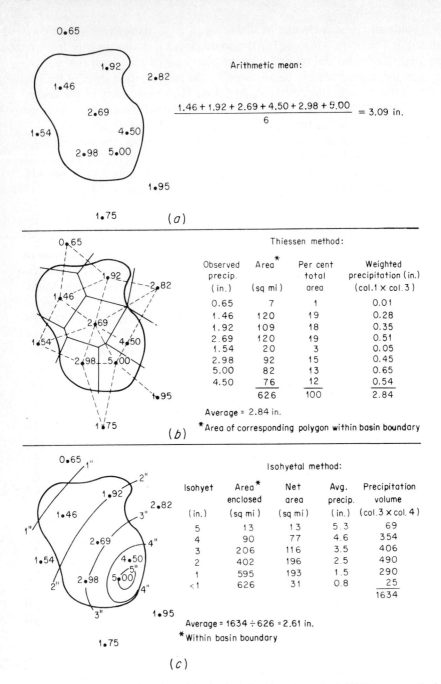

Arithmetic mean:

$$\frac{1.46 + 1.92 + 2.69 + 4.50 + 2.98 + 5.00}{6} = 3.09 \text{ in.}$$

(a)

Thiessen method:

Observed precip. (in.)	Area* (sq mi)	Per cent total area	Weighted precipitation (in.) (col.1 × col. 3)
0.65	7	1	0.01
1.46	120	19	0.28
1.92	109	18	0.35
2.69	120	19	0.51
1.54	20	3	0.05
2.98	92	15	0.45
5.00	82	13	0.65
4.50	76	12	0.54
	626	100	2.84

Average = 2.84 in.
*Area of corresponding polygon within basin boundary

(b)

Isohyetal method:

Isohyet (in.)	Area* enclosed (sq mi)	Net area (sq mi)	Avg. precip. (in.)	Precipitation volume (col.3 × col. 4)
5	13	13	5.3	69
4	90	77	4.6	354
3	206	116	3.5	406
2	402	196	2.5	490
1	595	193	1.5	290
<1	626	31	0.8	25
				1634

Average = 1634 ÷ 626 = 2.61 in.
*Within basin boundary

(c)

Figure 3-11 Areal averaging of precipitation by (a) arithmetic method, (b) Thiessen method, and (c) isohyetal method.

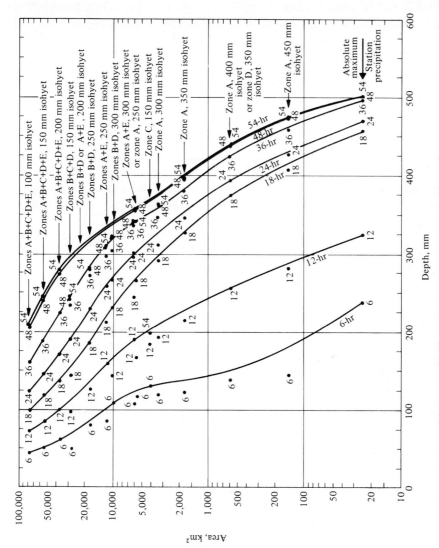

Figure 3-12 Maximum depth-area-duration curves for the southern United States storm of May 16–18, 1953. (*From* [61].)

computed by weighting the average precipitation between successive isohyets (usually taken as the average of the two isohyetal values) by the area between isohyets, totaling these products, and dividing by the total area.

The isohyetal method permits the use and interpretation of all available data and is well adapted to display and discussion. In constructing an isohyetal map, analysts can make full use of their knowledge of orographic effects and storm morphology, and in this case the final map should represent a more realistic precipitation pattern than could be obtained from the gaged amounts alone. If linear interpolation between stations is used, the results will be essentially the same as those obtained with the Thiessen method.

3-12 Depth-Area-Duration Analysis

Various hydrologic problems require an analysis of time as well as areal distribution of storm precipitation. Depth-area-duration analysis of a storm is performed to determine the maximum amounts of precipitation within various durations over areas of various sizes. The method [61] discussed here is widely used. The procedure has been computerized [62].

For a storm with a single major center, the isohyets are taken as boundaries of individual areas. The average storm precipitation within each isohyet is computed (Sec. 3-11). The storm total is distributed through successive increments of time (usually 6 hr) in accordance with the distribution recorded at nearby stations. When this has been done for each isohyet, data are available showing the time distribution of average rainfall over areas of various sizes. From these data, the maximum rainfall for various durations (6, 12, 18 hr, etc.) can be selected for each size of area. These maxima are plotted (Fig. 3-12), and an enveloping depth-area curve is drawn for each duration. Storms with multiple centers are divided into zones for analysis.

A summary [63] of depth-area-duration analyses for 853 storms in the contiguous United States presents the five greatest areal rainfall depths by seasons and $5 \times 5°$ sections for durations of 6 to 48 hr and areas from 100 to 10,000 mi^2 (259 to 25,900 km^2).

VARIATIONS IN PRECIPITATION

3-13 Geographic Variations

In general, precipitation is heaviest near the equator and decreases with increasing latitude. However, the irregularity and orientation of the isohyets on the mean-annual-precipitation maps of the world (Fig. 3-13) and the United States (Fig. 3-14) indicate that the geographic distribution of precipitation depends on factors other than distance from the equator.

The main source of moisture for precipitation is evaporation from the oceans. Therefore, precipitation tends to be heavier near coastlines. Distortions in the isohyets reflect orographic effects.

Since lifting of air masses accounts for almost all precipitation, amounts and frequency are generally greater on the windward side of mountain barriers. Conversely, since downslope motion of air results in decreased relative humidity, the lee sides of barriers usually experience relatively light precipitation. However, the continued rise of air immediately downwind from the ridge and the slanting fall of the precipitation can produce heavy amounts on the lee slopes near the crest.

The variation of precipitation with elevation and other topographic factors has been investigated [64] with somewhat varying conclusions. Perhaps the most detailed early study of orographic influences on precipitation is that by Spreen [65], who correlated mean seasonal precipitation with elevation, slope, orientation, and exposure for western Colorado. While elevation alone accounted for only 30 percent of the variation in precipitation, the four parameters together accounted for 85 percent. Relations of this type are very useful for constructing isohyetal maps in rugged areas having sparse data.

The development of a relation like Spreen's is a tedious and lengthy procedure. In a mountainous area having relatively homogeneous geographic features, elevation generally accounts for a large proportion of the variation in normal annual precipitation. Some investigators [66, 67] develop only the precipitation-elevation relation for such an area. Differences between observed station precipitation and precipitation estimated by the relation are determined and plotted on a map, and isoanomalous lines are then drawn. Precipitation for any ungaged point can then be obtained by using the elevation of the point and the precipitation-elevation relation to make a preliminary estimate and then adjusting the estimate as indicated by the anomaly map.

3-14 Time Variations

While portions of a precipitation record may suggest an increasing or decreasing trend, there is usually a tendency to return to the mean; abnormally wet periods tend to be balanced by dry periods. The regularity of these fluctuations has been repeatedly investigated. More than 100 apparent cycles, ranging in period from 1 to 744 years, have been propounded [68]. The bibliography lists a few reports on attempts to detect these variations. However, with the exception of diurnal and seasonal variations, no persistent regular cycles of any appreciable magnitude have been conclusively demonstrated [69].

The seasonal distribution of precipitation varies widely within the United States. Figure 3-15 shows typical seasonal distributions for selected stations.

The time distribution of rainfall within storms is important for estimating flood hydrographs. Distributions vary with storm type, intensity, and duration, and there is no typical distribution that is applicable to all situations. Studies of

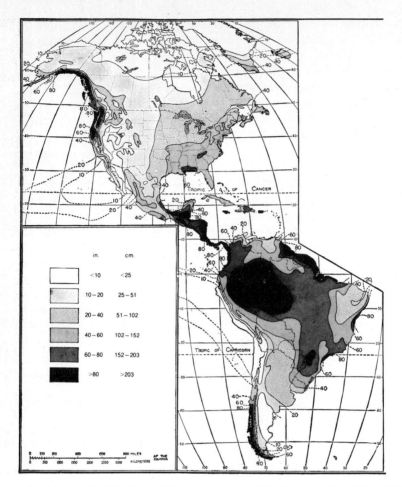

Figure 3-13 World distribution of mean annual precipitation, in inches. (*From G. T. Trewartha, "Introduction to Climate," 3d ed., copyright 1954, McGraw-Hill Book Company. Used with permission of McGraw-Hill Book Company.*)

time distribution within storms are listed in the bibliography section entitled, "Spatial and Temporal Distribution of Precipitation," at the end of this chapter.

3-15 Record Rainfalls

Table 3-2 lists the world's greatest observed point rainfalls. Values of Table 3-2 plotted on logarithmic paper define an enveloping curve closely approximating a straight line (Fig. 3-16).

The maximum rainfalls of record for durations up to 24 hr at five major United States cities are given in Table 3-3. Table 3-4 lists maximum depth-area-duration data for the United States and the storms producing them. They represent enveloping values for over 850 of the country's major storms ana-

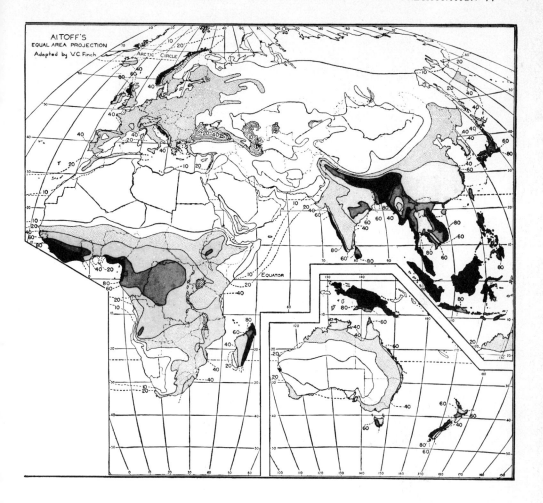

lyzed by the U.S. Army Corps of Engineers, the U.S. National Weather Service, and the U.S. Water and Power Resources Service (formerly the Bureau of Reclamation).

SNOWPACK AND SNOWFALL

3-16 Measurement

Snow is usually composed of ice crystals and liquid water; the amount of liquid water is referred to as the *water content* of the snow. The *quality of snow,* i.e., the percentage by weight which is ice, can be determined by a calorimetric

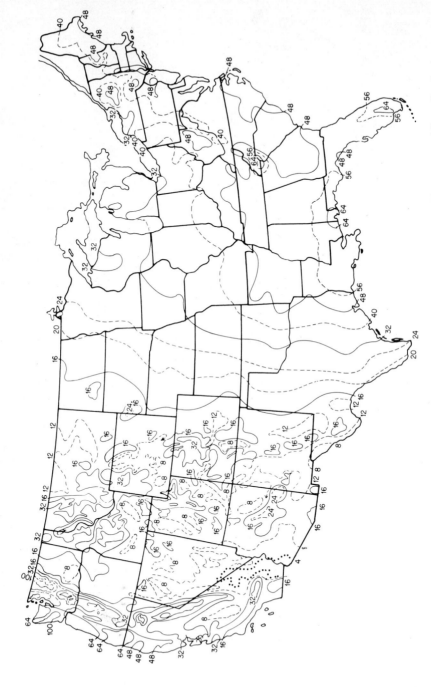

Figure 3-14 Mean annual precipitation in the United States, in inches (1 in = 25.4 mm). (*U.S. Environmental Data Service.*)

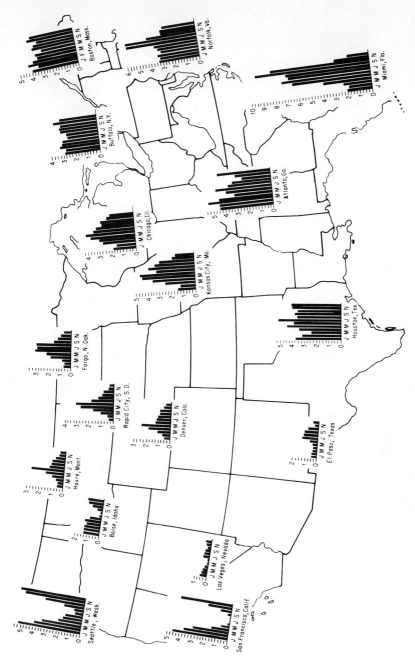

Figure 3-15 Normal monthly distribution of precipitation in the United States, in inches (1 in = 25.4 mm). (*U.S. Environmental Data Service.*)

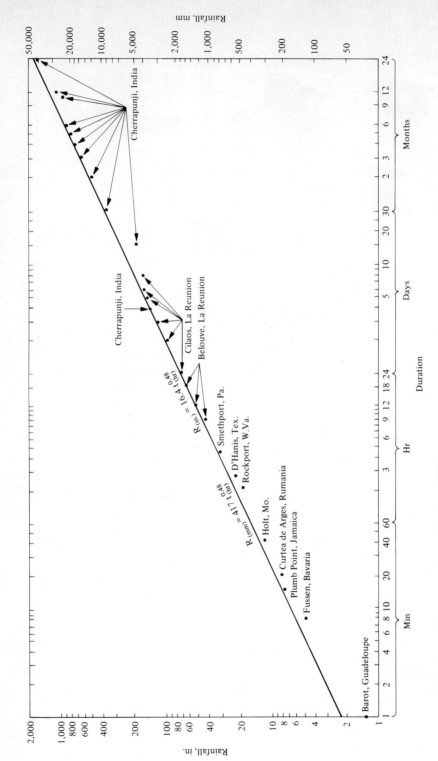

Figure 3-16 World's greatest observed point rainfalls.

Table 3-2 World's greatest observed point rainfalls

Duration	Depth in	Depth mm	Location	Date
1 min	1.50	38	Barot, Guadeloupe	Nov. 26, 1970
8 min	4.96	126	Füssen, Bavaria	May 25, 1920
15 min	7.80	198	Plumb Point, Jamaica	May 12, 1916
20 min	8.10	206	Curtea-de-Arges, Roumania	July 7, 1889
42 min	12.00	305	Holt, Mo.	June 22, 1947
2 hr 10 min	19.00	483	Rockport, W.Va.	July 18, 1889
2 hr 45 min	22.00	559	D'Hanis, Tex. (17 mi NNW)	May 31, 1935
4 hr 30 min	30.8+	782+	Smethport, Pa.	July 18, 1942
9 hr	42.79	1,087	Belouve, Réunion	Feb. 28, 1964
12 hr	52.76	1,340	Belouve, Réunion	Feb. 28–29, 1964
18 hr 30 min	66.49	1,689	Belouve, Réunion	Feb. 28–29, 1964
24 hr	73.62	1,870	Cilaos, Réunion	Mar. 15–16, 1952
2 days	98.42	2,500	Cilaos, Réunion	Mar. 15–17, 1952
3 days	127.56	3,240	Cilaos, Réunion	Mar. 15–18, 1952
4 days	146.50	3,721	Cherrapunji, India	Sept. 12–15, 1974
5 days	151.73	3,854	Cilaos, Réunion	Mar. 13–18, 1952
6 days	159.65	4,055	Cilaos, Réunion	Mar. 13–19, 1952
7 days	161.81	4,110	Cilaos, Réunion	Mar. 12–19, 1952
8 days	162.59	4,130	Cilaos, Réunion	Mar. 11–19, 1952
15 days	188.88	4,798	Cherrapunji, India	June 24–July 8, 1931
31 days	366.14	9,300	Cherrapunji, India	July 1861
2 mo	502.63	12,767	Cherrapunji, India	June–July 1861
3 mo	644.44	16,369	Cherrapunji, India	May–July 1861
4 mo	737.70	18,738	Cherrapunji, India	Apr.–July 1861
5 mo	803.62	20,412	Cherrapunji, India	Apr.–Aug. 1861
6 mo	884.03	22,454	Cherrapunji, India	Apr.–Sept. 1861
11 mo	905.12	22,990	Cherrapunji, India	Jan.–Nov. 1861
1 year	1041.78	26,461	Cherrapunji, India	Aug. 1860–July 1861
2 years	1605.05	40,768	Cherrapunji, India	1860–1861

process [70, 71]. Most evaluations of quality made to date indicate values of 90 percent or more, but values as low as 50 percent have been obtained at times of rapid melting.

Measurement of the depth of accumulated snow on the ground is a regular function of all U.S. National Weather Service observers. Where the accumulation is not large, the measurements are made with a rain-gage measuring stick. In regions where large accumulations are the rule, permanent *snow stakes* are used. *Aerial snow-depth markers,* a type of stake adapted for visual reading from low-flying aircraft, are used in some remote areas. All stakes for measuring snow depth should be installed where they will be least affected by blowing or drifting snow [72]. Snow depths are reported in inches in the United States, but most countries use centimeters.

The hydrologist is usually more interested in the water equivalent of the

Table 3-3 Maximum recorded rainfalls at five major United States cities, in inches and (millimeters)

	Duration					
	Minutes			Hours		
Station	5	15	30	1	6	24
New York, N.Y.	1.02 (26) 7/5/1973	1.63 (41) 7/10/1905	2.34 (59) 8/12/1926	2.97 (75) 8/26/1947	4.44 (113) 10/1/1913	9.55 (243) 10/8/1903
St. Louis, Mo.	0.88 (22) 3/5/1897	1.39 (35) 8/8/1923	2.56 (65) 8/8/1923	3.47 (88) 7/23/1933	5.82 (148) 7/9/1942	8.78 (223) 8/15/1946
New Orleans, La.	1.00 (25) 2/5/1955	1.90 (48) 4/25/1953	3.18 (81) 4/25/1953	4.71 (120) 4/25/1953	8.62 (219) 9/6/1929	14.01(356) 4/15/1927
Denver, Colo.	0.91 (23) 7/14/1912	1.57 (40) 7/25/1965	1.99 (51) 7/25/1965	2.20 (56) 8/23/1921	2.91 (74) 8/23/1921	6.53 (166) 5/21/1876
San Francisco, Calif.	0.33 (8) 11/25/1926	0.65 (17) 11/4/1918	0.83 (21) 3/4/1912	1.07 (27) 3/4/1912	2.34 (59) 10/11/1962	4.67 (119) 1/29/1881

snowpack than in its depth. The *water equivalent* of the snowpack, i.e., the depth of water that would result from melting, depends on snow density as well as on depth. *Snow density,* the ratio between the volume of meltwater from a sample of snow and the initial volume of the sample, has been observed to vary from 0.004 for freshly fallen snow at high latitudes to 0.91 for compacted snow in glaciers. An average density of 0.10 for freshly fallen snow is often assumed. In regions of heavy snow accumulation, densities of 0.4 to 0.6 are common by the time the spring thaw begins.

Measurements of water equivalent are usually made by sampling with a snow tube (Fig. 3-17). The tube is driven vertically into the snowpack, sections of tubing being added as required. The cutting edge on the leading section is designed to penetrate ice layers when the tube is rotated. When the bottom of the snowpack is reached, the snow depth is determined from the graduations on the tube. The tube and its contents are then withdrawn and weighed to determine the water equivalent.

Because of the highly variable snowpack resulting from drifting and nonuniform melting, a number of measurements are made along an established line, or *snow course* [73]. Sampling sites should be free from extensive wind effects and lateral drainage of meltwater. Even then the measurement for each course is considered to be only an index to areal snowpack.

Snow tubes tend to overmeasure water equivalent by 7 to 12 percent, the error increasing with snow density [74]. Chief disadvantages of snow surveys are that they are costly and disturb snow at the sampling sites so that natural changes in the snowpack at any particular site cannot be determined reliably.

The water equivalent of the snowpack can also be measured by *pressure*

Table 3-4 Maximum observed depth-area-duration data for the United States
Average rainfall in inches and (millimeters)

Area mi² (km²)	Duration, hr						
	6	12	18	24	36	48	72
10	24.7[a]	29.8[b]	36.3[c]	38.7[c]	41.8[c]	43.1[c]	45.2[c]
(26)	(627)	(757)	(922)	(983)	(1062)	(1095)	(1148)
100	19.6[b]	26.3[c]	32.5[c]	35.2[c]	37.9[c]	38.9[c]	40.6[c]
(259)	(498)	(668)	(826)	(894)	(963)	(988)	(1031)
200	17.9[b]	25.6[c]	31.4[c]	34.2[c]	36.7[c]	37.7[c]	39.2[c]
(518)	(455)	(650)	(798)	(869)	(932)	(958)	(996)
500	15.4[b]	24.6[c]	29.7[c]	32.7[c]	35.0[c]	36.0[c]	37.3[c]
(1,295)	(391)	(625)	(754)	(831)	(889)	(914)	(947)
1,000	13.4[b]	22.6[c]	27.4[c]	30.2[c]	32.9[c]	33.7[c]	34.9[c]
(2,590)	(340)	(574)	(696)	(767)	(836)	(856)	(886)
2,000	11.2[b]	17.7[c]	22.5[c]	24.8[c]	27.3[c]	28.4[c]	29.7[c]
(5,180)	(284)	(450)	(572)	(630)	(693)	(721)	(754)
5,000	8.1[bd]	11.1[b]	14.1[b]	15.5[c]	18.7[e]	20.7[e]	24.4[e]
(12,950)	(206)	(282)	(358)	(394)	(475)	(526)	(620)
10,000	5.7[d]	7.9[f]	10.1[g]	12.1[g]	15.1[e]	17.4[e]	21.3[e]
(25,900)	(145)	(201)	(257)	(307)	(384)	(442)	(541)
20,000	4.0[d]	6.0[f]	7.9[g]	9.6[g]	11.6[e]	13.8[e]	17.6[e]
(51,800)	(102)	(152)	(201)	(244)	(295)	(351)	(447)
50,000	2.5[gh]	4.2[i]	5.3[g]	6.3[g]	7.9[g]	9.9[l]	13.2[l]
(129,500)	(64)	(107)	(135)	(160)	(201)	(251)	(335)
100,000	1.7[h]	2.5[hk]	3.5[g]	4.3[g]	5.9[m]	6.6[j]	8.9[j]
(259,000)	(43)	(64)	(89)	(109)	(150)	(168)	(226)

Storm	Date	Location of center
a	July 17–18, 1942	Smethport, Pa.
b	Sept. 8–10, 1921	Thrall, Tex.
c	Sept. 3–7, 1950	Yankeetown, Fla.
d	June 27–July 4, 1936	Bebe, Tex.
e	June 27–July 1, 1899	Hearne, Tex.
f	Apr. 12–16, 1927	Jefferson Parish, La.
g	Mar. 13–15, 1929	Elba, Ala.
h	May 22–26, 1908	Chattanooga, Okla.
i	Apr. 15–18, 1900	Eutaw, Ala.
j	July 5–10, 1916	Bonifay, Fla.
k	Nov. 19–22, 1934	Millry, Ala.
l	Sept. 19–24, 1967	Sombreretillo, Mex.
m	Sept. 29–Oct. 3, 1929	Vernon, Fla.

pillows. The pillows are filled with a mixture of water and antifreeze. As the snow accumulates on the pillow, the internal pressure increases. The weight of the snow on the pillow is then determined by measuring the pressure with a manometer or pressure transducer.

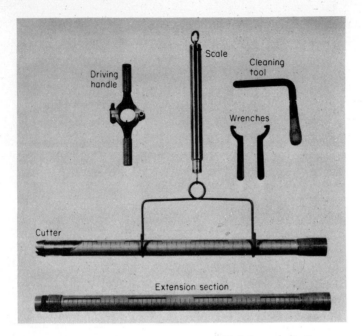

Figure 3-17 Mt. Rose snow sampler. (*Leupold & Stevens, Inc.*)

The first pressure pillows were made of thin butyl rubber, and were 5 to 12 ft (1.5 to 3.7 m) in diameter, accuracy increasing with size. Rubber pillows are easily damaged, and this led to the development of stainless-steel pillows 4 by 5 ft (1.2 by 1.5 m) and 0.5 in (13 mm) thick.

The chief advantage claimed for the pressure-pillow snow gage is that it is not subject to the wind errors of the usual type of precipitation gage so that the snow accumulating on it is more likely to be representative of the true snow cover. The disadvantages are that ice in the snow cover may form a bridge over the pillow, thus leading to false readings, and that the pillows are vulnerable to puncturing by vandals and animals. Instrumental problems include pinhole leaks in the pillow surface, leaks in the fittings, exposed fluid lines, air locks in the system, and air in the pressure transducer and pillows. The care required in the installation and maintenance of the pillows has been summarized by Cox et al. [75].

At U.S. National Weather Service stations the water equivalent is measured by cutting out a sample of the snow cover with the inverted overflow can of the standard 8-in (20.3-cm) rain gage. A piece of sheet metal or thin wood is slipped beneath the mouth of the can to hold the sample as the can is withdrawn. At some stations equipped with weighing scales, the water equivalent is determined by weighing. At most stations, however, the snow sample is melted by pouring a measured quantity of warm water into the can. After the snow is completely melted, the contents of the overflow can are poured into the measuring tube, and measured as for rainfall. The amount measured minus the known quantity of warm water added for melting the snow is the water equivalent [23].

Nuclear-radiation snow gages have been developed [76] for measuring water equivalent of the snowpack. Measurement depends on attenuation of the radiation beam by the snow. Early models had the radioisotope source at the ground and detector overhead, but positions were reversed in later models to minimize temperature effects on the detector. One such gage [77] has the source about 15 ft (5 m) above the detector at ground level. Nuclear-radiation gages are expensive, and few are in use. Also, changes in the physical state of the snow affect measurement accuracy.

Special types of radioactive snow gages have been developed [78, 79] to determine the variation of the water equivalent, or density, with depth in the snowpack. These gages, known generally as *profiling radioactive snow gages* or *twin-probe snow-density gages,* consist of a gamma photon source and detector moved synchronously through vertical tubes about 2 ft (60 cm) apart in the snowpack. The gages are used for studying temporal variations of the internal structure of the snowpack.

Areal measurements of the snowpack water equivalent can be made by aircraft flying at about 150 m (500 ft) above the ground. Natural gamma emissions from the soil are attenuated by the snow cover. The degree of attenuation is related to the mass of water in the snow. Gamma spectral and total counting rates are collected and recorded by an airborne system using sodium iodide scintillation crystals [80]. Corrections are made for soil moisture, background radiation, altitude, and air density. Comparison of results with extensive ground measurements indicates that gamma spectral data yield areal values of water equivalent accurate within 5 to 12 mm (0.2 to 0.5 in). Natural gamma radiation measurements of water equivalent have been used for many years in the Soviet Union, where depths of up to 300 mm (12 in) have been measured successfully [81].

The percentage of a drainage area that is covered by snow is an important factor in the prediction of snowmelt runoff. The extent of the snow cover can be mapped from satellite imagery [82] with an accuracy of 1 to 5 km (0.6 to 3 mi) depending on camera resolution and scan angle. This degree of accuracy provides more detailed mapping than can be obtained from the conventional snow-reporting station network. Near-infrared imagery, however, may sometimes fail to show snow revealed by video imagery. This deficiency is likely to occur during melting, when radiation in the near-infrared band is apparently absorbed by liquid water on or near the snow surface. The deficiency is an asset when video and near-infrared photographs are taken simultaneously, as melting areas are then identified [83].

Qualitative estimates of snow depth can also be obtained from satellite photographs. In nonforested areas, snow brightness, or reflectivity, increases rapidly with depth up to about 30 cm (1 ft), there being little further increase for greater depths [84]. Brightness estimates from satellite photography must be used cautiously, however, since anisotropy of snow seems to be a major factor affecting its appearance in photographs. No discernible relationship between reflectivity and snow depth is observed in forested areas.

Differentiation between snow and cloud is a major problem, but cloudiness can be determined from weather reports and charts. Since a cloud layer precludes photographs of snow cover, satellite observations cannot be made on a regular schedule. Indications of little or no cloudiness can be gained from topographic or pattern recognition of such features as lakes, dendritic stream patterns in mountains, dark forested areas, etc. Composite minimum brightness (CMB) charts [85] have proved useful in minimizing the cloudiness problem. These charts are computer-composited for a period of 5 to 10 days using only the minimum brightness observed for each array point during the compositing period.† Cloudiness is thus retained in the CMB chart only when present in a given area every day of the period. Since cloudiness is usually transitory, the bright areas in the CMB chart generally represent the relatively stationary boundaries of the snow and ice fields. Cloudiness persisting over an area for the entire compositing period cannot be filtered out by the technique, which is why a minimum period of 5 days is advisable. Brightness for such time-composited areas, however, is usually less than for composited snow and ice fields. The CMB technique is most effective in flat and unforested regions and less so in heavy coniferous-forest areas.

3-17 Variations

The distribution of mean annual snowfall in the United States is shown in Fig. 3-18. This map may be considerably in error in mountainous regions because of the paucity of measurements at high elevations. As should be expected, there is a gradual increase of snowfall with latitude and elevation. In the Sierra Nevada and Cascade Range, annual snowfalls of 400 in (1000 cm) are not uncommon. In general, maximum annual snowfall occurs at slightly higher elevations than maximum annual precipitation.

Maps of mean depth of snow on the ground or mean water equivalent for specific dates are not available. Snow depths would, of course, be much lower than annual snowfall amounts because of compaction of old snow, evaporation, and melting. Snow depth usually builds up rapidly early in the season and then remains relatively constant as compaction of old snow compensates for new falls. Maximum depths on the ground are usually less than one-half the annual snowfall at high elevations and are still less at lower elevations, where intermittent melting occurs. The mean water equivalent at the beginning of the melting season is a good index to the mean annual precipitation [86] in areas which experience little winter melting.

Because of drifting, considerable variation in snow depth and water equivalent may be observed within short distances. This variation can be intensified

---

† In 1979, the U.S. National Weather Service was using a 7-day sequence composited daily.

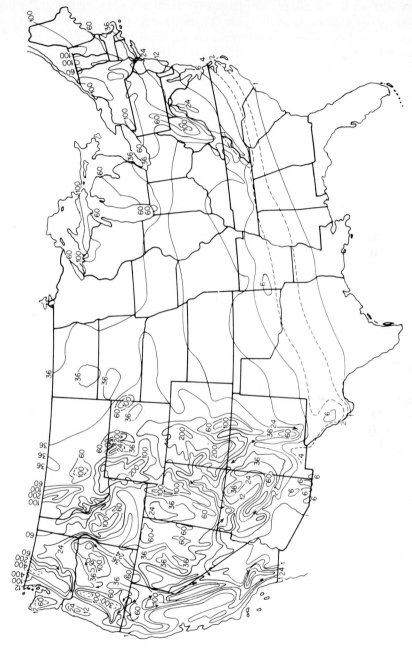

Figure 3-18 Mean annual snowfall in the United States, in inches (1 in = 2.54 cm). (*U.S. Environmental Data Service.*)

by differences in melting rates, which are generally greater on south slopes and in areas without forest cover. Nevertheless, there is a consistency in the distribution pattern from year to year as many important factors affecting melting—slope, aspect, elevation, forest cover, wind currents, etc.—normally undergo little change.

PROBLEMS

3-1 Assuming rain falling vertically, express the catch of a gage inclined 20° from the vertical as a percentage of the catch for the same gage installed vertically.

3-2 Shielded and unshielded gages at a station indicate a storm rainfall of 100 and 90 mm, respectively. Estimate the actual rainfall. Assume $b = 1.7$.

3-3 With a Z value of 300,000 mm^6/m^3, what are the rainfall rates, in mm/hr, indicated by Z-R relationships with a and b values of (a) 200 and 1.6, and (b) 300 and 1.4, respectively?

3-4 Comparison of rain-gage measurements with rainfall rates estimated by radar using the so-called standard Z-R relationship shows that the rates measured by the gage are 2.5 times those indicated by the radar. If the b exponent in the Z-R equation is assumed to be correct, what should the value of a be?

3-5 Precipitation station X was inoperative for part of a month during which a storm occurred. The respective storm totals at three surrounding stations, A, B, and C, were 98, 80, and 110 mm. The normal annual precipitation amounts at stations X, A, B, and C are, respectively, 880, 1008, 842, and 1080 mm. Estimate the storm precipitation for station X.

3-6 The annual precipitation at station X and the average annual precipitation at 15 surrounding stations are as shown in the following table.

   (a) Determine the consistency of the record at station X.

   (b) In what year is a change in regime indicated?

   (c) Compute the mean annual precipitation for station X for the entire 30-year period without adjustment.

   (d) Repeat part (c) for station X at its 1979 site with the data adjusted for the change in regime.

	Annual precipitation, cm			Annual precipitation, cm	
Year	Station X	15-Station average	Year	Station X	15-Station average
1950	47	29	1965	36	34
1951	24	21	1966	35	28
1952	42	36	1967	28	23
1953	27	26	1968	29	33
1954	25	23	1969	32	33
1955	35	30	1970	39	35
1956	29	26	1971	25	26
1957	36	26	1972	30	29
1958	37	26	1973	23	28
1959	35	28	1974	37	34
1960	58	40	1975	34	33
1961	41	26	1976	30	35
1962	34	24	1977	28	26
1963	20	22	1978	27	25
1964	26	25	1979	34	35

3-7 Plot the 15-station average precipitation of Prob. 3-6 as a time series. Also plot 5-year moving averages and accumulated annual departures from the 30-year mean. Are there any apparent cycles or time trends? Discuss.

3-8 The average annual precipitation for the four subbasins constituting a large river basin is 73, 85, 112, and 101 cm. The areas are 930, 710, 1090, and 1680 km^2, respectively. What is the average annual precipitation for the basin as a whole?

3-9 Compute the mean annual precipitation for some river basin selected by your instructor. Use the arithmetic average, Thiessen network, and isohyetal map, and compare the three values. Which do you feel is the most accurate? How consistent are the answers determined by each method among members of your class?

3-10 A calorimetric procedure for evaluating quality Q_t of melting snow consists of inserting into a thermos bottle or jug containing a known weight W_1 of warm water a sample of snow of known weight W_2. Write an equation for evaluating Q_t as a percentage when the initial and final temperatures of the water are, respectively, T_1 and T_2, in degrees Celsius, and the calorimeter constant is k.

3-11 Using the formula developed in Prob. 3-10, determine the snow quality when 100 g of snow inserted into a thermos bottle containing 350 g of water at 25°C lowers the temperature to 5°C. Assume a calorimeter constant of 50 g.

REFERENCES

1. McDonald, J. E.: The Evaporation-Precipitation Fallacy, *Weather,* vol. 17, pp. 168–177, May 1962.
2. Kumai, M.: Snow Crystals and the Identification of Nuclei in the Northern United States of America, *J. Meteorol.,* vol. 18, no. 2, pp. 139–150, April 1961.
3. Mason, B. J.: Precipitation of Rain and Drizzle by Coalescence in Stratiform Clouds, *Q. J. R. Meteorol. Soc.,* vol. 78, no. 337, pp. 377–386, July 1952.
4. Gunn, R.: Collision Characteristics of Freely Falling Water Drops, *Science,* vol. 150, no. 3697, pp. 695–791, Nov. 5, 1965.
5. Hamilton, P. M.: Vertical Profiles of Total Precipitation in Shower Situations, *Q. J. R. Meteorol. Soc.,* vol. 92, no. 392, pp. 346–362, April 1966.
6. Das, P.: Role of Condensed Water in the Life Cycle of a Convective Cloud, *J. Atmos. Sci.,* vol. 21, no. 4, pp. 404–418, July 1964.
7. Roys, G. P., and E. Kessler: Measurement by Aircraft of Condensed Water in Great Plains Thunderstorms, *ESSA Tech. Note* 49-NSSP-19, July 1966.
8. Reynolds, D. W., T. H. Vonder Haar, and L. O. Grant: Meteorological Satellites in Support of Weather Modification, *Bull. Am. Meteorol. Soc.,* vol. 59, no. 3, pp. 269–281, March 1978.
9. Vardiman, L., and J. A. Moore: Generalized Criteria for Seeding Winter Orographic Storms, *J Appl. Meteorol.,* vol. 17, no. 12, pp. 1769–1777, December 1978.
10. Henderson, T. J.: Kings River Weather Modification Program, *Special Regional Weather Modification Conference—Augmentation of Winter Orographic Precipitation in Western United States, San Francisco, Calif., Nov. 11-13, 1975,* pp. 100–105, American Meteorological Society, Boston.
11. Fleagle, R. G. (ed.): "Water Modification: Science and Public Policy," University of Washington Press, Seattle, pp. 31–32, 1969.
12. Biswas, K. R., and A. S. Dennis: Formation of a Rain Shower by Salt Seeding, *J. Appl. Meteorol.,* vol. 10, no. 4, pp. 780–784, August 1971.
13. Wexler, R.: Efficiency of Natural Rain, *Am. Geophys. Union Geophys. Monog.* 5, pp. 158–163, 1960.
14. The Management of Water Resources, Volume 1; Proposals for a National Policy and Program, Report to the Secretary of Commerce from the Weather Modification Advisory Board, p. 82, June 1978.
15. Neyman, J., E. L. Scott, and M. A. Wells: Statistics in Meteorology, *Rev. Int. Stat. Inst.,* vol. 37, no. 2, pp. 119–148, 1969.

16. Fleagle, R. G., J. A. Crutchfield, Jr., R. Johnson, and M. Abdo: "Weather Modification in the Public Interest," University of Washington Press, Seattle, 1974.
17. Sonka, S. T.: The Economics of Weather Modification: A Review and Suggestions for Future Economics, *J. Appl. Meteorol.*, vol. 17, no. 6, pp. 778–785, June 1978.
18. Thomas, W. A. (ed.): "Legal and Scientific Uncertainties of Weather Modification," Duke University Press, Durham, N.C., 1977.
19. Taubenfeld, H. J. (ed.): "Controlling the Weather: A Study of Law and Regulatory Procedures," Cambridge University Press, Cambridge, and Dunellen, New York, 1970.
20. Changnon, S. A., F. A. Huff, and C. F. Hsu: On the Need to Evaluate Operational Weather Modification Projects, *Bull. Am. Meteorol. Soc.*, vol. 60, no. 7, July 1979.
21. Huff, F. A.: Comparison between Standard and Small Orifice Raingages, *Trans. Am. Geophys. Union*, vol. 36, no. 4, pp. 689–694, August 1955.
22. Jones, D. M. A.: Effect of Housing Shape on the Catch of Recording Gages, *Mon. Weather Rev.*, vol. 97, no. 8, pp. 604–606, August 1969.
23. U.S. National Weather Service: Substation Observations, *Obs. Handb.*, no. 2, pp. 17–37, 1970, rev. December 1972.
24. Beausoleil, R. W., and K. W. Davis: Final Report of the Fischer and Porter Precipitation Gage Installation, *U.S. Natl. Weather Serv. Syst. Dev. Off. Rep.* 5, pp. 15–23, June 1970.
25. Parsons, D. A.: Calibration of a Weather Bureau Tipping-Bucket Gage, *Mon. Weather Rev.*, vol. 69, no. 7, p. 205, July 1941.
26. Struzer, L. R.: Method of Measuring the Correct Value of Solid Atmospheric Precipitation, *Sov. Hydrol. Sel. Pap.*, no. 6, pp. 560–565, 1969.
27. See Windshields in Bibliography.
28. Brown, M. J., and E. L. Peck: Reliability of Precipitation Measurements as Related to Exposure, *J. Appl. Meteorol.*, vol. 1, no. 2, pp. 203–207, June 1962.
29. Weiss, L. L.: Securing More Nearly True Precipitation Measurements, *J. Hydraul. Div. ASCE*, vol. 89, no. HY3, pp. 11–18, March 1963.
30. Hamon, W. R.: Computing Actual Precipitation, in Distribution of Precipitation in Mountainous Areas, WMO/OMM no. 326, pp. 159–173, World Meteorological Organization, Geneva, 1973.
31. Hanson, C. L., R. P. Morris, and D. L. Coon: A Note on the Dual-Gage and Wyoming Shield Precipitation Measurements Systems, *Water Resour. Res.*, vol. 15, no. 4, pp. 956–960, August 1979.
32. Hamilton, E. L.: Rainfall Sampling on Rugged Terrain, *U.S. Dept. Agric. Tech. Bull.* 1096, 1954.
33. Hoeck, E.: Report of the Committee on the Measurement of Precipitation, *Trans. Int. Assoc. Sci. Hydrol.*, vol. 3, pp. 81–93, 1952.
34. Serra, L.: The Correct Measurements of Precipitation, *Houille Blanche*, spec. no. A, pp. 152–158, March–April 1953.
35. See Network Design and Estimates of Areal Precipitation and Runoff in Bibliography.
36. Thunderstorm Rainfall, *Hydrometeorol. Rep.* 5, U.S. Weather Bureau in cooperation with Corps of Engineers, pp. 234–259, 1947.
37. World Meteorological Organization: Guide to Hydrometeorological Practices, 3d ed., WMO no. 168, *Tech. Pap.* 82, pp. 3.8–3.10, Geneva, 1974.
38. Eagleson, P. S.: Optimum Density of Rainfall Networks, *Water Resour. Res.*, vol. 3, no. 4, pp. 1021–1033, 1967.
39. Johanson, R. C.: Precipitation Network Requirements for Streamflow Estimation, *Stanford Univ. Dept. Civ. Eng. Tech. Rep.* 147, August 1971.
40. Kessler, E., and K. E. Wilk: The Radar Measurement of Precipitation for Hydrological Purposes, *Rep. WMO/IHD Proj.*, no. 5, World Meteorological Organization, Geneva, 1968.
41. Stout, G. E., and E. A. Mueller: Survey of Relationships between Rainfall Rate and Radar Reflectivity in the Measurement of Precipitation, *J. Appl. Meteorol.*, vol. 7, no. 3, pp. 465–474, June 1968.

42. Carlson, P. E., and J. S. Marshall: Measurement of Snowfall by Radar, *J. Appl. Meteorol.,* vol. 11, no. 3, pp. 494–500, April 1972.
43. Hudlow, M. D., and R. E. Arkell: Effect of Temporal and Spatial Sampling Errors and *Z-R* Variability on Accuracy of GATE Radar Rainfall Estimates, 18*th Radar Meteorology Conference, Atlanta, Ga., March* 28–31, 1978, pp. 342–349, American Meteorological Society, Boston.
44. Greene, D. R., and A. F. Flanders: Radar Hydrology—The State of the Art, *First Conference on Hydrometeorology, Ft. Worth, Tex., April* 20–22, 1976, pp. 66–71, American Meteorological Society, Boston.
45. Saffle, R. E.: *D/RADEX* Products and Field Operation, 17*th Radar Meteorology Conference, Seattle, Wash., October* 26–29, 1976, pp. 555–559, American Meteorological Society, Boston.
46. Brandes, E. A.: Optimizing Rainfall Estimates with the Aid of Radar, *J. Appl. Meteorol.,* vol. 14, no. 7, pp. 1339–1345, October 1975.
47. Cain, D. E., and P. L. Smith, Jr.: A Sequential Analysis Strategy for Adjusting Radar Rainfall Estimates on the Basis of Rain Gage Data in Real Time, *Second Conference on Hydrometeorology, Toronto, Canada, October* 25–27, 1977, pp. 280–285, American Meteorological Society, Boston.
48. Frederick, R. H., V. A. Myers, and E. P. Auciello; Storm Depth-Area Relations from Digitized Radar Returns, *Water Resour. Res.,* vol. 13, no. 3, pp. 675–679, June 1977.
49. Oliver, V. J., and R. A. Scofield: Estimation of Rainfall from Satellite Imagery, *First Conference on Hydrometeorology, Ft. Worth, Tex., April* 20–22, 1976, pp. 98–101, American Meteorological Society, Boston.
50. Wilheit, T. T., et al.: A Satellite Technique for Quantitatively Mapping Rainfall Rates over Oceans, *J. Appl. Meteorol.,* vol. 16, no. 5, pp. 551–560, May 1977.
51. Griffith, C. G., W. L. Woodley, and P. C. Grube: Rain Estimation from Geosynchronous Satellite Imagery—Visible and Infrared Studies, *Mon. Weather Rev.,* vol. 106, no. 8, pp. 1153–1171, August 1978.
52. Follansbee, W. A.: Estimation of Average Daily Rainfall from Satellite Cloud Photographs, *NOAA Tech. Memo. NESS* 44, January 1973.
53. Kilonsky, B. J., and C. S. Ramage: A Technique for Estimating Tropical Open-Ocean Rainfall from Satellite Observations, *J. Appl. Meteorol.,* vol. 15, no. 9, pp. 972–975, September 1976.
54. Follansbee, W. A.: Estimation of Daily Precipitation over China and the USSR Using Satellite Imagery, *NOAA Tech. Memo. NESS* 81, September 1976.
55. Greene, D. R., M. D. Hudlow, and R. K. Farnsworth: A Multiple Sensor Rainfall Analysis System, *Third Conference on Hydrometeorology, Bogota, Colombia, August* 20–24, 1979, preprint, American Meteorological Society, Boston.
56. Paulhus, J. L. H., and M. A. Kohler: Interpolation of Missing Precipitation Records, *Mon. Weather Rev.,* vol. 80, no. 8, pp. 129–133, August 1952.
57. U.S. National Weather Service: *National Weather Service River Forecast System. Forecast Procedures. NOAA Tech. Memo. NWS HYDRO* 14, pp. 3.1–3.14, December 1972.
58. Kohler, M. A.: Double-Mass Analysis for Testing Consistency of Records and for Making Required Adjustments, *Bull. Am. Meteorol. Soc.,* vol. 30, no. 5, pp. 188–189, May 1949.
59. Chang, M., and R. Lee: Objective Double-Mass Analysis, *Water Resour. Res.,* vol. 10, no. 6, pp. 1123–1126, December 1974.
60. Peck, E. L. Discussion of Problems in Measuring Precipitation in Mountainous Areas, *Symposium on Distribution of Precipitation in Mountainous Areas, Geilo, Norway, August* 1972, *WMO/OMN* no. 326, vol. 1, pp. 5–16, World Meteorological Organization, Geneva.
61. World Meteorological Organization: Manual for Depth-Area-Duration Analysis of Storm Precipitation, WMO no. 237, *Tech. Pap.* 129, pp. 1–31, Geneva, 1969.
62. See [61], pp. 32–56.

63. Shipe, A. P., and J. T. Riedel: Greatest Known Areal Storm Rainfall Depths for the Contiguous United States, *NOAA Tech. Memo. NWS HYDRO*-33, December 1976.
64. See Spatial and Temporal Distribution of Precipitation in Bibliography.
65. Spreen, W. C.: Determination of the Effect of Topography upon Precipitation, *Trans. Am. Geophys. Union,* vol. 28, no. 2, pp. 285–290, April 1947.
66. Dawdy, D. R., and W. B. Langbein: Mapping Mean Areal Precipitation, *Bull. Int. Assoc. Sci. Hydrol.,* vol. 5, no. 19, pp. 16–23, September 1960.
67. Peck, E. L., and M. J. Brown: An Approach to the Development of Isohyetal Maps for Mountainous Areas, *J. Geophys. Res.,* vol. 67, no. 2, pp. 681–694, February 1962.
68. Shaw, N.: "Manual of Meteorology," 2d ed., vol. 2, pp. 320–325, Cambridge University Press, London, 1942.
69. Mitchell, J. M., Jr.: A Critical Appraisal of Periodicities in Climate, *Iowa State Univ., Cent. Agric. Econom. Devel. Rep.* 20, pp. 189–227, 1964.
70. Bernard, M., and W. T. Wilson: A New Technique for the Determination of the Heat Necessary to Melt Snow, *Trans. Am. Geophys. Union,* vol. 22, pt. 1, pp. 178–181, 1941.
71. Radok, U., S. K. Stevens, and K. L. Sutherland: On the Calorimetric Determination of Snow Quality, *Int. Assoc. Sci. Hydrol. Publ.* 54, pp. 132–135, 1961.
72. Miller, R. W.: Aerial Snow Depth Marker Configuration and Installation Considerations, *West. Snow. Conf. Proc. 1962,* pp. 1–5.
73. U.S. Soil Conservation Service: Snow Survey and Water Supply Forecasting, "SCS National Engineering Handbook," sec. 22, April 1972.
74. Work, R. A., H. J. Stockwell, and R. T. Beaumont: Accuracy of Field Snow Surveys, *U.S. Army Corps Eng. Cold Regions Res. Eng. Lab. Tech. Rep.* 163, 1965.
75. Cox, L. M., et al.: The Care and Feeding of Snow Pillows, *46th Annual Meeting Western Snow Conference, Otter Rock, Oregon,* April 18–20, 1978, pp. 40–47.
76. Warnick, C. C., and V. E. Penton: New Methods of Measuring Water Equivalent of Snow Pack for Automatic Recording at Remote Mountain Locations, *J. Hydrol.,* vol. 13, pp. 201–215, 1971.
77. McKean, G. A.: A Nuclear Radiation Snow Gage, *Univ. Idaho Eng. Exp. Sta. Bull.* 13, August 1967.
78. Smith, J. L., H. G. Halverson, and R. A. Jones: The Profiling Radioactive Snow Gage, *Trans. Isotopic Snow Gage Info. Meet.,* Sun Valley, Idaho, Oct. 28, 1970, Idaho Nuclear Energy Commission and U.S. Soil Conservation Service.
79. Western Snow Conference, 44th Annual Meeting, Calgary, Alberta, Apr. 20–22, 1976, pp. 1–19.
80. Peck, E. L., and V. C. Bissell: Aerial Measurement of Snow Water Equivalent by Terrestrial Gamma Radiation Survey, *Bull. Int. Assoc. Hydrolog. Sci.,* vol. 18, no. 1, pp. 47–62, 1973.
81. Zotimov, N. V.: Investigation of a Method of Measuring Snow Storage by Using the Gamma Radiation of the Earth, *Sov. Hydrol.: Sel. Pap.,* no. 3, pp. 254–265, 1968.
82. Schneider, S. R., D. R. Wiesnet, and M. C. McMillan: River Basin Snow Mapping at the National Environmental Satellite Service, *NOAA Tech. Memo. NESS* 83, November 1976.
83. Strong, A. E., E. P. McClain, and D. F. McGinnis: Detection of Thawing Snow and Ice Packs through the Combined Use of Visible and Near-Infrared Measurements from Earth Satellites, *Mon. Weather Rev.,* vol. 99, no. 11, pp. 828–830, November 1971.
84. McGinnis, D. F., Jr., J. A. Pritchard, and D. R. Wiesnet: Determination of Snow Depth and Snow Extent from NOAA 2 Satellite Very High Resolution Radiometer Data, *Water Resour. Res.,* vol. 11, no. 6, pp. 897–902, December 1975.
85. McClain, E. P., and D. R. Baker: Experimental Large Scale Snow and Ice Mapping with Composite Brightness Charts, *ESSA Tech. Mem., NESCTM* 12, September 1969.
86. Paulhus, J. L. H., C. E. Erickson, and J. T. Riedel: Estimation of Mean Annual Precipitation from Snow Survey Data, *Trans. Am. Geophys. Union,* vol. 33, no. 5, pt. 1, pp. 763–767, October 1952.

BIBLIOGRAPHY

Artificial induction of precipitation

American Meteorological Society: Sixth Conference on Planned and Inadvertent Weather Modification, Oct. 10–13, 1977, Champaign-Urbana, Ill.

Dennis, A. S.: "Weather Modification by Cloud Seeding," Academic Press, New York, 1980.

Foote, G. B., and C. A. Knight (eds.): Hail: A Review of Hail Science and Hail Suppression, *Meteorol. Mono., Am. Meteorol. Soc.,* vol. 16, no. 38, December 1977.

Hess, W. N.: "Weather and Climate Modification," Wiley, New York, 1974.

Mason, B. J.: "Clouds, Rain and Rainmaking," 2d ed., Cambridge University Press, London, 1975.

Depth-area relations

Court, A.: Area-Depth Rainfall Formulas, *J. Geophys. Res.,* vol. 66, no. 6, pp. 1823–1832, June 1961.

Stout, G. E., and F. A. Huff: Studies of Severe Rainstorms in Illinois, *J. Hydraul. Div. ASCE,* vol. 88, no. HY7, pp. 129–146, July 1962.

Measurement of precipitation

Battan, L. J.: "Radar Observation of the Atmosphere," University of Chicago Press, Chicago, 1973.

Israelsen, C. E.: Reliability of Can-Type Precipitation Gage Measurements, *Utah State Univ., Utah Water Resour. Lab. Tech. Rep.* 2, July 1967.

Larson, L. W.: Approaches to Measuring "True" Snowfall, paper presented at 29*th East. Snow Conf., Oswego, N.Y., Feb.* 3–4, 1972.

Martin, D. W., and W. D. Scherer: Review of Satellite Rainfall Estimation Methods, *Bull. Am. Meteorol. Soc.,* vol. 54, no. 7, pp. 661–674, July 1973.

Middleton, W. E. K., and A. F. Spilhaus: "Meteorological Instruments," 3d ed., pp. 118–131, University of Toronto Press, Toronto, 1953.

Rodda, J. C.: Annotated Bibliography on Precipitation Measurement Instruments, *Rep. WHO/IHD Proj.,* no. 17, WMO no. 343, World Meteorological Organization, Geneva, 1973.

U.S. Weather Bureau: Excessive Precipitation Techniques, *Key Meteorol. Rec. Doc.,* no. 3.081, 1958.

———: History of Weather Bureau Precipitation Measurements, *Key Meteorol. Rec. Doc.,* no. 3.082, 1963.

Wilson, J. W., and E. A. Brandes: Radar Measurement of Rainfall—A Summary, *Bull. Am. Meteorol. Soc.,* vol. 60, no. 9, pp. 1048–1058, September 1979.

Network design and estimates of areal precipitation and runoff

Alvarez, F., and W. K. Henry: Rain Gage Spacing and Reported Rainfall, *Bull. Int. Assoc. Sci. Hydrol.,* vol. 15, no. 3, pp. 97–107, March 1970.

American Geophysical Union: Chapman Conference on Design of Hydrologic Networks, Tucson, Ariz., Dec. 11–14, 1978.

Hall, A. J., and P. A. Barclay: Methods of Determining Areal Rainfall from Observed Data, in T. G. Chapman and F. X. Dunin (eds.): "Prediction in Catchment Hydrology," Australian Academy of Science, Canberra, pp. 47–57, 1975.

Huff, F. A.: Sampling Errors in Measurement of Mean Precipitation, *J. Appl. Meteorol.,* vol. 9, no. 1, pp. 35–44, February 1970.

International Association of Hydrologic Sciences: Design of Hydrological Networks, Symposium of Quebec, *IASH Publ.* 67 and 68, 1965.

Lenton, R. L., and I. Rodriguez-Iturbe: Rainfall Network Systems Analysis: The Optimal Estimation of Total Area Storm Depth, *Water Resour. Res.,* vol. 13, no. 5, pp. 825–836, October 1977.

Němec, J.: A Contribution to the Design of Recording Gage Network, *Bull. Int. Assoc. Sci. Hydrol.,* vol. 10, no. 2, pp. 70–73, June 1965.

Nordenson, T. J.: Preparation of Co-ordinated Precipitation, Runoff and Evaporation Maps, *Rep. WMO/IHD Proj.,* no. 6, World Meteorological Organization, Geneva, 1967.

Rainbird, A. F.: Methods of Estimating Areal Average Precipitation, *Rep. WMO/IHD Proj.,* no. 3, World Meteorological Organization, Geneva, 1967.

Schickedanz, P. T.: Application of Dense Rainfall Data to Regions of Sparse Data Coverage, State Water Survey, Urbana, Ill., October 1971.

Stephenson, P. M.: Objective Assessment of Adequate Numbers of Raingauges for Estimating Areal Rainfall Depths, *Int. Assoc. Sci. Hydrol. Pub.* 78, pp. 252–264, 1968.

World Meteorological Organization: Hydrological Network Design and Information Transfer, *Opera. Hydrol. Rep.* 8, WMO no. 433, Geneva, 1976.

Precipitation cycles

Brier, G. W., R. Shapiro, and N. J. MacDonald: A Search for Rainfall Calendricities. *J. Atmos. Sci.,* vol. 20, no. 6, pp. 529–532, November 1963.

Lambor, J.: "Hydrologic Forecasting Methods," Scientific Publications Foreign Cooperation Center of the Central Institute for Scientific, Technical, and Economic Information, Warsaw, pp. 314–336, 1962; available from U.S. Dept. of Commerce National Technical Information Center, Springfield, Va.

Namias, J.: Long-Range Weather Forecasting History, Current Status and Outlook, *Bull. Am. Meteorol. Soc.,* vol. 49, no. 5, pt. 2, pp. 438–470, May 1968.

Precipitation forms and formation

Byers, H. R.: "Elements of Cloud Physics," University of Chicago Press, Chicago, 1965.

Fletcher, N. H.: "The Physics of Rain Clouds," Cambridge University Press, London, 1962.

Hardy, K. R.: The Development of Raindrop-Size Distributions and Implications Related to the Physics of Precipitation, *J. Atmos. Sci.,* vol. 20, no. 4, pp. 299–312. July 1963.

Mason, B. J.: "The Physics of Clouds," 2d ed., Oxford University Press, London, 1971.

Spatial and temporal distribution of precipitation

Changnon, S. A., Jr., R. G. Semonin, and F. A. Huff: A Hypothesis for Urban Rainfall Anomalies, *J. Appl. Meteorol.,* vol. 15, no. 6, pp. 544–560, June 1976.

Fedorov, S. F., and A. S. Burov: Influence of the Forest on Precipitation, *Sov. Hydrol.: Sel. Pap.,* no. 3, pp. 217–227, 1967.

Frederick, R. H.: Interduration Precipitation Relations for Storms—Southeast States, *NOAA Tech. Rep. NWS* 21, March 1979.

———: Interstorm Relations in Pacific Northwest, *J. Hydraul. Div. ASCE,* vol. 104, no. HY12, pp. 1577–1586, December 1978.

———: Time Distribution of Precipitation in 4- to 10-Day Storms—Arkansas-Canadian River Basins, *NOAA Tech. Memo. NWS HYDRO* 15, June 1973.

Griffiths, J. F.: Bibliography of the Urban Modification of the Atmospheric and Hydrologic Environment, *NOAA Tech. Memo. EDS*-21, February 1974.

Huff, F. A.: Spatial Distribution of Rainfall Rates, *Water Resour. Res.,* vol. 6, no. 1, pp. 254–260, February 1970.

———: Time Distribution Characteristics of Rainfall Rates, *Water Resour. Res.,* vol. 6, no. 2, pp. 447–454, April 1970.

Miller, J. F., and R. H. Frederick: Time Distribution of Precipitation in 4- to 10-Day Storms—Ohio River Basin, *NOAA Tech. Memo NWS HYDRO* 13, July 1972.

Windshields

Rechard, P. A., and L. W. Larson: Snow Fence Shielding of Precipitation Gages, *J. Hydraul. Div. ASCE,* vol. 97, no. HY9, pp. 1427–1440, September 1971.
Warnick, C. C.: Experiments with Wind Shields for Precipitation Gages, *Trans. Am. Geophys. Union,* vol. 34, no. 3, pp. 379–388, June 1953.
Weiss, L. L.: Relative Catches of Snow in Shielded and Unshielded Gages, *Mon. Weather Rev.,* vol. 89, no. 10, pp. 397–400, October 1961.

UNITED STATES DATA SOURCES

The main source of daily precipitation data is *Climatological Data* published monthly through 1965 by the Weather Bureau and then by the Environmental Data Service. Hourly rainfall intensities are found in *Hydrologic Bulletin* (1940–1948), *Climatological Data* (1948–1951), and since 1951 in *Hourly Precipitation Data.* Weekly and monthly maps of precipitation are found in *Weekly Weather and Crop Bulletin* and *Monthly Weather Review.* Monthly and annual data are summarized from beginning of record through 1950 in *Climatic Summary for the United States* and from 1951 through 1960 in *Climatography of the United States,* no. 86. Monthly data for about 1000 stations scattered throughout the world are published in *Monthly Climatic Data for the World,* which is sponsored by the World Meteorological Organization in cooperation with the Environmental Data Service.

The *Water Bulletin* of the International Boundary and Water Commission contains data for the Rio Grande Basin, and *Precipitation in the Tennessee River Basin,* published by the Tennessee Valley Authority, summarizes data in its service area. Many states and local groups publish data summaries which contain information on precipitation.

Results of snow surveys in California are published by the State Department of Water Resources in *Water Conditions in California.* Elsewhere in the West, snow-survey data are published by the U.S. Soil Conservation Service in the *Federal-State-Private Cooperative Snow Surveys and Water-Supply Outlook.* In the East snow-survey data are published by the U.S. National Weather Service in *Snow Cover Surveys by Eastern Snow Conference.*

Special summaries and data on precipitation normals, extremes, and frequencies may be found in:

"Climatological Data, National Summary," U.S. Environmental Data Service.
"Climatic Atlas of the United States," U.S. Environmental Data Service.
Monthly Normals of Temperature, Precipitation, and Heating and Cooling Degree Days, *Climatography of the United States,* no. 81 (by states), NOAA Environmental Data Service, August 1973.
Frederick, R. H., V. A. Myers, and E. O. Auciello: Five- to 60-Minute Precipitation Frequency for the Eastern and Central United States, *NOAA Tech. Memo. NWS HYDRO* 35, June 1977.
Hershfield, D. M.: Rainfall-Frequency Atlas of the United States for Durations from 30 Minutes to 24 Hours and Return Periods from 1 to 100 Years, *U.S. Weather Bur. Tech. Pap.* 40, 1961. (Similar data for durations up to 10 days are also available in the *Weather Bureau Technical Paper* series for the United States, including Alaska and Hawaii, Puerto Rico, and Virgin Islands.)
Ludlum, D. M.: Extremes of Snowfall in the United States, *Weatherwise,* vol. 15, no. 6, pp. 246–253, December 1962.
Miller, J. F., and R. H. Frederick: Normal Monthly Number of Days with Precipitation of 0.5, 1.0, 2.0, and 4.0 Inches or More in the Conterminous United States, *ESSA Tech. Pap.* 57, 1966.
——, ——, and R. J. Tracey: "Precipitation-Frequency Atlas for the Conterminous Western United States (by States)," *NOAA Atlas* 2 (11 vols.), 1974.

Thom, H. C. S.: Probabilities of One-Inch Snowfall Thresholds for the United States, *Mon. Weather Rev.,* vol. 85, no. 8, pp. 269–271, August 1957.

———: Distribution of Maximum Annual Water Equivalent of Snow on the Ground, *Mon. Weather Rev.,* vol. 94, no. 4, pp. 265–271, April 1966.

Detailed descriptions of most of the above and additional sources are given in Selective Guide to Climatic Data Sources, *Key Meteorolog. Rec. Doc.* 4.11, U.S. Environmental Data Service, 1969, and in *Annotated Bibliography of NOAA Publications of Hydrometeorological Interest,* which is updated at irregular intervals by the National Weather Service.

FOUR

STREAMFLOW

Most data used by hydrologists serve other purposes in the geophysical sciences. Streamflow data are gathered primarily by hydrologists for hydrologic studies. To the engineering hydrologist, streamflow is the dependent variable in most studies, since engineering hydrology is concerned mainly with estimating rates or volumes of flow or the changes in these values resulting from human activities. Because it is difficult to make a direct, continuous measurement of the rate of flow in a stream but relatively simple to obtain a continuous record of stage, the primary field data gathered at most streamflow measurement stations are river stage [1]. This approach is satisfactory only if there is an adequate correlation between stage and discharge (Sec. 4-5).

WATER STAGE

4-1 Manual Gages

River stage is the elevation above some arbitrary zero datum of the water surface at a station. The datum is sometimes taken as mean sea level but more often is slightly below the point of zero flow in the stream.

The simplest way to measure river stage is by means of a *staff gage,* a scale set so that a portion of it is immersed in the water at all times. The gage may consist of a vertical scale attached to a bridge pier, piling, wharf, or other structure that extends into the low-water channel of the stream. If no suitable structure exists in a location accessible at all stages, a *sectional staff gage* (Fig. 4-1) may be used. Short sections of staff are mounted on available structures or

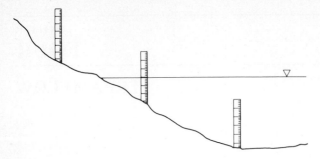

Figure 4-1 A sectional staff gage.

on specially constructed supports in such a way that one section is always accessible.

The gage scale may be painted on an existing structure or on a special gage board, usually in feet and tenths or in centimeters. Enameled metal sections are available where particularly accurate stage data are desired. If a stream carries a large amount of fine sediment or industrial waste, scale markings may be quickly obliterated. In this case, a serrated edge on the staff or raised marking symbols may be helpful.

In another type of manual gage a weight is lowered from a bridge or other overhead structure until it reaches the water surface. By subtracting the length of line paid out from the elevation of a fixed reference point on the structure, the water-surface elevation can be determined. The *wire-weight gage* has a drum with a circumference such that each revolution unwinds 1 ft of wire. A counter records the number of revolutions of the drum while a fixed reference point indicates hundredths of feet on a scale around the circumference.

4-2 Recording Gages

Manual gages are simple and inexpensive but must be read frequently to define the hydrograph adequately when stage is changing rapidly. In most situations water-stage recorders, in which the motion of a float is recorded on a chart, should be used. In a *continuous-chart recorder,* motion of the float moves a pen across a long strip chart. When the pen reaches the edge of the chart, it reverses direction and records in the other direction across the chart (Fig. 4-2a). Clocks may be weight-driven and will run as long as there is room for the clock weight to drop. Electric clocks that can operate for a year on a battery are also used.

Recorders which punch the stage at fixed intervals (usually 15 min) on paper tape are also used. The tape (Fig. 4-2b) can be read, verified, and converted to streamflow by electronic equipment. Since punched tapes do not offer ready visual inspection of the stage record for observing stage or detecting errors in the stage recording, chart recorders are preferred in some instances.

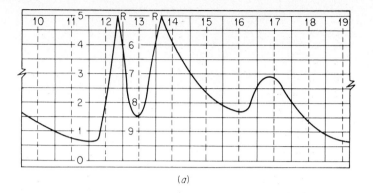

(a)

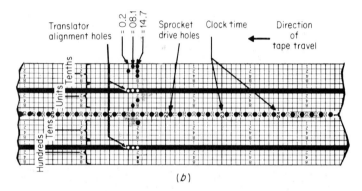

(b)

Figure 4-2 Recording-gage charts: (a) continuous-strip chart, and (b) punched tape. [*Part* (b) *Fischer & Porter Co.*]

Float-type water-stage recorders are generally installed in a shelter house and stilling well (Fig. 4-3). The stilling well serves to protect the float and counterweight cables from floating debris and (if the intakes are properly designed) suppress fluctuations from surface waves in the stream. Generally two or more intake pipes are placed from the stilling well into the stream so that at least one will admit water at all times. When the well is attached to a bridge pier and water enters through the open bottom of the well, an inverted cone may be placed over the bottom of the well to reduce the size of the opening and suppress the effect of surface waves. The open-bottom type of stilling well has the advantage of being less likely to fill with sediment. If a stilling well with a closed bottom is installed on a stream with a high sediment load, it is necessary to provide for removal of the sediment which accumulates in the well. It is customary to install staff gages inside and outside the well to check the performance of the recorder.

Bubbler gages [2] record the pressure required to maintain a small flow of

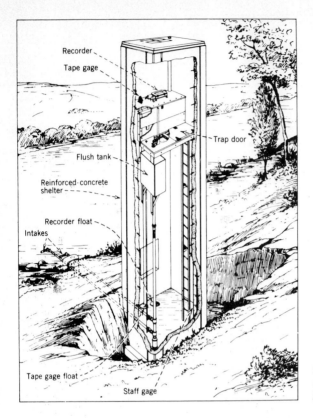

Recorder
Tape gage
Trap door
Flush tank
Reinforced concrete shelter
Recorder float
Intakes
Tape gage float
Staff gage

Figure 4-3 A typical water-stage recorder installation. (*U.S. Geological Survey.*)

gas from an orifice submerged in the stream. The aim is to eliminate the costly stilling well required with float-operated gages. It has proved difficult to develop gages which are accurate at low stages yet have the ruggedness and range required to record flood stages.

4-3 Crest-Stage Gages

Crest gages provide low-cost, supplementary records of crest stages at locations where recorders are not justified and where manually read staff gages are inadequate. A variety of such gages have been devised, including small floats which rise with stage but are restrained at the maximum level [3, 4] and water-soluble paints [5] on bridge piers where they are protected from rain and can indicate a definite high-water mark. The gage used by the U.S. Geological Survey consists of a length of pipe (Fig. 4-4) containing a graduated stick and a small amount of ground cork [6]. The cork floats as the water rises, and some adheres to the stick at the highest level reached by the water. The stick can be removed, the crest reading recorded, the cork wiped off, and the stick replaced ready for the next rise.

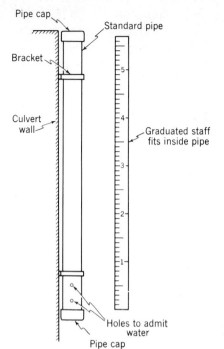

Figure 4-4 Crest-stage gage used by the U.S. Geological Survey.

4-4 Miscellaneous Stage Gages

Water- or mercury-filled manometers can be used to indicate reservoir water levels or to actuate recording devices. Remote recorders in which a system of selsyn motors is used to transmit water-level information from streamside to a recorder at a distance are available, as are numerous remote transmitting telephonic or radio gages. The latter gages use a coding device which converts stage to a signal transmitted as a series of impulses which can be counted, a change in frequency of oscillation which can be measured, or the time interval required for a sensing element to move from a zero point to the water surface at constant speed. Such remote recording devices are used primarily for flood forecasting or reservoir operation. Use of earth satellites as relay stations for transmission of data from remote stations eliminates surface relay stations.

4-5 Selection of Station Site

If a stream gage is solely to record water level for flood warning or as an aid to navigation, the prime factor in its location is accessibility. If the gage is to be used to obtain a record of discharge, the location should be carefully selected. The relation between stage and discharge is controlled by the physical features of the channel downstream from the gage. When controlling features are situated in a short length of channel, a *section control* exists. If the stage-discharge relation is governed by the slope, size, and roughness of the channel over a considerable distance, the station is under *channel control.* In many cases no single control is effective at all stages, but a complex of controlling elements function as stage varies.

The ideal low-water control is a section control consisting of rapids or falls where critical depth occurs. If this control is in rock, it will be reasonably permanent and once calibrated need be checked only infrequently. Where no such natural control exists, an artificial control consisting of a low concrete weir, sometimes with a shallow V notch, may be constructed to maintain a stable low-water rating. A channel control is more likely to change with time as a result of scour or deposition of sediment, and more frequent flow measurements are required to maintain an accurate stage-discharge relation.

Rapids may also be effective controls at high flows if the slope of the stream is steep, but where slopes are flat, the section control is likely to be submerged and ineffective at high flows. High-water controls are more likely to be channel controls, although in some cases the contraction at a bridge or the effect of a dam may control at high stages. It is advisable to avoid locations where varying backwater from a dam, an intersecting stream, or tidal action occurs. These situations require special ratings (Sec. 4-8) which are usually less accurate.

DISCHARGE

4-6 Current Meters

The stage record is transformed to a discharge record by calibration. Since the control rarely has a regular shape for which discharge can be computed, calibration is accomplished by relating field measurements of discharge with the simultaneous river stage (Secs. 4-8 and 4-9). Except in special situations, the discharge at a section is derived from point measurements of velocity.

The most common current meter in the United States, the *Price meter* (Fig. 4-5), consists of six conical cups rotating about a vertical axis [7, 8]. Electric contacts driven by the cups close a circuit through a battery and the wire of the supporting cable to cause a click for each revolution (or each fifth revolution) in headphones worn by the operator. Electrical counting devices are also used.

For measurements in deep water, the meter is suspended from a cable. Tail vanes keep the meter facing into the current, and a heavy weight keeps the meter cable as nearly vertical as possible. Special cranes are available to support the meter over a bridge rail, to simplify handling the heavy weights, and to permit measuring the length of cable paid out. In shallow water the meter is mounted on a rod, and the observer wades the stream.

Propeller-type current meters employ a propeller turning about a horizontal axis (Fig. 4-6). The contacting mechanism of a propeller meter is similar to that of a Price meter, and similar suspensions are used. The vertical-axis meter

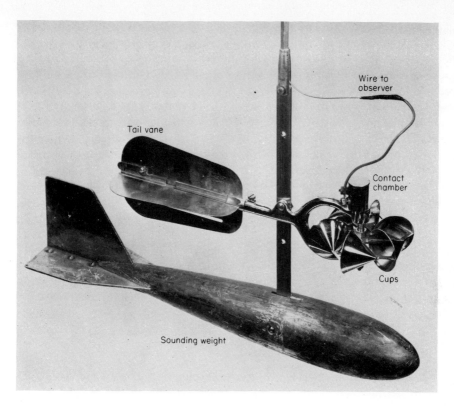

Figure 4-5 Price current meter and 30-lb C-type sounding weight. (*U.S. Geological Survey.*)

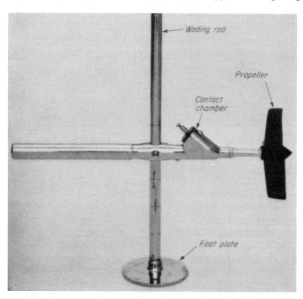

Figure 4-6 Propeller-type current meter on wading rod. (*U.S. Geological Survey.*)

has an important advantage in that the bearings supporting the shaft can be enclosed in inverted cups which trap air and prevent entrance of sediment-laden water. The bearings of propeller meters cannot be so protected and are exposed to damage by abrasion. On the other hand, vertical currents or upstream velocity components will rotate the cups of a vertical-axis meter in the same direction as downstream currents. A Price meter moved vertically in still water will indicate a positive velocity. Hence, it tends to overestimate stream velocity. If the measuring section is well chosen with the current flow nearly parallel to the channel axis and with a minimum of turbulence, the error is probably not more than 2 percent [9].

The relation between revolutions per second N of the meter cups and water velocity v is given by an equation of the form

$$v = a + bN \tag{4-1}$$

where b is the constant of proportionality and a is the starting velocity or velocity required to overcome mechanical friction. Some differences in these constants must be expected as a result of manufacturing variations, effects of wear, and accidental damage. Consequently each meter should be recalibrated periodically [10].

4-7 Current-Meter Measurements

A discharge measurement [11] requires determination of sufficient point velocities to permit computation of an average velocity in the stream. Cross-sectional area multiplied by average velocity gives the total discharge. The number of velocity determinations must be limited to those which can be made within a reasonable time, especially if stage is changing rapidly, since it is desirable to complete the measurement with a minimum change in stage.

The practical procedure involves dividing the stream into a number of vertical sections (Fig. 4-7). No section should include more than about 10 percent of the total flow; thus 20 to 30 vertical sections are typical, depending on the width of the stream. Velocity varies approximately as a parabola (Fig.

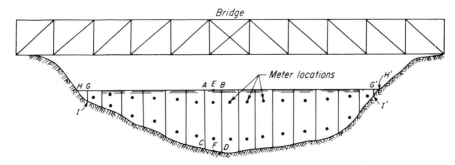

Figure 4-7 The procedure for current-meter measurement.

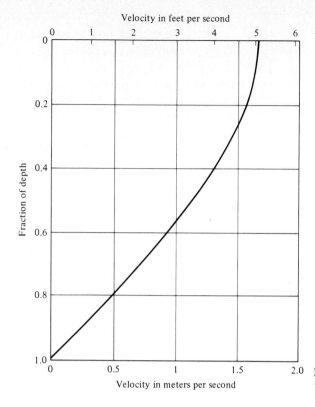

Figure 4-8 A typical vertical velocity profile in a stream.

4-8) from zero at the channel bottom to a maximum at (or near) the surface. On the basis of many field tests, the variation for most channels is such that the average of the velocities at two-tenths and eight-tenths depth below the surface equals the mean velocity in the vertical. The velocity at six-tenths depth below the surface also closely approximates the mean in the vertical. The adequacy of these assumptions for a particular stream can be tested by making a detailed vertical velocity traverse.

Sample current-meter field notes are shown in Fig. 4-9. Determination of the mean velocity in a vertical is as follows:

1. Measure total depth of water by sounding with the meter cable.
2. Raise meter to eight-tenths depth and measure velocity by starting a stop-watch on an impulse from the meter and stopping it on another impulse about 45 s later. The number of impulses counted (taking the first as zero) and the elapsed time permit calculation of velocity from the meter calibration.
3. Raise meter to two-tenths depth and repeat step 2.

In shallow water near the shore a single velocity determination at six-tenths depth may be used.

BIG CR. River at— DOGWOOD, VA.

9-275-F
Jan. 1952

UNITED STATES
DEPARTMENT OF THE INTERIOR
GEOLOGICAL SURVEY
WATER RESOURCES DIVISION

Meas. No. 270
Comp. by TJB.

DISCHARGE MEASUREMENT NOTES   Checked by ....

BIG CREEK NR DOGWOOD, VA.
Date SEP. 25 , 19 62 Party T. J. Buchanan
Width 70 Area 143 Vel. 0.53 G. H. 1.93 Disch. 76.4
Method 0.2+0.8 No. secs. 28 G. H. change −0.02 in 1 hrs. Susp. Rod
Method coef. 1 Hor. angle coef. Varies Susp. coef. 1 Meter No. 3684

	GAGE READINGS			Date rated 2-16-62 Used rating
Time	Recorder	Inside	Outside	for rod ✓ susp. Meter .... ft.
1310	1.94	1.94	1.92	above bottom of wt. Tags checked ...
1330	Start	(1.94)		Spin before meas. 2:45 after free.
1430	Finish	(1.92)		Meas. plots ...... % diff. from .... rating
1445		1.92	1.92	1.90	Wading, cable, ice, boat, upstr., downstr., side
				bridge 2.5 feet, mile, above, below

Weighted M. G. H. 1.93
G. H. correction ....
Correct M. G. H. ....

gage, and ....
Check-bar, chain found ....
changed to .... at ....
Correct ....
Levels obtained ....

Measurement rated excellent (2%), good (5%), fair (8%), poor (over 8%), based on following
conditions: Cross section Sand and gravel, fairly even
Flow Good distribution Weather Cloudy
Other Greatest flow in center Air 67 °F @ 1435
Gage .... Water 54 °F @ 1435
Record removed Yes Intake flushed ....
Observer Talked with ....

Control Clear

Remarks ....

G. H. of zero flow 1.92 −1.61 = 0.3 ft. ± 0.1 ft

16—58354-3

Angle coefficient	Dist. from initial point	Width	Depth	Observation depth	Revolutions	Time in seconds	Velocity At point	Velocity Mean in vertical	Adjusted for hor. angle or	Area	Discharge
	1	1.5	0							0	0
	4	3	0.95	.6	5	40		.31		2.8	.87
	7	3	1.4	.6	7	43		.40		4.2	1.68
	10	3	2.0	.6	7	52		.33		6.0	1.98
	13	3	2.1	.6	7	40		.42		6.3	2.65
	16	3	2.3	.6	10	55		.44		6.9	3.04
	19	3	2.25	.6	10	53		.46		6.8	3.13
	22	2.5	2.2	.6	10	48		.50		5.5	2.75
	24	2	2.5	.2	15	50	.70	.59		5.0	2.95
				.8	10	50	.48				
	26	2	2.8	.2	15	45	.78	.64		5.6	3.58
				.8	10	48	.50				
	28	2	3.0	.2	15	43	.82	.67		6.0	4.02
				.8	10	46	.52				
	30	2	2.95	.2	20	52	.90	.74		5.9	4.37
				.8	10	40	.59				
	32	2	3.1	.2	20	50	.93	.75		6.2	4.65
				.8	10	42	.57				
	34	2	3.2	.2	20	48	.97	.74		6.4	4.74
				.8	10	46	.52				
	36	2	3.05	.2	20	52	.90	.77		6.1	4.70
				.8	15	55	.64				
	38	2	3.1	.2	15	42	.83	.68		6.2	4.22
				.8	10	46	.52				

Figure 4-9 Sample current-meter field notes used by the U.S. Geological Survey.

If velocities are high, the meter and weight will not hang vertically below the point of suspension but will be carried downstream by the current (Fig. 4-10). Under these conditions the length of line paid out is greater than the true vertical depth, and the meter is higher than indicated. If the angle between the line and a vertical becomes large, it is necessary to apply a correction to the measured depths [12]. The actual correction depends on the relative lengths of line above and below the water surface, but at a vertical angle of 12° the error will be about 2 percent. A slight additional error is introduced if the current is not normal to the measuring section.

Computation of total discharge is made as follows [13] (Fig. 4-9):

1. Compute average velocity in each vertical by averaging velocities at two-tenths and eight-tenths depths.
2. Multiply the average velocity in a vertical by the area of a vertical section extending halfway to adjacent verticals (*ABCD*, Fig. 4-7). This area is taken

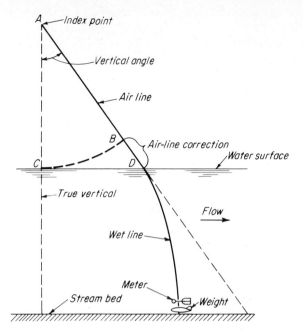

Figure 4-10 Position of sounding line in swift water.

as the measured depth at the vertical (*EF*) times the width of the section (*AB*).

3. Add the increments of discharge in the several verticals. Incremental discharge in the shore section (*GHI*, Fig. 4-7) is taken as zero.

Access to individual verticals of a section may be obtained by wading if the water is shallow. At high stages the meter must be lowered from an overhead support. Where possible, bridges are used as the measuring section. The measuring section need not be in the immediate vicinity of the control section provided that the intervening inflow is not large. Where no existing bridge is suitable, a special cableway may be used. The hydrographer rides in a small car suspended beneath the cable and lowers the meter through an opening in the floor of the car. Where neither bridge nor cableway is practical, measurements may be made from a boat. This is far less satisfactory because of the difficulty of maintaining position during a measurement and because either vertical or horizontal motion of the boat results in a positive velocity indication by a Price meter. In many countries cableways are used to carry the meter over the stream with the observer remaining on the shore.

4-8 Stage-Discharge Relations

Periodic meter measurements of flow and simultaneous stage observations provide data for a calibration curve called a *rating curve* or *stage-discharge relation* [14]. For most stations a simple plot of stage versus discharge (Fig. 4-11) is

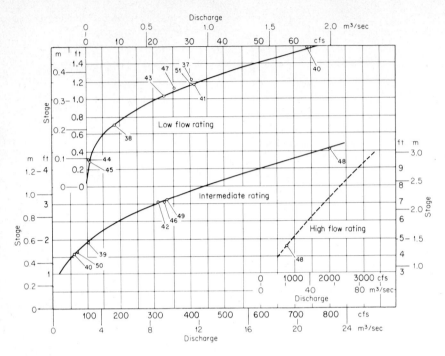

Figure 4-11 A simple stage-discharge relation.

satisfactory. Such a curve is approximately parabolic but may show some irregularities if the control changes within the range of flows experienced or if the cross section is irregular.

The dispersion of the measured data about the mean rating curve should be small (generally less than 2 percent). A larger dispersion indicates either (1) that the control shifts more or less continuously with scour and deposition in bed and banks or the growth of vegetation, (2) that the water-surface slope at the control varies as the result of varying backwater from tides, reservoir fluctuations, or variable tributary inflow downstream, or (3) that the measurements are not carefully made.

Under conditions of *shifting control*, discharge is usually estimated by noting the difference between the stage at the time of a discharge measurement and the stage on the mean rating curve which shows the same discharge. This difference is applied as a correction to all stages before entering the rating. If the correction changes between measurements, a linear variation with time is usually assumed.

If variable backwater is present, the rating curve must include slope as a parameter [15, 16]. The basic approach is represented by

$$\frac{q}{q_0} = \left(\frac{s}{s_0}\right)^m = \left(\frac{F}{F_0}\right)^k \tag{4-2}$$

In other words, discharge q is proportional to a power of water-surface slope s. From fluid mechanics the exponent m would be expected to be $\frac{1}{2}$. The fall F is the difference in water-surface elevation between two fixed sections and is usually measured by two conventional river gages. There is no assurance that the water-surface profile between these gages is a straight line, i.e., that $F/L = s$. Consequently the exponent k need not be $\frac{1}{2}$ and must be determined empirically.

A *slope-stage-discharge relation* requires a base gage and an auxiliary gage. The gages should be far enough apart for F to be at least 1 ft (30 cm) to minimize the effect of observational errors. The fall F applicable to each discharge measurement is determined, and if the observed falls do not vary greatly, an average value F_0 is selected. All measurements with values of $F \approx F_0$ are plotted as a simple stage-discharge relation, and a curve is fitted (Fig. 4-12). This is the q_0 curve representing the discharge when $F = F_0$. If $F \neq F_0$, the ratio F/F_0 is plotted against q/q_0 on an auxiliary chart. The discharge at any time may be computed by calculating the ratio F/F_0 and selecting a value of q/q_0 from the auxiliary curve. A value of q_0 corresponding to the existing stage is taken from the q_0 curve and multiplied by q/q_0 to give q. If the auxiliary curve plots as a straight line on logarithmic paper, the slope of the line is k in Eq. (4-2). The rating just described is known as a *constant-fall rating*, since the adopted mean fall F_0 is constant.

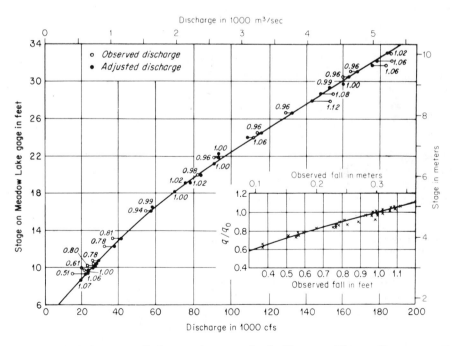

Figure 4-12 A slope-stage-discharge rating curve for the Tennessee River at Chattanooga, Tenn. (*U.S. Geological Survey.*)

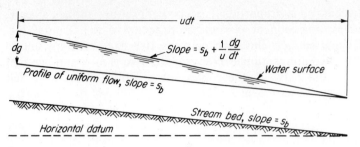

Figure 4-13 Profile of a flood wave.

If the fall varies over a wide range and is correlated with stage, a *normal-fall rating* may be used. In this case F_0 is determined from a curve or equation expressing the relation between fall and stage.

Stations not subject to variable slope because of backwater may still be materially affected because of variations in the rate of change in stage. Under this circumstance it is possible to develop a *change-in-stage* rating and thus eliminate the need for an auxiliary gage. Figure 4-13 shows the profile of a flood wave as it moves past a station. The slope is equal to $s_b + s_r$, where s_b is the slope of the channel bottom (or the slope of the water surface in uniform flow) and $s_r = dg/u\ dt$. Here dg/dt is the slope of the flood wave expressed as a rate of change of stage with time and u is the velocity of movement of the flood wave. Since s_r is the variable factor in the slope, a rating can be constructed by plotting stage against discharge for those measurements for which dg/dt is near zero. If u is assumed to be constant, a correction curve can then be established by plotting q/q_c versus dg/dt, where q_c is the discharge corresponding to the existing stage when $dg/dt = 0$. Theoretically, dg/dt is the tangent to the water-stage recorder trace, but in practice it is convenient to use Δg, the change in stage during a finite time period, usually 1 hr but sometimes much longer.

4-9 Extension of Rating Curves

There is no completely satisfactory method for extrapolating a rating curve beyond the highest measured discharge. It is often assumed that the equation of the rating curve is

$$q = k(g - a)^b \tag{4-3}$$

where g is gage height and a, b, and k are station constants. If the constants can be determined from the observed rating, Eq. (4-3) can be used to extend the rating. The constants can be determined by plotting q versus $(g - a)$ on semilog paper and trying various values of a until a straight line results. For three points on the observed rating such that $q_2 = \sqrt{q_1 q_3}$

$$a = \frac{g_1 g_3 - g_2^2}{g_1 + g_3 - 2g_2} \tag{4-4}$$

Values of b and k can be determined from Eq. (4-3). Since the constants rarely provide a good fit throughout the entire range of the rating, the three points should be selected in the middle and upper flow range. Even so, the procedure cannot account for any marked change in the geometry of the stream at high flows.

Illustrative example 4-1 Extend the rating of Fig. 4-11 by use of Eqs. (4-3) and (4-4).

Let

$$q_1 = 50 \text{ ft}^3/\text{s} \qquad g_1 = 1.41 \text{ ft}$$

and

$$q_3 = 800 \qquad g_3 = 4.60$$

then

$$q_2 = \sqrt{40,000} = 200 \qquad \text{and} \qquad g_2 = 2.55$$

From Eq. (4-4)

$$a = \frac{1.41(4.60) - 2.55^2}{1.41 + 4.60 - 2(2.55)} = -0.02$$

From Eq. (4-3)

$$\frac{q_1}{q_2} = \left(\frac{g_1 - a}{g_2 - a}\right)^b$$

hence

$$0.25 = \left(\frac{1.43}{2.57}\right)^b \qquad \text{and} \qquad b = 2.36$$

$$k = \frac{q_1}{(g - a)^b} = \frac{50}{(1.43)^{2.36}} = 21.5$$

The equation of the rating is $q = 21.5(g + 0.02)^{2.36}$. For $g = 9.00$ ft, $q = 21.5\,(9.02)^{2.36} = 3860 \text{ ft}^3/\text{s}$.

Another method [17, 18] of extending rating curves is based on the Chézy formula

$$q = AC\sqrt{Rs} \qquad (4-5)$$

where C is a roughness coefficient, s is the slope of the energy line, A is the cross-sectional area, and R is the *hydraulic radius* (A divided by the wetted perimeter). If $C\sqrt{s}$ is assumed to be constant for the station and D, the mean depth, is substituted for R,

$$q = ka\sqrt{D} \qquad (4-6)$$

Known values of q and $A\sqrt{D}$ are plotted on a graph and usually define something close to a straight line which can be extended. Values of $A\sqrt{D}$ for stages above the existing rating can be obtained by field measurement and used with the extended curve for estimates of q. An abrupt discontinuity may be observed at bankfull stage [18].

4-10 Effect of Ice on Streamflow

When ice covers a stream, a new friction surface is formed and the stream becomes a closed conduit with lower discharge because of the decreased hydraulic radius. The underside of the ice sheet may be extremely rough if ice cakes are tilted helter-skelter and then frozen together. Movement of the water under the ice gradually causes a smooth surface to develop. If the stage falls, leaving the ice as a bridge across the stream, the stage-discharge characteristics return to those of a free stream.

In turbulent streams the first ice to form is *frazil ice,* small crystals suspended in the turbulent flow. Frazil ice collecting on rocks on the streambed is called *anchor ice* and may cause a small increase in stage. If the turbulence is not sufficient to keep the frazil ice mixed in the stream, it rises to the surface to form sheet ice. Until a complete ice sheet is formed, small variations in the stage-discharge relation must be expected from time to time.

When ice conditions exist, it is necessary to make periodic measurements [19] through holes in the ice and to interpolate the discharge between these measurements in any manner which seems reasonable. The meter must be moved rapidly from hole to hole and should be kept in the water at all times (except when being moved) to prevent freezing. Fortunately, if the stream is frozen over, the flow is usually small since there will be little snowmelt or other source of runoff within the tributary area.

4-11 Other Methods of Obtaining Streamflow Data

The discharge at dams can be determined from calibration of the spillway, sluiceway, and turbine gates. If a record of gate and turbine operation is maintained, the discharge can be computed.

On small streams, flow measurements may be made with weirs or flumes [20]. These devices are commonly rated on the basis of laboratory calibration, although the rating may be checked in place with current meters. For small streams, a combination of a V-notch weir for low flows and a venturi flume for high flows may be necessary to assure accuracy. Large weirs are usually unsatisfactory because deposition of sediment in the upstream pool changes the discharge characteristics.

Highway culverts have frequently been suggested as flow-measuring devices, and in many instances this is feasible. However, the hydraulics of culverts is quite complex [21]. In flat terrain, care must be taken to avoid culverts affected by backwater unless both headwater and tailwater stages are measured. On steep slopes it is necessary to establish whether the culvert flows full with control in the barrel or partially full with control at the entrance. Temporary changes in discharge capacity may occur as the result of sediment or debris deposits.

Another method of estimating flow is by application of hydraulic principles. The procedure is often referred to as a *slope-area computation* [22, 23]. Sufficient high-water marks must be located along a reach of channel to permit determination of the water-surface slope at the time of peak. Cross sections of the channel may be determined by leveling or sounding, and the area and hydraulic radius calculated. The Chézy-Manning formula is ordinarily used to compute discharge:

$$q = \frac{1}{n} A R^{2/3} s^{1/2} \qquad (4\text{-}7)$$

in metric units. For English units (feet) the equation should be multiplied by 1.49. The main source of error in applying Eq. (4-7) is in estimating the roughness coefficient n (Appendix Table A-18). Since q is inversely proportional to n and the average value of n for natural streams is about 0.035, an error of 0.001 in n represents about 3 percent in discharge. Considerable doubt may exist whether the cross section measured after the flood is the section which existed at the time of peak. A stream often scours during rising stages and redeposits material during falling stages (Fig. 4-14). Under the most favorable conditions, an error of 10 percent may be expected in a slope-area estimate of flow.

Current-meter gaging is difficult and sometimes impossible in boulder-strewn mountain torrents, very small streams, etc. Here chemical gaging may prove useful. Common salt, fluorescein dye [24], a radioactive material, or any

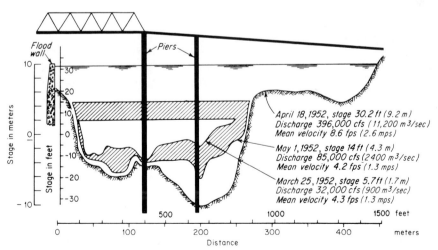

Figure 4-14 Cross sections of the Missouri River at Omaha, Neb., showing the progressive scour and filling during the passage of a flood wave.

easily measurable material not present in the stream and not likely to be lost by chemical combination with materials in the stream may be used. The tracer can be used to measure mean velocity through a reach by measuring the time between injection of a pulse and its arrival at a downstream point [25].

In the dilution method [26] a tracer of concentration c_t is injected into the stream at rate q_t. At a downstream point samples are taken, and after an equilibrium concentration c_e is reached, the discharge q is

$$q = \left(\frac{c_t}{c_e} - 1 \right) q_t \tag{4-8}$$

Complete mixing of the tracer in the flow and accurate determination of the initial and final concentrations are essential.

Rough measurements of discharge can be made by timing the speed of floats. A surface float travels with a velocity which is about 1.2 times the mean velocity. Objects floating with a greater submerged depth will travel at speeds closer to the mean velocity in the section. A float extending from the surface to mid-depth travels with a velocity about 1.1 times the mean velocity.

For very large streams conventional meter measurements may be too slow because of the many verticals that should be measured. In the *moving-boat method,* a boat traverses the stream at constant speed on a course normal to the flow. A special meter which operates continuously and indicates the instantaneous velocity is mounted in the bow. The meter has a trail vane which keeps it pointed along the resultant of the boat and stream velocities. An echo sounder measures the cross section of the stream during the traverse. With depth, meter angle, and water velocity measured at 30 to 40 points across the section, the discharge can be computed. Normally several traverses are made and averaged [27]. The method is well adapted to streams over 300 m in width.

The ultrasonic and electromagnetic methods can provide continuous discharge measurement. In the *ultrasonic method,* sonic pulses are emitted from transducers on opposite banks and located on a line about 45° from the direction of flow. One pulse has a component with the stream velocity and the other is opposed. The difference in pulse velocity can be related to mean water velocity at the level of the transducers. By using several transducer pairs at different levels and a water-level indicator, the discharge at the station can be computed and printed at any chosen time interval. The procedure is said to be capable of accuracies within plus or minus 2 percent [28, 29, 30].

When flowing water cuts the vertical component of the earth's magnetic field, an electromotive force is induced in the water and can be sensed by suitable electrodes [31]. This procedure has been used to measure the flow of large streams with reasonable success, but for small streams the noise of the systems prevents accurate measurement of the flow-induced emf. Pipe flow-meters are in use in which a field is produced by coils adjacent to the pipe [32] and research is underway in England in which a field coil is installed under a stream. Cost will restrict this method to small streams, but it may prove suitable for streams in which heavy weed growth prevents meter measurement [28].

4-12 Planning a Streamflow Network

How many gaging stations should be installed, and where should they be located? The design of a streamflow network is a problem both of statistical sampling over area and of sampling the locations where data are most likely to be needed.

It is convenient to recognize three types of stations. *Operational stations* are required for streamflow forecasting, project operation, water allocation, etc. They are located as required for the purposes they serve and are operated as long as the purposes exist. *Special stations* are installed to secure data for a project investigation, special studies, or research. Their location is determined by the special need, and they are operated until the study is completed. *Basic data stations* are operated to obtain data for future use. The time and nature of this future use are usually unknown when the station is established. Planning for such stations is the main problem of streamflow-network design.

The decision problem is how best to utilize a limited budget for capital costs and operating expenses of a network [33]. Since there is no rational evaluation of the value of the data, it seems logical to attempt to define a minimum size of drainage area for which reasonable estimates of streamflow should be available. This minimum will be determined by the level of regional development, the hydrologic characteristics of the region, and the probable information needs (culvert design, irrigation diversion, etc.). The minimum area might range from 10 mi^2 (26 km^2) in a well-developed region to 100 mi^2 (260 km^2) or more in developing areas.

Every major stream in the region should be gaged near its mouth and a number of tributaries as well. If it is obvious which streams will be developed, they are the ones which should be gaged. This includes those streams which may pose present or future flood problems as well as those streams which may be used for power, irrigation, etc. The network should sample catchments of all sizes larger than the specified minimum and should also sample the range of hydrologic and geologic characteristics in the region. It will be a practical impossibility to gage all streams at every site where data may be needed. A major function of engineering hydrologists is to estimate flow at ungaged locations. Their task is made easier and the results more reliable by a network which samples the regional characteristics effectively.

Because hydrologic design information is most often expressed in terms of probability (Chaps. 12 and 13), there is a considerable value in long flow records provided they are internally homogeneous, i.e., represent essentially similar basin characteristics throughout the entire period of record. Human activities are now so widespread that few streams remain unaffected. *Benchmark stations* [34] should be maintained permanently on all streams that are substantially unaffected by people.

Basic data stations can be discontinued as soon as data are sufficient for a synthetic record to be derived if needed. If the budget for data is severely constrained, there may be considerable advantage in terminating a data station after say 10 years and moving the equipment to another site to enlarge the sample of gaged basins [35]. If this policy is adopted, a suitable method for deriving a synthetic record should be available, and all the necessary data for its use should be collected. Monthly or annual streamflow volumes often correlate very well with those of a nearby station. Because of rainfall variations, flood peaks are sometimes better related to precipitation (Chaps. 8 and 12).

INTERPRETATION OF STREAMFLOW DATA

4-13 Units

The basic flow unit in the United States is the *cubic foot per second* (ft³/s), also called *second-foot* and abbreviated *cusec,* or *cfs.* Countries using the metric system usually express flow in *cubic meters per second* (m³/s). Volume of flow may be expressed in cubic feet or cubic meters, but since this leads to very large numbers, larger volume units are commonly used. In the United States the *cfs-day,* or the volume of water discharged in twenty-four hours, with a flow of one cubic foot per second is widely used; 1 cfs-day is $24 \times 60 \times 60 = 86,400$ ft³. The average flow in cubic feet per second for any 24-hr period is the volume of flow in cfs-days. The other common unit of volume is the *acre·foot,* the volume of water required to cover one acre to a depth of one foot. Hence, 1 acre·ft contains 43,560 ft³ and equals 0.504 cfs-day. Within an error of 1 percent, 1 acre·ft equals $\frac{1}{2}$ cfs-day. In some cases it is convenient to use cfs-hours ($\frac{1}{24}$ cfs-day) as a volume unit. In countries using the metric system, volume is usually expressed in millions of cubic meters, 10^6 m³ $= 409$ cfs-day (Appendix Table A-1).

Less common units of flow are the cubic foot per second per square mile (csm), the inch, and the miner's inch. Cubic feet per second per square mile (cubic meters per second per square kilometer) is a convenient unit for comparing rates of flow on streams with differing tributary area and is the flow rate divided by the drainage area. The *inch,* the amount of water required to cover the drainage area to a depth of one inch, is a useful unit for comparing streamflow with the precipitation which caused it. The inch is a unit of volume only when associated with a specific drainage area. In the metric system the millimeter or centimeter is used as the unit of runoff depth and as a unit of volume when associated with a specific drainage area. The *miner's inch* is an old unit used in the western United States and defined as the rate of discharge through an orifice one inch square under a specified head. By statute the miner's inch has been defined between 0.020 and 0.028 ft³/s, depending on the state (Appendix Table A-7).

It is desirable to treat annual streamflow data in such a way that the flood season is not divided between successive years. Various *water years* have been used for special purposes; the U.S. Geological Survey uses the water year, October 1 to September 30, for data publication. Water years are customarily designated by the ending year, i.e., water year 1975 ends September 30, 1975.

4-14 Hydrographs

A *hydrograph* is a graph of stage or discharge versus time. Many different methods of plotting are used, depending on the purpose of the chart. Monthly and annual mean or total flow is used to display the record of past runoff at a station (Fig. 4-21). The characteristics of a particular flood, however, usually cannot be shown successfully by plotting average flow for periods longer than 1 day. For detailed analysis, discharge hydrographs are plotted by computing instantaneous flow values from the water-stage recorder chart. The visual shape of the hydrograph is determined by the scales used, and in any particular study it is good practice to use the same scales for all floods on a given basin.

4-15 Mean Daily Flows

Streamflow data are usually published in the form of mean daily flows (Fig. 4-15) from midnight to midnight. Since 1945, the U.S. Geological Survey has included the magnitude and time of occurrence of all significant flood peaks. On large streams this form of publication is quite satisfactory, but on small streams it leaves something to be desired. The picture presented by mean daily flows depends on the chance relation between time of storm occurrence and fixed clock hours (Fig. 4-16). On large streams the maximum instantaneous flow may be only slightly higher than the maximum mean daily flow. On small streams the maximum instantaneous flow is usually much greater than the highest mean daily flow. Wherever possible, the hydrologist should secure copies of the recorder charts and work with a hydrograph of instantaneous flow when dealing with small basins. If this is not feasible, Fig. 4-17 may be used as a guide to estimate the peak flow and time of peak. The figure is an average relationship [36] developed from data for stations all over the United States and cannot be expected to give exact results.

4-16 Adjustment of Streamflow Data

Before publication, streamflow data should be carefully reviewed and adjusted for errors resulting from instrumental and observational deficiencies until they are as accurate a presentation of the flow as it is possible to make. For a number of reasons, the published flow may not represent the data actually required by the analyst. The location of the station may have changed during the period of record with a resultant change in drainage area and hence volume and rate of flow. In this case an adjustment of the record is possible by use of the double-mass curve (Sec. 3-10). The base for the double-mass curve may be flow at one or more gaging stations that have not been moved. The average precipitation at a number of stations in the area should not be used. The double-mass-curve method implies a relationship of the form $q = kP$, which is not correct if precipitation is used as a base. A more effective procedure is to develop a relationship between precipitation and runoff and make a double-mass curve of observed streamflow versus runoff as estimated from this relation (Sec. 8-12).

POTOMAC RIVER BASIN

01619500 Antietam Creek near Sharpsburg, Md.

LOCATION.--Lat 39°27'01", long 77°43'52", Washington County, on left bank 400 ft downstream from Burnside Bridge, 1 mile southeast of Sharpsburg, and 4 miles upstream from mouth.

DRAINAGE AREA.--281 sq mi.

PERIOD OF RECORD.--June 1897 to September 1905. August 1928 to current year. Monthly discharge only for some periods, published in WSP 1302.

GAGE.--Water-stage recorder. Concrete control since Mar. 29, 1934. Datum of gage is 311.00 ft above mean sea level, adjustment of 1912. June 24, 1897, to Aug. 25, 1905, nonrecording gage a few hundred feet downstream from Middle Bridge, 1.2 miles upstream at datum about 12 feet higher. Aug. 21, 1928, to July 13, 1933, nonrecording gage at Burnside Bridge at present datum.

AVERAGE DISCHARGE.--48 years (1897-1903, 1904-5, 1930-1971), 257 cfs (12.42 inches per year), adjusted for inflow since 1930.

EXTREMES.--Current year: Maximum discharge, 2,670 cfs Feb. 13 (gage height, 7.15 ft); minimum, 122 cfs part of each day Sept. 8-11 (gage height, 2.45 ft).
Period of record: Maximum discharge, 12,600 cfs July 20, 1956 (gage height, 16.73 ft), from rating curve extended above 4,300 cfs on basis of contracted-opening measurement of peak flow; minimum discharge, 9.4 cfs Nov. 22, 1957, result of regulation caused by construction work above station; minimum daily, 37 cfs Jan. 30, 1966.

REMARKS.--Records good. Some diurnal fluctuation caused by powerplant above station. Since 1928, records include pumpage from Potomac River for municipal supply of Hagerstown. This water later enters Antietam Creek above station as sewage. Records of chemical analyses and water temperatures for the water year 1971 are published in Part 2 of this report.

REVISIONS (WATER YEARS).--WSP 192: 1897-1905. WSP 726: Drainage area. WSP 1432: 1929-31(M), 1933, 1935(M), 1937(M), 1949(M), 1952(M).

DISCHARGE, IN CUBIC FEET PER SECOND, WATER YEAR OCTOBER 1970 TO SEPTEMBER 1971

DAY	OCT	NOV	DEC	JAN	FEB	MAR	APR	MAY	JUN	JUL	AUG	SEP
1	146	206	228	320	260	876	392	262	391	216	219	136
2	144	166	211	300	240	909	388	269	327	208	372	136
3	141	289	205	300	230	790	389	297	344	200	241	134
4	135	426	208	391	230	910	370	271	400	188	348	132
5	134	256	202	864	290	722	360	256	379	183	341	128
6	137	208	192	760	352	724	359	351	374	182	243	126
7	135	197	185	600	365	784	453	489	387	183	192	127
8	134	174	186	520	350	912	428	473	467	177	176	125
9	132	166	183	500	480	712	382	479	377	173	168	125
10	132	170	183	480	398	676	365	408	339	170	165	133
11	131	232	181	456	340	657	354	373	319	171	189	127
12	130	264	201	439	382	628	346	354	310	195	192	198
13	130	313	226	426	1,160	629	343	479	302	181	169	178
14	129	293	201	421	1,590	623	338	588	302	177	156	178
15	164	294	191	432	691	585	326	471	506	173	151	150
16	144	336	191	380	582	576	322	483	434	176	153	135
17	132	240	352	340	567	542	320	495	352	174	150	157
18	127	260	388	320	553	514	319	438	321	167	149	151
19	125	248	301	310	640	528	306	421	298	172	151	144
20	126	269	279	300	694	645	301	398	285	173	154	152
21	150	377	267	310	816	564	301	420	275	165	146	341
22	292	306	376	321	992	526	297	386	270	160	146	216
23	212	272	504	340	1,630	514	290	349	259	157	143	163
24	154	266	661	340	1,360	494	284	331	251	157	139	151
25	142	247	545	327	1,060	474	276	328	242	203	136	145
26	137	237	495	320	939	462	274	326	235	171	134	146
27	135	230	453	300	1,040	450	272	301	227	165	167	155
28	130	225	415	280	1,000	437	273	291	219	158	207	158
29	129	219	396	260	------	433	284	292	222	158	162	145
30	142	222	360	300	------	422	274	376	219	173	143	141
31	202	------	340	290	------	404	------	496	------	192	139	------
TOTAL	4,531	7,628	9,296	12,247	19,311	18,924	9,986	11,951	9,633	5,498	5,741	4,631
MEAN	146	254	300	395	690	607	333	386	321	177	185	154
MAX	292	426	661	864	1,630	912	453	588	506	216	372	341
MIN	125	166	181	260	230	404	272	256	219	157	134	125
(†)	-13.1	-11.0	-6.9	-5.7	-6.0	-5.5	-6.5	-5.5	-6.8	-10.8	-11.3	-13.0
MEAN ‡	133	243	293	389	684	602	326	380	314	166	174	141
CFSM ‡	.47	.86	1.04	1.38	2.43	2.14	1.16	1.35	1.12	.59	.62	.50
IN ‡	.54	.96	1.20	1.59	2.53	2.47	1.29	1.56	1.25	.68	.72	.56

CAL YR 1970 TOTAL 124,639 MEAN 341 MAX 1,810 MIN 110 MEAN‡ 333 CFSM‡ 1.19 IN‡ 16.15
WTR YR 1971 TOTAL 119,277 MEAN 327 MAX 1,630 MIN 125 MEAN‡ 318 CFSM‡ 1.13 IN‡ 15.34

† Pumpage, in cubic feet per second, from Potomac River for municipal supply of Hagerstown.
‡ Adjusted for pumpage.

PEAK DISCHARGE (BASE, 1,500 CFS)

DATE	TIME	G. H.	DISCHARGE	DATE	TIME	G. H.	DISCHARGE
2-13	2115	7.15	2,670	2-23	1800	5.67	1,680

Figure 4-15 Sample page from the U. S. Geological Survey, *Water Resources Data*.

Storage reservoirs, diversions, levees, etc., cause changes in either total flow volume or rate of flow or both. An analysis of the effects on the record at a given station requires a careful search to determine the number and size of reservoirs, the number and quantity of diversions, and the date of their construction. Many small diversions may be unmeasured, and estimates of the flow

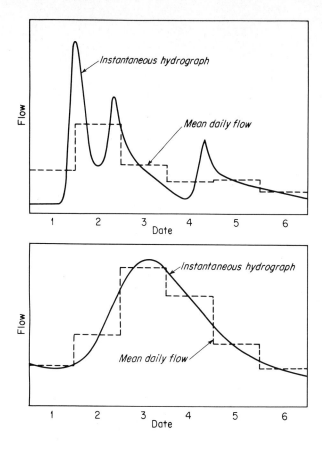

Figure 4-16 The relationship between instantaneous and mean daily flows (upper and lower charts typical of small and large streams, respectively).

diverted must be based on electric-power consumption of pumps, capacity of pump equipment, duration of pumping, or conduit capacities for gravity diversions. Diversions for irrigation may be estimated from known irrigable or irrigated acreage and estimated unit water requirements (Chap. 5). The adjustment of the streamflow record for the effect of reservoirs or diversions on flow volume requires the addition of the net change in storage and/or the total diversion to the reported total flow. It may be necessary to consider also channel losses and losses by evaporation from the reservoirs.

Adjustment of short-period or instantaneous flow rates for the effect of storage or diversion is a much more complex problem. Levees, channel improvement, and similar works may also affect flow rate. In some instances pumping of groundwater has markedly reduced low flows, as has the construction of stock ponds [37]. Correction for the effect of storage or diversion on flow rates is made by adding the rate of change of storage or the diversion rate to the observed flows. In addition, it may be necessary to use storage-routing techniques (Chap. 9) to correct for the effect of channel storage between the reservoir or diversion point and the gaging station. Channel improvement and

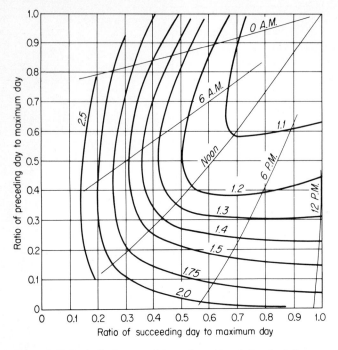

Figure 4-17 Peak discharge and time of crest in relation to daily mean discharge. Labels on curves indicate clock time of crest on maximum day and ratio of instantaneous peak flow to maximum daily discharge. (*U.S. Geological Survey.*)

levee work change flow through their effects on the channel storage of the stream. Unless it is possible to establish "before" and "after" correlations with some station outside the influence of the channel works, corrections must be made entirely by storage-routing methods.

Land-use changes, urbanization, deforestation, or reforestation affect streamflow and cause apparent shifts in the flow record. Unless the timing and areal extent of such changes are well documented, correction of the record is almost impossible. Even with good documentation of changes, the adjustment process is complex. The most direct solution may be the use of simulation (Chap. 11) to reconstruct a flow series from rainfall and other meteorological factors using parameters appropriate for current or expected future conditions.

4-17 Mean Annual Runoff

Figure 4-18 is a map of mean annual runoff in the United States. In addition to the problems of representing any highly variable element by isopleths drawn on the basis of limited data, there are other problems peculiar to streamflow. The record of flow at a gaging station represents the integrated runoff for the entire basin above the station. In most areas the production of runoff is not uniform

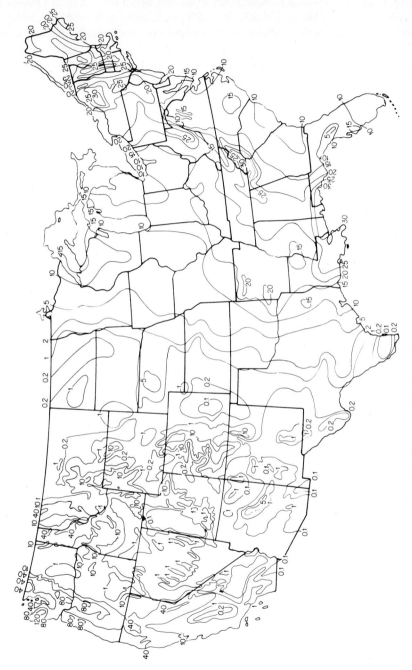

Figure 4-18 Average annual runoff (inches) in the United States. (*U.S. Geological Survey.*)

over the basin. The average annual runoff of the Missouri River at Omaha is 10,370,000 cfs-day (2540 × 10⁶ m³), or 1.20 in (30 mm) over the 322,800-mi² (836,000-km²) drainage area. The headwater areas produce much heavier runoff [Yellowstone River at Corwin Springs = 15.6 in (396 mm)], whereas other large areas must average below the mean for Omaha. In addition, storage and diversion further complicate the picture. Figure 4-18 was prepared by the Geological Survey [38] using all available information and represents the most reliable map available for the United States. Such a map is intended to present a general picture of geographical variations in runoff. It cannot possibly show fine detail and should be used only for general information and preliminary studies. It should not be used as a source of information for a specific design problem. Such information can be developed more reliably through use of actual streamflow data and the analytical techniques discussed in the chapters which follow. It is helpful to coordinate precipitation maps and evaporation maps with the runoff map [39].

4-18 Streamflow Variations

The normal or average values of runoff serve an important purpose, but they do not disclose all the pertinent information concerning the hydrology of an area. Especially significant are variations of streamflow about this normal. These variations include the following:

1. Variations in total runoff from year to year
2. Seasonal variations in runoff
3. Variations of daily rates of runoff throughout the year

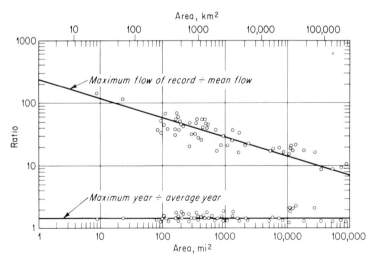

Figure 4-19 Ratios of maximum annual runoff to average annual runoff and maximum flow of record to mean flow for stations in the upper Ohio River Basin.

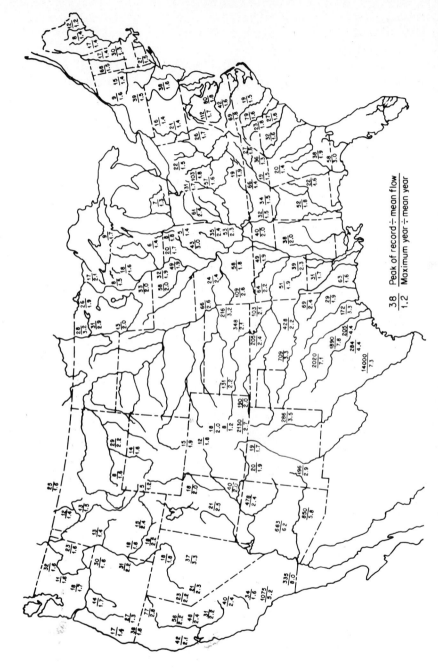

Peak of record ÷ mean flow
$\dfrac{38}{1.2}$ ——————— Maximum year ÷ mean year

Figure 4-20 Ratios of maximum annual flow and maximum peak flow to mean flow for selected United States stations with drainage areas between 1000 and 2000 mi² (2590 and 5180 km²).

123

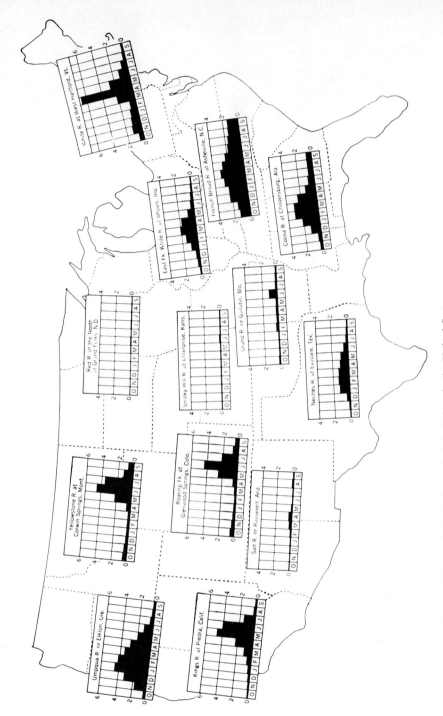

Figure 4-21 Median monthly runoff (inches) at selected stations in the United States.

Table 4-1 Selected record peak flows in the United States

Stream	Station	Peak flow ft³/s	Peak flow m³/s	Drainage area mi²	Drainage area km²	Flow per unit area ft³/s · mi²	Flow per unit area m³/s · km²	Date
		Northeastern States						
Lillibridge Cr.	‡Port Allegany, Pa.	16,000	453	6.7	17.4	2,380	26.0	7/42
Annin Cr.	†Turtlepoint, Pa.	24,000	679	11.4	29.5	2,110	23.1	7/42
Bull Run	†Catharpin, Va.	39,400	1,116	25.8	66.8	1,527	16.7	6/72
East Cr.	‡Rutland, Vt.	36,500	1,034	51	132	719	7.8	6/47
Bull Run	†Manassas, Va.	76,100	2,155	148	383	514	5.6	6/72
East Mahantango Cr.	†Dalmatia, Pa.	69,900	1,979	162	420	431	4.7	6/72
Esopus Cr.	‡Colbrook, N.Y.	59,600	1,688	192	497	310	3.4	3/51
Patapsco R.	‡Hollofield, Md.	80,600	2,282	285	738	283	3.1	6/72
Conestoga Cr.	‡Lancaster, Pa.	88,300	2,500	324	839	273	3.0	6/72
Loyalsock Cr.	‡Loyalsock, Pa.	88,700	2,512	443	1,147	200	2.2	6/72
White R.	‡West Hartford, Vt.	120,000	3,398	690	1,787	174	1.9	11/27
Tioga R.	†Lindley, N.Y.	128,000	3,625	771	1,997	166	1.8	6/72
Tioga R.	†Erwins, N.Y.	190,000	5,380	1,377	3,566	138	1.5	6/72
Chemung R.	†Big Flats, N.Y.	235,000	6,654	2,150	5,568	109	1.2	6/72
Potomac R.	†Hancock, Md.	340,000	9,628	4,073	10,549	83	0.9	3/36
Potomac R.	‡Point of Rocks, Md.	480,000	13,592	9,651	24,996	50	0.5	3/36
Susquehanna R.	‡Harrisburg, Pa.	1,020,000	28,883	24,100	62,419	42	0.5	6/72
		Southeastern States						
Ellerbe Cr.	†Durham, N.C.	3,350	95	2.9	7.5	1,170	12.8	8/67
N. Fork, Catawba R.	‡Ashford, N.C.	15,000	424	4.5	11.7	3,350	36.6	8/40
E. Fork, Globe Cr.	‡McKenzie School, Tenn.	16,300	461	6.6	17.1	2,470	27.0	6/39
Trail Cr.	‡Athens, Ga.	15,000	424	11.8	30.6	1,270	13.9	6/67
Wilson Cr.	†Adako, N.C.	99,000	2,802	65	168.3	1,520	16.6	8/40
Catawba Cr.	‡Marion, N.C.	71,400	2,022	171	443	418	4.6	8/40
Yadkin R.	‡Wilkesboro, N.C.	160,000	4,531	493	1,277	325	3.5	8/40

Table 4-1 (*Continued*)

Stream	Station	Peak flow		Drainage area		Flow per unit area		Date
		ft³/s	m³/s	mi²	km²	ft³/s · mi²	m³/s · km²	
		North Central States						
Triplett Cr.	‡Morehead, Ky.	44,000	1,245	47.5	123.0	926	10.1	7/39
Wautauga R.	‡Sugar Grove, N.C.	50,800	1,438	91	236	559	6.1	8/40
Platte R.	‡Rockville, Wis.	43,500	1,232	149	386	313	3.2	7/50
Buffalo R.	†Flat Woods, Tenn.	90,000	2,549	447	1,158	202	2.2	2/48
Cuivre R.	†Troy, Mo.	120,000	3,398	903	2,339	133	1.5	10/41
New R.	†Galax, Va.	141,000	3,993	1,131	2,929	125	1.4	8/40
Caney Fork	†Rock Island, Tenn.	210,000	5,947	1,678	4,346	125	1.4	3/29
Great Miami R.	‡Dayton, Ohio	250,000	7,079	2,511	6,503	100	1.1	3/13
Great Miami R.	‡Hamilton, Ohio	352,000	9,968	3,630	9,402	97	1.1	3/13
Ohio R.	‡Sewickly, Ohio	574,000	16,250	19,500	50,500	29	0.3	3/36
Ohio R.	‡Owensboro, Ohio	1,210,000	34,260	97,200	251,750	12	0.1	1/37
Ohio R.	‡Metropolis, Ill.	1,850,000	52,390	203,000	525,770	9	0.1	2/37
		Northern Plains						
Soldier Cr.	‡Goff, Kan.	7,080	200	2.1	5.3	3,440	37.6	5/70
Burnt Cr.	†Reese, Kan.	20,500	580	8.8	22.9	2,320	25.3	6/65
E. Fork, Fishing R.	†Excelsior Springs, Mo.	12,000	340	20	52	600	6.5	7/51
Boxelder Cr.	‡Rapid City, S.D.	51,600	1,460	117	303	441	4.8	6/72
Battle Cr.	‡Keystone, S.D.	26,200	742	66	171	397	4.4	6/72
Big Bull Cr.	†Hillsdale, Kan.	45,200	1,280	147	381	308	3.4	7/51
Little Nemaha R.	†Syracuse, Neb.	225,000	6,371	212	549	1,061	11.6	5/50
Little Nemaha R.	‡Auburn, Neb.	164,000	4,644	793	2,054	207	2.3	5/50
Marais des Cygnes R.	‡Ottawa, Kan.	142,000	4,021	1,250	3,237	114	1.2	7/51

Southern Plains

Boney Draw	†Rockport, Mo.	5,080	144	0.8	2.0	6,680	72.8	7/65
Cass Draw	‡Carlsbad, N.M.	32,500	920	9.3	24.1	3,490	38.0	5/65
E. Fork, James R.	‡Old Noxville, Tex.	105,000	2,970	61	158	1,730	18.8	7/32
Johnson Cr.	‡Ingram, Tex.	138,000	3,908	114	295	1,211	13.2	7/32
Salado Cr.	‡Salado, Tex.	143,000	4,047	148	383	966	10.5	9/21
Neosho R.	‡Parsons, Kan.	410,000	11,610	4,905	12,704	84	0.9	7/51
Devils R.	‡Del Rio, Tex.	597,000	16,895	4,305	11,150	139	1.5	9/32
Nueces R.	§Uvalde, Tex.	616,000	17,440	1,947	5,053	317	3.5	6/35

Colorado River Basin and Great Basin

Little Pinto Cr. trib.	‡Old Irontown, Utah	2,630	74.4	0.3	0.8	8,770	95.6	8/64
Dry Canyon	‡Cedar City, Utah	3,670	103.9	0.9	2.3	4,080	44.5	8/65
Bronco Cr.	‡Wikieup, Ariz.	73,500	2,165	19.0	49.2	3,870	42.2	8/71
Piccacho Wash	‡Yuma, Ariz.	37,000	1,047	41.5	107.5	892	9.7	9/39
Whitewater R.	‡Whitewater, Calif.	42,000	1,189	57	148	730	8.0	3/38
Deep Cr.	†Hesperia, Calif.	46,600	1,320	136	352	343	3.8	3/38
Mojave R.	‡Victorville, Calif.	70,600	1,999	514	1,331	137	1.5	3/38
San Pedro R.	‡Charleston, Ariz.	98,000	2,775	1,219	3,157	80	0.9	9/26
Bill Williams R.	†Planet, Ariz.	200,000	5,663	5,140	13,312	39	0.4	2/1891
Humboldt R.	†Imlay, Nev.	6,080	172	157,000	406,630	0.04	0.0004	5/52

South Pacific Coast

San Juan Cr. trib.	‡Elsinore, Calif.	2,130	60.3	0.4	1.0	5,460	59.5	2/69
Haines Cr.	‡Tujunga, Calif.	4,620	130.8	1.3	3.3	3,670	40.0	2/14
Cucamonga Cr.	‡Upland, Calif.	14,100	399	10	26	1,410	15.3	1/69
Arroyo de la Cruz	‡San Simeon, Calif.	35,200	996	41.2	107	854	9.3	12/66
Tujunga R.	†Sunland, Calif.	50,000	1,416	106	275	471	5.1	3/38
S. Fork, Eel R.	‡Miranda, Calif.	199,000	5,635	537	1,391	371	4.0	12/64
Smith R.	†Crescent City, Calif.	228,000	6,456	609	1,577	374	4.1	12/64
Eel R.	‡Scotia, Calif.	752,000	21,290	3,113	8,063	242	2.6	12/64

Table 4-1 (Continued)

Stream	Station	Peak flow		Drainage area		Flow per unit area		Date
		ft³/s	m³/s	mi²	km²	ft³/s · mi²	m³/s · km²	
		Pacific Northwest						
Knapp Coulee trib.	‡Chelan, Wash.	1,860	52.6	0.3	0.7	6,640	72.4	8/56
Highland Valley Gulch	‡Boise, Idaho	2,100	59.4	0.4	1.0	5,380	58.6	8/59
Squaw Cr.	‡Boise, Idaho	7,320	207.2	1.5	3.8	4,980	54.3	8/59
Maynard Gulch	‡Boise, Idaho	9,540	270	2.2	5.8	4,240	46.2	8/59
N. Fork, Skykomish R.	‡Hoodsport, Wash.	27,000	764	57	148	474	5.2	11/34
Willow Cr.	†Heppner, Oreg.	36,000	1,019	87	225	414	4.5	7/03
Lewis R.	‡Ariel, Wash.	129,000	3,653	731	1,893	177	1.9	12/33
Skagit R.	‡Concrete, Wash.	500,000	14,150	2,737	7,089	183	2.0	/1815
Willamette R.	Albany, Oreg.	340,000	9,622	4,840	12,540	70	0.8	12/1861
Willamette R.	Salem, Oreg.	500,000	14,160	7,280	18,860	69	0.8	12/1861
Columbia R.	Dalles, Oreg.	1,240,000	35,110	237,000	613,800	5	0.06	6/1894
		Hawaii						
Kalihi Stream	†Honolulu	12,400	351	2.6	6.7	4,769	52.4	11/30
N. Fork, Wailua R.	†Kapaa	53,200	1,506	18.7	48.4	2,845	31.1	11/55
S. Fork, Wailua R.	†Lihue	87,300	2,472	22.4	58	3,897	42.6	4/63
Waimea R.	†Waimea	37,100	1,051	58	150	640	7.0	2/49
Wailuku R.	‡Piihonua	63,400	1,795	125	324	507	5.5	8/40

† Near.   ‡ At.   § Below.

To some extent these variations are regional characteristics, but size of the drainage area is invariably a factor. Figure 4-19 shows the ratio of the maximum annual runoff to the mean annual runoff as a function of drainage area for a number of stations in the upper Ohio River basin having at least 20 years of record. On the same chart is a curve showing the ratio of the maximum flood of record to the average flow. In the first instance, little effect of area is evident, but in the second case substantially higher ratios are associated with the smaller basins. Basin shape, geology, and climatic exposure are responsible for most of the scatter in Fig. 4-19. The map of Fig. 4-20 indicates the variation of these two ratios over the United States. To minimize the effect of area, ratios are given for stations having drainage areas between 1000 and 2000 mi^2 (2590 to 5180 km^2). High ratios generally occur in arid regions with low normal annual runoff.

Median monthly runoff for selected stations is shown in Fig. 4-21 (p. 124). Here again there are important regional differences in magnitude and seasonal distribution. The hydrographs for Kings River, Roaring Fork, and Yellowstone River reflect primarily the melting of the mountain snowpacks from April through July. The Umpqua River, on the other hand, shows the effect of heavy winter rains (November through March) and some snowmelt thereafter. The Red, Grand, and Salt Rivers show very low runoff with a spring snowmelt period and a summer rain period. The remaining stations show a dominant winter rain and snow period but with some rain in all months. Figures 4-21 and 3-15 should be compared to see the interrelationships.

Table 4-1 summarizes some of the largest flood peaks in the United States.

PROBLEMS

4-1 Compute the streamflow for the measurement data below. Take the meter rating from Eq. (4-1) with $a = 0.1$ and $b = 2.2$ for v in ft/s. Note that the rule against more than 10 percent of the flow in any vertical section is violated in this problem to reduce computation.

Distance from bank, ft	Depth, ft	Meter depth, ft	Revolu- tions	Time, s
2	1	0.6	10	50
4	3.5	2.8	22	55
		0.7	35	52
6	5.2	4.2	28	53
		1.0	40	58
9	6.3	5.0	32	58
		1.3	45	60
11	4.4	3.5	28	45
		0.9	33	46
13	2.2	1.3	22	50
15	0.8	0.5	12	49
17	0			

4-2 The table that follows gives discharge, base stage, and stage at an auxiliary gage 2000 ft downstream. Develop a slope-stage-discharge relationship from these data. Compute the average error of the rating, using the tabulated data. What is the estimated discharge for base and auxiliary stages of 25.00 and 24.50 ft, respectively?

Base stage, ft	Discharge, ft³/s	Auxiliary stage, ft
14.02	2,400	13.00
23.80	29,600	23.25
17.70	21,200	16.60
24.60	85,500	23.55
20.40	28,200	19.55
17.00	7,400	16.40
18.65	34,000	17.50
26.40	55,000	25.70
22.20	74,200	21.00
16.20	9,550	15.30
21.10	43,500	20.13
25.60	84,000	24.60
23.20	93,500	21.95

4-3 Given below are data for a station rating curve. Extend the relation and estimate the flow at a stage of 14.5 ft by both the logarithmic and $A\sqrt{D}$ methods.

Stage, ft	Area, ft²	Depth, ft	Discharge, ft³/s
1.72	263	1.5	1,020
3.47	1,200	2.1	4,900
4.26	1,790	3.2	7,700
5.61	2,380	4.6	10,700
6.70	3,280	5.2	15,100
7.80	3,960	5.7	19,000
9.21	5,000	6.1	25,000
14.50	8,200	9.0	
2.50	674	1.8	2,700
4.02	1,570	2.8	6,600
5.08	2,150	3.9	9,450
5.98	2,910	4.9	13,100
6.83	3,420	5.4	16,100
8.75	4,820	6.0	24,100
9.90	5,250	6.5	27,300

4-4 What volume is represented by 1.43 in of runoff from a basin of 254 mi²? Give answer in cubic feet, cfs-days, acre·feet, and millions of cubic meters.

4-5 Given below are the daily mean flows in cubic feet per second at a gaging station for a period of 5 days. What is the mean flow rate for the period in cubic feet per second? What is the total discharge during the period in cfs-days? Acre·feet? If the drainage area is 756 mi², what is the runoff volume in inches?

Day	1	2	3	4	5
Flow, ft³/s	700	4800	3100	2020	1310

4-6 Using the data of Prob. 4-5 and the relationship of Fig. 4-17, estimate the peak flow and the time of peak.

4-7 Obtain a copy of a *U.S. Geological Survey Water-Supply Paper* or *Water Resources Data,* and for some stream in your area determine the maximum flow of record in cubic feet per second and in cubic feet per second per square mile. Find the average annual runoff in acre·feet, cfs-days, and inches.

4-8 For some stream selected by your instructor determine the mean flow for each month on the basis of 10 years of record. What percent of the total annual flow occurs each month? Compare these percentages with the monthly distribution of precipitation. Can you explain the differences?

4-9 For three basins in your area find the ratio of the maximum peak flow of record to the average daily flow rate. Is there an apparent relationship with drainage area? Can you explain the differences?

4-10 Compare the data in Table 4-1 for the several regions of the country. What regional differences do you see? Look for such things as date of floods, relative magnitude of flood peaks, apparent areal extent of floods, etc. Can you explain the differences?

4-11 What volume of runoff in cubic meters is represented by a depth of 37 mm on a basin of 600 km²? How many hectares can be irrigated with this volume if 60 cm of water is required for irrigation?

4-12 Taking the flow data of Prob. 4-5 in cubic meters per second, what is the total discharge in cubic meters? If the tributary area is 100,000 km², what is the equivalent runoff depth in millimeters?

REFERENCES

1. Buchanan, T. J., and W. P. Somers: Stage Measurement at Gaging Stations, *U.S. Geol. Surv. Tech. Water Resour. Inv.,* bk. 3, chap. A7, 1968.
2. Barron, E. G.: New instruments for Surface-Water Investigations, in Selected Techniques for Water-Resources Investigations, *U.S. Geol. Surv. Water-Supply Pap.* 1692-Z, pp. Z-4–Z-8, 1963.
3. Collet, M. H.: Crest-Stage Meter for Measuring Static Heads, *Civ. Eng.,* vol. 12, p. 396, 1942.
4. Stevens, J. C.: Device for Measuring Static Heads, *Civ. Eng.,* vol. 12, p. 461, 1942.
5. Doran, F. J.: High Water Gaging, *Civ. Eng.,* vol. 12, pp. 103–104, 1942.
6. Ferguson, G. E.: Gage to Measure Crest Stages of Streams, *Civ. Eng.,* vol. 12, pp. 570–571, 1942.
7. For some history of the development of cup-type meters see A. H. Frazier: Daniel Farrand Henry's Cup Type "Telegraphic" Current Meter, *Tech. Cult.,* vol. 5, pp. 541–565, 1964.
8. Frazier, A. H.: William Gunn Price and the Price Current Meters, *U.S. Natl. Mus. Bull.,* vol. 252, pp. 37–68, 1967.
9. O'Brien, M. P., and R. G. Folsom: Notes on the Design of Current Meters, *Trans. Am. Geophys. Union,* vol. 29, pp. 243–250, April 1948.

10. Smoot, G. F., and C. E. Novak: Calibration and Maintenance of Vertical-Axis Type Current Meters, *U.S. Geol. Surv. Tech. Water Resour. Inv.,* bk. 3, chap. A8, 1968.
11. Buchanan, T. J., and W. P. Somers: Discharge Measurements of Gaging Stations, *U.S. Geol. Surv. Tech. Water Resour. Inv.,* bk. 3, chap. A8, 1969.
12. Corbett, D. M., and others: Stream-Gaging Procedure, *U.S. Geol. Surv. Water-Supply Pap.* 888, pp. 43–51, 1945.
13. Young, K. B.: "A Comparative Study of Methods for Computation of Discharge," U.S. Geological Survey, February 1950.
14. Carter, R. W., and J. Davidian: Discharge Ratings at Gaging Stations, *U.S. Geol. Surv. Surface-Water Tech.,* bk. 1, chap. 12, 1965.
15. Boyer, M. C.: Determining Discharge at Gaging Stations Affected by Variable Slope, *Civ. Eng.,* vol. 9, p. 556, 1939.
16. Mitchell, W. D.: Stage-Fall-Discharge Relations for Steady Flow in Prismatic Channels, *U.S. Geol. Surv. Water-Supply Pap.* 1164, 1954.
17. Stevens, J. C.: A Method of Estimating Stream Discharge from a Limited Number of Gagings, *Eng. News,* pp. 52–53, July 18, 1907.
18. Sittner, W. T.: Extension of Rating Curves by Field Surveys. *J. Hydraul. Div., ASCE,* vol. 89, pp. 1–9, March 1963.
19. Hoyt, W. G.: The Effect of Ice on Stream Flow, *U.S. Geol. Surv. Water-Supply Pap.* 337, pp. 24–30, 1913.
20. U.S. Bureau of Reclamation: "Water Measurement Manual," 1953.
21. Bodhaine, C. L.: Measurement of Peak Discharge of Culverts by Indirect Measurements, *U.S. Geol. Surv. Tech. Water-Resour. Inv.,* bk. 3, chap. A3, 1968.
22. Benson, M. A.: Measurement of Peak Discharge by Indirect Methods, *World Meteorol. Org. Tech. Note* 90, Geneva, 1968.
23. Dalrymple, T., and M. A. Benson: Measurement of Peak Discharge by the Slope-Area Method, *U.S. Geol. Surv. Tech. Water-Resour. Inv.,* bk. 3, chap. A2, 1967.
24. Wilson, J. F.: Fluorometric Procedures for Dye Tracing, *U.S. Geol. Surv. Tech. Water Resour. Inv.,* bk. 3, chap. A11, 1968.
25. Allen, C. M., and E. A. Taylor: The Salt Velocity Method of Water Measurement, *Trans. ASME,* vol. 45, p. 285, 1923.
26. Addison, H.: "Applied Hydraulics," 4th ed., pp. 583–584, Wiley, New York, 1954.
27. Smoot, G. F., and C. E. Novak: Measurement of Discharge by the Moving Boat Method, *U.S. Geol. Surv. Tech. Water Resour. Inv.,* bk. 3, chap. A11, 1969.
28. Herschy, R. W.: New Methods of River Gaging, Chap. 5 in J. C. Rodda (ed.): "Facets of Hydrology," Wiley, New York, 1976.
29. Herschy, R. W., and W. R. Loosemore: The Ultrasonic Method of River Flow Measurement, Symposium on River Gaging by Ultrasonic and Electromagnetic Methods, University of Reading, England, 1974.
30. Schuster, J. C.: Measuring Water Velocity by Ultrasonic Flowmeter, *J. Hydraul. Div. ASCE,* vol. 101, pp. 1503–1517, 1975.
31. Faraday, M.: Experimental Researches *Phil. Trans. R. Soc.,* p. 175, 1832.
32. Balls, B. W., and K. J. Brown: The Magnetic Flowmeter, *Trans. Soc. Inst. Tech.,* vol. 11, pp. 123–130, 1959.
33. Kohler, M. A.: Design of Hydrological Networks, *World Meteorological Organization Tech. Note,* no. 25, 1958.
34. Langbein, W. B.: Hydrological Bench Marks, *Rep. WMO/IHD Proj.,* no. 8, World Meteorological Organization, Geneva, 1968.
35. Langbein, W. B.: Stream Gaging Networks, *Publ. 38, Int. Assoc. Hydrol. Sci.,* General Assembly, Rome, 1954.
36. Langbein, W. B.: Peak Discharge from Daily Records, *U.S. Geol. Surv. Water-Resour. Bull.,* p. 145, Aug. 10, 1944.
37. Culler, R. C., and H. V. Peterson: Effect of Stock Reservoirs on Runoff in the Cheyenne River Basin above Angostura Dam, *U.S. Geol. Surv. Circ.* 223, 1953.

38. Langbein, W. B., and others: Annual Runoff in the United States, *U.S. Geol. Surv. Circ.* 52, June 1949.
39. Nordenson, T. J.: Preparation of Coordinated Maps of Precipitation, Runoff, and Evaporation, *Rep. WMO/IHD Proj.,* no. 6, World Meteorological Organization, Geneva, 1968.

BIBLIOGRAPHY

Benson, M. A., and T. Dalrymple: General Field and Office Procedures for Indirect Measurements, *U.S. Geol. Surv. Tech. Water-Resour. Inv.,* bk. 3, chap. A1, 1967.
Carter, R. W., and J. Davidian: General Procedure for Gaging Streams, *U.S. Geol. Surv. Tech. Water-Resour. Inv.,* bk. 3, chap. A6, 1968.
International Association of Hydrological Sciences: "Design of Hydrologic Networks," Symposium of Quebec, *IAHS Publs.* 67 and 68, 1965.
King, H. W., and E. F. Brater: "Handbook of Hydraulics," 5th ed., McGraw-Hill, New York, 1963.
Schaefer, H.: Report on Hydrometry, *Hydrol. Sci. Bull.,* vol. 17, pp. 145–166, July 1972.
World Meteorological Organization: Guide to Hydrological Practices, 3d ed., *WMO Publ.* 168, Geneva, 1974.
———: Casebook on Hydrological Network Design Practices, *WMO Publ.* 324, Geneva, 1972.
———: Technical Regulations Vol. III, Operational Hydrology, *WMO Publ.* 49, Geneva, 1971.

UNITED STATES DATA SOURCES

The main sources of streamflow data are the publications of the U.S. Geological Survey. More detailed information may be obtained by writing to the agency or one of its field offices, located in principal cities throughout the country.

The U.S. Corps of Engineers, U.S. Water and Power Resources Service (formerly Bureau of Reclamation), U.S. National Weather Service, and U.S. Soil Conservation Service all make occasional observations of stage and/or flow. Information from these sources can usually be obtained from the nearest field office. Many states have agencies which make or cooperate in streamflow measurements.

FIVE

EVAPORATION AND TRANSPIRATION

This chapter discusses that phase of the hydrologic cycle in which precipitation reaching the earth's surface is returned to the atmosphere as vapor. Of the precipitation falling earthward, a portion evaporates before reaching the ground. Since precipitation is normally measured near the ground, evaporation from raindrops is of no practical concern in hydrology, except in the interpretation of radar reflectivity as a measure of precipitation intensity (Sec. 3-7). Likewise, evaporation from the oceans lies beyond the hydrologist's scope of direct interest. Precipitation deposited on vegetation eventually evaporates, and the quantity of water actually reaching the soil surface is correspondingly reduced below that observed in a precipitation gage. Other evaporative mechanisms to be considered are transpiration by plants and evaporation from soil, snow, and free-water surfaces (lakes, reservoirs, streams, and depressions).

Anticipated evaporation is a decisive element in design of reservoirs to be constructed in arid regions. Ten reservoirs the size of Lake Mead would evaporate virtually the entire flow of the Colorado River in a normal year, and normal evaporation from Lake Mead alone is equivalent to almost one-third of the minimum annual inflow to the reservoir. Evaporation and transpiration are indicative of changes in the moisture deficiency of a basin and, in this capacity, are sometimes used to estimate storm runoff in the preparation of river forecasts. Estimates of these factors are also used in determining water-supply requirements of proposed irrigation projects.

EVAPORATION

Although there is always continuous exchange of water molecules to and from the atmosphere, the hydrologic definition of *evaporation* is restricted to the net rate of vapor transport to the atmosphere. This change in state requires an exchange of approximately 600 cal for each gram of water evaporated [Eq. (2-1)]. If the temperature of the surface is to be maintained, these large quantities of heat must be supplied by radiation and conduction from the overlying air or at the expense of energy stored below the surface.

5-1 Factors Controlling the Evaporation Process

Rates of evaporation vary, depending on meteorological factors and the nature of the evaporating surface. Much of the ensuing discussion of meteorological factors is couched in terms of evaporation from a free-water surface. The extent to which the material is applicable to other surfaces of interest in hydrology is believed to be self-evident.

Meteorological factors. If natural evaporation is viewed as an energy-exchange process, it can be demonstrated that radiation is by far the most important single factor and that the term *solar evaporation* is basically applicable. On the other hand, theory and wind-tunnel experiments have shown that the rate of evaporation from water of specified temperature is proportional to wind speed and is highly dependent on the vapor pressure of the overlying air. How are these two conclusions to be reconciled? In essence, it can be said that water temperature is not independent of wind speed and vapor pressure. If radiation exchange and all other meteorological elements were to remain constant over a shallow lake for an appreciable time, the water temperature and the evaporation rate would become constant. If the wind speed were then suddenly doubled, the evaporation rate would also double momentarily. This increased rate of evaporation would immediately begin to extract heat from the water at a more rapid rate than what could be replaced by radiation and conduction. The water temperature would approach a new, lower equilibrium value, and evaporation would diminish accordingly. On a long-term basis, a change of 10 percent in wind speed will change evaporation only 1 to 3 percent, depending on other meteorological factors. In deep lakes with capacity for considerable heat storage, sudden changes in wind and humidity have longer-lasting effects; heat into or from storage assists in balancing the energy demands. The use of stored energy by excessive evaporation during a dry, windy period can thus result in reduced evaporation during subsequent weeks.

The relative effect of pertinent meteorological factors is difficult to evaluate at best, and any conclusions must be qualified in terms of the time period considered. The rate of evaporation is influenced by solar radiation, air temperature, vapor pressure, wind, and minimally by atmospheric pressure. The manner in which these factors influence the evaporative process becomes evident with an understanding of Secs. 5-3 and 5-4. Since solar radiation is an important factor, evaporation also varies with latitude, season, time of day, and sky condition.

Nature of evaporating surface. All surfaces exposed to precipitation, such as vegetation, buildings, and paved streets, are potentially evaporation surfaces. Since the rate of evaporation during rainy periods is small, the quantity of storm rainfall disposed of in this manner is essentially limited to that required to saturate the surface. Although this evaporation is appreciable on an annual basis, it is seldom evaluated separately but is considered part of overall evaporation and transpiration.

The rate of evaporation from a saturated soil surface is approximately the same as that from an adjacent water surface of the same temperature. As the soil begins to dry, evaporation decreases and its temperature rises to maintain the energy balance. Eventually, evaporation virtually ceases, since there is no effective mechanism for transporting water from appreciable depths. Thus the rate of evaporation from soil surfaces is limited by the availability of water, or *evaporation opportunity*.

Evaporation from snow and ice constitutes a special problem since the melting point lies within the range of temperatures normally experienced. Evaporation can occur only when the vapor pressure of the air is less than that of the snow surface [Eq. (5-11)], i.e., only when the dewpoint is lower than the temperature of the snow. The vapor pressure of the surface film on melting snow is 6.11 millibars (0.611 kilopascal), and this value represents the maximum possible air-to-snow vapor-pressure difference under evaporating conditions. The maximum rate of evaporation from snow is only about one-fourth that from a water surface at 30°C (86°F) when the dewpoint is 15°C (59°F) under the same wind conditions. These and other considerations lead to the conclusion that with temperatures much above freezing, the rate of snowmelt must exceed evaporation unless a large part of the area consists of exposed, wet soil [1]. The impression that a chinook, or foehn, is conducive to excessive evaporation from the snow cover is fallacious unless the wind is strong enough to result in blowing snow. The dewpoint increases downslope, and evaporation from the snow must cease when the dewpoint rises to the freezing point. Reasonable assumptions yield an upper limit of about 5 mm (0.2 in) water equivalent per day for evaporation from a snow surface. Considerable quantities of snow can disappear during periods when the wind is strong enough to whip the snow into the overrunning air [2] and the accumulated evaporation during chinook periods over a season can be appreciable [3].

Effects of water quality. The effect of salinity, or dissolved solids, is brought about by the reduced vapor pressure of the solution. The vapor pressure of seawater (35,000 ppm dissolved salts) is about 2 percent less than that of pure water at the same temperature. The reduction in evaporation is less than that indicated by the change in vapor pressure [Eq. (5-11)] because with reduced evaporation there is an increase in water temperature which partially offsets the vapor-pressure reduction [4, 5]. Even for seawater, the reduction in evaporation is never in excess of a few percent over an extended period of time, so that salinity effects can be neglected in the estimation of reservoir evaporation. Any foreign material which tends to seal the water surface or change its vapor pressure or albedo (Sec. 5-8) will affect the evaporation.

5-2 Water-Budget Determinations of Reservoir Evaporation

The direct measurement of evaporation under field conditions is not feasible, at least in the sense that one is able to measure river stage, rainfall, etc. As a consequence, a variety of techniques have been derived for determining or estimating vapor transport from water surfaces. The most obvious approach involves the maintenance of a water budget. Assuming that storage S, surface inflow I, surface outflow O, subsurface seepage O_g, and precipitation P can be measured, evaporation E can be computed from the continuity equation

$$E = (S_1 - S_2) + I + P - O - O_g \qquad (5\text{-}1)$$

This approach is simple in theory, but application rarely produces reliable results since all errors in measuring precipitation, inflow, outflow, and change in storage are reflected directly in the computed evaporation.

Seepage is usually the most difficult factor to evaluate since it must be estimated indirectly from measurements of groundwater levels, permeability, etc. If seepage approaches or exceeds evaporation, reliable evaporation determinations by this method are usually not possible. Under some circumstances, both seepage and evaporation can be evaluated by simultaneous solution of Eqs. (5-1) and (5-11) for periods of insignificant inflow and outflow [6]. The water budget can then be applied on a continuing basis through application of a stage-seepage relation.

The determination of rainfall generally does not represent a major obstacle, provided the average of on-shore measurements is representative of the reservoir. Difficulties in this respect may be expected when the surrounding topography is of high relief and for very large lakes which modify local weather. Errors in the measurement of snowfall can be large during periods of high wind. In addition to the usual deficiency of gage catch, a small reservoir can trap considerable quantities of blowing snow.

Water-stage recorders are sufficiently precise for determining the storage changes provided that the stage-area relationship is accurately established. Variations in bank storage are sometimes an important source of error in monthly computations but can usually be neglected in estimates of mean annual or annual evaporation. Similarly, expansion or contraction of stored water with large temperature changes can introduce appreciable errors. At Lake Hefner, Oklahoma [7], corrections as large as 10 mm (0.4 in) per month were required for changes in density.

The relative effect of errors in the surface inflow and outflow terms varies considerably from lake to lake, depending upon the extent of ungaged areas, the reliability of the rating curves, and the relative magnitude of flows with respect to evaporation. Determinations of streamflow to within 5 percent are normally considered excellent, and corresponding evaporation errors may be expected in an off-channel reservoir without appreciable outflow. If the quantity of water passing through a reservoir is large in comparison with evaporation losses, water-budget results are of questionable accuracy.

Under somewhat idealized conditions, it was found that daily evaporation from Lake Hefner, Oklahoma, could be reliably computed from a water budget; results were considered to be within 5 percent one-third of the time and within 10 percent two-thirds of the time. It should be emphasized that Lake Hefner was selected, after a survey of more than 100 lakes and reservoirs [8], as one of the three or four best meeting water-budget requirements. The requirements are not nearly so stringent for estimates of annual or mean annual reservoir evaporation, and satisfactory estimates have been made for many reservoirs.

5-3 Energy-Budget Determinations of Reservoir Evaporation

The energy-budget approach, like the water budget, employs a continuity equation and solves for evaporation as the residual required to maintain a balance. Although the continuity equation in this case is one of energy, an approximate water budget is required as well, since inflow, outflow, and storage of water represent energy values which must be considered in conjunction with the respective temperatures [9]. Although application of the energy budget had been attempted by numerous investigators, with cases selected to minimize the effect of terms that could not be evaluated, the Lake Hefner experiment is believed to constitute the first test of the method with adequate control. The energy-budget approach is receiving increasing application to special studies but is not likely to be used on a broad-scale, continuing basis until instrumentation is improved.

The energy budget for a lake or reservoir may be expressed as

$$Q_n - Q_h - Q_e = Q_\theta - Q_v \qquad (5\text{-}2)$$

where Q_n is the net (all-wave) radiation absorbed by the water body, Q_h the sensible-heat transfer (conduction) to the atmosphere, Q_e the energy used for evaporation, Q_θ the increase in energy stored in the water body, and Q_v the *advected energy* (net energy content of inflow and outflow elements), all expressed in equivalent energy units per unit of surface area.† Letting H_v represent the latent heat of vaporization [Eq. (2-1)] and R the ratio of heat loss by conduction to heat loss by evaporation (*Bowen ratio*), Eq. (5-2) becomes

$$E = \frac{Q_n + Q_v - Q_\theta}{\rho H_v(1 + R)} \qquad (5\text{-}3)$$

---

† Usually expressed as calories (15°C) per square centimeter or megajoules per square meter [10] for daily and mean daily computations, and watts per square meter in terms of flux (1 cal/cm² = 0.04186 MJm⁻² and 1 W = 1 J/s).

where E is the evaporation in centimeters and ρ is the density of water (cal, cm, g units). The Bowen ratio [11] can be computed from the equation

$$R = 0.66 \frac{T_0 - T_a}{e_0 - e_a} \frac{p}{1000} \tag{5-4}$$

where p is the atmospheric pressure, T_a the temperature of the air, e_a the vapor pressure of the air, T_0 the water-surface temperature, and e_0 the saturation vapor pressure corresponding to T_0; all temperatures and pressures are in degrees Celsius and millibars.

Sensible-heat transfer cannot readily be observed or computed, and the Bowen ratio was conceived as a means of eliminating this term from the energy-budget equation. The validity of the constant in Eq. (5-4) has been the subject of much discussion [12]. Bowen found limiting values of 0.58 and 0.66, depending on the stability of the atmosphere, and concluded that 0.61 is applicable under normal atmospheric conditions. Using an independent approach, Pritchard [13] derived values of 0.57 and 0.66 for smooth and rough surfaces, respectively. At Lake Hefner, Oklahoma, monthly values of the ratio [computed from Eq. (5-4)] were found to vary from -0.32 in February to 0.25 in November, while the annual value was -0.03. It is obvious that one need not be concerned over variations in the constant of Eq. (5-4) for annual computations.

It will be seen from Eq. (5-3) that evaporation is indeterminate when $R = -1.0$ and that it becomes critically dependent upon R as this value of the ratio is approached. A similar situation occurs when the vapor-pressure difference [Eq. (5-4)] approaches zero. In a practical sense, such situations do not greatly affect the accuracy of the computed evaporation for periods in excess of 10 days.

In the application of Eq. (5-3), it is important that the net radiation exchange be accurately evaluated. The net radiation can be expressed in terms of its five components

$$Q_n = Q_s - Q_r + Q_a - Q_{ar} - Q_0 \tag{5-5}$$

where Q_s is sun and sky (global) shortwave radiation incident upon the water surface, Q_r reflected shortwave radiation, Q_a incident atmospheric longwave radiation, Q_{ar} reflected longwave radiation, and Q_0 emitted longwave radiation. Most routine observations from established networks provide measurements of incident shortwave radiation only, and prior to the development of longwave and all-wave radiometers, radiation exchange was necessarily estimated from empirical relations [14, 15]. Radiometers can be designed to measure either the total incoming or net radiation (Sec. 2-4). Net radiometers must be exposed over the water at one or more points constituting a representative surface temperature. Because of the difficulties of maintaining observations over a lake, net radiometers are sometimes exposed over a tank of water. Assuming that the emissivity ε and reflectivity are the same for the water in the tank and lake, the incident minus reflected all-wave radiation for the tank $\hat{Q}_{ir}$ and the adjacent lake Q_{ir} can be obtained from the net radiation $\hat{Q}_n$ and absolute temperature $\hat{T}_0$ of the tank water surface:

$$\hat{Q}_{ir} = Q_{ir} = \hat{Q}_n + \varepsilon\sigma(\hat{I}_0)^4 \qquad (5\text{-}6)$$

where σ is the *Stefan-Boltzmann* constant $(11.71 - 10^{-8}\,\text{cal cm}^{-2}\,\text{K}^{-4}\,\text{day}^{-1})$ and ε may be taken as 0.97. The net radiation for the lake can then be obtained from

$$Q_n = Q_{ir} - \varepsilon\sigma(T_0)^4 \qquad (5\text{-}7)$$

or

$$Q_n = \hat{Q}_n + \varepsilon\sigma(\hat{T}_0 - T_0)^4 \qquad (5\text{-}8)$$

Another approach to the determination of net radiation involves the application of the energy budget to an insulated evaporation pan (*radiation integrator*) [16, 17]. The assumption is again made that Q_{ir} is the same for the pan and an adjacent lake, and values of this term are computed from Eqs. (5-2) and (5-6).

The energy budget has also been applied to small lakes using onshore measurements of incoming shortwave Q_s and total hemispherical radiation $Q_s + Q_a$. In this case, Q_{ir} is computed assuming average reflectivities $[Q_{ar} = 0.03\,Q_a$ and $Q_r = (0.07 \pm 0.02)Q_s$, depending on solar altitude; see discussion of "albedo" in Sec. 2-2].

The energy advection and storage terms $(Q_v - Q_\theta)$ of Eq. (5-3) are computed from an approximate water budget and temperatures of the respective water volumes. Equation (5-1) can be written

$$S_2 - S_1 = I + P - O - O_g - E \qquad (5\text{-}9)$$

Variations in density are usually neglected, and all terms are expressed in cubic centimeters. The energy content per gram of water (with respect to 0°C) is the product of its specific heat and Celsius temperature. Assuming unity for the values of density and specific heat gives

$$Q_v - Q_\theta = \frac{1}{A}\,(IT_I + PT_P - OT_O - O_gT_g - ET_E + S_1T_1 - S_2T_2) \qquad (5\text{-}10)$$

where T_I, T_P, . . . are the Celsius temperatures of the respective volumes of water, and the surface area A of the lake is introduced to convert energies into units of calories per square centimeter. Equation (5-9) should be balanced before solving Eq. (5-10), although approximate values of the individual terms will suffice. The temperature of precipitation can be taken as the wet-bulb temperature, seepage temperature as that of the water in the lowest levels of the lake, and T_E as the lake-surface temperature. Advected energy and change in energy storage tend to balance for most lakes, particularly over long periods of time, and are frequently assumed to cancel when considering annual or mean annual evaporation.

5-4 Aerodynamic Determination of Reservoir Evaporation

The theoretical development of turbulent-transport equations has followed two basic approaches: the discontinuous, or mixing-length, concept introduced by

Prandtl and Schmidt, and the continuous-mixing concept of Taylor. An extensive physical and mathematical review [18] of the two approaches was prepared in advance of the Lake Hefner experiment, and a number of the equations were tested at Lake Hefner and Lake Mead [19]. Equations derived by Sverdrup [20] and Sutton [21] gave good results at Lake Hefner but were considered inadequate when applied to Lake Mead.

Numerous empirical formulas [22, 23] have been derived which express evaporation as a function of atmospheric elements and which parallel the turbulent-transport approach in some respects. Many such equations are of the Dalton type [24] and may be written in the form

$$E = (e_0 - e_a)(a + bv) \tag{5-11}$$

where e_0 and e_a are vapor pressures of the water surface and at some fixed height in the overrunning air, respectively, and v is the wind speed (also at some fixed height). The saturation vapor pressure at the temperature of the air e_s is sometimes used instead of e_0.

Several empirical equations have been derived from the data collected at Lake Hefner:

$$E = 0.122(e_0 - e_2)v_4 \qquad e_2 \text{ and } v_4 \text{ over lake} \tag{5-12}$$

$$E = 0.097(e_0 - e_8)v_8 \qquad e_8 \text{ and } v_8 \text{ over lake} \tag{5-13}$$

$$E = 0.109(e_0 - e_2)v_4 \qquad e_2 \text{ upwind and } v_4 \text{ over lake} \tag{5-14}$$

where E is lake evaporation in millimeters per day, vapor pressures are in millibars, wind is in meters per second, and numerical subscripts designate heights above the surface in meters. With vapor pressures in inches of mercury and wind speeds in miles per day, the constants in Eqs. (5-12) to (5-14) become 0.00304, 0.00241, and 0.00270, respectively, for evaporation in inches per day. Equation (5-12) was found to yield excellent results for Lake Mead, and has given satisfactory results on other lakes as well [25, 26]. Equation (5-13) yielded a satisfactory value of annual evaporation for Lake Mead, but there was a seasonal bias which appeared to be correlated with atmospheric stability.

Vapor pressure of the air increases downwind across an open-water surface; hence turbulent-transport concepts have led to the belief that point evaporation decreases downwind. Sutton [27] concluded that average depth of evaporation from a circular water surface is proportional to the -0.11 power of its diameter under adiabatic conditions, and this functional relation has been verified in wind-tunnel experiments [28]. The theory assumes that water temperature and wind are unchanging downwind, and this condition tends to prevail in a wind tunnel, where solar radiation is not a factor. Observations show that wind speed increases downwind from the upwind edge of a lake, however, and a consideration of energy conservation requires that any immediate reduction in evaporation rate brought about by decreased vapor-pressure gradient be followed by an increase in water temperature. Although experimental data are insufficient to determine the magnitude of the *size effect,* it is believed that Eq. (5-12) can be applied to lakes ranging up to several hundred square kilometers without appreciable error in this respect if all observations are well centered in the lake.

Using studies of many reservoirs up to 120 km² (29,000 acres) in area, Harbeck [29] found that the data fit Eq. (5-11) such that $a = 0$ and b is given by

$$b = 0.29 A^{-0.05} \tag{5-15}$$

where E is in millimeters per day, A is the area of the reservoir in square kilometers, v_2 is in meters per second, and e_a is measured in the unmodified air (upwind). The coefficient in Eq. (5-15) becomes 0.00014 for E in inches per day, A in acres, v_2 in miles per day, and vapor pressures in millibars. It should be stressed that Harbeck's results do not demonstrate a size effect in actual evaporation or in Eq. (5-12). The relation between e_2 over the water and e_a is a function of lake size, and this would assure a size effect in Eq. (5-14) if Eq. (5-12) were entirely independent of lake area.

5-5 Combination Methods of Estimating Reservoir Evaporation

The simultaneous solution of aerodynamic and water-balance equations to estimate evaporation and seepage is suggested in Sec. 5-2. A similar approach can be used to eliminate the need for water-surface temperature observations by simultaneous solution of aerodynamic and energy-budget equations [30–33]. Assuming a thin free-water surface (without heat storage or conduction from below), Penman derived the equation

$$E = \frac{\Delta}{\Delta + \gamma} Q_n + \frac{\gamma}{\Delta + \gamma} E_a \tag{5-16}$$

where Δ is the slope of the saturation-vapor-pressure versus temperature curve at the air temperature T_a, E_a is the evaporation given by Eq. (5-11) assuming the water-surface temperature $T_0 = T_a$, Q_n is the net radiation exchange expressed in the same units as E, and γ is defined by the Bowen ratio equation:

$$R = \gamma \frac{T_0 - T_a}{e_0 - e_a} \tag{5-17}$$

The value of γ depends to some extent on atmospheric pressure [Eq. (5-4)], but the variation is appreciable only with changes in elevation. The dimensionless ratios $\Delta/(\Delta + \gamma)$ and $\gamma/(\Delta + \gamma)$ in Eq. (5-16) constitute weighting factors, their sum being equal to unity. Equations for estimating the ratios are presented in Sec. 5-18.

Figure 5-1 is based on an application [34] of the Penman equation to the Class A pan (Fig. 5-4) modified [35] to reflect differences in γ for the pan and an extended free-water surface, and assuming that the class A pan coefficient is 0.70 when pan-water temperature is the same as air temperature (Sec. 5-6). Figure 5-1 provides a convenient means of estimating reservoir evaporation when net advection is not appreciable; otherwise, adjustment should be made using Eq. (5-20).

Inherent in the relationship of Fig. 5-1 are graphical correlations for es-

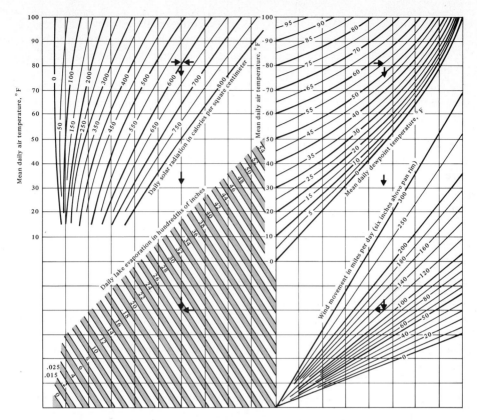

Figure 5-1 Shallow-lake evaporation as a function of solar radiation, air temperature, dewpoint, and wind movement.

timating Q_n from air temperature and solar radiation, and E_a from wind movement and vapor-pressure difference. To facilitate computer solution, equations for estimating Q_n and E_a are given in Sec. 5-18. Entering Eq. (5-30) with these values and the Δ, γ ratios will yield essentially the same evaporation as that obtained from Fig. 5-1.

> **Illustrative example 5-1** Given the sea-level observations $T_a = 70°F$, $T_d = 45°F$, $Q_s = 530$ cal cm^{-2} day^{-1}, and $v_p = 140$ mi/day, compute the daily evaporation from Fig. 5-1 and from the equations given in Sec. 5-18. Entering Fig. 5-1 with the values provided yields $E = 0.22$ in/day.
> From Eq. (5-25b), $\Delta/(\Delta + \gamma) = 0.697$
> From Eq. (5-26), $\gamma/(\Delta + \gamma) = 0.303$
> From Eq. (5-27b), $Q_n = 0.250$
> From Eq. (5-28b), $e_s - e_a = 0.4385$
> From Eq. (5-29b), $E_a = 0.457$
> From Eq. (5-30), $E = 0.22$ in/day

Equation (5-16) assumes that Q_n is representative of exchange at the water surface; E_a is based on an aerodynamic equation which yields correct values of E when used with observed vapor pressure of the water surface; and Δ at T_a is a good approximation of its average value between T_a and T_0. Net radiation must be measured over the water surface or estimated so as to reflect the water temperature (Sec. 5-3). While Penman's derivation takes into account the effect of any difference between T_0 and T_a on evaporation and convective heat transfer, his method of estimating Q_n assumes that the two temperatures are equal. This assumption can result in an appreciable overestimate of evaporation under calm, humid conditions and a corresponding underestimate for dry, windy conditions.

By using Eq. (5-7), T_0 can be eliminated from the radiation term in Penman's derivation [36], yielding

$$E = \frac{(Q_{ir} - \varepsilon\sigma T_a^4)\,\Delta + E_a\,[\gamma + 4\varepsilon\sigma T_a^3/f(v)]}{\Delta + [\gamma + 4\varepsilon\sigma T_a^3/f(v)]} \tag{5-18}$$

where T_a is in kelvins and $f(v)$ is the wind function in a suitable aerodynamic equation. The units of E_a and Q_{ir} must be the same as E, and σ must be consistent with these units. The similarity between Eqs. (5-16) and (5-18) is apparent: the term appearing in brackets with γ corrects for the difference between emitted longwave radiation at T_0 and T_a. Methods of obtaining values of Q_{ir} are discussed in Sec. 5-3.

Equation (5-18) assumes that net radiation is entirely dissipated through evaporation and sensible-heat exchange with the atmosphere, and it is therefore necessary to consider the effects of heat exchange within the water body as they are related to energy storage and advection.

It can be shown that the effects of advected energy (inflow and outflow of water) are unimportant except when flows are large relative to the rate of evaporation. Even then, the inflow and outflow temperatures must be appreciably different. The effects of changes in energy storage, on the other hand, may be relatively large, depending on the period of computation and the depth of the lake. Energy storage can be safely neglected in the computation of mean annual evaporation in all cases and in the computation of annual evaporation from relatively shallow lakes.

The effects of advection and/or changes in energy storage are brought about through a change in water-surface temperature and are thus distributed among sensible transfer, evaporation, and emitted radiation. It can be shown [37] that the portion of such energy affecting evaporation is

$$\alpha = \frac{\Delta}{\Delta + \gamma + 4\varepsilon\sigma T_0^3/f(v)} \tag{5-19}$$

The ratio α is analogous to the Bowen ratio. Figure 5-2 provides a convenient solution to Eq. (5-19). Equations for deriving α taking into account the effects of pressure are given in Sec. 5-18.

When it becomes necessary to take advection and energy storage into

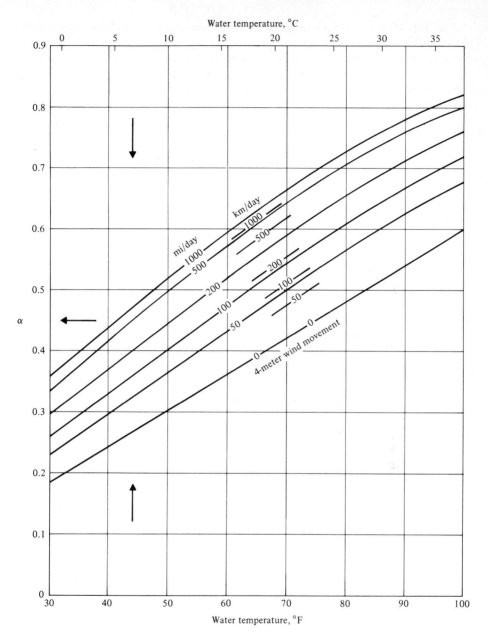

Figure 5-2 Variation of α with water temperature and wind movement at sea-level elevation.

account, the adjusted lake evaporation E_L can be computed from

$$E_L = E + \alpha(Q_v - Q_\theta) \qquad (5\text{-}20)$$

where E is evaporation computed from Eq. (5-18) and the *net advection,* $Q_v - Q_\theta$ [Eq. (5-10)], is expressed in equivalent units of evaporation. Thus, while Eq. (5-18) provides generalized estimates of evaporation from a very shallow lake, it may be necessary to use Eq. (5-20) when comparing the results with evaporation from specific lakes. Approximate adjustments for net advection may also be feasible when considering reservoir design. Although annual net advection is negligibly small for most lakes, this is not necessarily true—it is equivalent to about 30 cm of evaporation for Lake Mead as a result of low-level outflow which is on the average colder than inflow.

5-6 Estimation of Reservoir Evaporation from Pan Evaporation and Related Meteorological Data

The pan is undoubtedly the most widely used evaporation instrument today, and its application in hydrologic design and operation is of long standing. Although criticism of the pan may be justified on theoretical grounds, for some types of pans the ratio of annual lake-to-pan evaporation (*pan coefficient*) is quite consistent, year by year, and does not vary excessively from region to region.

Pan observations. There are three types of exposures employed for pan installations—sunken, floating, and surface—and divergent views on the best exposure persist. Burying the pan tends to eliminate objectionable boundary effects, such as radiation on the side walls and heat exchange between the atmosphere and the pan itself, but creates observational problems. Sunken pans collect more trash; they are difficult to install, clean, and repair; leaks are not easily detected; and height of vegetation adjacent to the pan is quite critical. Moreover, appreciable heat exchange does take place between the pan and the soil [38], depending on such factors as soil type, moisture content, and vegetative cover. Heat exchange with the soil can readily change the annual evaporation from a 2-m pan 10 percent and a 5-m pan 7 percent. It is therefore obvious that such heat exchange will produce large climatic variation in the pan coefficient of a small, sunken, uninsulated pan.

The evaporation from a pan floating in a lake more nearly approximates evaporation from the lake than that from an on-shore installation, but even so, the boundary effects are appreciable. Observational difficulties are prevalent with floating pans (splashing frequently renders the data unreliable), and installation and operational expense is excessive. Since relatively few such installations are now in existence, floating pans are not considered in the subsequent discussion.

Pans exposed above ground experience greater evaporation than sunken pans, primarily because of the radiant energy intercepted by the side walls, and heat exchange through the pan produces unrealistic effects which must be taken into account. Both deficiencies can be minimized by insulating the pan. The principal advantages of surface exposure are economy and ease of installation, operation, and maintenance.

Of the various sunken pans used, only three have gained prominence in the

United States: the Young screened pan, the Colorado pan, and the Bureau of Plant Industry (BPI) pan. The Young pan [39] is 2 ft (61 cm) in diameter and 3 ft (91.5 cm) deep and is covered with ¼-in-mesh (6-mm) hardware cloth. The screen modifies the pan coefficient to near unity, on an average, but the small size of the pan leads to an unstable coefficient and the overall effect of screening may be adverse. The Colorado pan is 3 ft (91.5 cm) square and 18 in (46 cm) deep. The BPI pan, 6 ft (183 cm) in diameter by 2 ft (61 cm) deep, provides by far the best index to lake evaporation because of its size.

The standard National Weather Service Class A pan is the most widely used evaporation pan in the United States; in 1979 records were published for about 480 stations. It is of unpainted galvanized iron 4 ft (122 cm) in diameter by 10 in (25.4 cm) deep and is exposed on a wood frame to let air circulate beneath the pan (Fig. 5-3). It is filled to a depth of 8 in (20 cm), and instructions [40] require that it be refilled when the depth has fallen to 7 in (18 cm). Water-surface level is measured daily with a hook gage in a stilling well, and evaporation is computed as the difference between observed levels, adjusted for any precipitation measured in a standard rain gage. Alternatively, water is added each day to bring the level up to a fixed point in the stilling well. This method assures proper water level at all times.

Many other types of pans are in use in different parts of the world, and the need for international standardization has long been recognized by the World Meteorological Organization. Pending standardization, many intercomparisons of the various types of pans have been made [41] which show that pan-to-pan ratios display appreciable geographic (climatic) variation. The two most widely used pans are the Class A and the GGI-3000, the standard pan in the U.S.S.R. [42]. The latter pan is circular, 3000 cm² in area (61.8 cm or 24.3 in in diameter). The depth is 60 cm (23.6 in) at the wall and somewhat greater at the center. It is fabricated of galvanized sheet iron. The pan and a precipitation gage of similar dimensions are both sunk into the ground.

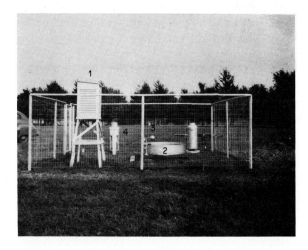

Figure 5-3 Class A evaporation station: 1 = instrument shelter, 2 = evaporation pan, 3 = anemometer, 4 = standard 8-in precipitation gage, and 5 = weighing-type recording precipitation gage. (*U.S. National Weather Service.*)

The value of a pan as an index to lake evaporaton must depend on energy-exchange considerations rather than aerodynamic similarities. As a "calorimeter," there is obviously much to be gained by insulating the pan, and if this is done, it seems that the disadvantages of a sunken exposure should be avoided. Experiments with an insulated pan of approximately the same dimensions as the GGI-3000 are extremely promising [17]. The geographic variation of the pan coefficient for such a pan is appreciably less than for any small pan in general use (see Table 5-2 on page 151).

In some localities, it is necessary to screen evaporation pans to eliminate loss of water due to birds and animals drinking from the pan. It should be realized that screening increases the pan coefficient in addition to the elimination of drinking. The unintended reduction in pan evaporation depends on the wire size and its albedo, the density of mesh, and the method of mounting the screen [41, 43]. Great care must be exercised to distinguish between the effects of the animals and the screen in any experimental tests, and to make certain that any effects of the screen on precipitation catch are taken into account.

Pan evaporation and meteorological factors. Many attempts have been made to derive reliable relations between pan evaporation and meteorological factors [44]. Obvious purposes to be served by such relations are as follows:

1. To increase our knowledge of evaporation.
2. To estimate missing pan records (pans are not operated during winter in areas where ice cover would occur much of the time, and records for days with snow or heavy rain are frequently fallacious).
3. To estimate data for stations at which pan observations are not made.
4. To test the reliability and representativeness of observed data.
5. To aid the study of lake-pan relations.

Some of the relations developed involve the substitution of air temperature for water temperature with a resultant seasonal and geographic bias.

The Penman approach [Eq. (5-16)] was applied to the composite record from a number of stations over the United States to derive the coaxial relation [35] shown in Fig. 5-4. To compute pan evaporation, E_a is first determined from the upper left-hand chart by entering with values of air temperature, dewpoint, and wind. Equation (5-16) is solved by entering the other three curve families of Fig. 5-4 with air temperature, solar radiation, E_a, and air temperature. Although the correlation was based on daily data, experience has shown that only minor errors result when monthly evaporation, i.e., mean daily value for the month, is computed from monthly averages of the daily values of the variables.

There are only a limited number of locations where all the data required for application of Fig. 5-4 are available. In 1974 solar radiation was observed at only about 85 stations in the United States, and the data from only 38 were being published as of 1979. There are reasonably reliable means of estimating this factor [45, 46], however, and the other required elements can usually be esti-

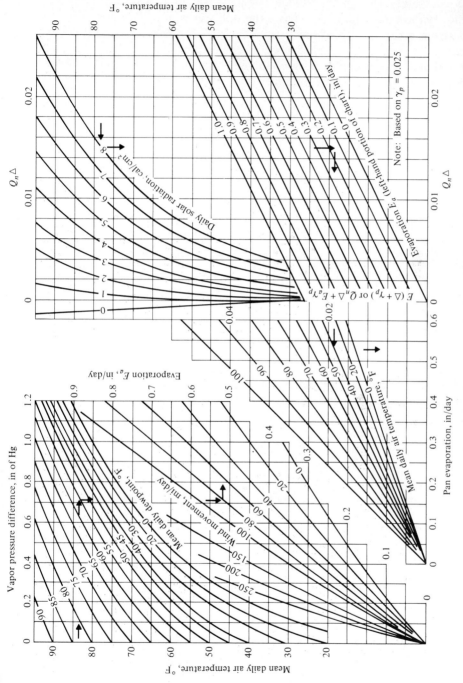

Figure 5-4 Evaporation from Class A pan as a function of solar radiation, air temperature, dewpoint, and wind movement. (*U.S. National Weather Service.*)

mated accurately enough to provide adequate values of monthly or normal monthly evaporation. Table 5-1 shows the magnitude of evaporation error resulting from error in each of the various factors, for selected meteorological situations. In view of the high degree of correlation found to exist between computed and observed evaporation, Table 5-1 is also indicative of the relative influence of the various elements under typical meteorological situations.

Pan coefficients. Water-budget, energy-budget, and aerodynamic techniques can be used to estimate evaporation from existing reservoirs and lakes. However, these methods are not directly applicable to design problems, since water-temperature observations are required in their use. The combination methods (Sec. 5-5) are beginning to come into use, but most estimates of reservoir evaporation, both for design and operation, have been made by applying a coefficient (Table 5-2) to observed or derived pan evaporation. Although too few determinations have been made to appraise the approach accurately, assuming an annual Class A pan coefficient of 0.70 for the lakes included in the table would result in a maximum difference of 15 percent. Part-year coefficients are more variable because energy storage in the lake can be appreciably different at the beginning and end of the period and changes in heat storage cause pronounced variations in monthly coefficients which must be taken into account [47]. Equation (5-20) should be used to adjust for net advection (in this case, E is pan evaporation reduced by an appropriate coefficient).

Effects of advected energy on pan evaporation. Observations demonstrate that the sensible-heat transfer through the pan can be appreciable and may flow in either direction, depending upon water and air temperatures. Since annual heat transfer through the bottom of the reservoir is essentially zero, pan data require adjustment.

Table 5-1 Variation of Class A pan evaporation with related factors

Case no.	T_a, °F	% error/ deg change in T_a	T_d, °F	% error/ deg change in T_d	Q_s, cal cm^{-2} day^{-1}	% error/ % change in Q_s	v, mi/day	% error/ % change in v	E, in/day
1	91	1.8	41	0.4	700	0.7	50	0.2	0.51
2	91	2.3	63	0.8	700	0.8	50	0.1	0.46
3	84	3.8	75	2.7	600	0.9	50	0.1	0.28
4	66	6.0	55	4.0	300	0.6	50	0.2	0.12
5	45	6.3	28	2.9	250	0.3	50	0.3	0.09
6	91	1.8	41	0.3	700	0.6	100	0.3	0.60
7	91	2.3	63	1.0	700	0.7	100	0.2	0.52
8	84	4.4	75	3.1	600	0.8	100	0.2	0.31
9	66	6.2	55	4.6	300	0.5	100	0.4	0.15
10	45	6.2	28	3.1	250	0.3	100	0.5	0.11

Table 5-2 Summary of pan coefficients

See text for description of pans

Location	Years of record	Period	Surface exposure		Sunken pans			
			Class A	X-3[a]	BPI	Young	Colo.	GGI-3000
Davis Calif.[b]	1966–69	Ann.	0.72	0.71	...	...	...	0.94
Denver, Colo.[c]	1915–16	Ann.	0.67					
	1916	June–Oct.	...	...	0.94			
Felt Lake, Calif.	1955	Ann.	0.77	...	0.91	0.99	0.85	
Ft. Collins, Colo.	1926–28	Apr.–Nov.	0.70	...	...	...	0.79	
Fullerton, Calif.[c]	1936–39	Ann.	0.77	...	0.94	0.98	0.89	
Lake Colorado City, Tex.	1954–55	Ann.	0.72					
Lake Elsinore, Calif.	1939–41	Ann.	0.77	...	...	0.98		
Lake Hefner, Okla.	1950–51	Ann.	0.69	...	0.91	0.91	0.83	
Lake Mead, Ariz.–Nev.[b]	1966–69	Ann.	0.66[d]	0.73[d]	...	...	...	0.71[d]
Lake Okeechobee, Fla.	1940–46	Ann.	0.81	...	...	...	0.98	
Red Bluff Res., Tex.	1939–47	Ann.	0.68					
Salton Sea, Calif.	1967–69	Ann.	0.64	0.64				
Silver Hill, Md.[e]	1955–60	Apr.–Nov.	0.74	...	1.05	...	0.97	
Sterling, Va.[b]	1965–68	Apr.–Nov.	0.69[f]	0.71[f]	...	...	...	1.11[f]
Australia[g]								
Lake Albacutya	...	Ann.	0.79					
Lake Cawndilla	...	Ann.	0.71					
Lake Hindmarsh	...	Ann.	0.74					
Lake Menindee	...	Ann.	0.71					
Lake Pamamaroo	...	Ann.	0.66					
Stephens Cr. Resv.	...	Ann.	0.69					
India								
Poona[b]	1965–68	Ann.	0.69	...	...	...	...	0.78
Israel								
Lod Airport[c]	1954–60	Ann.	0.74					
Sudan								
Khartoum[c]	1960–61	Ann.	0.65					
United Kingdom								
London	1956–62	Ann.	0.70					
U.S.S.R								
Dubovka[b]	1957–59	May–Oct.	0.64	...	...	...	...	0.91
	1962–67	May–Oct.	0.64	...	...	...	...	0.84
Valdai[b]	1949–53	May–Sept.	0.82	...	...	...	...	0.93
	1958–63	May–Sept.	0.67	...	...	...	...	0.98

Source: References 16, pp. 127–148; 17; 19; 22; 35; 38; 39; 41; and 43; also unpublished material.

[a] Insulated; dimensions approximately the same as for the GGI-3000 pan.

[b] Assuming that evaporation from a tank 5 m (16.5 ft) in diameter is equivalent to lake evaporation.

[c] Assuming that evaporation from a tank 12 ft (3.65 m) in diameter is equivalent to lake evaporation.

[d] Correction for heat flow from soil into 5-m tank would reduce coefficient at least 5 percent.

[e] Assuming that evaporation (adjusted for heat flow to the soil) from a tank 15 ft (4.57 m) in diameter is equivalent to lake evaporation.

[f] Correction for heat flow from 5-m tank to the soil would increase coefficient a few percent.

[g] The listed coefficients are based on water-budget determinations. Heat-budget determinations for eight additional lakes are cited in the reference [43]; their average value is appreciably higher.

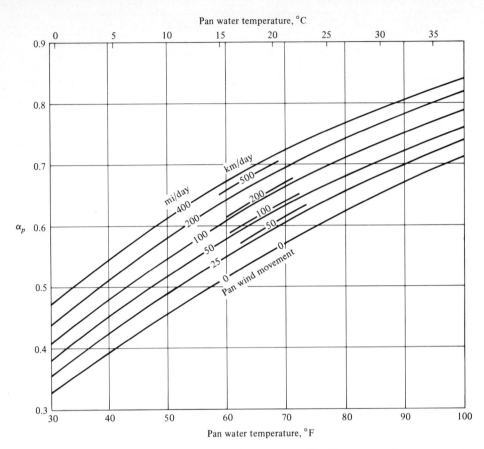

Figure 5-5 Portion of advected energy (into a Class A pan) utilized for evaporation at sea level. (*U.S. National Weather Service.*)

Figure 5-5, derived in much the same manner as Fig. 5-2, shows the relative effect α_p of net advection on Class A pan evaporation [35]. Equations for estimating α_p are given in Sec. 5-18. Advection by means of water added to the pan is normally unimportant, but advection of sensible heat through the pan walls is sufficient to produce moderate variation in the pan coefficient under varying climatic regimes. From the Bowen ratio concept and the empirical relation of Fig. 5-4, a function can be derived for estimating heat transfer through the Class A pan from observations of air and water temperature, wind movement, and atmospheric pressure. Making the further assumption that the pan coefficient is 0.7 when air and pan water temperatures are equal:

$$E = 0.7[E_p \pm 0.00051 \, p\alpha_p \, (0.37 + 0.0041 \, v_p)|T_0 - T_a|^{0.88}] \quad (5\text{-}21a)$$

where E_p is pan evaporation in inches per day, p is in inches of mercury, v_p is wind movement 6 in above the pan rim in miles per day, and temperatures are in degrees Fahrenheit. The sign following E_p is $(+)$ when $T_0 > T_a$ and $(-)$ when $T_0 < T_a$. Equation $(5\text{-}21a)$ becomes

$$E = 0.7[E_p \pm 0.00064p\alpha_p (0.37 + 0.00255 v_p)|T_0 - T_a|^{0.88}] \quad (5\text{-}21b)$$

when evaporation is in millimeters per day, p is in millibars, v_p is in kilometers per day, and temperatures are in degrees Celsius. Lake evaporation computed from Eq. (5-21) should be adjusted for any appreciable net advection to the lake.

5-7 Summary and Appraisal of Techniques for Estimating Reservoir Evaporation

There are relatively few reservoirs for which reliable evaporation estimates can be derived from a water budget on a continuous basis, but values for selected periods may often serve to calibrate or check other techniques. When conditions are such that satisfactory results cannot be obtained by applying a water budget, evaporation from an existing reservoir can be determined by either the empirical aerodynamic or energy-budget approach. Both instrumentation and maintenance of continuing observations are expensive for these two approaches, and their widespread use may not be economically feasible for some time. However, the purposes to be served may justify their application for a short period to calibrate a less costly method.

The operation of a pan station (near the reservoir, but not close enough to be materially affected by it) is relatively inexpensive and should provide reasonably good estimates of annual reservoir evaporation [Eq. (5-21)]. Some reliability would be gained if net reservoir advection were estimated, but this item is rarely of much importance unless monthly or seasonal distribution of annual evaporation is required. Even though evaporation pans are normally inoperative during freezing weather, pan evaporation is small during such periods and can usually be estimated with sufficient accuracy. Reservoir evaporation can be relatively large during early winter, however, because of changes in energy storage.

For reservoir-design studies, Fig. 5-1 or Eqs. (5-18) or (5-21) may be applied if representative data are available. All relevant data for the area should be analyzed using all techniques for which the data are suitable if the economic aspects of design so justify. There is seldom justifiable reason for constructing a major reservoir prior to the collection of at least 1 or 2 years of pan and related meteorological data at the site for checking purposes. Figure 5-6, which presents generalized estimates of mean annual evaporation from shallow lakes for the conterminous United States [34], is considered sufficiently reliable for preliminary estimates and for design of projects in which evaporation is not important. Similar maps are available for a number of other countries [48–50], and Fig. 2-3 is a world map of heat used by evaporation.

In some design problems, it is necessary to estimate the monthly distribution of annual evaporation. While several of the techniques discussed will yield

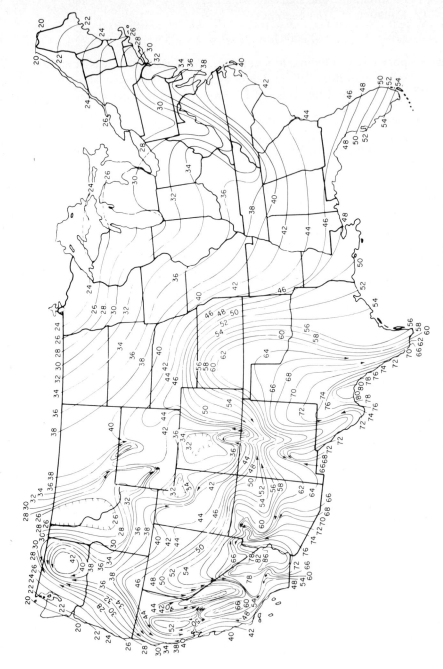

Figure 5-6 Average annual evaporation (inches) from shallow lakes. (*U.S. National Weather Service.*)

monthly values of free-water evaporation, adjustment is required for net advection (or heat-storage changes, at least) [51–54]. The estimation of monthly pan coefficients by applying numerical lake modeling techniques is also a possibility [55].

In reservoir design, the engineer is really concerned with the increased loss over the reservoir site resulting from the construction of the dam, i.e., reservoir evaporation less evapotranspiration under natural conditions. In humid areas construction of a dam causes only a nominal increase in water loss.

5-8 Increased Water Supplies through Reduced Evaporation

Any steps which can be taken to reduce reservoir evaporation per unit of storage provide a corresponding increase in usable water supply. Selecting the site and design yielding a minimum of reservoir area per unit of storage is advantageous. Designing the outlet works so that the warmer surface water can be released to meet demands will reduce evaporation from a reservoir. Such an operation will not provide equivalent increased supplies at a distant point downstream, since increased evaporation will take place in the channel below the dam.

Small reservoirs are sometimes entirely covered to reduce evaporation losses. The use of floating covers [56] and floating granular materials has also been advocated [57]. Such methods are effective but expensive to apply. Although windbreaks have been repeatedly recommended, they obviously are effective only on very small reservoirs: a 25 percent reduction in wind speed will normally reduce evaporation by only about 5 percent on a sustained basis.

Extensive research has been undertaken in the application of monomolecular films to reduce evaporation [58–60]. Despite early optimism, the approach is little used. Extremely small quantities of cetyl alcohol will reduce evaporation from a small pan as much as 40 percent, but it is seldom possible to maintain more than 10 to 20 percent coverage over a reservoir on a continuing basis. Moreover, any reduction in evaporation is accompanied by an increase in water temperature. Much of the excess heat can be dissipated through the pan, but this is not the case for a reservoir [61]. It appears that any hope for achieving appreciable, practical reductions in evaporation from large reservoirs lies in finding a material which will effectively increase the reflectivity of the water surface without causing undesirable side effects.

TRANSPIRATION

Only minute portions of the water absorbed by the root systems of plants remain in the plant tissues; virtually all is discharged to the atmosphere as vapor through *transpiration*. This process constitutes an important phase of the hydrologic cycle since it is the principal mechanism by which the precipitation falling on land areas is returned to the atmosphere. In studying the water balance of a drainage basin, it is usually impracticable to separate evaporation and transpiration, and the practicing engineer therefore treats the two factors as a single item. Nevertheless, a knowledge of each process is required to assure that the techniques employed are consistent with physical reality.

5-9 Factors Affecting Transpiration

The difference in concentration between the sap in the root cells of a plant and the soil water causes an osmotic pressure which moves soil water through the root membrane into the root cells. High salinity of the soil solution and/or high moisture tension in the soil may prevent or greatly reduce the osmotic transfer. Once inside the root, the water is transferred [62] through the plant to the *intercellular space* within the leaves (Fig. 5-7). Air enters the leaf through the *stomata,* openings in the leaf surface, and the *chloroplasts* within the leaf use carbon dioxide from the air and a small portion of the available water to manufacture carbohydrates for plant growth (*photosynthesis*). As air enters the leaf, water escapes through the open stomata; this is the process of *transpiration*. The ratio of the water transpired to that used in forming plant matter is very large—up to 800 or more.

Plants serve to move water to the surface from depths penetrated by the root systems. The rate of transpiration is largely independent of plant type, provided there is adequate soil water and the surface is entirely covered by vegetation. Since photosynthesis is highly dependent on the radiation received, about 95 percent of daily transpiration occurs during daylight hours [63], compared with 75 to 90 percent for soil evaporation [64]. Plant growth normally ceases when temperatures drop to near 4°C (40°F), and transpiration becomes very small.

Transpiration is limited by the rate at which moisture becomes available to the plants. Although there is little doubt that the rate of soil evaporation, under fixed meteorological conditions, decreases quasi-exponentially with time, divergent views persist with respect to transpiration (see Sec. 5-15). It is believed that the controversy and apparent discrepancies can be attributed to the varied methods of deriving supporting data and the nondescript terminology used to describe the results. Some investigators believe that transpiration is indepen-

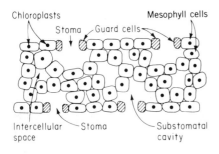

Figure 5-7 Internal structure of a leaf.

dent of available moisture until it has receded to the *wilting point* (moisture content at which permanent wilting of plants occurs), while others assume that transpiration is roughly proportional to the moisture remaining in the soil and available to the plants (Sec. 6-3). Available water varies with soil type, ranging from about 0.5 mm/cm (0.5 in/ft) of depth for sand to over 2 mm/cm (2 in/ft) of depth for clay loams.

Plant type becomes an important factor in controlling transpiration when available soil moisture is limited. As the upper layers of the soil dry out, shallow-rooted species can no longer obtain water and wilt, while deep-rooted species continue to transpire until the soil moisture at greater depths is reduced to the wilting point. Thus deep-rooted vegetation transpires more water during sustained dry periods than shallow-rooted species. Transpiration per unit area also depends on the density of vegetative cover. With widely spaced plants (low cover density), not all solar radiation reaches the plants, and some of it is absorbed at the soil surface. The relative transpiration is not proportional to cover density, however, for two reasons: (1) an isolated plant receives radiation on the side facing the sun which would fall on an adjacent plant were there solid cover, and (2) a portion of the radiation reaching the ground is subsequently transmitted to the plants (*oasis effect*).

Plant type also influences transpiration during drought conditions, even with specified soil-moisture conditions. *Xerophytes,* desert species, which have fewer stomata per unit area and less surface area exposed to radiation, transpire relatively little water. *Phreatophytes,* on the other hand, have root systems reaching to the water table and transpire at rates largely independent of moisture content in the zone of aeration. All plants can control stomatal opening to some extent, and thus even *mesophytes,* plants of the temperate zones, have some ability to reduce transpiration during periods of drought. Note, however, that the ability to control is only to reduce transpiration. Even aquatic plants, *hydrophytes,* cannot pump water into the atmosphere at rates in excess of those controlled by available radiant and sensible energy. A pond covered with aquatic plants does not lose water at a rate appreciably different from a pond free of vegetation. Any difference in sensible-heat transfer brought about by increased surface roughness tends to be counterbalanced by the increase in albedo.

Rainfall intercepted by vegetation is subsequently evaporated and thereby utilizes some of the energy otherwise available for transpiration. Experiments with grass cover [65] indicate that the transpiration reduction may be equivalent to the interception, while forestry studies show that the reduction may be much less than interception losses [66–68]. Energy required for the increased evapotranspiration is apparently provided through downward flux of sensible heat.

5-10 Measurement of Transpiration

Since it is not possible to measure transpiration loss from an appreciable area under natural conditions, determinations are restricted to studies of small samples under laboratory conditions. One method involves placing one or more potted plants in a closed container and computing transpiration as the increase in moisture content of the confined space. Most measurements are made with a *phytometer,* a large vessel filled with soil in which one or more plants are rooted. The only escape of moisture is by transpiration (the soil surface is sealed to prevent evaporation), which can be determined by weighing the plant and container at desired intervals of time. By providing aeration and additional water, a phytometer study can be carried through the entire life cycle of a plant. Since it is virtually impossible to simulate natural conditions, the results of phytometer observations are mostly of academic interest to the hydrologist, constituting little more than an index to water use by a crop under field conditions.

Ceramic and Piche atmometers have been widely used in transpiration studies. Such instruments [41] automatically feed water from a reservoir to an exposed, wetted surface. The change in content of the reservoir serves as an index to transpiration. With proper care and exposure, atmometers are useful in experimental work for estimating temporal and spatial variations in potential transpiration.

EVAPOTRANSPIRATION

In studying the hydrologic balance for a catchment area, one is usually concerned only with the *total evaporation* (or *evapotranspiration*), the evaporation from all water, soil, snow, ice, vegetation, and other surfaces plus transpiration. *Consumptive use* is the total evaporation from an area plus the water used directly in building plant tissue. The distinction between the two terms is largely academic, falling well within the error of measurement, and they are now generally treated as synonymous [69]. When used with respect to a specific crop, consumptive use is the evapotranspiration experienced if water supply is adequate at all times (see Sec. 5-16).

On the assumption that any reduction in evapotranspiration brought about by a deficiency of soil moisture is independent of meteorological conditions, the concept of *potential evapotranspiration* introduced by Thornthwaite [70] is widely used. He defined the term as "the water loss which will occur if at no time there is a deficiency of water in the soil for the use of vegetation." It has since been found that evapotranspiration depends on the density of cover and its stage of development. To serve the purpose, potential evapotranspiration must either be independent of the nature and condition of the surface, except with respect to available moisture, or it must be defined in terms of a particular surface. Penman [71] suggested that the original definition be modified to include the stipulation that the surface be fully covered by green vegetation. This modified definition is generally satisfactory, but it becomes meaningless during winter in higher latitudes.

In the interest of clarity and reproducible results, there is good reason to consider potential evapotranspiration to be equivalent to the evaporation from a free-water surface of extended proportions but with negligible heat-storage capacity [72]. Potential evapotranspiration as defined by Thornthwaite approaches free-water evaporation provided there is complete vegetal cover, and the effects of meteorological factors on the two are sufficiently alike to be converted into actual evapotranspiration in the same manner. Priestly and Taylor [73] have proposed that potential evapotranspiration be considered equivalent to the first term of the Penman equation [Eq. (5-16)] multiplied by the factor 1.26 and taking into account the heat flux into the soil (see Sec. 5-14).

There are numerous approaches to estimation of actual and potential evapotranspiration, none of which is generally applicable for all purposes. The type of data required depends on the intended use. In some hydrologic studies, mean basin evapotranspiration is required, while in other cases we are interested in water use of a particular crop cover or the change in water use resulting from changed vegetal cover.

5-11 Water-Budget Determination of Mean Basin Evapotranspiration

Assuming that storage and all items of inflow and outflow except evapotranspiration can be measured, the volume of water (usually expressed in units of depth) required to balance the continuity equation for a basin represents evapotranspiration. The reliability of a water-budget computation hinges largely on the time increments considered. As a rule, normal annual evapotranspiration can be reliably computed as the difference between long-time averages of precipitation and streamflow, since the change in storage over a long period of years is inconsequential [74]. Any deficiencies in such computations are usually attributable either to inadequate precipitation or runoff data or to subterranean flow into or out of the basin. Estimates of annual evapotranspiration are subject to appreciable errors if changes in storage are neglected, except where moisture storage in a basin is nearly the same on a given date each year. Generally it is necessary to evaluate soil moisture, groundwater, and surface storage at the beginning of each year.

The water-budget method can also be applied to selected short time periods [75]. In Fig. 5-8, heavy rain fell in a 3-day period, resulting in the rise of June 21, and 117 mm (4.60 in) more fell by June 29. The total runoff (Sec. 7-5) produced by the second storm was 60 mm (2.37 in). If it is assumed that the soil was equally near saturation at the end of rain on June 21 and 29, the evapotranspiration during the period was 57 mm (117 − 60), or 7 mm/day (0.28 in/day). The error in computations for periods as short as a week can be appreciable, and it is advisable to use longer periods when feasible. If the computations are carried through July 16, the estimated evapotranspiration averages about 6 mm/day (0.23 in/day), undoubtedly a more reliable value. This procedure is best adapted to regions where depth to groundwater is relatively small and precipitation is evenly distributed throughout the year. Although evapotranspiration estimates derived in the manner described must be geared to the fortuitous occurrence of large storms or relatively wet periods, sufficient determinations can be made

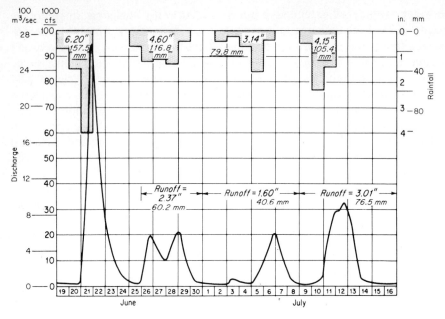

Figure 5-8 Derivation of short-period evapotranspiration.

from many years of record to define the seasonal distribution. If the resulting curve is to represent normal annual evapotranspiration, the computations must be carried on a continuous basis, since omitting dry periods will bias the results. If the curve is to represent potential evapotranspiration, computations should be made only for those periods during which potential conditions existed.

5-12 Field-Plot Determination of Evapotranspiration

Application of a water budget to field plots produces satisfactory results only under ideal conditions, which are rarely attained. Precise measurement of percolation is not possible, and consequent errors tend to be cumulative. If the groundwater table lies at great depth, accretions may be inconsequential but not necessarily so. If accretions are inconsequential, soil-moisture measurements become the principal source of error—random in nature but large enough to preclude computation of short-period evapotranspiration. Reasonable seasonal estimates may be feasible, however [76, 77].

The energy budget can be applied to the determination of evapotranspiration from a field plot much as for a lake (Sec. 5-3). Instead of being concerned with heat storage in a mass of water, however, we must compute that stored in the soil profile. In lieu of heat storage [78], it is usually advantageous to measure the soil heat flux G. The evapotranspiration rate is then given by

$$E_T = \frac{Q_n - G - M}{H_v(1 + R)} \tag{5-22}$$

where H_v is the latent heat of vaporization and M is the rate of heat storage in the layer above the soil surface and below the level where Q_n is observed (significant only for a forest canopy). The introduction of H_v yields E_T in units of depth when Q_n, G, and M are expressed in energy units. In applying Bowen's ratio [Eq. (5-4)] to a lake, temperature and vapor pressure of the surface are used. Measuring the vapor pressure of a vegetated surface presents quite another problem, and it becomes necessary to measure the temperature and vapor-pressure gradients between two levels above the surface [79].

The Thornthwaite-Holzman turbulent-transport equation has been used to estimate evapotranspiration [80, 81], although there is some question whether it has been adequately tested. Instrumental requirements are not easily satisfied under field conditions, since computed evapotranspiration is proportional to the differences in measured wind and vapor pressure at two levels near the surface under investigation.

The energy-budget and turbulent-transport techniques are also suited to the determination of potential evapotranspiration, the only added requirement being that the area under observation have sufficient water at all times. Since extreme care is required in the application of these techniques, their use has been largely limited to experimental purposes.

5-13 Lysimeter Determination of Evapotranspiration

Many observations of evapotranspiration are made in soil containers [82–87], variously known as *tanks, evapotranspirometers,* and *lysimeters.* The first two terms customarily refer to containers with sealed bottoms, while there has been an attempt to restrict the word lysimeter to containers with pervious bottoms or with a mechanism for maintaining negative pressure at the bottom. Evapotranspiration is computed by maintaining a water budget for the container.

Like evaporation pans, small evapotranspirometers provide only an index to potential evapotranspiration. Accordingly, instrumental and operational standardization is of extreme importance. In summarizing the results of worldwide observations, Mather [82] states:

> The evapotranspirometer, when properly operated, i.e., when watered sufficiently so that there is no moisture deficiency and no appreciable moisture surplus in the soil of the tank, and when exposed homogeneously within a protective buffer area of the proper size to eliminate the effect of moisture advection, is an instrument which should give reasonably reliable values of potential evapotranspiration. Great care must be taken in the operation of the instrument, and standardized soil, vegetation, cultivation, and watering practices must be maintained on the tanks in order to insure comparable results from one installation to another.

Reliable tank observations of actual evapotranspiration (when appreciably less than potential) are seldom attained, since it is virtually impossible to maintain comparable soil moisture and vegetal cover in and adjacent to the tank under such conditions. Experimental results indicate [88] that reliable measurements of evapotranspiration can be made with large lysimeters 5 m (15 ft) or more in diameter if provision is made for applying a suction force at the base which is comparable with that in the natural soil profile. It is necessary that root development not be inhibited by limited dimensions of the lysimeter, and cover characteristics (density, height, and vigor) must be the same over and adjacent to the lysimeter.

5-14 Estimating Potential Evapotranspiration from Meteorological Data

Several empirical techniques have been developed for estimating potential evapotranspiration from readily available climatological data and latitude (duration of possible sunshine). Thornthwaite [89] has derived a somewhat involved procedure using only temperature and duration of possible sunshine. Blaney's approach [69] involves the same two factors but was designed primarily to transpose observed consumptive-use data for irrigated areas to other localities on the basis of derived coefficients (see Sec. 5-16). Using average yearly data, Lowry and Johnson [90] found high correlations between consumptive use and accumulated degree-days during the growing season. Procedures which rely on temperature as the sole index to heat supply at a particular latitude and which neglect cloudiness, humidity, wind, and other factors are subject to rather large errors under adverse circumstances.

Potential evapotranspiration and evaporation from a thin free-water surface are affected by the same meteorological factors: radiation, humidity, wind, and temperature. Even though the differences in surface roughness, albedo, and possibly other factors are involved, free-water evaporation should be a better index to potential evapotranspiration than air temperature is. Thus it would seem that Eq. (5-16) or (5-18) constitutes the best approach to computing potential evapotranspiration, even though there remains a question concerning the need for a reduction factor.

Conflicting experimental results for reduction factors appear in the literature, and the definition of potential evapotranspiration becomes involved (see Sec. 5-13). Using observations from one of the largest, most reliable lysimeters in the United States, Pruitt and Lourence [91] found the average annual potential evapotranspiration from fescue grass to be 68 in (173 cm). The evaporation from a 20-m² sunken evaporation pan for the same 3-year period was actually 2 percent less. These and other results [92, 93] lead to the conclusion that the annual reduction factor is much nearer unity than the 0.75 found by Penman [14], and pending further research an assumed value of unity is perhaps satisfactory when considering catchment areas of varied vegetal cover. Any of the techniques for estimating free-water evaporation may therefore be applied, including the pan-coefficient approach.

Priestly and Taylor [73] showed that for several well-watered experimental sites (and in the absence of advection), the potential rate of evapotranspiration $E_{\hat{T}}$ was approximately equal to the equilibrium rate [32] multiplied by 1.26:

$$E_{\hat{T}} = 1.26E_{\tilde{T}} = 1.26 \, \frac{[\Delta/(\Delta + \gamma)](Q_n - G - M)}{H_v} \tag{5-23}$$

It will be seen by comparing Eqs. (5-16) and (5-23) that the equilibrium evapotranspiration, used by many micrometeorologists, is the evaporation rate when $E_a = 0$ (i.e., when air and dewpoint temperatures are equal). There has been support for the conclusion of Priestly and Taylor [94–95], but other studies [96–99] have found the ratio to be both higher and lower than 1.26. If the ratio is constant under potential conditions, it follows that the Bowen ratio is a function of temperature only. Priestly and Taylor stress that the value 1.26 is probably satisfactory as a basis for apportioning the net radiation over substantial saturated land areas (their intent), but not over water.

5-15 Estimating Actual Evapotranspiration from Potential

The effect of moisture deficiency on the relation between actual and potential evapotranspiration has long been a subject for debate [100–106]. Some investigators contend that evapotranspiration from a homogeneous plot continues at an undiminished rate until moisture content throughout the root zone is reduced to near the wilting point; others cite experimental results to show that the rate (relative to potential) is approximately proportional to the remaining available water; a third point of view is that the rate is a complex function of available water (only) but limited to the potential rate. Regardless of the functional relationship for a homogeneous plot, the rate of depletion from an initially saturated heterogeneous catchment does soon decrease with time (for constant potential evapotranspiration) because of variations in root-zone capacity and other relevant factors.

The assumption that the ratio of actual to potential evapotranspiration is proportional to the remaining available water would perhaps be satisfactory for basin accounting if each storm saturated the soil. Unfortunately, this simple function cannot adequately provide for the increased evapotranspiration immediately following a moderate storm on relatively dry soil. This difficulty can be overcome by arbitrary separation of moisture storage into two categories [107]. In this approach it is visualized that *upper-zone* moisture is always depleted at the potential rate and that any deficiency in this zone must be satisfied before rainfall begins to recharge the *lower zone*. Depletion from the lower zone occurs only when there is no remaining moisture in the upper zone, in which case the rate of evapotranspiration is assumed to be proportional to the available moisture in the lower zone. By applying this simple model to observations of precipitation and total runoff (Sec. 7-5) daily values of evapotranspiration can be derived by accounting procedures. It should be pointed out that the more complex models used for simulation of streamflow also provide derived values of evapotranspiration (Chap. 11).

5-16 Irrigation Water Requirements

A key element in the design of any irrigation system is the determination of the total water requirement—the consumptive use less the expected contribution from precipitation plus any losses associated with delivery and application of the water. In some cases it is also necessary to provide for excess irrigation water to leach accumulating salts from the soil, since removal by the evapotranspiration process is minimal.

The estimation of reduced irrigation requirements resulting from precipitation on the cropped area is beyond the scope of this chapter, but it should be noted that only a portion of the precipitation is effective in this respect [108, 109]. Precipitation is effective only when it remains in the soil and is available to plants, or otherwise offsets a requirement for leaching water. Local runoff is not available to crops, and thus the effectiveness of rainfall is dependent on its fortuitous occurrence following irrigation. Simulation (Chap. 12) offers a possible approach to estimating the effectiveness of precipitation.

The most widely used techniques for estimating consumptive use rely largely on the transposition of data derived from tanks, field plots, or irrigated valleys, taking into account differences in temperature regime [69, 90, 109, 110]. It is believed that improved results can be expected by modifying a more reliable estimate of potential evapotranspiration, which is the accepted upper limit. Actual consumption will be somewhat less, depending on the crop and its stage of development.

5-17 Controlling Evapotranspiration

Following reported success in reducing evaporation from water surfaces by the monomolecular film technique (Sec. 5-8), experiments were undertaken to reduce transpiration from plants by mixing fatty alcohols into the soil [111]. Some positive results have been reported, while other experiments have indicated no significant effects [112] or even increased transpiration. Detailed analysis of many independent experiments leads to the conclusions [113] that concentrations of fatty alcohols sufficient to reduce transpiration also reduce plant growth and that such materials are not suitable as antitranspirants. Spraying a cedar-hemlock catchment in Idaho with an aqueous emulsion of silicone oil is reported to have increased summer runoff by 12 percent [114].

A major and continuing research effort has been under way since the early part of this century to determine and predict the hydrologic effects of land-use changes. There is little doubt that land-use changes can have an appreciable effect on annual evapotranspiration and on its seasonal distribution as well [115–117]. Differences in albedo, aerodynamic roughness, and plant behavior have some effect, but primary factors are those pertaining to the availability of water supply and extent to which the area is covered by freely growing vegetation.

The availability of water supply is largely determined by the extent of the root zone and the climatic regime. If extended periods without rainfall during the growing season are characteristic of an area, then a deep-rooted forest cover will be transpiring freely much of the time when the supply to shallow-rooted plants has been exhausted. In areas where climatic conditions are such that shallow-rooted plants have adequate supply most of the time, evapotranspiration is less affected by the extent of the root zone. Changes in land use reflecting differences in duration of the growing season also have an effect.

Attempting to reduce evapotranspiration through changing land use should be undertaken only after thorough study of possible related side effects. Clear-cut logging will decrease evapotranspiration and increase streamflow but may result in unacceptable rates of erosion [118] and increased peak flows [119].

5-18 Equations for Evaporation Computations

Interest in evaporation and evapotranspiration spans a number of fields, including hydrology, meteorology, several branches of engineering, and agriculture. As might be expected under these circumstances, a variety of units are in continuing use. The psychrometric equation relating temperature and saturation vapor pressure, which is so basic to evaporation processes, and many other relationships used are empirical and/or graphical in nature. This section is included to accommodate the different units and to facilitate computer solution of certain graphical procedures.

Differentiating Eq. (2-5) and neglecting a term which is inconsequential when the temperature is above $-25°C$:

$$\frac{de_s}{dT_a} = \Delta = (0.00815T_a + 0.8912)^7 \qquad T_a \geq -25°C \qquad (5\text{-}24a)$$

where T_a is the air temperature in degrees Celsius and e_s is the saturation vapor pressure in millibars. If T_a is expressed in degrees Fahrenheit and e_s in inches of mercury,

$$\Delta = (0.00252T_a + 0.4149)^7 \qquad T_a \geq -13°F \qquad (5\text{-}24b)$$

From Eqs. (5-4) and (5-17), it is seen that $\gamma = 0.00066\,p$ mb/°C and $0.000367p$ in Hg/°F. At sea level ($p = 1013$ mb, or 29.92 in Hg, standard atmosphere), $\gamma = 0.66$ mb/°C, or 0.011 in Hg/°F. Noting also that $\Delta/(\Delta + \gamma)$ is equivalent to $(1 + \gamma/\Delta)^{-1}$,

$$\frac{\Delta}{\Delta + \gamma} = \left[1 + \frac{0.66}{(0.00815T_a + 0.8912)^7}\right]^{-1} \qquad (5\text{-}25a)$$

where T_a is in degrees Celsius, and

$$\frac{\Delta}{\Delta + \gamma} = \left[1 + \frac{0.011}{(0.00252T_a + 0.4149)^7}\right]^{-1} \qquad (5\text{-}25b)$$

where T_a is in degrees Fahrenheit. Equations (5-25a) and (5-25b) are sufficiently accurate for the purpose throughout the range of natural weather conditions.

The other dimensionless ratio in the Penman equation can be computed from

$$\frac{\gamma}{\Delta + \gamma} = 1 - \frac{\Delta}{\Delta + \gamma} \qquad (5\text{-}26)$$

The increase of γ with atmospheric pressure is sufficient to result in appreciable variation of the two ratios with elevation, but sea-level values are customarily applied without adjustment. Since the ratios define relative weights for two terms which are usually of the same order of magnitude, the resulting error is less than might be expected.

The effective net radiation as used in Figs. 5-1 and 5-4 is a derived variable obtained through the graphical correlation of Class A pan data. It can be estimated from

$$Q_n = 7.14 \times 10^{-3}Q_s + 5.26 \times 10^{-6}Q_s(T_a + 17.8)^{1.87} + 3.94 \times 10^{-6}Q_s^2$$
$$- 2.39 \times 10^{-9}Q_s^2(T_a - 7.2)^2 - 1.02 \quad (5\text{-}27a)$$

where Q_n is in equivalent millimeters of evaporation per day, Q_s is the daily solar radiation in calories per square centimeter, and T_a is in degrees Celsius, or from

$$Q_n = 2.81 \times 10^{-4}Q_s + 6.90 \times 10^{-8}Q_s(T_a)^{1.87} + 1.55 \times 10^{-7}Q_s^2$$
$$- 3.14 \times 10^{-11}Q_s^2(T_a - 45)^2 - 0.040 \quad (5\text{-}27b)$$

where Q_n is in equivalent inches of evaporation per day, the units of Q_s are unchanged, and T_a is in degrees Fahrenheit. If Q_s is expressed in megajoules per square meter-day, then the first four constants in Eq. (5-27a) become 0.171, 1.26×10^{-4}, 2.25×10^{-3}, and 1.36×10^{-6}, respectively.

Based on Eq. (2-5), and dropping an inconsequential term, the vapor-pressure difference can be computed from [120]

$$e_s - e_a = 33.86[(0.00738T_a + 0.8072)^8 - (0.00738T_d + 0.8072)^8]$$
$$T_d \geq -27°C \quad (5\text{-}28a)$$

where vapor pressures are in millibars and the dewpoint T_d and air temperature are in degrees Celsius. With vapor pressures in inches of mercury and temperatures in Fahrenheit, Eq. (5-28a) becomes

$$e_s - e_a = (0.0041T_a + 0.676)^8 - (0.0041T_d + 0.676)^8 \quad T_d \geq 16°F \quad (5\text{-}28b)$$

If vapor pressures are expressed in kilopascals, the factor preceding the brackets in Eq. (5-28a) becomes 3.386.

The computed Class A pan evaporation assuming that air and water temperatures are equal, E_a in Fig. 5-4, is given by

$$E_a = (e_s - e_a)^{0.88}(0.42 + 0.0029v_p) \quad (5\text{-}29a)$$

where E_a is in millimeters per day, $e_s - e_a$ is determined by Eq. (5-28a), and v_p is the wind movement 150 mm above the pan rim in kilometers per day. With E_a in inches per day, v_p in miles per day, and the vapor-pressure difference from Eq. (5-28b),

$$E_a = (e_s - e_a)^{0.88}(0.37 + 0.0041v_p) \quad (5\text{-}29b)$$

Daily lake evaporation (Fig. 5-1) can be computed from

$$E = 0.7\left[\frac{\Delta}{\Delta + \gamma} Q_n + \frac{\gamma}{\Delta + \gamma} E_a\right] \tag{5-30}$$

where the units of E are determined by the units of Q_n and E_a as derived from the above equations.

The following equation for α, the portion of net advected energy contributing to evaporation, can be derived from Eqs. (5-19) and (5-24a) by assuming a suitable wind function [26]:

$$\alpha = \left[1 + \frac{0.00066p + (T_0 + 273)^3 \times 10^{-8}/(0.177 + 0.00143v_4)}{(0.00815T_0 + 0.8912)^7}\right]^{-1} \tag{5-31a}$$

where the atmospheric pressure p is in millibars, T_0 is the water temperature of the lake surface in degrees Celsius, and v_4 is the 4-m wind movement (upwind of the lake) in kilometers per day. For p in inches of mercury, T_0 in degrees Fahrenheit, and v_4 in miles per day,

$$\alpha = \left[1 + \frac{0.000367p + (T_0 + 459.7)^3 \times 10^{-10}/(0.629 + 0.00817v_4)}{(0.00252T_0 + 0.4149)^7}\right]^{-1}$$

$$\tag{5-31b}$$

The derivation of an equation for α_p applicable to the Class A pan involves additional assumptions [28], and the result is not particularly suited to computer use. The following equation is an adequate approximation:

$$\alpha_p = 0.34 + 0.0117T_0 - 3.5 \times 10^{-7}(T_0 + 17.8)^3 + 0.0135(v_p)^{0.36} \tag{5-32a}$$

where T_0 is in degrees Celsius and v_p is the wind movement 150 mm above the pan rim in kilometers per day. In degrees Fahrenheit and miles per day

$$\alpha_p = 0.13 + 0.0065T_0 - 6.0 \times 10^{-8}T_0^3 + 0.016(v_p)^{0.36} \tag{5-32b}$$

The effects of atmospheric pressure are almost inconsequential—the increase in α_p is the order of 0.01 per 600 m (2000 ft) increase in elevation.

PROBLEMS

5-1 Entering the upper left-hand relation of Fig. 5-4 with evaporation, wind movement, and dewpoint in reverse of the order indicated provides an estimate of surface-water temperature in the pan (on the axis labeled "air temperature"). In this manner, compute water temperature for each of the 10 meteorological conditions enumerated in Table 5-1. Describe those conditions which result in a water temperature higher than that of the overlying air, and vice versa.

5-2 Using Fig. 5-1, compute lake evaporation (thus neglecting advection and changes in heat storage) for each set of data presented in Table 5-1. Also compute the pan coefficient for each case. Is the range of coefficient displayed here indicative of the range in annual coefficients to be expected over the United States? Why?

5-3 An indication of the gross variation of lake evaporation with elevation can be obtained from Fig. 5-1 for specified conditions if values of the respective variables are known for two elevations. Considering the data in Table 5-1 to constitute a series of sea-level observations, compute the evaporation at 3000 ft, mean sea level, using the following elevation gradients (per 1000 ft): air temperature, −3 Fahrenheit degrees; dewpoint, −1 Fahrenheit degree; wind, 10 percent; and radiation, 2 percent. If these results are compared with those of Prob. 5-2, is the elevation effect reasonably constant for these selected circumstances? Discuss.

5-4 Given the data tabulated below, compute monthly and annual lake evaporation from Fig. 5-1. Assuming that a proposed reservoir will experience this computed amount of evaporation per year (and noting that precipitation less runoff is natural evapotranspiration), what would be the net anticipated loss from the reservoir in inches? On the basis of the computed lake evaporation and that provided for a Class A pan, compute monthly, mean monthly, and annual pan coefficients. Discuss qualitatively the effects of heat storage in a deep reservoir on the derived monthly pan coefficients.

Period	Air temperature, °F	Dewpoint, °F	Wind, mi/day	Radiation, cal cm^{-2}/day	Pan evaporation, in	Precipitation, in	Runoff, in
Oct.	57	46	52	290	3.2	2.7	0.8
Nov.	46	35	59	200	1.7	2.4	1.1
Dec.	37	28	68	150	1.0	3.1	1.3
Jan.	34	25	81	150	0.8	3.4	1.6
Feb.	36	24	90	230	1.0	2.9	1.7
Mar.	44	30	92	310	2.8	3.5	1.8
Apr.	54	39	83	400	4.8	3.6	1.7
May	65	51	66	480	6.0	3.7	1.4
June	72	60	50	510	6.4	3.9	1.2
July	77	65	44	500	7.1	4.0	0.8
Aug.	75	64	41	440	5.8	4.6	1.0
Sept.	69	58	44	370	4.2	3.6	0.7
Sum	...	...	...	...	44.8	41.4	15.1
Mean	56	44	64	336			

5-5 Using data published in *Water-Supply Papers* and *Climatological Data,* plot hydrographs of mean daily flow and bar charts of daily precipitation (several years) for a selected small basin. The basin and period should be selected to include reasonably saturated conditions on a number of occasions. Compute mean daily evapotranspiration for several periods delineated by times of assumed basin saturation (Fig. 5-8). Which of the periods analyzed do you believe to be indicative of potential conditions?

5-6 Estimate the Class A pan coefficient, at sea level, for a period during which the average observed daily pan evaporation is 6.0 mm, wind movement is 200 km/day, water temperature is 20°C, and air temperature is 25°C. Use Fig. 5-5 and Eq. (5-21b), and assume that net advection and change in energy storage for the lake are insignificant.

5-7 Given the data tabulated below, compute the adjusted monthly and annual lake evaporation in millimeters using Eq. (5-20). Assume $H_v = 590$ cal/cm^3 and $p = 1000$ mb.

	J	F	M	A	M	J	J	A	S	O	N	D
T_0, °C	13.5	12.7	13.7	16.0	17.5	21.8	27.5	28.0	27.5	25.0	19.7	15.5
v_4, m/s	3.35	3.73	3.73	4.28	5.03	4.28	3.91	4.10	2.61	2.05	4.10	3.91
Q_v, cal cm^{-2}/day	−70	−80	−50	−25	15	495	165	70	−20	−120	−120	−140
Q_θ, cal cm^{-2}/day	−165	−215	70	160	135	740	320	85	−165	−295	−665	−490
E, mm/day	2.5	3.1	5.3	6.8	8.7	10.0	8.7	9.1	5.3	2.7	2.2	1.3

REFERENCES

1. Hutchinson, B. A.: A Comparison of Evaporation from Snow and Soil Surfaces, *Bull. Int. Assoc. Sci. Hydrol.,* vol. 11, pp. 34–42, March 1966.
2. Tabler, R. D.: Design of a Watershed Snow Fence System and First-Year Snow Accumulation, *Proc. West, Snow Conf., 39th Ann. Meet.,* pp. 50–55, 1971.
3. Golding, D. L.: Snowpack Evaporation during Chinooks along the East Slopes of the Rocky Mountains, "Preprints," Second American Meteorological Society Conference on Hydrometeorology (Toronto, Oct. 25–27, 1977), pp. 251–254.
4. Harbeck, G. E.: The Effect of Salinity on Evaporation, *U.S. Geol. Surv. Prof. Pap.* 272-A, 1955.
5. Turk, L. J.: Evaporation of Brine: A Field Study on the Bonneville Salt Flats, Utah, *Water Resour. Res.,* vol. 6, pp. 1209–1215, 1970.
6. Langbein, W. B., C. H. Hains, and R. C. Culler: Hydrology of Stock-Water Reservoirs in Arizona, *U.S. Geol. Surv. Circ.* 110, 1951.
7. Harbeck, G. E., and F. W. Kennon: The Water-Budget Control, in Water-Loss Investigations: Lake Hefner Studies, Technical Report, *U.S. Geol. Surv. Prof. Pap.* 269, p. 34, 1954.
8. Harbeck, G. E., and others: Utility of Selected Western Lakes and Reservoirs for Water-Loss Studies, *U.S. Geol. Surv. Circ.* 103, 1951.
9. Anderson, E. R., L. J. Anderson, and J. J. Marciano: A Review of Evaporation Theory and Development of Instrumentation, *U.S. Navy Electron. Lab. Rep.* 159, pp. 41–42, 1950.
10. American Society for Testing and Materials: "Metric Practice Guide," ASTM E 380-72e, June 1972.
11. Bowen, I. S.: The Ratio of Heat Losses by Conduction and by Evaporation from Any Water Surface, *Phys. Rev.,* vol. 27, pp. 779–787, 1926.
12. See [9], pp. 43–48.
13. See [9], p. 45.
14. Penman, H. L.: Natural Evaporation from Open Water, Bare Soil, and Grass, *Proc. R. Soc. London, ser.* A, vol. 193, pp. 120–145, April 1948.
15. Anderson, E. A., and D. R. Baker: Estimating Incident Terrestrial Radiation under All Atmospheric Conditions, *Water Resour. Res.,* vol. 3, pp. 975–988, 1967.
16. Harbeck, G. E.: Cummings Radiation Integrator, in Water-Loss Investigations: Lake Hefner Studies, Technical Report, *U.S. Geol. Surv. Prof. Pap.* 269, pp. 120–126, 1954.
17. Peck, E. L., and R. Farnsworth: A Dual-Purpose Evaporimeter, U.S. National Weather Service.
18. See [9], pp. 3–37.
19. Harbeck, G. E.: Water-Loss Investigations: Lake Mead Studies, *U.S. Geol. Surv. Prof. Pap.* 298, pp. 29–37, 1958.
20. Sverdrup, H. U.: On the Evaporation from the Oceans, *J. Marine Res.,* vol. 1, no. 1, pp. 3–14, 1937–1938.
21. Sutton, O. G.: Wind Structure and Evaporation in a Turbulent Atmosphere, *Proc. R. Soc. London, ser.* A, vol. 146, no. 858, pp. 701–722, October 1934.
22. Linsley, R. K., M. A. Kohler, and J. L. H. Paulhus: "Applied Hydrology," pp. 165–169, McGraw-Hill, New York, 1949.
23. Phillips, D. W.: Evaluation of Evaporation from Lake Ontario during IFYGL by a Modified Mass Transfer Equation, *Water Resour. Res.,* vol. 14, no. 2, pp. 196–205, April 1978.
24. Dalton, J.: Experimental Essays on the Constitution of Mixed Gases; on the Force of Steam or Vapor from Waters and Other Liquids, Both in a Torricellian Vacuum and in Air; on Evaporation; and on the Expansion of Gases by Heat, *Mem. Proc. Manch. Lit. Phil. Soc.,* vol. 5, pp. 535–602, 1802.
25. Webb, E. K.: Evaporation from Lake Eucumbene, *Div. Meteorol. Phys. Tech. Pap.* 10, C.S.I.R.O., Australia, 1960.
26. Derecki, J.: Multiple Estimates of Lake Erie Evaporation, *J. Great Lakes Res.,* vol. 2, no. 1, pp. 124–129, 1976.

27. See [9], p. 10.

28. Lettau, H., and F. Dorffel: Der Wasserdampfübergang von einer nassen Platte an strömende Luft, *Ann. Hydrogr. Marit. Meteorol.,* vol. 64, pp. 342, 504, 1936.

29. Harbeck, G. E.: A Practical Field Technique for Measuring Reservoir Evaporation Utilizing Mass-Transfer Theory, *U.S. Geol. Surv. Prof. Pap.* 272-E, pp. 101–105, 1962.

30. See [14], pp. 125–126.

31. Ferguson, J.: The Rate of Natural Evaporation from Small Ponds, *Aust. J. Sci. Res.,* vol. 5. pp. 315–330, 1952.

32. Slatyer, R. O., and I. C. McIlroy: "Practical Microclimatology," UNESCO, Paris, 1961.

33. Kohler, M. A., and L. H. Parmele: Generalized Estimates of Free-Water Evaporation, *Water Resour. Res.,* vol. 3, pp. 997–1005, 1967.

34. Kohler, M. A., T. J. Nordenson, and D. R. Baker: Evaporation Maps for the United States, *U.S. Weather Bur. Tech. Pap.* 37, 1959.

35. Kohler, M. A., T. J. Nordenson, and W. E. Fox: Evaporation from Pans and Lakes, *U.S. Weather Bur. Res. Pap.* 38, 1955.

36. See [33], p. 999.

37. See [33], p. 1001.

38. Nordenson, T. J., and D. R. Baker: Comparative Evaluation of Evaporation Instruments, *J. Geophys. Res.,* vol. 67, pp. 671–679, 1962.

39. Young, A. A.: Some Recent Evaporation Investigations, *Trans. Am. Geophys. Union,* vol. 28, pp. 279–284, April 1947.

40. U.S. National Weather Service Substation Observations, *Obs. Handb.,* no. 2, pp. 38–55, 1970; rev. December 1972.

41. World Meteorological Organization, Measurement and Estimation of Evaporation and Evapotranspiration, *WMO* no. 201, *Tech. Note* 83, 1966.

42. Instructions for Hydrometeorological Stations and Posts, no. 7, pt. II, Hydrometeorological Publishing House, Leningrad, 1963.

43. Hoy, R. D., and S. K. Stephens: Field Study of Lake Evaporation—Analysis of Field Data from Phase 2 Storages and Summary of Phase 1 and Phase 2, *Australian Water Resour. Council Tech. Pap.* 41, 1979.

44. See [22], pp. 168–169.

45. Hamon, R. W., L. L. Weiss, and W. T. Wilson: Insolation as an Empirical Function of Daily Sunshine Duration, *Mon. Weather Rev.,* vol. 82, pp. 141–146, June 1954.

46. Fritz, S., and T. H. MacDonald: Average Solar Radiation in the United States, *Heat. Vent.,* vol. 46, pp. 61–64, July 1949.

47. Nordenson, T. J.: Appraisal of Seasonal Variation in Pan Coefficients, *Int. Assoc. Sci. Hydrol. Publ.* 67, pp. 279–286, 1965.

48. Ferguson, H. L., A. D. J. O'Neill, and H. F. Cork: Mean Evaporation over Canada, *Water Resour. Res.,* vol. 6, pp. 1618–1633, 1970.

49. Stoenescu, V., La Répartition de l'évaporation à la surface de l'eau sur la territorie de la Roumanie (Distribution of Water Surface Evaporation in Romania), *Int. Assoc. Sci. Hydrol. Publ.* 78, pp. 316–337, 1968.

50. Němec, J.: "Engineering Hydrology," p. 207, McGraw-Hill, London, 1972.

51. Beard, L. R., and R. G. Willey: An Approach to Reservoir Temperature Analysis, *Water Resour. Res.,* vol. 6, pp. 1335–1345, 1970.

52. Goodling, J. S., and T. G. Arnold: Deep Reservoir Thermal Stratification Model, *Water Resour. Bull.,* vol. 8, no. 4, pp. 745–749, August 1972.

53. Rahman, M., and N. Marcotte: On Thermal Stratification in Large Bodies of Water, *Water Resour. Res.,* vol. 10, no. 6, pp. 1143–1147, December 1974.

54. Burt, W. V.: Verification of Water Temperature Forecasts for Deep Stratified Reservoirs, *Water Resour. Res.,* vol. 10, no. 1, pp. 93–97, February 1974.

55. Garrett, D. R., and R. D. Hoy: A Study of Monthly Lake to Pan Coefficients Using a Numerical Lake Model, *Proc. Hydrology Symposium,* Canberra, Sept. 5–6, 1978, pp. 145–149, 1978.

56. Cooley, K. R.: Energy Relationships in the Design of Floating Covers for Evaporation Reduction, *Water Resour. Res.,* vol. 6, pp. 717–727, 1970.
57. Myers, L. E., and G. W. Frasier: Evaporation Reduction with Floating Granular Materials, *J. Irrig. Drain Div. ASCE,* vol. 96, pp. 425–436, December 1970.
58. Mansfield, W. W.: Influence of Monolayers on the Natural Rate of Evaporation of Water, *Nature,* vol. 175, p. 247, 1955.
59. Gunaji, N. N.: Evaporation Investigations at Elephant Butte Reservoir in New Mexico, *Int. Assoc. Sci. Hydrol. Publ.* 78, pp. 308–325, 1968.
60. Pushkarev, V. F., and G. P. Levchenko: Use of Monomolecular Films to Reduce Evaporation from the Surface of Bodies of Water, *Tr. GGI* 142, pp. 84–107, 1967 (*Sov. Hydrol.: Sel. Pap.,* no. 3, pp. 253–272, 1967).
61. Bartholic, J. F., J. R. Runkels, and E. B. Stenmark: Effects of a Monolayer on Reservoir Temperature and Evaporation, *Water Resour. Res.,* vol. 3, pp. 173–180, 1967.
62. Hendricks, D. W., and V. E. Hansen: Mechanics of Transpiration, *J. Irrig. Drain Div. ASCE,* vol. 88, pp. 67–82, June 1962.
63. Lee, C. H.: Transpiration and Total Evaporation, chap. 8, p. 280, in O. E. Meinzer (ed.): "Hydrology," McGraw-Hill, New York, 1942.
64. Landsberg, H.: "Physical Climatology," 2d ed., p. 190, Pennsylvania State University, University Park, Pa., 1958.
65. McMillan, W. D., and R. H. Burgy: Interception Loss from Grass, *J. Geophys. Res.,* vol. 65, pp. 2389–2394, 1960.
66. Thorud, D. B.: The Effect of Applied Interception on Transpiration Rates of Potted Ponderosa Pine, *Water Resour. Res.,* vol. 3, pp. 443–450, 1967.
67. Stewart, J. B.: Evaporation from the Wet Canopy of a Pine Forest, *Water Resour. Res.,* vol. 13, no. 6, pp. 915–921, December 1977.
68. Singh, B., and G. Szeicz: The Effect of Intercepted Rainfall on the Water Balance of a Hardwood Forest, *Water Resour. Res.,* vol. 15, no. 1, pp. 131–138, February 1979.
69. Blaney, H. F.: Consumptive Use of Water, *Trans. ASCE,* vol. 117, pp. 949–973, 1952.
70. Thornthwaite, C. W.: Report of the Committee on Transpiration and Evaporation, 1943–1944, *Trans. Am. Geophys. Union,* vol. 25, pt. 5, p. 687, 1944.
71. Penman, H. L.: Estimating Evaporation, *Trans. Am. Geophys. Union,* vol. 37, pp. 43–50, 1956.
72. Kohler, M. A., and M. M. Richards: Multicapacity Basin Accounting for Predicting Runoff from Storm Precipitation, *J. Geophys. Res.,* vol. 67, pp. 5187–5197, 1962.
73. Priestly, C. H. B., and R. J. Taylor: On the Assessment of Surface Heat Flux and Evaporation Using Large Scale Parameters, *Mon. Weather Rev.,* vol. 100, pp. 81–92, February 1972.
74. Knox, C. E., and T. J. Nordenson: Average Annual Runoff and Precipitation in the New England–New York Area, *U.S. Geol. Surv. Hydrol. Invest. Atlas* HA-7, undated.
75. Fox, W. E.: Computation of Potential and Actual Evapotranspiration, U.S. Weather Bureau (processed), 1956.
76. Johnston, R. S.: Evapotranspiration from Bare, Herbaceous, and Aspen Plots: A Check on a Former Study, *Water Resour. Res.,* vol. 6, pp. 324–327, 1970.
77. Brown, H. E., and J. A. Thompson: Summer Water Use by Aspen, Spruce, and Grassland in Western Colorado, *J. For.,* vol. 63, pp. 756–760, 1965.
78. Tan, C. S., and T. A. Black: Factors Affecting the Canopy Resistance of a Douglas-Fir Forest, *Boundary Layer Meteorol.,* vol. 10, pp. 475–488, 1976.
79. See [67], p. 916.
80. Thornthwaite, C. W., and B. Holzman: Measurement of Evaporation from Land Water Surfaces, *U.S. Dept. Agric. Tech. Bull.* 817, 1942.
81. Rider, N. E.: Evaporation from an Oat Field, *Q. J. R. Meteorol. Soc.,* vol. 80, pp. 198–211, April 1954.
82. Mather, J. R. (ed.): The Measurement of Potential Evapotranspiration, *Johns Hopkins Univ. Lab. Climatol., Seabrook, N.J., Publ. Climatol.,* vol. 7, no. 1, 1954.

83. Harrold, L. L., and F. R. Dreibelbis: Agricultural Hydrology as Evaluated by Monolith Lysimeters, *U.S. Dept. Agric. Tech. Bull.* 1050, 1951.
84. Pruitt, W. O., and D. E. Angus: Large Weighing Lysimeter for Measuring Evapotranspiration, *Trans. Am. Soc. Agric. Eng.,* vol. 3, pp. 13–18, 1960.
85. Popov, O. V.: Lysimeters and Hydraulic Soil Evaporimeters, *Int. Assoc. Sci. Hydrol. Publ.* 49, pp. 26–37, 1959.
86. McIlroy, I. C., and D. E. Angus: The Aspendale Multiple Weighed Lysimeter Installation, *Div. Meteorol. Phys. Tech. Pap.* 14, C.S.I.R.O., Australia, 1963.
87. See [50], pp. 62–66.
88. Pruitt, W. O., and F. J. Lourence: Tests of Aerodynamic, Energy Budget, and Other Evaporation Equations over a Grass Surface, in "Investigations of Energy, Momentum, and Mass Transfer near the Ground," U.S. Army Electronics Command Atmospheric Laboratory, Ft. Huachuca, Ariz., pp. 37–63, 1966.
89. Thornthwaite, C. W.: An Approach toward a Rational Classification of Climate, *Geograph. Rev.,* vol. 38, pp. 55–94, 1948.
90. Lowry, R. L., and A. F. Johnson: Consumptive Use of Water for Agriculture, *Trans. ASCE,* vol. 107, p. 1252, 1942.
91. Pruitt, W. O., and F. J. Lourence: Correlation of Climatological Data with Water Requirements of Crops, *Univ. Calif. Water Sci. Eng. Pap.* 9001, Davis, Calif., June 1968.
92. Van Bavel, C. H. M.: Potential Evaporation: The Combination Concept and Its Experimental Verification, *Water Resour. Res.,* vol. 2, pp. 455–467, 1966.
93. Blaney, H. F.: Discussion of paper by H. L. Penman, Estimating Evaporation, *Trans. Am. Geophys. Union,* vol. 37, pp. 46–48, February 1956.
94. Davies, J. A., and C. D. Allen: Equilibrium, Potential, and Actual Evaporation from Cropped Surfaces in Southern Ontario, *J. Appl. Meteorol.,* vol. 12, no. 4, pp. 649–657, June 1973.
95. Stewart, R. B., and W. R. Rouse: Substantiation of the Priestly and Taylor Parameter $\alpha = 1.26$ for Potential Evaporation at High Latitudes, *J. Appl. Meteorol.,* vol. 16, pp. 649–650, June 1977.
96. Jury, W. A., and C. B. Tanner: A Modification of the Priestly and Taylor Evapotranspiration Formula, *Agron. J.,* vol. 67, pp. 840–842, 1975.
97. McNaughton, K. G., and T. A. Black: A Study of Evapotranspiration from a Douglas-Fir Forest Using the Energy Balance Approach, *Water Resour. Res.,* vol. 9, no. 6, pp. 1579–1590, December 1973.
98. Black, T. A.: Evapotranspiration from Douglas Fir Stands Exposed to Soil Water Deficits, *Water Resour. Res.,* vol. 15, no. 1, pp. 164–170, February 1979.
99. Munro, D. S.: Daytime Energy Exchange and Evaporation from a Wooded Swamp, *Water Resour. Res.,* vol. 15, no. 5, pp. 1259–1265, October 1979.
100. Veihmeyer, F. J., and A. H. Hendrickson: Does Transpiration Decrease as Soil Moisture Decreases?, *Trans. Am. Geophys. Union,* vol. 36, pp. 425–448, 1955.
101. Lemon, E. R.: Some Aspects of the Relationship of Soil, Plant, and Meteorological Factors to Evapotranspiration, *Soil Sci. Soc. Am. Proc.,* vol. 21, pp. 464–468, 1957.
102. Pruitt, W. O.: Correlation of Climatological Data with Water Requirements of Crops, *Univ. Calif. (Davis) Dept. Irrig. 1959–1960 Ann. Rep.,* p. 91, September 1960.
103. Gardner, W. R., and D. I. Hillel: The Relation of External Evaporative Conditions to the Drying of Soils, *J. Geophys. Res.,* vol. 67, pp. 4319–4325, October 1962.
104. Philip, J. R.: Evaporation, and Moisture and Heat Fields in the Soil, *J. Meteorol.,* vol. 14, pp. 354–366, 1957.
105. Marlatt, W. E., A. V. Havens, N. A. Willits, and G. D. Brill: A Comparison of Computed and Measured Soil Moisture under Snap Beans, *J. Geophys. Res.,* vol. 66, pp. 535–541, February, 1961.
106. Molz, F. J., I. Remson, A. A. Fungaroli, and R. L. Drake: Soil Moisture Availability for Transpiration, *Water Resour. Res.,* vol. 4, pp. 1161–1170, December 1968.
107. Kohler, M. A.: Meteorological Aspects of Evaporation Phenomena, *Gen. Assemb. Int. Assoc. Sci. Hydrol., Toronto,* vol. 3, pp. 421–436, 1957.

108. Linsley, R. K., and J. B. Franzini: "Water Resources Engineering," 3d ed., pp. 376–381, McGraw-Hill, New York, 1979.
109. Jensen, M. E. (ed.): "Consumptive Use of Water and Irrigation Water Requirements," American Society of Civil Engineers, New York, 1974.
110. U.S. Bureau of Reclamation Manual, vol. IV, Water Studies, Chap. 4.1, 1951.
111. Roberts, W. J.: Reduction of Transpiration, *J. Geophys. Res.,* vol. 66, pp. 3309–3312, 1961.
112. Anderson, H. W., A. J. West, R. R. Zeimer, and F. R. Adams: Evaporative Loss from Soil, Native Vegetation, and Snow as Affected by Hexadecanol, *Int. Assoc. Sci. Hydrol. Publ. 62,* pp. 7–12, 1963.
113. Gale, J., E. B. Roberts, and R. M. Hagan: High Alcohols as Antitranspirants, *Water Resour. Res.,* vol. 3, no. 2, pp. 437–441, 1967.
114. Belt, G. H., J. G. King, and H. F. Haupt: Augmenting Summer Streamflow by Use of a Silicone Antitranspirant, *Water Resour. Res.,* vol. 13, no. 2, pp. 267–272, April 1977.
115. Hibbert, A. R.: Forest Treatment Effects on Water Yields, in W. E. Sopper and H. W. Lull (eds.): "International Symposium on Forest Hydrology," pp. 527–544, Pergamon, New York, 1967.
116. Pereira, H. C.: The Influence of Man on the Hydrological Cycle, in "World Water Balance," *Proc. Reading Symp., July 1970,* pp. 553–569, IASH-UNESCO-WMO, Gentbrugge, Paris, Geneva, 1972.
117. Rich, L. T., and G. J. Gottfried: Water Yields Resulting from Treatments on the Workman Creek Experimental Watersheds in Central Arizona, *Water Resour. Res.,* vol. 12, no. 5, pp. 1053–1060, 1976.
118. Dryness, C. T.: Erosion Potential of Forest Watersheds, in W. E. Sopper and H. W. Hull (eds.): "International Symposium on Forest Hydrology," pp. 599–610, Pergamon, New York, 1967.
119. Nakano, H.: Effects of Changes of Forest Conditions on Water Yield, Peak Flow, and Direct Runoff of Small Watersheds in Japan, in W. E. Sopper and H. W. Lull (eds.): "International Symposium on Forest Hydrology," pp. 551–564, Pergamon, New York, 1967.
120. Lamoreux, W. W.: Modern Evaporation Formulae Adapted to Computer Use, *Mon. Weather Rev.,* vol. 90, pp. 26–28, January 1962.

BIBLIOGRAPHY

Australian Water Resources Council: Evaporation from Water Storages, *Dept. Natl. Dev., Canberra, Hydrol. Ser.* 4, 1970.
Hare, F. K., and J. E. Hay: Anomalies in the Large-Scale Annual Water Balance over Northern North America, *Canadian Geographer,* vol. XV, no. 2, pp. 79–94, 1971.
Hounam, C. E.: Problems of Evaporation Assessment in the Water Balance, *Rep. WMO/IHD Proj.,* no. 13, World Meteorological Organization, Geneva, 1971.
Jones, F. E.: Evaporation of Water: A Review of Pertinent Laboratory Research, *Natl. Bur. Stand. Rep.* 10 235, May 1970.
Konstantinov, A. R.: "Evaporation in Nature," Israel Program for Scientific Translations, Jerusalem, 1966; available from U.S. Dept. of Commerce National Technical Information Center, Springfield, Va.
McKay, G. A., and M. K. Thomas: Mapping of Climatological Elements, *Canadian Cartographer,* vol. 8, no. 1, pp. 27–40, 1971.
Nordenson, T. J.: Preparation of Coordinated Precipitation, Runoff and Evaporation Maps, *Rep. WMO/IHD Proj.* no. 6, World Meteorological Organization, Geneva, 1968.
Peck, E. L., and A. Hely: Precipitation, Runoff and Water Loss in the Lower Colorado River–Salton Sea Area, *U.S. Geol. Surv. Prof. Pap.* 486-B, 1964.
Priestly, C. H. B.: "Turbulent Transfer in the Lower Atmosphere," The University of Chicago Press, Chicago, 1959.

Rijtema, P. E.: "Analysis of Actual Evapotranspiration," Wageningen, Netherlands, 1965.
Rodda, J. C. (ed): "Facets of Hydrology," Wiley, London, 1976.
Thornthwaite, C. W., and J. R. Mather: The Water Balance, *Drexel Inst. Lab. Climatol. Centerton, N.J., Publ. Climatol.,* vol. 8, no. 1, 1955.
World Meteorological Organization: "Guide to Hydrological Practices," 3d ed., WMO no. 168, Geneva, 1974.

UNITED STATES DATA SOURCES

Observations of pan evaporation and other data for the station are published in monthly and annual issues of *Climatological Data* (NOAA, Environmental Data Service). The data are also available on magnetic tape from the National Climatic Center, Federal Building, Asheville, N.C. 28801.

SUBSURFACE WATER

About one-fourth of the total fresh water withdrawn in the United States (exclusive of hydroelectric power generation) is drawn from groundwater. Subsurface water is relatively free of pollution and is especially useful for domestic use in small towns and isolated farms. In arid regions groundwater is often the only reliable source of water for irrigation. Groundwater temperatures are usually relatively low, and large quantities are used for cooling in warm regions.

Aside from its direct use, groundwater is also an important phase of the hydrologic cycle. Most of the flow of perennial streams originates from subsurface water, while a large part of the flow of ephemeral streams may percolate beneath the surface. Thus, no treatment of surface water hydrology can ignore subsurface processes. Since the occurrence and movement of subsurface water are intimately related to geologic structure, an understanding of geologic controls is a prerequisite to a comprehension of groundwater hydrology. This chapter stresses interrelations between surface and subsurface water and presumes only an elementary knowledge of geology.

6-1 Occurrence of Subsurface Water

Figure 6-1 is a schematic cross section of the upper portion of the earth's crust with an idealized column showing a suggested classification of subsurface water [1]. The two major subsurface zones are divided by an irregular surface called the *water table*. The water table is the locus of points (in unconfined material) where hydrostatic pressure equals atmospheric pressure. Above the water table, in the *vadose zone*, soil pores may contain either air or water; hence it is

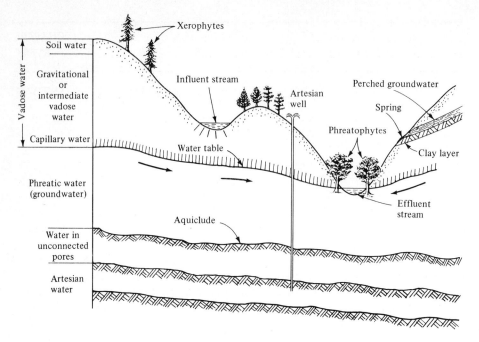

Figure 6-1 Schematic cross section showing the occurrence of groundwater.

sometimes called *zone of aeration*. In the *phreatic zone*, below the water table, interstices are filled with water; sometimes this is called the *zone of saturation*. The phreatic zone may extend to considerable depth, but as depth increases, the weight of overburden tends to close pore spaces and relatively little water is found at depths greater than 3 km (10,000 ft).

Local saturated zones sometimes exist as *perched groundwater* above an impervious layer of limited extent. Sometimes groundwater is overlain by an impervious stratum to form *confined*, or *artesian*, water. Confined groundwater is usually under pressure because of the weight of the overburden and the hydrostatic head. If a well penetrates the confining layer, water will rise to the *piezometric level*, the artesian equivalent of the water table. If the piezometric level is above ground level, the well discharges as a *flowing well*.

MOISTURE IN THE VADOSE ZONE

Three moisture regions can be identified within the vadose zone. In the region penetrated by roots of vegetation, ranging to 30 ft (10 m) below the soil surface, is the *soil water*, which fluctuates in amount as vegetation removes moisture between rains. Above the water table, moisture is raised by capillarity into the *capillary fringe*, which may have a vertical extent of a few inches to several feet depending on the pore sizes of the material. If the water table is close to the ground surface, the capillary fringe and the soil-moisture region may overlap, but where the water table is deep, an *intermediate* region exists where moisture levels remain constant at the field capacity of the soil and rock of the region.

6-2 Soil-Water Relationships

Soil moisture may be present as *gravity water* in transit in the larger pore spaces, as *capillary water* in the smaller pores (Fig. 6-2), as *hygroscopic moisture* adhering in a thin film to soil grains, and as water vapor. Gravity water is in a transient state. After a rain, water may move downward in the larger pores, but this water must either be dispersed into capillary pores or pass through the vadose zone to the groundwater or to a stream channel. Hygroscopic water, on the other hand, is held by molecular attraction and is not normally removed from the soil under usual climatic conditions. The important variable element of soil moisture, therefore, is capillary water.

If a soil-filled tube is placed with its open lower end in a container of water, some water will move up into the soil. The rate of upward movement becomes progressively smaller with time, until it eventually approaches zero. Measurement of the moisture content of the soil at various levels will show that moisture in the soil column decreases with height above the water surface (Fig. 6-3). If a sample of soil is saturated with water and then subjected to successively greater negative pressures and the moisture content is noted after each change in pressure, a similar curve (Fig. 6-4) results.

Buckingham [2] first proposed characterizing soil-moisture phenomena on the basis of energy relationships. He introduced the concept of capillary potential to describe the attraction of soil for water. With a free-water surface taken as reference, *capillary potential* is defined as the work required to move a unit mass of water from the reference plane to any point in the soil column. Thus, capillary potential is the potential energy per unit mass of water. By definition, capillary potential is negative since water will move upward by capillarity without external work. Capillary potential ψ is related to the acceleration of gravity g and height above datum y (negative) by the equation

$$\psi = gy \tag{6-1}$$

Curves like Fig. 6-3 provide a basis for relating capillary potential and moisture content for a particular soil.

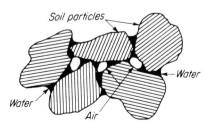

Figure 6-2 The occurrence of capillary moisture in soil.

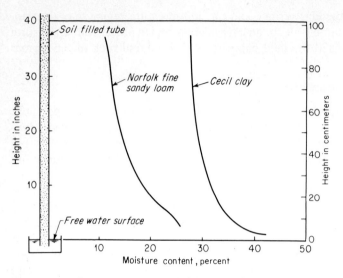

Figure 6-3 Moisture content versus height curves for two soils. (*From E. Buckingham, Studies on the Movement of Soil Moisture, U.S. Dept. Agric. Bur. Soils Bull* **38**, *1907.*)

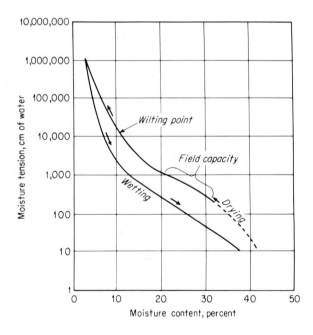

Figure 6-4 Moisture-tension curves for a typical soil. (*From R. K. Schofield, The pF of Water in Soil, Trans. 3d Int. Cong. Soil Sci., vol.* 2, *pp. 37–48, 1935.*)

6-3 Equilibrium Points

Visualizing several states of water in soil, early soil scientists tried to define limits of these states by *equilibrium points*. Figures 6-3 and 6-4 indicate that no clear-cut boundaries exist, but equilibrium points are convenient for discussing soil moisture. The two of greatest interest are field capacity and wilting point. *Field capacity* is defined as the moisture content of soil after gravity drainage is complete. Colman [3] has shown that field capacity is essentially the water retained in soil at a tension of $\frac{1}{3}$ atm. Veihmeyer and Hendrickson [4] found that the *moisture equivalent*, water retained in a soil sample, $\frac{3}{8}$ in (9.5 mm) deep after being centrifuged for 30 min at a speed equivalent to a force of $1000g$, was also nearly the field capacity of fine-grained soils.

The *wilting point* represents the soil-moisture level when plants cannot extract water from soil. It is the moisture held at a tension equivalent to the osmotic pressure in the plant roots. For years the wilting point was determined by growing sunflower seedlings in a soil sample, but it is now commonly assumed to be equivalent to the moisture content at a tension of 15 atm. The difference between the moisture content at field capacity and at wilting point is called *available moisture*. It represents the useful storage capacity of the soil (Secs. 5-9 and 5-15) and the maximum water available to plants. Typical values of moisture content at field capacity and wilting point and available moisture are given in Table 6-1.

6-4 Measurement of Soil Moisture

The standard determination of soil moisture is the loss in weight when a soil sample is oven-dried. A *tensiometer* (Fig. 6-5) consists of a porous ceramic cup which is inserted in the soil, filled with water, and connected to a manometer. The tensiometer remains in equilibrium with the soil. If the soil moisture falls below saturation, water is drawn from the cup and a negative pressure is indi-

Table 6-1 Typical moisture values for various soil types

Soil type	Percent dry weight of soil			Specific weight lb/ft³ dry	Density, kg/m³ dry
	Field capacity	Wilting point	Available water		
Sand	5	2	3	95	1520
Sandy loam	12	5	7	90	1440
Loam	19	10	9	85	1360
Silt loam	22	13	9	80	1280
Clay loam	24	15	9	80	1280
Clay	36	20	16	75	1200
Peat	140	75	65	25	400

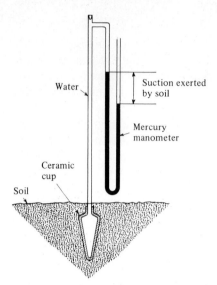

Water

Suction exerted
by soil

Mercury
manometer

Ceramic
cup

Soil

Figure 6-5 Schematic drawing of a simple tensiometer.

cated by the manometer. A tensiometer [5] can indicate soil-moisture tension from saturation to a tension of about 1 atm.

In the resistivity [6] method, a pair of electrodes embedded in a porous dielectric (plaster of paris, nylon, Fiberglas) is buried in the soil. The dielectric maintains a moisture equilibrium with the soil, and the resistance between the electrodes varies with the moisture content of the material. Resistance is measured with an alternating current bridge to avoid polarizing the element. The block must be in intimate contact with the soil. Calibration is best achieved by taking periodic soil samples from the area surrounding the installation and correlating moisture content of the samples with concurrent resistivity readings.

The neutron-scattering method of soil-moisture measurement [7] uses a source of fast neutrons which is lowered into an aluminum access tube in the soil. Fast neutrons lose energy and are converted to slow neutrons through collision with atoms of low atomic weight. The hydrogen in water is normally the only atom of low atomic weight in soil. A high count of slow neutrons indicates a high moisture content. Once calibrated, the device is easily and quickly used. However, it measures moisture content in a sphere of soil surrounding the source and is less accurate near the surface, where the sample volume is distorted. Organic matter in soil also causes errors in this method.

Aerial observation of natural gamma radiation from the soil may prove useful in determining soil-moisture variations over large areas. Satellite observations are also able to detect soil moisture levels [8].

6-5 Movement of Soil Moisture

Infiltration is the movement of water through the soil surface into the soil as distinguished from *percolation,* the movement of water through the soil. When water is first applied to the soil surface, gravity water moves down through the larger soil openings while the smaller surface pores take in water by capillarity. The downward-moving gravity water is also taken in by capillary pores. As capillary pores at the surface are filled and intake capacity is reduced, the infiltration rate decreases. In homogeneous soil, infiltration decreases gradually until the zone of aeration is saturated. Normally, the soil is stratified, and subsoil layers are often less permeable than the surface soil. In this case, the infiltration rate is eventually limited to the rate of percolation through the least pervious subsoil stratum.

Infiltration from rainfall usually occurs with very shallow depths of water on the soil surface. Quantities of water infiltrated are usually only a few inches per day and rarely sufficient to saturate a great depth of soil. When rain stops, gravity water remaining in the soil continues to move downward and at the same time is taken up in capillary-pore spaces. Usually the infiltrated water is distributed within the upper few feet of soil, with little or no contribution to groundwater unless the soil is highly permeable or the vadose zone very thin. Infiltration from rainfall is discussed in greater detail in Sec. 8-2.

For irrigation and artificial recharge of groundwater (Sec. 6-17), water is ponded to a considerable depth over limited areas for long periods of time. The aim of recharge operations is to saturate the soil down to the water table. Under these conditions the time variation of infiltration is complex, with temporary increases in rate superimposed on a gradually declining trend. Escape of soil air around the infiltration basin, bacterial action, changes in water temperature, changes in soil structure, and other factors influence these variations.

Movement of moisture in soil is governed by the moisture potential following the equation

$$q = -K \frac{\partial \Lambda}{\partial x} \tag{6-2}$$

where q is flow per unit time through unit area normal to the direction of flow, x is distance along the line of flow, K is conductivity, and Λ is potential. After gravity water has left the soil, the principal component of total potential is the capillary potential. Equation (6-2) states that flow is from a region of high potential to a region of lower potential. Conductivity increases with moisture content and decreases with pore size. Thus capillary movement decreases as soil dries and is least in fine-grained soil. Fortunately, a qualitative understanding of these phenomena is normally sufficient for engineering hydrology.

Vapor pressure in the soil is controlled by temperature. Vapor movement is from high temperature (high vapor pressure) to low temperature. Vapor transport is an important factor in moisture movement when the moisture content is lowered to the point where capillary moisture is discontinuous. Under this condition, however, moisture-content and temperature gradients are usually so small that the quantity of moisture moved is negligible. When the surface soil is frozen, the vapor-pressure gradient is upward and is accentuated by the lower vapor pressure of ice relative to water at the same temperature. Thus when frozen soil thaws, its moisture content may be greater than at the time of freezing. Conversely, during summer, vapor-pressure gradients would be downward were it not for evaporation and transpiration.

Electrical and chemical factors also contribute to total potential, but under natural conditions these factors are hydrologically unimportant.

MOISTURE IN THE PHREATIC ZONE

Within the phreatic zone all pore spaces are filled with water, and the different states of moisture, moisture tension, etc., are of little concern. Interest is centered on the amount of water present, the amount which can be removed, and the movement of this water.

6-6 Aquifers

A geologic formation which contains water and transmits it from one point to another in quantities sufficient to permit economic development is called an *aquifer*. In contrast, an *aquiclude* is a formation which contains water but cannot transmit it rapidly enough to furnish a significant supply to a well or spring. An *aquifuge* has no interconnected openings and cannot hold or transmit water. The ratio of the pore volume to the total volume of the formation is called *porosity*. The *original porosity* of a material is that which existed at the time the material was formed. *Secondary porosity* results from fractures and solution channels.

Secondary porosity cannot be measured without an impossibly large sample. Original porosity is usually measured by oven-drying an undisturbed sample and weighing it. It is then saturated with some liquid and weighed again. Finally, the saturated sample is immersed in the same liquid, and the weight of displaced liquid is noted. The weight of liquid required to saturate the sample divided by the weight of liquid displaced is the porosity as a decimal. If the material is fine-grained, the liquid may have to be forced into the sample under pressure to assure complete saturation.

High porosity does not necessarily indicate a productive aquifer, since much of the water may be retained in small pore spaces under capillary tension as the material is dewatered. The *specific yield* of an aquifer is the ratio of the water which will drain freely from the material to the total volume of the formation and is always less than the porosity. The relation between specific yield and porosity depends on the size of particles in the formation. Specific yield of a fine-grained aquifer will be small, whereas coarse-grained material will yield a greater amount of its contained water. Table 6-2 lists approximate average values of porosity and specific yield for some typical materials. Large variations from these average values must be expected. Note that although clay has a high porosity, it has a very low specific yield. Sand and gravel, which make up most of the more productive aquifers in the United States, yield about 80 percent of their total water content.

Table 6-2 Approximate average porosity, specific yield, and permeability of various materials

Material	Porosity, %	Specific yield, %	Permeability		Intrinsic permeability, D
			Meinzer units	m³ × day⁻¹ m⁻²	

Material	Porosity, %	Specific yield, %	Meinzer units	m³ × day⁻¹ m⁻²	Intrinsic permeability, D
Clay	45	3	0.01	0.0004	0.0005
Sand	35	25	1,000	41	50
Gravel	25	22	100,000	4,100	5,000
Gravel and sand	20	16	10,000	410	500
Sandstone	15	8	100	4.1	5
Dense limestone and shale	5	2	1	0.041	0.05
Quartzite, granite	1	0.5	0.01	0.0004	0.0005

6-7 Movement of Groundwater

In 1856 Darcy [9] confirmed the applicability of principles of fluid flow in capillary tubes, developed several years earlier by Hagen [10] and Poiseuille [11], to the flow of water in permeable media. *Darcy's law* is

$$v = \kappa s \tag{6-3}$$

where v is the velocity of flow, s is the slope of the hydraulic gradient, and κ is a coefficient having the units of v (feet per day or meters per day). The discharge q is the product of area A and velocity. The effective area is the gross area times the porosity p of the medium. Hence

$$q = \kappa p A s = K A s \tag{6-4}$$

The coefficient K is called the *coefficient of permeability* or the *hydraulic conductivity*. It is dependent on properties of the fluid and the medium, and can be expressed as

$$K = k \frac{w}{\mu} = C d^2 \frac{w}{\mu} \tag{6-5}$$

where k is the *intrinsic permeability* of the medium, w is the specific weight of the fluid, μ is its absolute viscosity, C is a factor involving the shape, packing, porosity, and other characteristics of the medium, and d is the average pore size of the medium.

The intrinsic permeability is expressed in *darcys,* which have the dimensions of area.† The hydraulic conductivity K has the dimensions of velocity and is stated in a variety of units by different disciplines and in different countries. In the United States for hydrologic purposes K is often given in *Meinzer units,*‡

---

† One darcy (D) = 1.062×10^{-11} ft² = 0.987×10^{-8} cm².
‡ One Meinzer unit = 0.0408 m³ day⁻¹ m⁻² with unit gradient.

the flow in gallons per day through an area of one square foot under a gradient of one foot per foot at 60°F (Table 6-2).

It is convenient to use the *transmissibility* T to represent the flow rate per day through a section 1 ft wide and the thickness of the aquifer under a unit head (slope of 1 ft/ft):

$$T = KY \tag{6-6}$$

where Y is the saturated thickness of the aquifer. With this coefficient Eq. (6-4) becomes

$$q = TBs \tag{6-7}$$

where B is the width of the aquifer.

6-8 Determination of Permeability

Laboratory measurements of permeability are made with *permeameters* (Fig. 6-6). A sample of the material is subjected to water under a known head, and the flow through the sample in a known time is measured. Such tests have limited value because of the difficulty of placing samples of unconsolidated materials in the permeameter in their natural state and the uncertainty whether a sample is truly representative of the aquifer. Flow in solution cavities or rock fractures and the effect of large boulders in gravel aquifers cannot be duplicated in a permeameter.

The earliest field techniques for determining permeability involved introducing salt into the aquifer at one well and timing its movement to a downstream well [12]. Fluorescein dye [13], detectable at a concentration of

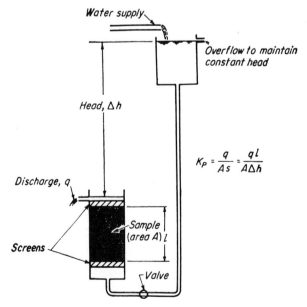

Figure 6-6 A simple upward-flow permeameter.

0.03 ppm by the unaided eye and in concentrations as low as 0.0001 ppm under ultraviolet light, has also been used as a tracer. More recently, radioactive materials have been used [14]. Tracer techniques have encountered numerous difficulties [14, 15]. Chemical reactions between the tracer elements and the formation sometimes occur. Because of diffusion, tests must be conducted over short distances in order to have detectable concentrations at the downstream well, and even then it is difficult to determine a representative time of arrival. Tracers are most useful for determining path of flow, e.g., when it is necessary to locate a source of pollution.

Today permeability is most commonly determined by pumping tests. By using the principles of well hydraulics (Secs. 6-11 and 6-12) it is possible to estimate average permeability of an aquifer for a large distance around the test well.

6-9 Sources of Groundwater

Almost all groundwater is *meteoric water* derived from precipitation. *Connate water* was present in the rock at its formation and is frequently highly saline. *Juvenile water,* formed chemically within the earth and brought to the surface in intrusive rocks, occurs in small quantities. Connate and juvenile waters are sometimes important sources of undesirable minerals in groundwater. Groundwater in the San Joaquin Valley, California, contains boron brought to the surface from great depths.

Water from precipitation reaches groundwater by infiltration and percolation (Fig. 6-7). Direct percolation is most effective in recharging groundwater

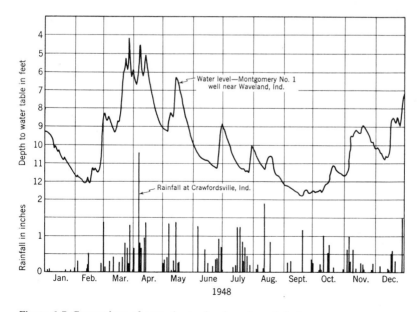

Figure 6-7 Comparison of groundwater levels and precipitation.

where the soil is highly permeable or the water table is close to the surface. Direct recharge is high through the permeable basaltic lavas of northern California, eastern Oregon, southern Idaho, and Hawaii, and in the southern Appalachian region, where thin soil cover overlies cavernous limestone.

Where annual rainfall is relatively low and the water table is hundreds of feet below the surface, little or no recharge from rain can be expected. In such areas irrigation water may provide for some recharge, but seepage from lakes and stream channels into permeable gravels is likely to be the main source of recharge. Streams contributing to groundwater are called *influent streams*. Such streams are frequently *ephemeral;* i.e., they go dry during protracted rainless periods when percolation depletes all flow. Streams are rarely influent throughout their entire length. Most of the percolation occurs in short reaches where bed materials are highly permeable. Considerable recharge often occurs from channels crossing coarse gravels in an alluvial fan.

In areas of artesian groundwater, the overlying aquiclude prevents appreciable direct recharge; the recharge area (*forebay*) may be far removed from the artesian area.

6-10 Discharge of Groundwater

Without interference by human beings, a groundwater basin fills and discharges excess water by several routes until a quasiequilibrium is reached. Streams intersecting the water table and receiving groundwater flow are called *effluent streams*. Perennial streams are generally effluent through at least a portion of their length.

Where an aquifer intersects the earth's surface, a spring or seep will form (Fig. 6-8). There may be a concentrated flow constituting the source of a small stream or merely seepage which evaporates from the ground surface. Most springs are small and of little hydrologic significance, although even a small spring may provide water for a single farmstead. First-magnitude springs [16] discharge 2.8 m³/s (100 ft³/s) and, according to Meinzer [17], there are 65 such springs in the United States: 38 in volcanic rocks of California, Oregon, and Idaho; 24 in limestone in the Ozarks, the Balcones Fault area of Texas, and Florida; and 3 sandstone springs in Montana. The Fontaine de Vaucluse in France has a discharge often exceeding 110 m³/s (4000 ft³/s). It is the world's largest spring and is in a limestone formation.

Where the water table is close to the surface, groundwater may be discharged by direct evaporation or by transpiration from the capillary fringe. Plants deriving their water from groundwater, called *phreatophytes,* often have root systems extending to depths of 12 m (40 ft) or more. This invisible evapotranspiration loss may be quite large. At a rate of 1 m/year the loss would be 10^6m³/km² per year. (Similarly 3 ft/year = 1920 acre · ft/mi² per year.)

The various channels of groundwater discharge may be viewed as spillways of the groundwater reservoir. When groundwater is high, discharge through natural spillways tends to maintain a balance between inflow and outflow.

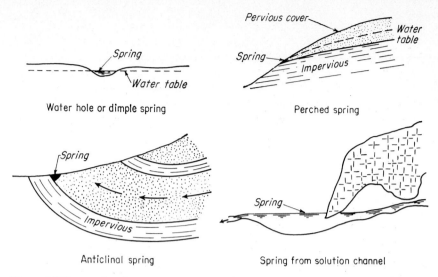

Figure 6-8 Types of springs.

During dry periods natural discharge is reduced as groundwater levels fall, and outflow may even cease. Artesian aquifers may not reflect this natural balance as rapidly as water-table aquifers, but sustained drought will decrease water levels in the recharge area and decrease discharge from the aquifer.

6-11 Equilibrium Hydraulics of Wells

Figure 6-9 shows a well in a homogeneous aquifer of infinite extent with an initially horizontal water table. For flow to occur to the well there must be a gradient toward the well. The resulting water-table form is called a *cone of depression*. If the decrease in water level at the well (*drawdown*) is small with respect to the total thickness of the aquifer, if the well completely penetrates the aquifer, and assuming equilibrium, a formula relating well discharge and aquifer characteristics can be derived.

Flow toward the well through a cylindrical surface at radius x must equal the discharge of the well, and from Darcy's law [Eq. (6-4)]

$$q = 2\pi xyK \frac{dy}{dx} \tag{6-8}$$

where $2\pi xy$ is the area of the cylinder and dy/dx is the slope of the water table. Integrating with respect to x from r_1 to r_2 and y from h_1 to h_2 yields

$$q = \frac{\pi K(h_1^2 - h_2^2)}{\ln (r_1/r_2)} \tag{6-9}$$

where h is the height of the water table above the base of the aquifer at distance r from the pumped well and ln is the logarithm to the base e. Since we have

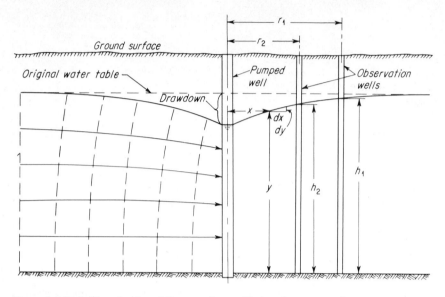

Figure 6-9 Definition sketch and flow net for equilibrium flow to a well.

assumed the drawdown Z to be small compared with the saturated thickness $(h_1 \approx h_2 \approx y)$, Eq. (6-9) can be written

$$T = \frac{q \ln (r_1/r_2)}{2\pi(Z_2 - Z_1)} \qquad (6\text{-}9a)$$

Equation (6-9) was first derived by Dupuit [18] and subsequently modified by Thiem [19]. Equations (6-9) and (6-9a) can be used to estimate T or K given q and Z, provided that the assumption of equilibrium is satisfied.

6-12 Nonequilibrium Hydraulics of Wells

During the initial period of pumping from a new well, much of the discharge is derived from storage in the portion of the aquifer unwatered as the cone of depression develops. Equilibrium analysis indicates a permeability which is too high, because only part of the discharge comes from flow through the aquifer to the well. This leads to an overestimate of the potential yield of the well.

In 1935 Theis [20] presented a formula based on the heat-flow analogy which accounts for the effect of time and the storage characteristics of the aquifer. His formula is

$$Z_r = \frac{q}{4\pi T} \int_u^{\infty} \frac{e^{-u}}{u}\, du \qquad (6\text{-}10)$$

where Z_r is the drawdown in an observation well at distance r from the pumped well, q is the flow in cubic feet per day, T is transmissibility in cubic feet per day per foot, and u is given by

$$u = \frac{r^2 S_c}{4Tt} \tag{6-11}$$

In Eq. (6-11) t is the time in days since pumping began, and S_c is the storage constant of the aquifer, or the volume of water removed from a column of aquifer 1 ft square when the water table or piezometric surface is lowered 1 ft. For water-table aquifers it is essentially the specific yield. The integral in Eq. (6-10), commonly written $W(u)$ and called the *well function of u,* can be evaluated from the series

$$W(u) = -0.5772 - \ln u + u - \frac{u^2}{2 \cdot 2!} + \frac{u^3}{3 \cdot 3!} \cdots \tag{6-12}$$

Values of $W(u)$ for various values of u are given in Table 6-3.

Equation (6-10) can be solved graphically by plotting a *type curve* of u versus $W(u)$ on logarithmic paper (Fig. 6-10). From Eq. (6-11),

$$\frac{r^2}{t} = \frac{4T}{S_c} u \tag{6-13}$$

If q is constant, Eq. (6-10) indicates that Z_r equals a constant times $W(u)$. Thus a curve of r^2/t versus Z_r should be similar to the type curve of u versus $W(u)$. After the field observations are plotted, the two curves are superimposed with their axes parallel and adjusted until some portions of the two curves coincide. The coordinates of a common point taken from the region where the curves coincide are used to solve for T and S_c, using Eqs. (6-10) and (6-13). Values of Z_r and r^2/t may come from one well with various values of t, from several wells with different values of r, or a combination of both. Metric units may be used in these equations without change in the constants.

Illustrative example 6-1 A 12-in-diameter well is pumped at a uniform rate of 1.5 ft³/s while observations of drawdown are made in a well 100 ft distant. Values of t and Z as observed and computed values of r^2/t are given below. Find T and S_c for the aquifer and estimate the drawdown in the observation well at the end of 30 days of pumping.

t, hr	1	2	3	4	5	6	8	10	12	18	24
Z, ft	0.6	1.4	2.4	2.9	3.3	4.0	5.2	6.2	7.5	9.1	10.5
r^2/t, (ft²/day) $\times 10^{-5}$	2.4	1.2	0.8	0.6	0.5	0.4	0.3	0.24	0.2	0.13	0.1

The relations between r^2/t and Z and between u and $W(u)$ are plotted as shown in Fig. 6-10. The match point coordinates are

Type curve: $\qquad u = 0.4 \qquad W(u) = 0.7$

Data curve: $\qquad Z = 3.4$ ft $\qquad \dfrac{r^2}{t} = 5.3 \times 10^4$ ft²/day

Table 6-3 Values of $W(u)$ for various values of u

u	1.0	2.0	3.0	4.0	5.0	6.0	7.0	8.0	9.0
×1	0.219	0.049	0.013	0.0038	0.00114	0.00036	0.00012	0.000038	0.000012
×10⁻¹	1.82	1.22	0.91	0.70	0.56	0.45	0.37	0.31	0.26
×10⁻²	4.04	3.35	2.96	2.68	2.48	2.30	2.15	2.03	1.94
×10⁻³	6.33	5.64	5.23	4.95	4.73	4.54	4.39	4.26	4.14
×10⁻⁴	8.63	7.94	7.53	7.25	7.02	6.84	6.69	6.55	6.44
×10⁻⁵	10.95	10.24	9.84	9.55	9.33	9.14	8.99	8.86	8.74
×10⁻⁶	13.24	12.55	12.14	11.85	11.63	11.45	11.29	11.16	11.04
×10⁻⁷	15.54	14.85	14.44	14.15	13.93	13.75	13.60	13.46	13.34
×10⁻⁸	17.84	17.15	16.74	16.46	16.23	16.05	15.90	15.76	15.65
×10⁻⁹	20.15	19.45	19.05	18.76	18.54	18.35	18.20	18.07	17.95
×10⁻¹⁰	22.45	21.76	21.35	21.06	20.84	20.66	20.50	20.37	20.25
×10⁻¹¹	24.75	24.06	23.65	23.36	23.14	22.95	22.81	22.67	22.55
×10⁻¹²	27.05	26.36	25.95	25.66	25.44	25.26	25.11	24.97	24.86
×11⁻¹³	29.36	28.66	28.26	27.97	27.75	27.56	27.41	27.28	27.16
×10⁻¹⁴	31.66	30.97	30.56	30.27	30.05	29.87	29.71	29.58	29.46
×10⁻¹⁵	33.96	33.27	32.86	32.58	32.35	32.17	32.02	31.88	31.76

Source: Adapted from [21].

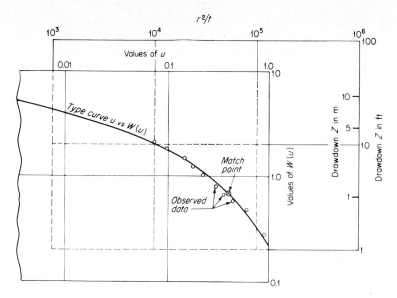

Figure 6-10 Using the Theis method to solve a well problem. Match point coordinates $u = 0.4$, $W(u) = 0.7$, $Z = 3.4$, $r^2/t = 5.3 \times 10^6$.

Substituting these values in Eqs. (6-10) and (6-13) and noting that q in ft^3/day is $1.5 \times 86,400 = 129,600$,

$$T = \frac{qW(u)}{4\pi Z} = \frac{129,600 \times 0.7}{12.56 \times 3.4} = 2124 \text{ ft}^2/\text{day}$$

$$S_c = \frac{4uT}{r^2/t} = \frac{4 \times 0.4 \times 2124}{.5.3 \times 10^4} = 0.064$$

From Eq. (6-11) at the end of 30 days

$$u = \frac{r^2 S_c}{4Tt} = \frac{10,000 \times 0.064}{4 \times 2124 \times 30} = 0.0025$$

From Table 6-3, $W(u) = 5.44$ which, substituted in Eq. (6-10), yields

$$Z = \frac{qW(u)}{4\pi T} = \frac{129,600 \times 5.44}{12.57 \times 2124} = 26.4 \text{ ft}$$

When u is small, the terms of Eq. (6-12) following $\ln u$ are small and may be neglected. Equation (6-11) indicates that u will be small when t is large, and in this case a modified solution of the Theis method is possible [22] by writing

$$T = \frac{2.3q}{4\pi \Delta Z} \log \frac{t_2}{t_1} \tag{6-14}$$

where ΔZ is the change in drawdown between times t_1 and t_2. The drawdown Z is plotted on an arithmetic scale against time t on a logarithmic scale (Fig. 6-11). If ΔZ is taken as the change in drawdown during one log cycle, $\log_{10}(t_2/t_1) = 1$, and T is determined from Eq. (6-14). When $Z = 0$,

$$S_c = \frac{2.25 T t_0}{r^2} \tag{6-15}$$

where t_0 is the intercept (in days) obtained if the straight-line portion of the curve is extended to $Z = 0$.

As in the Thiem equation, Theis assumes small drawdown and full penetration of the well. While Theis adjusts for the effect of storage in the aquifer, he does assume instantaneous unwatering of the aquifer material as the water table drops. These conditions are reasonably well satisfied in artesian aquifers. However, the procedure should be used with caution in thin or poorly permeable water-table aquifers.

Illustrative example 6-2 Using the modified Theis method, find the transmissibility and storage constant for the data of Illustrative example 6-1.

The time-drawdown curve for these data is plotted in Fig. 6-11. Between $t = 3$ hr and $t = 30$ hr, $\Delta Z = 11.0$ ft. Hence,

$$T = \frac{2.3 \times 129{,}600}{12.57 \times 11} = 2156 \text{ ft}^2/\text{day}$$

$$t_0 = 2.7 \text{ hr} = 0.112 \text{ day}$$

$$S_c = \frac{2.25 \times 2156 \times 0.112}{10{,}000} = 0.0546$$

6-13 Boundary Effects

The assumption of a symmetrical cone of depression implies a homogeneous aquifer of great extent. Such an ideal aquifer is rarely encountered, although in many cases the condition is approximated closely enough for reasonable accuracy. When several wells are close together, their cones of depression may overlap, or *interfere,* and the water table appears as in Fig. 6-12. Where the cones of depression overlap, the drawdown at a point is the sum of the drawdowns caused by the individual wells. The two-dimensional analysis is oversimplified, but it shows that when wells are located too close together, the flow from the wells is impaired and the drawdowns increased.

Figure 6-13 shows an aquifer with a positive boundary in the form of an intersecting surface stream. The gradient from the stream to the well causes influent seepage from the stream. If streamflow exceeds seepage, so that flow continues in the stream, the cone of depression of the well must coincide with the water surface in the stream. A rigorous analysis would require that the channel be the full depth of the aquifer to avoid vertical flow components. However, if the well is not too close to the stream, no serious error is intro-

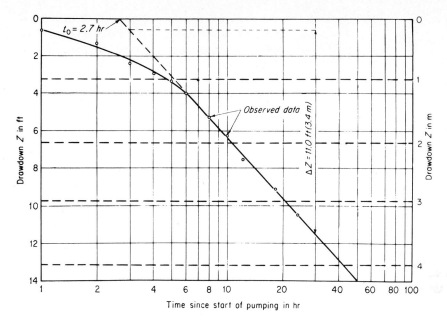

Figure 6-11 Use of the modified Theis method.

duced if this condition is not satisfied. The method of images devised by Lord Kelvin for electrostatic theory is a convenient way to treat boundary problems. An image well is assumed to have all the properties of the real well but to be located on the opposite side of the stream and at the same distance from it as the real well. Since the stream adds water to the aquifer, the image well is assumed to be a recharge well, i.e., one that adds water to the aquifer. Its cone of depression is the same as that of the real well but is inverted (Fig. 6-13). The resultant cone of depression for the real well is found by subtracting the drawdown caused by the image well from that caused by the real well (assuming no boundary). The corrected water table between the real well and the stream is

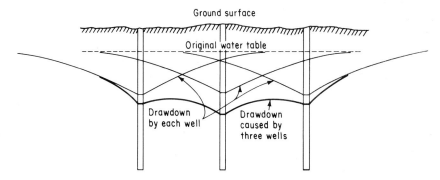

Figure 6-12. Effect of interference between wells.

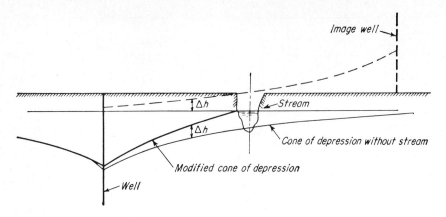

Figure 6-13 Image well stimulating the effect of seepage from a stream on water levels adjacent to a pumped well.

therefore higher than without the effect of the stream. At the stream, the two drawdowns are equal, and the new drawdown is zero.

Boundaries across which no flow is transmitted, such as faults, can be represented by pumping wells. Multiple boundaries may require multiple image wells. With judicious choice of image wells, fairly complicated problems may be analyzed quickly.

6-14 Aquifer Analysis

Techniques of the previous sections are suitable for analysis of single wells or a small well field, but study of a large aquifer generally requires more efficient computational systems. A Hele-Shaw apparatus consisting of closely spaced glass plates with a viscous fluid between them is often convenient for solving two-dimensional groundwater-flow problems [1]. The equations governing groundwater flow are the same as those for the viscous flow of the analog.

Three-dimensional problems are commonly treated with a digital or analog computer. The analog computer consists of a network of resistors and capacitors. Current is analogous to flow and voltage to potential; permeability is simulated by the reciprocal of resistance. Elaborate analogs representing large aquifers have been constructed [23]. They can indicate water-table changes, provide water-table maps, and evaluate the effect of pumping or changes in the recharge pattern.

The electrical analog solves the basic differential equation of groundwater flow

$$\frac{\partial^2 h}{\partial x^2} + \frac{\partial^2 h}{\partial y^2} = \frac{S_c}{T} \frac{\partial h}{\partial t} \tag{6-16}$$

In finite-difference form using the grid notation of Fig. 6-14 this becomes

$$\frac{h_2 + h_4 - 2h_1}{a^2} + \frac{h_3 + h_5 - 2h_1}{a^2} = \frac{h_2 + h_4 + h_3 + h_5 - 4h_1}{a^2}$$

$$= \frac{S_c}{T} \frac{h}{\partial t} \qquad (6\text{-}17)$$

where a is the grid size. This can also be solved with a digital computer [24] and without constructing an elaborate analog. If repeated solutions are desired over a long period of time, the analog may be the most effective device. If a relatively limited analysis is expected to suffice, the digital computer will generally prove more efficient. Grid sizes for both analog and digital solutions will vary depending on the nature of the problem. The solutions can be expected to be as precise as the aquifer description used. Aquifer properties are required at each grid point (thickness, permeability, and coefficient of storage). If they are not accurately defined, the solution may be in error.

POTENTIAL OF A GROUNDWATER RESERVOIR

A basic problem in engineering groundwater studies is the question of the permissible rate of withdrawal from a groundwater basin. This quantity, commonly called the *safe yield,* is defined by Meinzer [16] as "the rate at which water can be withdrawn for human use without depleting the supply to such an extent that withdrawal at this rate is no longer economically feasible." Many other definitions of safe yield have been suggested, and alternative terms such as sustained yield, feasible rate of withdrawal, and optimum yield have been proposed. The concept of safe yield has received considerable criticism. Kazmann [25] has suggested that it be abandoned because of its frequent interpretation as a permanent limitation on the permissible withdrawal. Safe yield must be

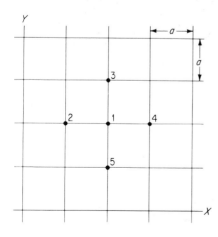

Figure 6-14 Grid notation for Eq. (6-17).

recognized as a quantity determined for a specific set of controlling conditions and subject to change as a result of changing economic or physical conditions. It should also be recognized that the concept can be applied only to a complete groundwater unit. The possible withdrawal from a single well or group of wells in a field is affected by a variety of factors such as size, construction, and spacing of wells as well as by any controls on the flow of groundwater toward the particular field.

6-15 Safe Yield

The safe yield of a groundwater basin is governed by many factors, one of the most important being the quantity of water available. This hydrologic limitation is often expressed by the equation

$$G = P - Q_S - E_T + Q_g - \Delta S_g - \Delta S_s \qquad (6\text{-}18)$$

where G is safe yield, P is precipitation on the area tributary to the aquifer, Q_S is surface streamflow from the same area, E_T is evapotranspiration, Q_g is net groundwater inflow to the area, ΔS_g is change in groundwater storage, and ΔS_s is change in surface storage. If the equation is evaluated on a mean annual basis, ΔS_s will usually be zero.

All terms of Eq. (6-18) are subject to artificial change, and G can be computed only by assuming the conditions regarding each item. Artificial-recharge operations can reduce Q_S. Irrigation diversion from influent streams may increase evapotranspiration. Lowering the water table by pumping may increase groundwater inflow (or reduce groundwater outflow) and may make otherwise effluent streams into influent streams.

The factors which control the assumptions on which Eq. (6-18) is solved are primarily economic. The feasibility of artificial recharge or surface diversion is usually determined by economics. If water levels in the aquifer are lowered, pumping costs are increased. Theoretically, there is a water-table elevation at which pumping costs equal the value of the water pumped and below which water levels should not be lowered. Practically, the increased cost is often passed on to the ultimate consumer, and the minimum level is never attained. Excessive lowering of the water table may result in contamination of the groundwater by inflow of undesirable waters. This hazard is especially prevalent near seacoasts, where seawater intrusion (Sec. 6-16) may occur. A similar problem may develop wherever an aquifer is adjacent to a source of saline groundwater.

The permanent withdrawal of groundwater from storage is called *mining,* the term being used in the same sense as for mineral resources. In 1975, 20 percent of all groundwater withdrawn in the United States was being mined [26], a rate of 80×10^6 m³/day or 65,000 acre · ft/day. If the storage in the aquifer is small, excessive mining may be disastrous to any economy dependent on the aquifer for water. On the other hand, many groundwater basins contain vast reserves of water, and planned withdrawal of this water at a rate that can be sustained over a long period may be a wise use of this resource. The annual increment of mined water, ΔS_g of Eq. (6-18), increases the yield. Thus Eq. (6-18) cannot properly be considered an equilibrium equation or solved in terms of mean annual values. It can be solved correctly only on the basis of specified assumptions for a stated period of years.

Transmissibility of an aquifer may also place a limit on safe yield. Although Eq. (6-18) may indicate a potentially large draft, this can be realized only if the aquifer is capable of transmitting the water from the source area to the wells at a rate high enough to sustain the draft. This problem is especially likely to occur in artesian aquifers which transmit water over long distances.

6-16 Seawater Intrusion

Since seawater (specific gravity about 1.025) is heavier than fresh water, the groundwater under a uniformly permeable circular island would appear as shown in Fig. 6-15. The lens of fresh water floating on salt water is known as a *Ghyben-Herzberg lens,* after the codiscoverers [27] of the principle. About 1/40 unit of fresh water is required above sea level for each unit of fresh water below sea level to maintain hydrostatic equilibrium. True hydrostatic equilibrium does not exist with a sloping water table since flow must occur. Thus, there is

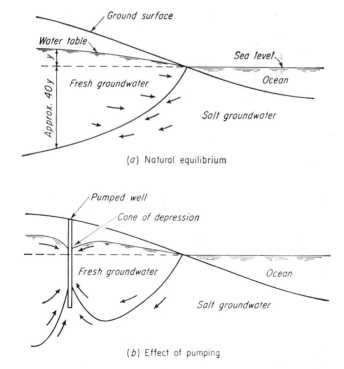

Figure 6-15 Saltwater-freshwater relations adjacent to a coastline.

likely to be a seepage face for freshwater flow to the ocean and a zone of mixing along the saltwater-freshwater interface. Areally variable recharge, pumping of wells, and tidal action also disturb the equilibrium. A hydrodynamic balance governs the form of the interface. If velocities are low, the 1/40 ratio may be a reasonable first approximation, but more adequate methods of analysis are available [28, 29].

When a cone of depression is formed about a pumping well in the fresh water, an inverted cone of salt water will rise into the fresh water (Fig. 6-15b). A saltwater rise of approximately 40 m/m (40 ft/ft) of freshwater drawdown may occur, depending on the local situation. Horizontal skimming wells are commonly used to avoid this effect.

6-17 Artificial Recharge

If transmissibility is not a problem, the yield of an aquifer may be increased artificially by introducing water into it. In most cases this is equivalent to reducing the surface runoff from the area [Q_s, Eq. (6-18)]. The methods employed for artificial recharge are controlled by the geologic situation of an area and by economic considerations. Some possible methods include:

1. Storing floodwaters in reservoirs constructed over permeable areas
2. Storing floodwaters in reservoirs for later release into the stream channel at rates approximating the percolation capacity of the channel
3. Diverting streamflow to spreading areas located in a highly permeable formation
4. Excavating recharge basins to reach permeable formations
5. Pumping water through recharge wells into the aquifer
6. Overirrigating in areas of high permeability
7. Construction of wells adjacent to a stream to induce percolation from streamflow

Where conditions are favorable, use of an aquifer as a reservoir may eliminate evaporation losses, protect against pollution, provide a low-cost distribution system, and result in a cost saving compared with a surface reservoir. Artificial recharge with wells has also been employed to create a groundwater mound along a coast as a barrier to saltwater intrusion and as a means of disposing of waste materials. The latter process must be used with great caution to avoid polluting useful aquifers [30]. Todd [29] discusses methods and results of recharge operations in some detail.

6-18 Artesian Aquifers

If the permeability of the aquiclude confining an artesian aquifer is 0.04 m/day and the hydraulic gradient is unity, the daily seepage would amount to 40,900 m³/km² (28 million gal/mi² or 86 acre · ft/mi²). A quantity of this magnitude could be quite significant in the groundwater exchange of an aquifer. Hantush [31] has demonstrated a procedure which accounts for such leakage in the analysis of pumping tests on artesian aquifers.

Artesian aquifers demonstrate considerable compressibility. There are cases where fluctuations in tide level, barometric pressure, or even the superimposed load of trains are reflected in fluctuations of water level in wells penetrating an aquifer. If the pressure in an artesian aquifer is relieved locally by removal of water, compression of the aquifer may result, with subsidence of the ground above it. Such subsidence has been observed [32] in areas subject to heavy withdrawal of groundwater, with ground-surface elevations declining more than 3.0 m (10 ft). Aside from the disrupting effects of the surface subsidence, pumping tests on such aquifers may be misleading because of the flow derived from storage as a result of the compression. Although the small fluctuations appear to exhibit elastic behavior, there is no evidence that the ground levels in regions of pronounced subsidence will recover if the aquifers are repressurized.

6-19 Time Effects in Groundwater

Flow rates in the groundwater are normally extremely slow, and considerable time may be involved in groundwater phenomena. A critical lowering of the water table adjacent to a coast may not bring immediate saltwater intrusion because of the time required for the salt water to move inland. Werner [33] suggests that several hundred years might be required for a sudden increase in water level in the recharge area of an extensive artesian aquifer to be transmitted through the aquifer. Jacob [34] found that water levels on Long Island were related to an effective precipitation which was the sum of the rainfalls for the previous 25 years, each weighted by a factor which decreased with time. McDonald and Langbein [35] found long-term fluctuations in streamflow in the Columbia basin which they believe are related to groundwater fluctuations. Thus in interpreting groundwater data it is important to give full weight to the influence of time. Observed variations in groundwater levels must be correctly related to causal factors if serious misconceptions are not to result.

PROBLEMS

6-1 An undisturbed rock sample has an oven-dry weight of 652.47 g. After saturation with kerosene its weight is 731.51 g. It is then immersed in kerosene and displaces 300.66 g. What is the porosity of the sample?

6-2 At station A the water-table elevation is 642 ft above sea level, and at B the elevation is 629 ft. The stations are 1100 ft apart. The aquifer has a permeability of 300 Meinzer units and a porosity of 14 percent. What is the actual velocity of flow in the aquifer?

6-3 If the root zone in clay-loam is 1m thick, what quantity of available moisture (in millimeters depth) should it hold? Use Table 6-1.

6-4 A soil sample has a coefficient of permeability of 250 Meinzer units. What would its permeability be at 50°F? What is its instrinsic permeability? What is its permeability in m/day? Feet/day?

6-5 A 12-in diameter well penetrates 80 ft below the static water table. After 24 hr of pumping at 1100 gal/min the water level in a test well at 320 ft is lowered 1.77 ft, and in a well 110 ft away the drawdown is 3.65 ft. What is the transmissibility of the aquifer? Use Eq. (6-9).

6-6 The time-drawdown data for an observation well 296 ft from a pumped well (500 gal/min) are tabulated below. Find the transmissibility and storage constant of the aquifer. Use the Theis method.

Time, hr	Drawdown, ft
1.9	0.28
2.1	0.30
2.4	0.37
2.9	0.42
3.7	0.50
4.9	0.61
7.3	0.82
9.8	1.09
12.2	1.25
14.7	1.40
16.3	1.50
18.4	1.60
21.0	1.70
24.4	1.80

6-7 Tabulated below are the time-drawdown data for an observation well 150 ft from a well pumped at 350 gal/min. Find the transmissibility and storage constant by the modified Theis method.

Time, hr	1.8	2.7	5.4	9.0	18.0	54.0
Drawdown, ft	1.8	2.4	3.6	4.3	5.8	8.1

6-8 A well 250 ft deep is planned in an aquifer having a transmissibility of 10,000 gal/day per foot width and a storage coefficient of 0.010. The well is expected to yield 500 gal/min and will have a 12-in diameter. If the static water level is 50 ft below the ground surface, estimate the pumping lift at the end of 1 and 3 years of operation.

6-9 After pumping a new 12-in well for 24 hr at 150 gal/min, the drawdowns in a number of nearby observation wells are as given below. Find the storage coefficient and transmissibility of the aquifer.

Well no.	Distance, ft	Drawdown, ft
1	100	10.5
2	141	7.5
3	190	6.2
4	200	4.0
5	283	2.4
6	347	1.4
7	490	0.6

6-10 An 18-in well is in an aquifer with a transmissibility of 8000 gal/day per foot width and a storage coefficient of 0.05. Draw a profile of the cone of depression after 1 year of pumping at 1.1 2 years will not exceed 20 ft?

6-11 A 24-in well is in an aquifer with a transmissibility of 10,000 gal/day per foot width and a storage coefficient of 0.05. Draw a profile of the cone of depression after 1 year of pumping at 1.1 ft³/s. If a fault is located 1000 ft from this well, what would be the profile of the cone of depression?

6-12 A well is 30 cm in diameter and penetrates 50 m below the static water table. After 36 hr of pumping at 4 m³/min [use Eq. (6-9)] the water level in a test well 200 m distant is lowered by 1.2 m and in a well 40 m away the drawdown is 2.7 m. What is the transmissibility of the aquifer?

6-13 Using the data for Prob. 6-6 but assuming the drawdowns to be given in meters, find the transmissibility and storage constant of the aquifer by the Theis method. The observation well is 100 m from the pumped well and the pumping rate is 2000 L/min.

6-14 Using the data from Prob. 6-7, determine the transmissibility and storage constant of the aquifer if the indicated drawdowns are in meters. The observation well is at a distance of 50 m, and the pumping rate is 2800 L/min. Use the modified Theis method.

6-15 The well of Prob. 6-11 is 800 ft from a stream which flows all year. How much is the drawdown midway between the well and the stream decreased because of this seepage?

6-16 Using data from the *Water-Supply Papers* or other source, find out what you can about the trend of groundwater levels in your area. What explanation can you see for the observed trends? What is the source of the groundwater? Is an overdraft of the available supply indicated? Are there any possible ways of improving the yield?

6-17 For a basin selected by your instructor, make an estimate of the safe yield, assuming no change in present conditions.

6-18 A new well is 30 cm in diameter. After 24 hr of pumping at a rate of 6000 L/min the drawdowns are as given for Prob. 6-9 (distances and drawdowns in meters). What are the storage constant and transmissibility of the aquifer?

REFERENCES

1. Davis, Stanley N., and R. J. DeWiest: "Hydrogeology," p. 39, Wiley, New York, 1966.
2. Buckingham, E.: Studies on the Movement of Soil Moisture, *U.S. Dept. Agric. Bur. Soils Bull.* 38, 1907.
3. Colman, E. A.: A Laboratory Procedure for Determining the Field Capacity of Soils, *Soil Sci.,* vol. 63, p. 277, 1947.
4. Veihmeyer, F. J., and A. H. Hendrickson: The Moisture-Equivalent as a Measure of the Field Capacity of Soils, *Soil Sci.,* vol. 32, pp. 181–193, 1931.
5. Watson, K. K.: Some Operating Characteristics of a Rapid Response Tensiometer System, *Water Resour. Res.,* vol. 1, pp. 577–586, 1965.
6. Broadfoot, W. M., and others: Some Field, Laboratory and Office Procedures for Soil-Moisture Measurement, *South Forest. Exp. Stn. Occ. Pap.* 135, 1954.
7. International Atomic Energy Agency: Neutron Moisture Gages, *Tech. Rep.* 112, Vienna, 1970.
8. Wiesnet, D. R.: "Remote Sensing and Its Application to Hydrology," chap. 2 in J. C. Rodda (ed.): "Facets of Hydrology," Wiley, New York, 1976.
9. Darcy, H.: "Les Fontaines Publiques de La Ville de Dijon," Dalmont, Paris, 1856.
10. Hagen, G.: Bewegung des Wassers in engen cylindrischen Rohren, *Pogg. Ann.,* vol. 47, 1839.
11. Poiseuille, J. M.: Experimental Investigations on the Flow of Liquids in Tubes of Very Small Diameters, *Proc. R. Acad. Sci. (France), Math. Phys. Sci. Mem.,* 1846.
12. Slichter, C. S.: Field Measurements of Rate of Movement of Underground Water, *U.S. Geol. Surv. Water-Supply Pap.* 140, 1905.
13. Dole, R. B.: Use of Fluorescein in Study of Underground Waters, *U.S. Geol. Surv. Water-Supply Pap.* 160, pp. 73–85, 1906.

14. International Atomic Energy Agency: Guide Book on Nuclear Techniques in Hydrology, *Tech. Rep.* 91, Vienna, 1968.
15. Kaufman, W. J., and G. T. Orlob: An Evaluation of Ground-Water Tracers, *Trans. Am. Geophys. Union,* vol. 37, pp. 297–306, June 1956.
16. Meinzer, O. E.: Outline of Groundwater Hydrology, *U.S. Geol. Survey Water-Supply Pap.* 494, 1923.
17. Meinzer, O. E.: Large Springs in the United States, *U.S. Geol. Survey Water-Supply Pap.* 557, 1927.
18. Dupuit, J.: "Etudes theoriques et practiques sur le mouvement des eaux," 2d ed., Paris, 1863.
19. Thiem, G.: "Hydrologische Methoden," Gebhardt, Leipzig, 1906.
20. Theis, C. V.: The Relation between the Lowering of the Piezometric Surface and the Rate and Duration of Discharge of a Well Using Ground-Water Storage, *Trans. Am. Geophys. Union,* vol. 16, pp. 519–524, 1935.
21. Wenzel, L. K.: Methods for Determining the Permeability of Water-Bearing Materials, *U.S. Geol. Surv. Water-Supply Pap.* 887, 1942.
22. Jacob, C. E.: Drawdown Test to Determine the Effective Radius of Artesian Well, *Trans. ASCE,* vol. 112, pp. 1047–1070, 1947.
23. Walton, W. C.: "Groundwater Resource Evaluation," pp. 518–598, McGraw-Hill, New York, 1970.
24. Prickett, T. A., and C. G. Lonnquist: Selected Digital Computer Techniques for Groundwater Resource Evaluation, *Ill. State Water Surv. Urbana, Ill., Bull.* 55, 1971.
25. Kazmann, R. G.: "Safe Yield" in Ground-Water Development, Reality or Illusion?, *J. Irrig. Drain. Div. ASCE,* vol. 82, November 1956; see also discussion by McGuinness, Ferris, and Kramsky, in ibid., vol. 82, May 1957.
26. U.S. Water Resources Council: "The Nation's Water Resources," Washington, D.C., 1979.
27. Ghyben, W. Badon: Nota in Verband met de Voorgenomen Putboring naby Amsterdam, *Tijdschr. K. Inst. Ing. (The Hague),* 1888–1889, p. 21.
28. Cooper, H. H.: A Hypothesis Concerning the Dynamic Balance of Fresh Water and Salt Water in a Coastal Aquifer, *J. Geophys. Res.,* vol. 64, pp. 461–468, April 1959.
29. Todd, D. K.: "Ground-Water Hydrology," Wiley, New York, 2d ed., 1980.
30. Jacob, C. E.: Full Utilization of Groundwater Reservoirs, *Trans. Am. Geophys. Union,* vol. 38, p. 417, June 1957.
31. Hantush, M. S.: Analysis of Data from Pumping Tests in Leaky Aquifers, *Trans. Am. Geophys. Union,* vol. 37, pp. 702–714, December 1956.
32. Poland, J. F., and G. H. Davis: Subsidence of the Land Surface in the Tulare-Wasco (Delano) and Los Banos–Kettleman City Area, San Joaquin Valley, California, *Trans. Am. Geophys. Union,* vol. 37, pp. 287–296, June 1956.
33. Werner, P. W.: Notes on Flow-Time Effects in the Great Artesian Aquifers of the Earth, *Trans. Am. Geophys. Union,* vol. 27, pp. 687–708, October 1946.
34. Jacob, C. E.: Correlation of Ground-Water Levels and Precipitation on Long Island, New York, *Trans. Am. Geophys. Union,* vol. 24, pt. 2, pp. 564–580, 1943, vol. 25, pt. 6, pp. 928–939, 1944.
35. McDonald, C. C., and W. B. Langbein: Trends in Runoff in the Pacific Northwest, *Trans. Am. Geophys. Union,* vol. 29, pp. 387–397, June 1948.

BIBLIOGRAPHY

Bear, J.: "Dynamics of Fluids in Porous Media," American Elsevier, New York, 1972.
———: "Hydraulics of Groundwater," McGraw-Hill, New York, 1979.
Bouwer, H.: "Groundwater Hydrology," McGraw-Hill, New York, 1978.
Domenico, A.: "Concepts and Models in Groundwater Hydrology," McGraw-Hill, New York, 1972.

Hantush, M. S.: Hydraulics of Wells, *Advan. Hydrosci.,* vol. 1, 1964.

Hillel, D.: "Fundamentals of Soil Physics," Academic Press, New York, 1980.

Hubbert, M. K.: The Theory of Ground-Water Motion, *J. Geol.,* vol. 48, pp. 785–944, 1940.

Meinzer, O. E.: Ground Water, chap. 10 in O. E. Meinzer (ed.): "Hydrology," vol. 9, Physics of the Earth Series, McGraw-Hill, New York, 1942; reprinted Dover, New York, 1949.

Polubarinova-Kochina, P. Ya.: "Theory of Groundwater Movements," trans. from Russian, Princeton University Press, Princeton, 1972.

Remson, I., G. Hornberger, and F. Molz: "Numerical Methods in Subsurface Hydrology," Wiley Interscience, New York, 1971.

SEVEN

STREAMFLOW HYDROGRAPHS

Engineering hydrology is concerned primarily with three characteristics of streamflow: monthly and annual volumes available for storage and use, low-flow rates which restrict in-stream uses of water, and floods. Detailed analysis of flood hydrographs is usually important in flood damage mitigation, flood forecasting, or establishing design flows for the many structures which must convey floodwaters.

CHARACTERISTICS OF THE HYDROGRAPH

The water which constitutes streamflow may reach the stream channel by any of several paths from the point where it first reaches the earth as precipitation. Some water flows over the soil surface, reaching the stream soon after its occurrence as rainfall. Other water infiltrates through the soil surface and flows beneath the surface to the stream. This water moves more slowly than the surface runoff and contributes to the sustained flow of the stream during periods of dry weather. In hydrologic studies involving rate of flow in streams it is necessary to distinguish between these components of total flow. The first step in traditional studies is to divide observed hydrographs of streamflow into components before analyzing the relation between rainfall and runoff (Chap. 8) or determining the characteristic shape of hydrographs for a basin.

7-1 Components of Runoff

The route followed by a water particle from the time it reaches the ground until it enters a stream channel is devious. It is convenient to visualize three main routes of travel: overland flow, interflow, and groundwater flow.

Overland flow, or *surface runoff,*† is that water which travels over the ground surface to a channel. The word channel as used here refers to any depression which may carry a small rivulet of water in turbulent flow during a rain and for a short while after. Such channels are numerous, and the distance water must travel as overland flow is relatively short, rarely more than a few hundred feet. Therefore overland flow soon reaches a channel, and if it occurs in sufficient quantity, is an important element in the formation of flood peaks. The amount of surface runoff may be quite small, however, since surface flow over a permeable soil surface can occur only when the rainfall rate exceeds the local infiltration capacity (Chap. 8). In many small and moderate storms, surface runoff may occur only from impermeable and saturated areas within the basin or from precipitation which falls directly on water surfaces. Except in urban areas, the extent of such surfaces is usually a small part of the basin area. Hence, surface runoff is usually an important factor in streamflow only as the result of heavy or high-intensity rains.

Some of the water which infiltrates the soil surface may move laterally through the upper soil layers until it enters a stream channel. This water, called *interflow* or *subsurface storm flow,* moves more slowly than the surface runoff and reaches the streams later [1]. The proportion of total runoff which occurs as interflow depends on the physical features of the basin. A thin soil cover overlying rock, hardpan, or plowbed a short distance below the soil surface favors substantial quantities of interflow, whereas uniformly permeable soil encourages downward percolation to groundwater. Although traveling more slowly than overland flow, interflow may be much larger in quantity, especially in storms of moderate intensity, and hence may be the principal factor in the smaller rises of streamflow.

Some precipitation may percolate downward until it reaches the water table (Chap. 6). This groundwater accretion may eventually discharge into the streams as *groundwater flow* (also called *base flow* and *dry-weather flow*) if the water table intersects the stream channels of the basin. The groundwater contribution to streamflow cannot fluctuate rapidly because of its very low flow velocity. In some cases, more than 2 years is required [2] for a given accretion to groundwater to be discharged into the streams. On the other hand, water precipitated adjacent to a channel intersected by the water table may contribute to streamflow relatively quickly [3].

Basins having permeable surface soils and large, effluent groundwater bodies show sustained high flow throughout the year, with a relatively small ratio between flood flow and mean flow. Basins with surface soils of low permeability or influent groundwater bodies have higher ratios of peak to average flows and very low or zero flows between floods. Hydrographs for each type of basin are shown in Fig. 7-1. Hat Creek drains volcanic terrain with a large

---

† Surface runoff includes precipitation falling on the stream system whereas, strictly speaking, overland flow does not.

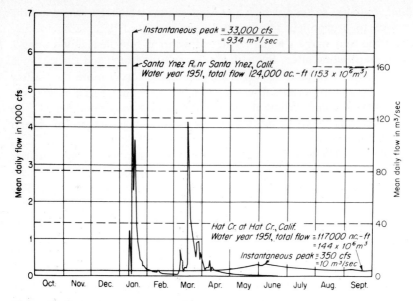

Figure 7-1 Comparison of hydrographs from two streams of differing geologic characteristics.

groundwater contribution, while the Santa Ynez River is influent throughout most of its length.

The distinctions drawn between the three components of flow are arbitrary. Water may start out as overland flow, infiltrate, and complete its trip to the stream as subsurface storm flow; or infiltrated water may surface where a relatively impervious stratum intersects a hillside, and finish its journey to the stream as overland flow. In limestone terrain, groundwater frequently moves at relatively high velocities through solution channels and fractures. For convenience it has been customary to consider the total flow to be divided into only two parts: *storm,* or *direct, runoff* and *base flow.* The distinction is actually on the basis of time of arrival in the stream rather than on the path followed. Direct runoff is presumed to consist of surface runoff and a substantial portion of the interflow, whereas base flow is considered to be largely groundwater. Computer simulation techniques (Chap. 12) commonly use all components.

7-2 Streamflow Recessions

A typical hydrograph resulting from an isolated period of rainfall (Fig. 7-2) consists of a *rising limb, crest segment,* and *falling limb,* or *recession.* The shape of the rising limb is influenced mainly by the character of the storm which caused the rise. The point of inflection on the falling side of the hydrograph is commonly assumed to mark the time at which surface inflow to the channel system ceases. Thereafter, the recession curve represents withdrawal of water from storage within the basin. The shape of the recession is largely independent

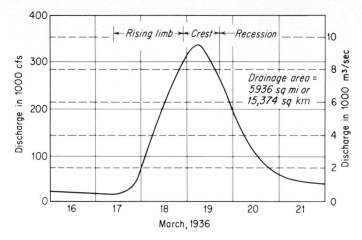

Figure 7-2 A typical hydrograph (Potomac River at Shepherdstown, W.Va.).

of the characteristics of the storm causing the rise. On large basins subject to runoff-producing rainfall over only a part of the basin, the recession may vary from storm to storm, depending on the particular area of runoff generation. If rainfall occurs while the recession from a previous storm is in progress, the recession will naturally be distorted. However, the recession curve for a basin is a useful tool in hydrology.

A number of functions have been used to describe the *recession curve,* or *depletion curve* [4], the one in general use being

$$q_1 = q_0 K_r \tag{7-1}$$

where q_0 is the flow at any time, q_1 is the flow one time unit later, and K_r is a recession constant which is less than unity. Equation (7-1) can be written in the more general forms as

$$q_t = q_0 K_r^t = q_0 e^{-\alpha t} \tag{7-2}$$

where q_t is the flow t time units after q_0, e is the napierian base, and $\alpha = -\ln K_r$. The time unit is frequently taken as 24 hr, although on small basins a shorter unit may be necessary. The numerical value of K_r depends on the time unit selected. Integrating Eq. (7-2) and remembering that the volume of water discharged during time dt is $q\, dt$ and is equal to the decrease in storage $-dS$ during the same interval, we see that the storage S_t remaining in the basin at time t is

$$S_t = -\frac{q_t}{\ln K_r} = \frac{q_t}{\alpha} \tag{7-3}$$

Equation (7-2) will plot as a straight line on semilogarithmic paper with q on the logarithmic scale. If the recession of a stream rise is plotted on semilogarithmic paper (Fig. 7-3), the result is usually not a straight line but a

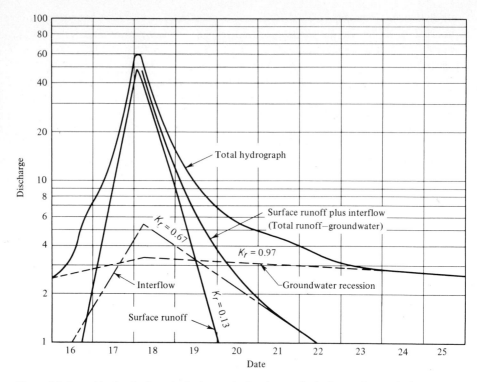

Figure 7-3 Logarithmic plotting of a hydrograph showing method of recession analysis.

curve with gradually decreasing slope, i.e., increasing values of K_r. The reason for this is that the water is coming from three different types of storage—stream channels, surface soil, and the groundwater—each having different lag characteristics. Barnes [5] suggests that the recession can be approximated by three straight lines on a semilogarithmic plot. The transition from one line to the next is often so gradual that it is difficult to select the points of change in slope. Considering the heterogeneity of the typical catchment, this is not surprising. Several aquifers may be contributing groundwater while influent seepage is occurring at other points in the stream. In most cases runoff occurs in varying amounts over the catchment.

The slope of the last portion of the recession should represent the characteristic K_r for groundwater since, presumably, both interflow and surface runoff have ceased. By projecting this slope backward in time (Fig. 7-3) and replotting the difference between the projected line and the total hydrograph, a recession which for a time consists largely of interflow is obtained. With the slope applicable to interflow thus determined, the process can be repeated to establish the recession characteristics of surface runoff.

The above technique represents a degree of refinement rarely used for engineering problems. A recession curve may be developed [6] by plotting values of q_0 against q_t some fixed time t later (Fig. 7-4). If Eq. (7-1) were strictly correct, the plotted data would indicate a straight line. Normally, however, a curve indicating a gradual change in the value of K_r results. This curve becomes asymptotic to a 45° line as q approaches zero.

The method illustrated in Fig. 7-4 may be used to construct recessions for base flow or direct runoff. For a base-flow recession, data should be selected from periods several days after the peak of a flood so that it is reasonably certain that no direct runoff is included. After the base-flow recession has been established, it can be projected back under the hydrograph immediately following a flood peak and the difference between projected base flow and the total hydrograph used to develop a direct-runoff recession curve. It is customary to

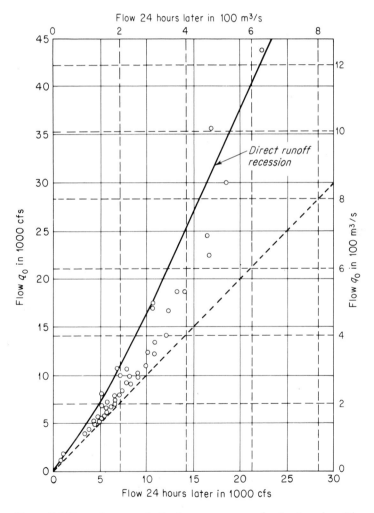

Figure 7-4 Recession curve in the form q_0 versus q_1 for the American River at Fair Oaks, Calif.

draw the base-flow curve to envelop the plotted data on the right, because such a curve represents the slowest recession (high K_r) and points deviating to the left may include direct runoff. By a similar argument, data for the direct-runoff recession are usually enveloped on the left.

7-3 Hydrograph Separation

Division of a hydrograph into direct and groundwater runoff as a basis for subsequent analysis is known as *hydrograph separation* or *hydrograph analysis.* Since there is no ready basis for distinguishing between direct and groundwater flow in a stream at any instant, and since definitions of these two components are relatively arbitrary, the method of separation is equally arbitrary.

For application of the unit hydrograph concept, the method of separation should be such that the time base of direct runoff remains relatively constant from storm to storm. This is usually provided by terminating the direct runoff at a fixed time after the peak of the hydrograph. As a rule of thumb, the time in days N may be approximated by

$$N = bA^{0.2} \qquad (7\text{-}4)$$

where A is the drainage area and b is a coefficient. The value of b can be taken as 0.8 when A is in square kilometers and unity when A is in square miles. With experience, N is probably better determined by inspection of a number of hydrographs, keeping in mind that the total time base should not be excessively long and the rise of the groundwater should not be too great. Figure 7-5 illustrates some reasonable and unreasonable assumptions regarding N.

The most widely used separation procedure consists of extending the recession existing before the storm to a point under the peak of the hydrograph (*AB*, Fig. 7-6). From this point a straight line is drawn to the hydrograph at a point N days after the peak. The reasoning behind this procedure is that as the stream rises, there is flow from the stream into the banks (Fig. 7-7). Hence, base flow should decrease until stages in the stream begin to drop and bank storage returns to the channel. While there is some support for this reasoning, there is no justification for assuming that decrease in base flow conforms to the usual recession. Actually, if the increment of bank storage is greater than inflow from groundwater, base flow is effectively negative.

The line *AC* (Fig. 7-6) connecting the point of rise to the hydrograph N days after the peak illustrates another arbitrary method of separation sometimes used. Differences in the results when applying methods *AC* and *ABC* are small and probably unimportant so long as one method is used consistently and it does not become necessary to analyze complex hydrographs.

7-4 Analysis of Complex Hydrographs

The discussion of hydrograph separation in Sec. 7-3 assumed an isolated streamflow event without subsequent rainfall until after direct runoff had left

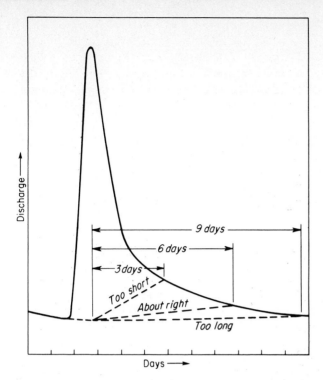

Figure 7-5 Selection of the time base for the direct runoff hydrograph.

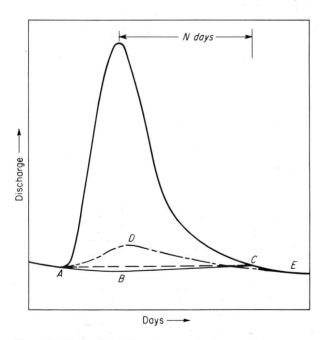

Figure 7-6 Some simple base-flow separation procedures.

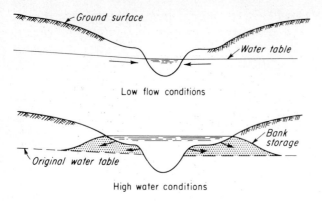

Figure 7-7 Bank-storage variations during a flood.

the basin. This type of event is easier to analyze than the complex hydrographs resulting from two or more closely spaced bursts of rainfall (Fig. 7-8). Often, however, analysis of the more complex cases cannot be avoided. In these cases it is necessary to separate the runoff caused by individual bursts of rainfall in addition to separating direct runoff from base flow.

If a simple base-flow separation such as line *ABC* of Fig. 7-6 is to be used, division between bursts of rain is usually accomplished by projecting the small segment of recession between peaks, using a total-flow recession curve for the basin (line *AB,* Fig. 7-8). Base-flow separation is then completed by drawing *CDB* and *EF.* Direct runoff for the two periods of rain is given by shaded areas I

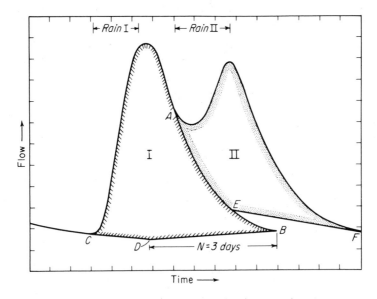

Figure 7-8 Separation of complex hydrograph using recession curve.

and II. A separation of this type is impracticable unless there are two clearly defined peaks with a short segment of recession following the first. If such a separation is in error, its consequences are usually only compensating errors in estimated runoff volume for the two events. It will be seen that the *AC* type of separation (Fig. 7-6) is not suitable for complex hydrographs; connecting points *A* and *F* in Fig. 7-8 would result in an unreasonable separation for the second rise.

Other methods have been developed for analyzing hydrographs (similar to *ADE* in Fig. 7-6) which are particularly useful for basins where the period between storms is frequently short [7, 8]. They may also have some advantages where groundwater is a relatively important component of runoff and reaches the stream fairly quickly.

7-5 Determination of Total Runoff

In some types of analysis (see Sec. 5-11) there is need to determine the total streamflow (direct runoff plus groundwater) resulting from a particular storm or group of storms. This can be done by computing the total volume of flow occurring during a period beginning and ending with the same discharge and encompassing the rise under consideration, making certain that groundwater-recession conditions prevail at both times. The runoff values in Fig. 5-8 were derived in this manner. Implicit in the technique is the assumption that base flow is indicative of the total water stored in the basin, principally groundwater and soil moisture.

HYDROGRAPH SYNTHESIS

The earliest method of estimating peak flows for design was by using empirical formulas considered unacceptable for engineering applications today, and river forecasts were based on gage relations. Efforts to synthesize the hydrograph were intensified in the early 1930s, the period in which Horton [9] first published his infiltration concepts and Sherman [10] introduced the unit hydrograph as a tool for estimating hydrograph shape. The unit hydrograph has been the mainstay of the flood hydrologist, but flood-routing methods (Chap. 9) offer greater flexibility and accuracy in many applications. Methods of estimating runoff volume are discussed in Chap. 8, and the more sophisticated techniques of hydrograph synthesis are presented in Chap. 12.

7-6 The Elemental Hydrograph

If a small, impervious area is subjected to rainfall at a constant rate, the resulting runoff hydrograph will appear much as in Fig. 7-9. Since sheet flow over the surface cannot occur without a finite depth of water on the surface, some of the rainfall goes into temporary storage, or *surface detention*. At any instant the

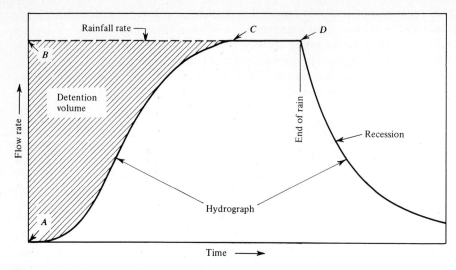

Figure 7-9 The elemental hydrograph.

quantity of water in such storage is equal to the difference between total inflow to the area (rain) and total outflow from the area. When equilibrium has been reached, rate of outflow equals rate of inflow (point C) and the volume in detention is ABC. The water is in constant motion, and any given element of rainfall may pass through the system in a fairly short time, but the volume difference between inflow and outflow remains constant.

When rainfall ends (point D), there is no further inflow to sustain detention, and rate of outflow and detention volume decrease. The outflow follows a recession with flow decreasing at a decreasing rate; that is, d^2q/dt^2 is negative.

Theoretically, an infinite time is required both for the rising portion of the hydrograph to reach equilibrium and for the recession to return to zero. Practically, both the rising and falling curves approach their limits rapidly. The results of experimental study of the hydrographs from small areas are discussed in Sec. 7-13.

7-7 The Unit-Hydrograph Concept

The hydrograph of outflow from a small basin is the sum of the elemental hydrographs from all the subareas of the basin modified by the effect of transit time through the basin and storage in the stream channels. Since the physical characteristics of the basin—shape, size, slope, etc.—are constant, one might expect considerable similarity in the shape of hydrographs from storms of similar rainfall characteristics. This is the essence of the unit hydrograph as proposed by Sherman [11]. The unit hydrograph is a typical hydrograph for the basin. It is called a unit hydrograph because, for convenience, the runoff volume under the hydrograph is commonly adjusted to 1 cm (or 1 mm or 1 in) equivalent depth over the catchment.

It would be wrong to imply that one typical hydrograph would suffice for any basin. Although the physical characteristics of the basin remain relatively constant, the variable characteristics of storms cause variations in the shape of the resulting hydrographs. The storm characteristics are rainfall duration, time-intensity pattern, areal distribution of rainfall, and amount of rainfall. Their effects are discussed below.

Duration of rain. Theoretically, the ideal unit hydrograph has a duration approaching zero, the *instantaneous unit hydrograph* (Sec. 7-10). More commonly the unit hydrograph has been derived for a finite duration. A unit hydrograph may be developed for a short duration (say 1 hr) and all storms treated by dividing the rainfall excess into similar intervals. An alternate approach is to derive a series of unit hydrographs covering the range of durations experienced on the catchment. The effect of small differences in duration is slight, and a tolerance of ±25 percent in duration is usually acceptable. Thus only a few unit hydrographs are actually required. Where a computer solution is used, a short-duration unit hydrograph is preferable.

Time-intensity pattern. If one attempted to derive a separate unit hydrograph for each possible time-intensity pattern, an infinite number of unit hydrographs would be required. Practically, unit hydrographs can be based only on an assumption of uniform intensity of runoff. However, large variations in rain intensity (and hence runoff rate) during a storm are reflected in the shape of the resulting hydrograph. The time scale of intensity variations that are critical depends mainly on basin size. Rainfall bursts lasting only a few minutes may cause clearly defined peaks in the hydrograph from a basin of a few hectares, while intensity changes lasting for hours are required to cause appreciable effects on the hydrograph from basins of several hundred square kilometers. If the unit hydrographs for a basin are applicable to storms of shorter duration than the critical time for the basin, hydrographs of longer storms can be synthesized quite easily (Sec. 7-10). A basic duration of about one-fourth of the basin lag† is generally satisfactory.

Areal distribution of runoff. The areal pattern of runoff can cause variations in hydrograph shape. If the area of high runoff is near the basin outlet, a rapid rise and sharp peak usually result. Higher runoff in the upstream portion of the basin produces a slow rise and a lower, broader peak. Unit hydrographs have been developed for specific runoff patterns, e.g., heavy upstream, uniform, or

---

† The *basin lag* is usually defined as the time from the centroid of rainfall to the hydrograph peak. A more rigorous definition is the time difference from the centroid of rainfall to the centroid of runoff, but the added difficulties in applying this definition are seldom justified.

heavy downstream. This is not wholly satisfactory because of the subjectivity of classification. A better solution is to apply the unit-hydrograph method only to basins small enough to ensure that the usual areal variations will not be great enough to cause major changes in hydrograph shape. The limiting basin size is fixed by the accuracy desired and regional climatic characteristics. Generally, however, unit hydrographs should not be used for basins much over 5000 km² (2000 mi²) unless reduced accuracy is acceptable. Where convective rainfall predominates, the acceptable limit is much smaller. What has been said does not apply to rainfall variations caused by topographic controls. Such rainfall patterns are relatively fixed characteristics of the basin. It is departures from the normal pattern that cause trouble.

Amount of runoff. Inherent in the unit hydrograph concept is the assumption that ordinates of flow are proportional to volume of runoff for all storms of a given duration and that the time bases of all such hydrographs are equal. This assumption is obviously not completely valid, since from the character of recession curves, duration of recession must be a function of peak flow. Moreover, unit hydrographs for storms of the same duration but different magnitudes do not always agree. Peaks of unit hydrographs derived from very small events are commonly lower than those derived from larger storms. This may be because the smaller events contain less surface runoff and relatively more interflow and groundwater than the larger events or because channel flow time is longer at low flows.

In the range of floods frequently experienced on a catchment, it is relatively simple to verify the adequacy of the assumption of linearity by comparing hydrographs from storms of various magnitudes. If important nonlinearity exists, derived unit hydrographs should be used only for reconstructing events of similar magnitude. A series of unit hydrographs covering appropriate ranges of magnitude would be required for each duration. If one is estimating an extreme event exceeding any which has occurred on the catchment, there is no way of obtaining empirical evidence about the changes in unit hydrograph peak. Many hydrologists increase the peaks of unit hydrographs derived from ordinary floods by 5 to 20 percent before using them for estimation of extreme floods. This increase is based on the belief that channel-flow time shortens as flood magnitude increases. However, if extreme floods overflow onto floodplains, the opposite effect might result. Caution should be exercised in using the unit hydrograph to extrapolate extreme events.

In the light of the foregoing discussion, the *unit hydrograph* can be defined as *the hydrograph of one centimeter, millimeter, or inch of direct runoff from a storm of specified duration.* For a storm of the same duration but with a different amount of runoff, the hydrograph of direct runoff is assumed to have the same time base as the unit hydrograph and ordinates of flow approximately proportional to the runoff volume. The duration assigned to a unit hydrograph should be the duration of rainfall producing significant runoff, determined by inspection of hourly rainfall data (Fig. 7-10).

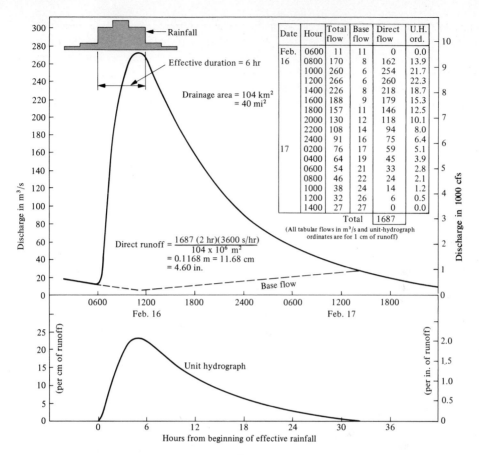

The data table in the figure:

Date	Hour	Total flow	Base flow	Direct flow	U.H. ord.
Feb. 16	0600	11	11	0	0.0
	0800	170	8	162	13.9
	1000	260	6	254	21.7
	1200	266	6	260	22.3
	1400	226	8	218	18.7
	1600	188	9	179	15.3
	1800	157	11	146	12.5
	2000	130	12	118	10.1
	2200	108	14	94	8.0
	2400	91	16	75	6.4
17	0200	76	17	59	5.1
	0400	64	19	45	3.9
	0600	54	21	33	2.8
	0800	46	22	24	2.1
	1000	38	24	14	1.2
	1200	32	26	6	0.5
	1400	27	27	0	0.0
			Total	1687	

(All tabular flows in m^3/s and unit-hydrograph ordinates are for 1 cm of runoff)

Drainage area = 104 km^2 = 40 mi^2

Effective duration = 6 hr

$$\text{Direct runoff} = \frac{1687\,(2\ hr)(3600\ s/hr)}{104 \times 10^6\ m^2}$$
$$= 0.1168\ m = 11.68\ cm$$
$$= 4.60\ in.$$

Figure 7-10 Development of a unit hydrograph.

7-8 Derivation of Unit Hydrographs

The unit hydrograph is best derived from the hydrograph of a storm of reasonably uniform intensity, duration of desired length, and a relatively large runoff volume. The first step (Fig. 7-10) is to separate the base flow from direct runoff. The volume of direct runoff is then determined, and the ordinates of the direct-runoff hydrograph are divided by observed runoff depth. The adjusted ordinates form a unit hydrograph.

A unit hydrograph derived from a single storm may not be representative, and it is therefore desirable to average unit hydrographs from several storms of about the same duration. This should not be an arithmetic average of superimposed ordinates, since if peaks do not occur at the same time, the peak so determined will be lower than the individual peaks. The proper procedure is to compute average peak flow and time to peak. The average unit hydrograph is then sketched to conform to the shape of the other graphs, passing through the computed average peak, and having the required unit volume (Fig. 7-11).

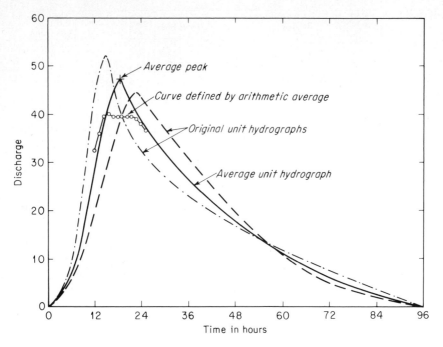

Figure 7-11 Construction of an average unit hydrograph.

7-9 Derivation of Unit Hydrograph from Complex Storms

The simple approach outlined in Sec. 7-8 often proves inadequate because no suitable storm is available in the record. It is then necessary to develop the unit hydrograph from a complex storm. If individual bursts of rain in the storm result in well-defined peaks, it is possible to separate the hydrographs of several bursts (Sec. 7-4) and to treat these hydrographs as independent storms. If the resulting unit hydrographs are averaged, errors in the separation are minimized.

A unit hydrograph can also be developed by successive approximations. A unit hydrograph is assumed and used to reconstruct the storm hydrograph (Fig. 7-17). If the reconstructed hydrograph does not agree with the observed hydrograph, the assumed unit hydrograph is modified and the process repeated until a unit hydrograph which seems to give the best fit is determined.

7-10 The Conversion of Unit-Hydrograph Duration

There is frequently a need to convert an existing unit hydrograph for one storm duration to another—shorter to better cope with spatial and intensity variations, or longer to reduce required computations and possibly in recognition of the coarseness of available data. If two t_R-hr (duration) unit hydrographs, one lagged by t_R hr with respect to the other, are added as shown in Fig. 7-12, the result is the characteristic hydrograph for 2 units of rainfall excess and $2 \cdot t_R$-hr duration. Dividing the ordinates by 2 yields the $2 \cdot t_R$-hr unit hydrograph. In other words, the $n \cdot t_R$-hr unit hydrograph is the average of n t_R-hr unit hydrographs, each lagged t_R hr with respect to the previous one. Unfortunately, it is not as simple to convert to shorter durations and to longer ones that are not multiples of t_R.

S-curve method. A unit hydrograph can be converted to one of shorter (or longer) duration by application of the S-curve, or summation-curve, method. The S *curve* is the hydrograph that would result from an infinite series of unit runoff increments. Thus, each S curve applies to a specific duration within which each unit of runoff is generated. The S curve is constructed by adding together a series of unit hydrographs, each lagged t_R hr with respect to the preceding one (Fig. 7-13). If the time base of the unit hydrograph is T hr, then a continuous rainfall producing one unit of runoff every period would develop a constant outflow at the end of T hr. Thus only T/t_R unit hydrographs need be combined to produce an S curve which should reach equilibrium at flow q_e:

$$q_e = \frac{2.78A}{t_R} \qquad (7\text{-}5)$$

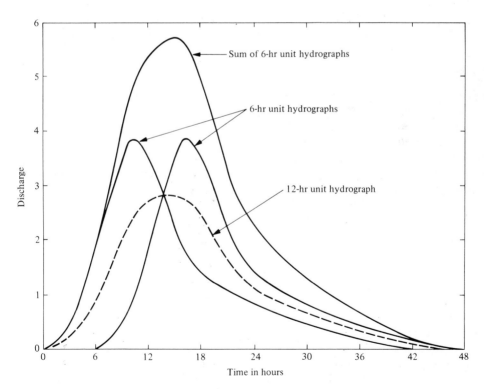

Figure 7-12 Construction of a unit hydrograph for duration 2 t.

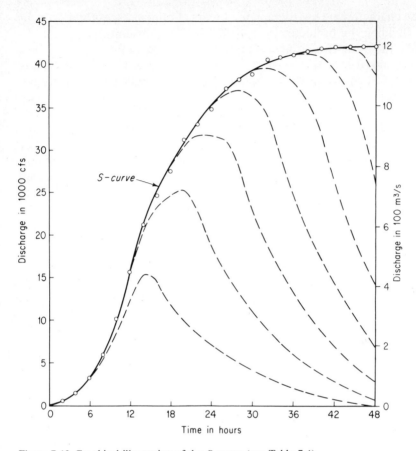

Figure 7-13 Graphical illustration of the S curve (see Table 7-1).

where q_e is in cubic meters per second, A is the drainage area in square kilometers, the runoff is in centimeters, and t_R is again the unit duration. In English units, 1 in/hr will produce an equilibrium flow of 645.3 ft³/s · mi².

Commonly, the S curve tends to fluctuate about the equilibrium flow. This means that the initial unit hydrograph does not actually represent runoff at a uniform rate over time t_R. If a uniform rate of runoff is applied to a basin, equilibrium flow at the rate given by Eq. (7-5) must eventually develop. If the actual effective duration of runoff associated with the original unit hydrograph is not t_R hr, the summation process results in a runoff diagram with either periodic gaps or periodic increases to a rate of 2 units in t_R hr (Fig. 7-14). Thus the S curve serves as an approximate check on the assumed duration of effective rainfall for the unit hydrograph. A duration which results in minimum fluctuation of the S curve can be found by trial. However, fluctuation of the S curve can also result from nonuniform runoff generation during the t_R hr, unusual areal distribution of rain, or errors in basic data. For this reason, the S curve can indicate only an approximate duration.

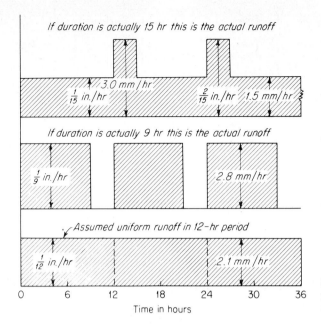

Figure 7-14 Influence of nonuniform rate of runoff on the S curve.

Construction of an S curve does not require tabulating and adding T/t_R unit hydrographs with successive lags of t_R hr. Table 7-1 illustrates the computation of an S curve, starting with an initial unit hydrograph for which $t_R = 6$ hr. For the first 6 hr the unit hydrograph and S curve are identical (columns 2 and 4). The S-curve additions (column 3) are the ordinates of the S curve set ahead 6 hr. Since an S-curve ordinate is the sum of all concurrent unit-hydrograph ordinates, combining the S-curve additions with the initial unit hydrograph is the same as adding all previous unit hydrographs.

The difference between two S curves with initial points displaced by t'_R hr gives a hydrograph for the new duration t'_R hr. Since the S curve represents runoff production at a rate of one unit in t_R hr, the runoff volume represented by this new hydrograph will be t'_R/t_R units. Thus the ordinates of the unit hydrograph for t'_R hr are computed by multiplying the S-curve differences by the ratio t_R/t'_R.

Instantaneous unit hydrograph (IUH). As the duration of the unit hydrograph approaches zero, the *instantaneous unit hydrograph* by definition, the flow sequence represents the outflow from the instantaneous application of unit rainfall excess over the catchment. Mathematically, the rate of direct runoff at time t is given by

$$q_t = \int_0^t f(\tau)i_e(t - \tau)d\tau \qquad (7\text{-}6)$$

where $f(\tau)$ is the IUH ordinate at time τ, i_e is the intensity of rainfall excess at time $(t - \tau)$, and τ is time in the past. In other words, the flow is determined by

Table 7-1 Application of S-curve method

Time, hr (1)	6-hr unit graph (2)	S-curve additions (3)	S curve (2) + (3) (4)	Lagged S curve (5)	(4) − (5) (6)	2-hr unit graph (7)
0	0	. . .	0	. . .	0	0
2	400	. . .	400	0	400	1,200
4	1,400	. . .	1,400	400	1,000	3,000
6	3,100	0	3,100	1,400	1,700	5,100
8	5,400	400	5,800	3,100	2,700	8,100
10	8,600	1,400	10,000	5,800	4,200	12,600
12	12,600	3,100	15,700	10,000	5,700	17,100
14	15,400	5,800	21,200	15,700	5,500	16,500
16	14,600	10,000	24,600	21,200	3,400	10,200
18	11,800	15,700	27,500	24,600	2,900	8,700
20	9,900	21,200	31,100	27,500	2,400†	7,200
22	8,400	24,600	33,000	31,100	2,000†	6,000
24	7,200	27,500	34,700	33,000	1,800†	5,400
26	6,000	31,100	37,100	34,700	1,600†	4,800
28	5,100	33,000	38,100	37,100	1,400†	4,200
30	4,200	34,700	38,900	38,100	1,200†	3,600
32	3,400	37,100	40,500	38,900	1,000†	3,000
34	2,700	38,100	40,800	40,500	800†	2,400
36	2,100	38,900	41,000	40,800	600†	1,800
38	1,600	40,500	41,500†	41,000	400†	1,200
40	1,100	40,800	41,900	41,500	200†	600
42	700	41,000	42,000†	41,900	100	300
44	400	41,500†	42,000†	42,000	0	0
46	200	41,900	42,100	42,000	0†	
48	0	42,000	42,100†	42,100	0	

† Adjusted value.

weighting the antecedent rainfall excess, where the weight applied to rainfall occurring τ hr ago is the IUH ordinate τ hr after the beginning of rainfall [12].

The IUH can be derived from an S curve [13] or by routing the time-area diagram of the catchment (Sec. 9-10). By analogy with the S-curve technique for deriving a short-duration unit hydrograph, it will be seen that the IUH ordinates are a function of the slope of the S curve and its duration:

$$f(\tau) = t_R \frac{dq}{dt} \tag{7-7}$$

Thus the IUH ordinate for any specific time can be determined graphically by placing a straightedge tangent to the S curve and reading the change in discharge for a time increment of t_R. Moreover, the S curve can be constructed by fitting a curve to a plot of the accumulated sum of the ordinates (e.g., the 6-hr entries in column 4 of Table 7-1 are the summation of those in column 2).

The IUH can be converted readily to one of any other duration. The required ordinate at any time is simply the average flow during the previous t_R hr.

The average flow can be taken as the average of flows at the beginning and end of the period, except when these flows bridge the peak.

7-11 Synthetic Unit Hydrographs

Unit hydrographs can be derived as described in previous sections only if records are available. Since only a relatively small portion of catchments are gaged, some means of deriving unit hydrographs for ungaged catchments is necessary. This requires a relation between the physical geometry of the area and the resulting hydrographs. Three approaches have been used: formulas relating hydrograph features to basin characteristics, transposition of unit hydrographs, and storage routing (Sec. 9-10). Basin characteristics formulas usually pertain to time of peak, peak flow, and time base of the unit hydrograph. When these features are established, the hydrograph can be sketched to provide the necessary unit volume. In a study of basins in the Appalachian Mountain region, Snyder [14] found the basin lag (in hours) to be a function of basin size and shape:

$$t_p = C_t(LL_c)^{0.3} \tag{7-8}$$

where L is the main stream distance from outlet to divide and L_c is the stream distance from outlet to a point opposite the basin centroid. With distances in miles, the coefficient C_t varied from 1.8 to 2.2 (about 1.4 to 1.7 with kilometers). Snyder found that the unit-hydrograph peak q_p could be estimated from

$$q_p = \frac{C_p A}{t_p} \tag{7-9}$$

where A is the drainage area. The coefficient C_p ranged from about 360 to 440 with A in square miles and q_p in cubic feet per second (about 0.15 to 0.19 for square kilometers and cubic meters per second for 1 mm of runoff). Snyder adopted as the time base of the unit hydrograph (days)

$$T = 3 + 3 \, \frac{t_p}{24} \tag{7-10}$$

The constants of Eq. (7-10) are dependant on the procedure used to separate base flow from direct runoff. Equations (7-8) to (7-10) define the three factors necessary to construct the unit hydrograph for duration t_r, which Snyder took as $0.18t_p$. For any other duration t_R he used an adjusted lag

$$t_{pR} = t_p + \frac{t_R - t_r}{4} \tag{7-11}$$

in Eqs. (7-9) and (7-10).

Snyder's formulas have been tried elsewhere with varying success, and other investigators have proposed similar formulas. However, the best way to apply such methods is to derive coefficients from gaged streams in the vicinity of the problem basin and use these for the ungaged stream, i.e., transposition of unit hydrographs.

One technique for transposing unit hydrographs makes use of a dimensionless unit hydrograph [15–17] such as one of those shown in Fig. 7-15. The dimensionless unit hydrograph masks the effect of basin size and essentially eliminates the effect of shape, except as they are reflected in the estimate of basin lag t_p and runoff volume. From Eq. (7-8) a general expression for basin lag might be expected to take the form

$$t_p = C_t \left(\frac{LL_c}{\sqrt{s}}\right)^n \tag{7-12}$$

where s is a weighted channel slope (Sec. 11-2). If known values of lag are plotted against $LL_c/\sqrt{s}$ on logarithmic paper (Fig. 7-16), the resulting plot should define a straight line for basins of similar hydrologic characteristics. Such a relation can be used to estimate the lag for an ungaged catchment in applying a dimensionless unit hydrograph.

Illustrative example 7-1 Given a 500-km² basin with lag of 10 hr (centroid of rainfall to time of peak flow) derive the 4-hr unit hydrograph using the Soil Conservation Service curve from Fig. 7-15.

Assuming constant intensity of rainfall excess, the centroid occurs 2 hr after beginning of rainfall and the time to peak $(t/t_p = 1)$ is 12 hr (10 + 2). The area under the dimensionless hydrograph (Fig. 7-15) is 1.33 $q_p t_p$, and this volume Q is equivalent to 1 cm of runoff over a 500-km² area A:

$$1.33 \ q_p t_p = Q = \tfrac{1}{100} \times 500 \times 10^6 \ \mathrm{m^3} = 5 \times 10^6 \ \mathrm{m^3}$$

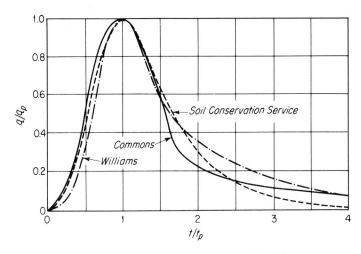

Figure 7-15 Dimensionless unit hydrographs. (*See* [15–17]).

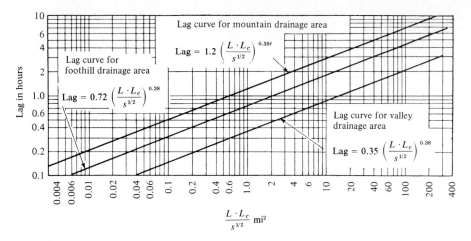

Figure 7-16 Relationship between basin lag (beginning of rain to $\int q\, dt = Q/2$) and basin characteristics in the vicinity of Los Angeles, Calif. (*U.S. Corps of Engineers.*)

Since $t_p = 12$ hr $= 12 \times 60 \times 60$ s $= 4.32 \times 10^4$ s,

$$q_p = \frac{5 \times 10^6}{1.33 \times 4.32 \times 10^4} = 87 \text{ m}^3/\text{s}$$

The ordinates from Fig. 7-15 (at intervals $t/t_p = 0.33$) multiplied by 87 constitute the required 4-hr unit hydrograph.

7-12 Application of Unit Hydrographs

Figure 7-17 illustrates the use of a 3-hr unit hydrograph to synthesize the storm hydrograph from a series of rainfall periods with varying intensity. The increments of runoff for successive 3-hr periods are computed using runoff relations (Chap. 8). The hydrograph of direct runoff resulting from each 3-hr increment is given by multiplying the unit hydrograph by the period runoff. The total hydrograph is the sum of all the incremental hydrographs and estimated base flow.

As stated earlier, the unit hydrograph has been a mainstay of the hydrologist even though some of the techniques discussed in Chap. 12 may offer more flexibility and accuracy in many applications. There are a number of reasons for this, but care should be taken not to apply the unit hydrograph without considering the advantages and disadvantages of other techniques. Perhaps because of assumptions and approximations inherent in the concept, unit hydrographs derived by synthetic and transposition methods seem to be of limited value.

7-13 Hydrograph of Overland Flow

Overland flow is supplied by rainfall and depleted by infiltration. As a result of variations in these factors, coupled with those of roughness and topographic

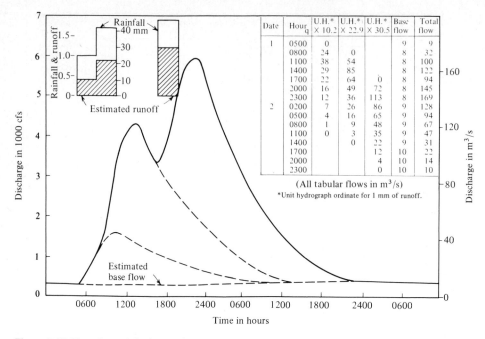

Date	Hour	U.H.* q × 10.2	U.H.* × 22.9	U.H.* × 30.5	Base flow	Total flow
1	0500	0			9	9
	0800	24	0		8	32
	1100	38	54		8	100
	1400	29	85		8	122
	1700	22	64	0	8	94
	2000	16	49	72	8	145
	2300	12	36	113	8	169
2	0200	7	26	86	9	128
	0500	4	16	65	9	94
	0800	1	9	48	9	67
	1100	0	3	35	9	47
	1400		0	22	9	31
	1700			12	10	22
	2000			4	10	14
	2300			0	10	10

(All tabular flows in m³/s)

*Unit hydrograph ordinate for 1 mm of runoff.

Figure 7-17 Use of a unit hydrograph to synthesize a streamflow hydrograph.

elements, overland flow is spatially varied and unsteady; it may be turbulent, laminar, or a combination; and depths may be subcritical or supercritical. The problem is further complicated when raindrops impact the surface of the flowing sheet.

Horton [18] postulated mixed flow conditions on natural slopes, with laminar and turbulent flow interspersed. Although the depth of flow D in the overland sheet is quite small, the quantity of water temporarily detained in this sheet (*surface detention*) can be relatively great. If the flow is laminar (Fig. 7-18)

$$\rho g(D - y)s = \mu \frac{dv}{dy} \tag{7-13}$$

where ρ is density, g is gravity, and μ is absolute viscosity. The assumption is made that the slope is so small that the sine and tangent are equal. Since μ/ρ is equal to kinematic viscosity v,

$$dv = \frac{gs}{v} (D - y)dy \tag{7-14}$$

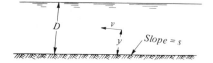

Figure 7-18 Definition sketch for laminar sheet flow.

Integrating and noting that $v = 0$ when $y = 0$, we have

$$v = \frac{gs}{\nu}\left(yD - \frac{y^2}{2}\right) \tag{7-15}$$

Integrating from $y = 0$ to $y = D$ and dividing by D gives the mean velocity

$$v_m = \frac{gsD^2}{3\nu} \tag{7-16}$$

and the discharge per unit width $v_m D$ or

$$q = bD^3 \tag{7-17}$$

where b is a coefficient involving slope and viscosity.

From the Chézy-Manning formula for turbulent flow [Eq. (4-7)]

$$q = b'D^{1.67} \tag{7-18}$$

where b' is a function of slope and the roughness coefficient. Thus the general equation of flow is

$$q = kD^m \tag{7-19}$$

where $1.67 < m < 3.0$ for mixed flow. It will be noted from Eqs. (7-17) and (7-18) that depth increases more rapidly to accommodate increased flow in the case of turbulent flow.

The most extensive early experiments on overland flow were those of Izzard [19]. His tests on long flumes at various slopes and with various surfaces showed that the time to equilibrium is

$$t_e = \frac{2V_e}{60q_e} \tag{7-20}$$

where t_e is defined as the time in minutes when flow is 97 percent of the supply rate and V_e is the volume of water in surface detention at equilibrium. The units of V_e and the equilibrium flow q_e must be consistent. From a 1-ft strip

$$q_e = \frac{iL}{43,200} \tag{7-21}$$

where i is the rainfall rate (or the rate of rainfall excess if the surface is pervious) and L is the distance of overland flow. The constant 43,200 gives q_e in cubic feet per second when i is in inches per hour and L is in feet. With i in millimeters per hour and L in meters the denominator becomes 3.6×10^6 to give q_e in cubic meters per second from a 1-m strip. When average depth on the strip V_e/L is substituted for outflow depth, Eq. (7-17) becomes

$$\frac{V_e}{L} = kq_e^{1/3} \tag{7-22}$$

Combining Eqs. (7-21) and (7-22),

Table 7-2 Retardance coefficient c in Eq. (7-24) [19]

Very smooth asphalt pavement	0.007
Tar and sand pavement	0.0075
Crushed-slate roofing paper	0.0082
Concrete	0.012
Tar and gravel pavement	0.017
Closely clipped sod	0.046
Dense bluegrass turf	0.060

$$V_e = \frac{kL^{4/3} i^{1/3}}{35.1} \tag{7-23}$$

where V_e is the volume of detention (in cubic feet) on the 1-ft strip at equilibrium. In metric units the denominator becomes 228 (for a 1-m strip and with the magnitude of k the same as for English units). Experimentally k was found to be given by

$$k = \frac{0.0007i + c}{s^{1/3}} \tag{7-24}$$

where s is the surface slope and the retardance coefficient c is given in Table 7-2. In metric units the multiplier for i is 2.76×10^{-5}.

Izzard found that the form of the overland-flow hydrograph can be presented as a dimensionless graph (Fig. 7-19). With t_e and q_e known, the q/q_e curve permits plotting of the rising limb of the overland-flow hydrograph. The dimensionless recession curve of Fig. 7-19b defines the shape of the receding limb. At any time t_a minutes after the end of rain, the factor β is

$$\beta = \frac{60q_e t_a}{V_0} \tag{7-25}$$

where V_0 is the detention given by Eqs. (7-22) and (7-24), taking $i = 0$.

Subsequent to Izzard's investigations, others have dealt with the overland-flow problem [20–22], demonstrating that equivalent results can be obtained using turbulent-flow equations. Based on extensive laboratory and field investigations, Emmett [23] found that:

1. With uniform flow (without simulated rain) in the laboratory, smooth or rough surface, flow was laminar up to some critical Reynolds number[†] (1500 to 6000, increasing with slope) and turbulent above. The values of m were about as expected from the above equations.
2. Upon applying simulated rain in the laboratory, some characteristics of turbulent flow were observed at relatively lower discharges, but the most important effect of impacting drops was to retard flow and increase depth for a given discharge.

[†] The nondimensional ratio of inertial and viscous forces, DV/ν.

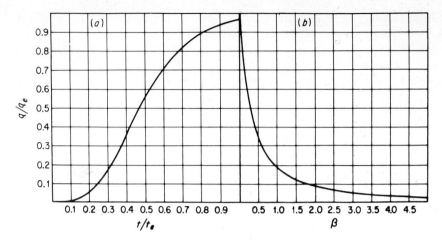

Figure 7-19 Dimensionless hydrograph of overland flow. (*From* [19].)

3. In field investigations, *m* ranged from 1.0 to 2.50, averaging 1.45 compared with 2.1 for rough-surface tests in the laboratory. Depths of flow were relatively greater in the field tests, most likely as a result of topographic irregularities.

The field investigations of Emmett further demonstrate the difficulties in attempting to simulate natural surfaces of meaningful extent in the laboratory.

PROBLEMS

7-1 Using the sample page from a *Water-Supply Paper* which appears as Fig. 4-15, construct direct- and base-flow recession curves of the type illustrated by Fig. 7-4. Are the values of K_r constant for these curves? What are the average values of K_r for direct and base flow? Using an average value of K_r, find the volume of groundwater storage when the flow is 500 ft³/s.

7-2 Tabulated below are ordinates at 24-hr intervals for a hydrograph. Separate the base flow from the direct runoff by the *AC* and *AB* methods illustrated in Fig. 7-6. Compute the volume of direct runoff in each case.

Time, days	Flow, m³/s	Time, days	Flow, m³/s
1	2,340	8	3,230
2	34,300	9	2,760
3	25,000	10	2,390
4	14,000	11	2,060
5	8,960	12	1,770
6	5,740	13	1,520
7	4,300	14	1,320

7-3 Plot the data of Prob. 7-2 on semilogarithmic paper and determine recession constants for surface runoff, interflow, and groundwater flow. What volume of each of the three components is present?

7-4 Tabulated below are the flows on a stream draining 2047 km². Using base-flow separation *ABC* of Fig. 7-6, determine the equivalent depth of direct runoff.

Time, days	Flow, m³/s	Time, days	Flow, m³/s
1	122	8	285
2	1137	9	223
3	950	10	185
4	627	11	161
5	531	12	149
6	429	13	140
7	347	14	132

7-5 Neglecting storage and assuming a linear rise and recession of the elemental hydrograph, sketch the outflow hydrograph from a basin in the shape of a 60° sector of a circle with outflow at the apex. Assume travel time proportional to distance and rainfall duration equal to time of concentration.

7-6 Repeat Prob. 7-5 for a semicircle with outflow at the midpoint of the boundary diameter. What change in shape would result if the rainfall (runoff) were to occur at unit rate for half the duration and 3 times that rate for the remainder?

7-7 How would the hydrograph of Prob. 7-6 be affected if runoff were to occur only from the outer half of the area?

7-8 Given below are the observed flows from a storm of 3-hr duration on a stream with a drainage area of 122 mi². Derive the unit hydrograph. Assume constant base flow = 600 ft³/s.

Hour	Day 1	Day 2	Day 3	Hour	Day 1	Day 2	Day 3
3 a.m.	600	4600	1700	3 p.m.	8000	2700	900
6 a.m.	550	4000	1500	6 p.m.	7000	2400	800
9 a.m.	6000	3500	1300	9 p.m.	6100	2100	700
Noon	9500	3100	1100	Midnight	5300	1900	600

7-9 Given below are three unit hydrographs derived from separate storms on a small basin. All are considered to have resulted from 4-hr rains. Find the average unit hydrograph.

Hours	Storm 1	Storm 2	Storm 3	Hours	Storm 1	Storm 2	Storm 3
0	0	0	0	8	130	170	215
1	110	25	16	9	95	130	165
2	365	125	58	10	65	90	122
3	500	358	173	11	40	60	90
4	390	465	337	12	22	35	60
5	310	405	440	13	10	20	35
6	235	305	400	14	5	8	16
7	175	220	285	15	0	0	0

7-10 Using storm 2 of Prob. 7-9, construct the S curve and find the 2- and 6-hr unit hydrographs. Smooth the S curve as required.

7-11 Take the flows given in Prob. 7-8 as being in cubic meters per second and the drainage area as 2231 km². Derive the unit hydrograph.

7-12 Given below is the 4-hr unit hydrograph for a basin of 84 mi². Derive the S curve and find the 2- and 6-hr unit hydrographs.

Time, hr	Flow, ft³/s	Time, hr	Flow, ft³/s
0	0	11	2700
1	400	12	2200
2	2500	13	1800
3	4400	14	1400
4	6000	15	1100
5	7000	16	800
6	6100	17	600
7	5200	18	400
8	4500	19	200
9	3800	20	100
10	3200	21	0

7-13 Given below is a 12-hr unit hydrograph. Construct the 12-hr S curve by plotting the summation of the given 12-hr ordinates, and derive both the instantaneous and 6-hr unit hydrographs.

Time, hr	0	12	24	36	48	60	72	84	96	108	120
Flow	0	103	279	165	78	36	20	11	5	3	0

7-14 Using the unit hydrograph of storm 3, Prob. 7-9, find the peak flow resulting from four successive 4-hr periods of rainfall producing 0.35, 0.87, 1.39, and 0.77 in of runoff, respectively. Ignore base flow.

7-15 For a drainage basin selected by your instructor, derive a synthetic unit hydrograph using Snyder's method with $C_t = 2.0$ and $C_p = 400$.

7-16 A basin of 360 km² has $L = 25$ km, $L_c = 10$ km. Using Snyder's method, with $C_t = 1.5$, $C_p = 0.17$, and $t_R = 3$ hr, derive the unit hydrograph.

7-17 Using actual streamflow data for a basin assigned by your instructor, find the unit hydrograph for a storm. What values of C_t and C_p for Snyder's method are indicated by the data?

7-18 A parking lot 150 ft long in the direction of the slope and 80 ft wide has a tar and gravel pavement on a slope of 0.0025. Assuming a uniform rainfall intensity of 2.75 in/hr for 30 min, construct the outflow hydrograph, using Izzard's method.

7-19 A city lot 200 ft deep and 100 ft wide has a slope of 0.005 toward the street. The street is 60 ft wide and has a 6-in crown. Assuming a rainfall intensity of 1.8 in/hr, $c = 0.040$ for the lot and 0.007 for the street, and a rainfall duration of 60 min, find the peak flow into the gutter. What will the peak flow be if the rainfall duration is 10 min?

7-20 A paved area 50 m long by 80 m wide has a concrete surface on a slope of 0.005. What will the outflow hydrograph be for a uniform rainfall intensity of 60 mm/hr for 30 min?

REFERENCES

1. Pilgrim, D. H., D. D. Huff, and T. D. Steele: Use of Specific Conductance and Contact Time Relations for Separating Flow Components in Storm Runoff, *Water Resour. Res.*, vol. 15, no. 2, pp. 329–339, April 1979.
2. McDonald, C. C., and W. B. Langbein: Trends in Runoff in the Pacific Northwest, *Trans. Am. Geophys. Union*, vol. 29, no. 3, pp. 387–397, June 1948.
3. Dunne, T.: Field Studies of Hillslope Flow Processes chap. 7 in M. J. Kirkby (ed.): "Hillslope Hydrology," pp. 247–251, Wiley, Chichester, 1978.
4. Hall, F. R.: Base-Flow Recessions—A Review, *Water Resour. Res.* vol. 4, no. 5, pp. 973–983, October 1968.
5. Barnes, B. S.: Discussion of Analysis of Runoff Characteristics, *Trans. ASCE*, vol. 105, p. 106, 1940.
6. Langbein, W. B.: Some Channel Storage and Unit Hydrograph Studies, *Trans. Am. Geophys. Union*, vol. 21, pt. 2, pp. 620–627, 1940.
7. Linsley, R. K., and W. C. Ackermann: A Method of Predicting the Runoff from Rainfall, *Trans. ASCE*, vol. 107, pp. 825–835, 1942.
8. Sittner, W. T., C. E. Schauss, and J. C. Monro: Continuous Hydrograph Synthesis with an API-Type Hydrological Model, *Water Resour. Res.*, vol. 5, no. 5, pp. 1007–1022, October 1969.
9. Horton, R. E.: The Role of Infiltration in the Hydrologic Cycle, *Trans. Am. Geophys. Union*, vol. 14, pp. 446–460, 1933.
10. Sherman, L. K.: Streamflow from Rainfall by the Unit-Graph Method, *Eng. News-Rec.*, vol. 108, pp. 501–505, 1932.
11. Sherman, L. K.: The Unit Hydrograph, chap. 11E in O. E. Meinzer (ed.): "Hydrology," McGraw-Hill, New York, 1942; reprinted Dover, New York, 1949.
12. Schaake, J. C., Jr.: Synthesis of the Inlet Hydrograph, *Tech. Rep.* no. 3, Department of Sanitary Engineering and Water Resources, The John Hopkins University, Baltimore, Md., 1965.
13. Wilson, E. M.: "Engineering Hydrology," pp. 122–123, Macmillan, London, 1969.
14. Snyder, F. F.: Synthetic Unit Hydrographs, *Trans. Am. Geophys. Union*, vol. 19, pt. 1, pp. 447–454, 1938.
15. Commons, G. G.: Flood Hydrographs, *Civil Eng.*, vol. 12, p. 571, 1942.
16. Williams, H. M.: Discussion of "Military Airfields," *Trans. ASCE*, vol. 110, p. 820, 1945.
17. Mockus, V.: "Use of Storm and Watershed Characteristics in Synthetic Hydrograph Analysis and Application," U.S. Soil Conservation Service, 1957.
18. Horton, R. E.: Erosional Development of Streams and Their Drainage Basins; Hydrophysical Approach to Quantitative Morphology, *Bull. Geol. Soc. Am.*, vol. 56, pp. 275–370, 1945.
19. Izzard, C. F.: Hydraulics of Runoff from Developed Surfaces, *Proc. High Res. Board*, vol. 26, pp. 129–150, 1946.
20. Henderson, F. M., and R. A. Wooding: Overland Flow and Groundwater Flow from a Steady Rainfall of Finite Duration, *J. Geophys. Res.*, vol. 69, pp. 1531–1540, April 1964.
21. Morgali, J. R., and R. K. Linsley: Computer Analysis of Overland Flow, *J. Hydraul. Div. ASCE*, vol. 91, pp. 81–100, May 1965.
22. Woo, D. C., and E. F. Brater: Spatially Varied Flow from Controlled Rainfall, *Proc. ASCE*, vol. 88, pp. 31–56, November 1962.
23. Emmett, W. W.: The Hydraulics of Overland Flow on Hillslopes, *U.S. Geol. Surv. Prof. Pap.* 662A, 1970.

BIBLIOGRAPHY

Chow, V. T.: Runoff, sec. 14 in V. T. Chow (ed.): "Handbook of Applied Hydrology," McGraw-Hill, New York, 1964.

Eagleson, P. S.: "Dynamic Hydrology," McGraw-Hill, New York, 1970.

Linsley, R. K., M. A. Kohler, and J. L. H. Paulhus: "Applied Hydrology," McGraw-Hill, New York, 1949.

Kirkby, M. J. (ed.): "Hillslope Hydrology," Wiley, Chichester, 1978.

Viessman, W., Jr., J. W. Knapp, G. L. Lewis, and T. E. Harbaugh: "Introduction to Hydrology," 2d ed., pp. 102–143, IEP, New York, 1977.

EIGHT

RELATIONS BETWEEN PRECIPITATION AND RUNOFF

The flow of streams is controlled primarily by variations in precipitation. Relationships between precipitation and runoff and techniques for distributing the runoff through time (Chap. 7) are a basis for efficient forecasting for the operation of hydraulic projects, for the extension of flow records on gaged streams, and for estimating the flow of ungaged streams.

THE PHENOMENA OF RUNOFF

Section 7-1 discussed the movement of runoff to the streams. The discussion which follows will be limited to processes which retain water in the catchment until it is removed by evapotranspiration.

8-1 Surface Retention

Much of the rain falling during the first part of a storm is stored on the vegetal cover as *interception* and, as rain continues, in surface puddles as *depression storage*. That part of storm precipitation retained on or above the ground surface is *surface retention*. In other words, surface retention includes interception, depression storage, and evaporation during the storm but does not include water temporarily detained en route to the streams.

Although the effect of vegetal cover is unimportant in major floods, interception by some types of cover may be a considerable portion of the annual rainfall. Interception-storage capacity is usually satisfied early in a storm so that a large percentage of the rain in the numerous small storms is intercepted. After the vegetation is saturated, interception would cease were it not for the fact that some water evaporates from the wetted surface of the foliage. Once interception storage is filled, the amount of water reaching the soil surface is equal to the rainfall less evaporation from the vegetation. Interception-storage capacity is reduced as wind speed increases, but the rate of evaporation is increased. Apparently, high wind speeds tend to augment total interception during a long storm and to decrease it for a short storm.

Extensive experimental interception data have been accumulated by numerous investigators [1–5], but evaluation and application of the data to specific problems are made difficult because of varied experimental techniques employed. Data for forest cover, relatively more plentiful than those for crops and other low-level vegetation, are usually obtained by placing several rain gages on the ground under the canopy and comparing the average of their catch with that in the open. Sometimes the gages (*interceptometers*) are placed at random in an attempt to measure average interception for an area. In other cases, the interceptometers are placed at carefully selected points under the tree crown to measure interception on its projected area. In either case, application of the data requires detailed knowledge of the cover density over the area of interest. Most data for forests have been collected by placing interceptometers free of underbrush, grass, etc., and little is known of total interception as required in most hydrologic problems.

Trimble and Weitzman [6] found that mixed hardwood about 50 years old and typical of considerable area in the southern Appalachian Mountains intercepts about 20 percent of the rainfall both in summer and winter. No measurements were made of flow down the tree trunks, but it is likely that net interception is nearer 18 percent for storms with rainfall in the order of 0.5 in (13 mm). Qualitatively, it can be said that annual interception by a well-developed forest canopy is about 10 to 20 percent of the rainfall and that the storage capacity of the canopy ranges from 0.03 to 0.06 in (0.8 to 1.5 mm).

Horton [7] derived a series of empirical formulas for estimating interception (per storm) by various types of vegetal cover. Applying these formulas to 25-mm (1-in) storms and assuming normal cover density give the values of interception in Table 8-1. Assuming sufficient rainfall to satisfy the interception-storage capacity, an equation for total-storm interception V_i can be written as

$$V_i = S_i + Et_R \tag{8-1}$$

where S_i is storage capacity per unit of projected area, E is the evaporation rate, and t_R is the duration of rainfall. Assuming that the interception given by Eq. (8-1) is approached exponentially as the rainfall increases from zero to some high value, then

$$V_i = (S_i + Et_R)(1 - e^{-kP}) \tag{8-2}$$

Table 8-1 Values of interception as computed with Horton's equations for 25-mm (1-in) storms

Crop	Height		Interception	
	ft	m	in	mm
Corn	6	1.8	0.03	0.8
Cotton	4	1.2	0.33	8.4
Tobacco	4	1.2	0.07	1.8
Small grains	3	0.9	0.16	4.1
Meadow grass	1	0.3	0.08	2.0
Alfalfa	1	0.3	0.11	2.8

where e is the napierian base, P is the amount of rain, and k is equal to $1/(S_i + Et_r)$.

Rainwater retained in puddles, ditches, and other depressions in the surface is termed *depression storage*. These depressions vary widely in area and depth; their size depends to a considerable degree on the definition of a depression. As soon as rainfall intensity exceeds the local infiltration capacity (Sec. 8-2), the rainfall excess begins to fill surface depressions. Each depression has its own storage capacity, and when it is filled, further inflow is balanced by outflow plus infiltration and evaporation. Depressions of various sizes are both superimposed and interconnected. Almost immediately after the beginning of rainfall excess, the smallest depressions become filled and overland flow begins. Most of this water in turn fills larger depressions, but some of it follows an unobstructed path to the stream channel. This chain of events continues, with successively larger portions of overland flow contributing to the streams. Water held in depressions at the end of rain either evaporates or contributes to soil moisture and/or subsurface flow following infiltration.

Meaningful observations of the magnitude of depression storage cannot be obtained easily, and values are highly dependent on definition of terms. Individual depressions of appreciable area relative to the drainage basin under consideration are usually called *blind drainage* and excluded from hydrologic analysis. The remaining depression storage is usually lumped with interception and treated as *initial loss* with respect to storm runoff. Nevertheless, depression storage may be of considerable magnitude and may play an important role in the hydrologic cycle. Even impervious areas such as roads and parking lots may have significant depression storage. Stock ponds, terraces, and contour farming all tend to moderate the flood hydrograph by increasing depression storage, while land leveling and drainage may reduce depression storage. Because most basins have some deep depressions, it is likely that the depression-storage capacity is never completely filled.

The volume of water in depression storage V_s can be expressed as

$$V_s = S_d(1 - e^{-kP_e})$$ (8-3)

where S_d is the depression-storage capacity of the basin, k has the value $1/S_d$, and P_e is the volume of precipitation in excess of interception and infiltration. Experience suggests that values of S_d for most basins lie between 10 and 50 mm (0.5 and 2.0 in). The equation neglects evaporation from depression storage during a storm, a factor which is usually unimportant.

8-2 Runoff Mechanisms

Following Horton's presentation of the infiltration theory [8], his concept that surface runoff is generated where and when the rainfall intensity exceeds the rate at which water can enter the soil, most hydrologists considered that all storm runoff was generated by this mechanism. Recent studies have demonstrated, however, that "Hortonian overland flow" is one of several mechanisms, and that it is not necessarily the dominant one [9].

Infiltration. The passage of water through the soil surface into the soil is termed *infiltration*. Although a distinction is made between infiltration and *percolation*, the gravity flow of water within the soil, the two phenomena are closely related since infiltration cannot continue unimpeded unless percolation removes infiltered water from the surface soil. The soil is permeated by noncapillary channels through which gravity water flows downward toward the groundwater, following the path of least resistance. Capillary forces continuously divert gravity water into capillary-pore spaces, so that the quantity of gravity water passing successively lower horizons is steadily diminished. This leads to increasing resistance to gravity flow in the surface layer and a decreasing rate of infiltration as a storm progresses. The rate of infiltration in the early phases of a storm is less if the capillary pores are filled from a previous storm.

The maximum rate at which water can enter the soil at a particular point under a given set of conditions is called the *infiltration capacity*. The actual infiltration rate f_i equals the infiltration capacity f_p only when the *supply rate* i_s (rainfall intensity less rate of retention) equals or exceeds f_p. Theoretical concepts presume that actual infiltration rates equal the supply rate when $i_s \leq f_p$ and are otherwise at the capacity rate (Fig. 8-1). The value of f_p is at a maximum f_0 at the beginning of a storm and approaches a low, constant rate f_c as the soil profile becomes saturated. The limiting value is controlled by subsoil permeability. Horton [8] found that infiltration-capacity curves approximate the form

$$f_p = f_c + (f_0 - f_c)e^{-kt}$$ (8-4)

where e is the napierian base, k is an empirical constant, and t is time from beginning of rainfall. The equation is applicable only when $i_s \geq f_p$ throughout the storm. Philip [10] suggested the equation

$$f_p = \frac{bt^{-1/2}}{2} + a$$ (8-5)

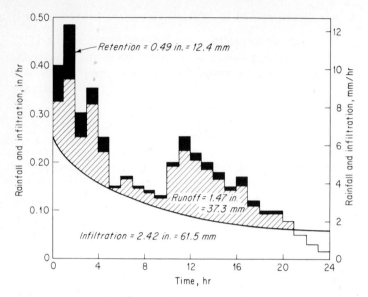

Figure 8-1 Simple separation of infiltration and surface runoff using rainfall data, estimated retention, and an infiltration-capacity curve.

Integrating Eq. (8-5) with respect to time gives the cumulative infiltration F at time t as

$$F = bt^{1/2} + at \qquad (8-6)$$

In Eqs. (8-5) and (8-6), a and b are parameters to be determined experimentally. Other equations for infiltration have been suggested, both physically based and empirical [11, 12].

Infiltration capacity depends on many factors such as soil type, moisture content, organic matter, vegetative cover, and season. Of the soil characteristics affecting infiltration, noncapillary porosity is perhaps the most important. Porosity determines storage capacity and also affects resistance to flow. Thus, infiltration tends to increase with porosity. An increase in organic matter also results in increased infiltration capacity, largely because of a corresponding change in porosity [13].

Figure 8-2 demonstrates the effect of initial moisture content and the variations to be expected from soil to soil. The effect of vegetation on infiltration capacity is difficult to determine, for it also influences interception. Nevertheless, vegetal cover does increase infiltration compared with barren soil because (1) it retards surface flow, giving the water additional time to enter the soil; (2) the root systems make the soil more pervious; and (3) the foliage shields the soil from raindrop impact and reduces rain packing of the surface soil.

Most data on infiltration rates are derived from infiltrometer tests. An *infiltrometer* is a tube or other boundary designed to isolate a section of soil. The effective area varies from less than 1000 cm² to several square meters. Although

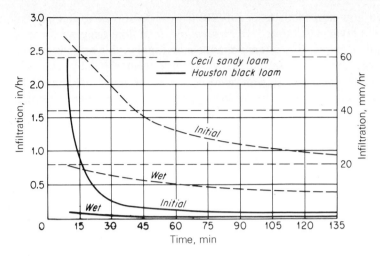

Figure 8-2 Comparative infiltration rates during initial and wet runs. (*From* [13].)

many of the earlier tests were made by flooding the infiltrometer, this method is no longer recommended and has been supplanted by sprinkling techniques which can better simulate heavy and uniform rainfall.† Since it is impossible to measure directly the quantity of water penetrating the soil surface, infiltration is computed by assuming it to equal the difference between water applied and measured surface runoff. In addition to the difficulties inherent in simulating raindrop size and velocity of fall with sprinklers, experiments using artificial rainfall have other features tending to cause higher infiltration rates in tests than under natural conditions.

Attempts have been made to derive infiltration data by analyzing rainfall and runoff data from small drainage basins with homogeneous soils [14], and Horton [15], Sherman [16], and others have presented techniques for deriving average "equivalent" infiltration-capacity data from records of heterogeneous basins. On natural basins allowance must be made for subsurface flow and surface retention. Since the infiltration capacity at a specific time and point depends on many factors (antecedent precipitation, soil characteristics, vegetal cover, slope, and aspect of the surface, among others), it must be highly variable, even within a small area [17]. Betson [18] used the concept of spatial variability of infiltration capacity and interpreted a regression coefficient as a measure of contributing area. The concept is also embodied in the Stanford Watershed Model [19].

Saturation overland flow. Rain falling on the stream surfaces constitutes an effective mechanism for flow generation even though it is not usually a major

---

† Tests to determine infiltration rates during flood or furrow irrigation are customarily made by flooding.

factor. Overland flow also tends to occur [20] if water is forced to the surface where the surface layer is saturated. *Saturation overland flow* occurs primarily at the base of slopes marginal to stream channels, in topographic hollows where flow lines converge, and in localized areas with thin soils underlain by relatively impervious strata. Streamflow so generated is said to be non-Hortonian [21], since it postulates saturation as the criterion for runoff rather than a rainfall intensity in excess of the local infiltration capacity.

The distinction between Hortonian and saturation overland flow is perhaps somewhat academic since the uppermost minute layer of the soil must be saturated at any point where overland flow is present, and saturation of a surface layer does not preclude infiltration. It has been proposed [22] that the distinction be clarified by specifying that overland flow is the result of a rising shallow water table or lateral subsurface storm inflow (e.g., "surface saturation from below"). Academic or not, there are catchments for which the principal mechanism of surface runoff is not entirely in accord with the concepts of Horton or Betson. Rainfall intensity over some forested upland catchments seldom exceeds the infiltration capacity, except for small characteristic areas in valley bottoms and hollows [23, 24]. The *variable source concept* postulates that these contributing (saturated) areas shrink and expand, depending on antecedent conditions and storm rainfall.

Subsurface storm flow. Interflow as a component of runoff is discussed in Chap. 7. For subsurface storm flow (interflow) to provide a major contribution to the total storm runoff requires the existence of a shallow layer of high permeability at the surface. Even then, the effectiveness of the mechanism has been questioned [22]. Mosley [9] reports to have found that the subsurface storm flow mechanism is dominant for a small forested catchment (0.3 ha) with steep slopes and shallow soil on an impermeable base.

8-3 The Runoff Cycle

The *runoff cycle* is the descriptive term applied to that portion of the hydrologic cycle between incident precipitation over land areas and subsequent discharge of this water through stream channels or evapotranspiration. Hoyt [25] has presented a comprehensive description of the hydrologic phenomena occurring at selected times during the runoff cycle by considering an idealized cross section of a basin.

Figure 8-3 shows schematically the time variations of the hydrologic factors during an extensive, uniform storm on a relatively dry basin of idealized spatial homogeneity. The shaded area of the figure represents the portion of total precipitation which eventually becomes streamflow measured at the basin outlet. Channel precipitation is the only increment of streamflow during the initial period of rainfall in the absence of any impervious areas. As streams rise, surface area and consequently the volume rate of channel precipitation increase.

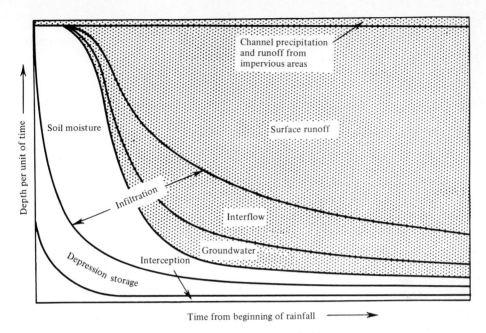

Figure 8-3 Schematic diagram of the disposition of storm rainfall.

The rate of interception is high at the beginning of rain, especially during summer and with dense vegetal cover. However, the available storage capacity is depleted rather quickly, so that the interception rate decreases to that required to replace water evaporated from the vegetation. The rate at which depression storage is filled also decreases rapidly from a high initial value as the smaller depressions become filled and approaches zero at a relatively high value of total-storm rainfall.

Except in very intense storms, the greater portion of the local soil-moisture deficiency is satisfied before appreciable surface runoff takes place. However, some of the rain occurring late in the storm undoubtedly becomes soil moisture, since the downward movement of this water is relatively slow.

Water infiltrating the soil surface and not retained as soil moisture either moves to the stream as interflow or penetrates to the water table and eventually reaches the stream as groundwater. The rate of surface runoff starts at zero, increases slowly at first and then more rapidly, eventually approaching a relatively constant percentage of the rainfall rate. Both the percentage and the rate of runoff depend upon rainfall intensity.

Figure 8-3 illustrates only one of an infinite number of possible cases. A change in rainfall intensity would change the relative magnitude of all the factors. Further complications are introduced by varying rainfall intensity during the storm or by occurrence of snow or frozen ground. To appreciate further the complexity of the process in a natural basin, remember that all the factors of Fig. 8-3 vary from point to point within the basin during a storm. Nevertheless,

the foregoing description should aid in understanding the relative time variations of hydrologic phenomena which are important in considering the runoff
relations discussed later in the chapter.

ESTIMATING THE VOLUME OF STORM RUNOFF

Despite the complex nature of the rainfall-runoff process, the practice of estimating runoff as a fixed percentage of rainfall is the most commonly used
method in design of urban storm-drainage facilities, highway culverts, and
many small water-control structures. The method can be correct only when
dealing with a surface which is completely impervious so that the applicable
runoff coefficient is near 1.00.

Computer simulation techniques (Chap. 12) offer the most reliable method
of computing runoff from rainfall because they permit a relatively detailed
analysis using short time intervals. The type of analysis used in computer
simulation would be virtually impossible to carry through by hand because of
the detailed computations required. The constraints of hand calculation led to
methods using longer time intervals and a correspondingly less rigorous model.
The following sections discuss some of the more successful approaches.

8-4 Initial Moisture Conditions

The quantity of runoff from a storm depends on (1) the moisture conditions of
the catchment at the onset of the storm and (2) the storm characteristics—
rainfall amount, intensity, and duration. The storm characteristics are defined
by the data from the precipitation-gage network, but no single observation
serves to define the antecedent moisture conditions. Much of the investigation
of rainfall-runoff relations was directed at finding a simple index of basin moisture conditions.

In humid areas, where streams flow continuously, groundwater discharge
at the beginning of the storm has been found to be a good index to initial
moisture conditions. In a study of the Valley River, North Carolina, Linsley
and Ackermann [26] found that field-moisture deficiency at any time was approximately equal to 90 percent of the total Class A pan evaporation since the
ground was last saturated less any additions made to field moisture by intervening rains. Basin-accounting techniques (Sec. 5-15) applied on a daily basis
provide a reasonably accurate estimate of moisture deficiency which can be
used as an index to runoff [27].

The most common index is based on antecedent precipitation. The rate at
which moisture is depleted from a particular basin under specified meteorological conditions is roughly proportional to the amount in storage (Sec. 5-15). In
other words, the soil moisture should decrease logarithmically with time during
periods of no precipitation [28],

$$I_t = I_0 k^t \tag{8-7}$$

where I_0 is the initial value of the *antecedent-precipitation index*, I_t is the reduced value t days later, and k is a recession factor ranging normally between 0.85 and 0.98. Letting t equal one gives

$$I_1 = kI_0 \qquad (8\text{-}8)$$

Thus, the index for any day is equal to that of the previous day multiplied by the factor k. If rain occurs on any day, the amount of rain is added to the index (Fig. 8-4). Since storm runoff does not add to the residual moisture of the basin, an index of precipitation minus runoff, i.e., *basin recharge,* should be more satisfactory than the precipitation index alone. Commonly, however, the minor improvements gained do not justify the added computation.

Equation (8-7) assumes that the daily depletion of soil moisture (primarily evapotranspiration) is

$$I_0 - I_1 = I_0(1 - k) \qquad (8\text{-}9)$$

Since actual evapotranspiration is a function of the potential value and the available moisture (I_0), k should be a function of potential evapotranspiration. The variation in potential evapotranspiration is largely seasonal, and Eq. (8-7) has been found to be reasonably satisfactory when used jointly with calendar date (Sec. 8-6). There is an added advantage in using the date as a parameter because it also reflects variations in surface conditions related to farming practices, stage of plant growth, etc.

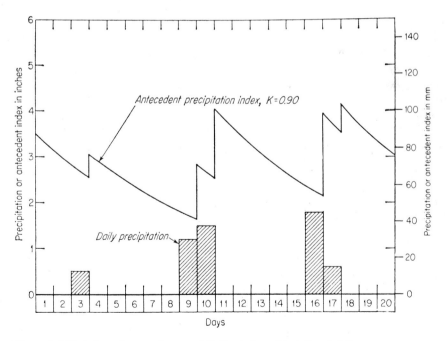

Figure 8-4 Variation of antecedent precipitation index with daily rainfall.

The value of the index on any day theoretically depends on precipitation over an infinite antecedent period, but if a reasonable initial value is assumed, the computed index will closely approach the true value within a few weeks. The index value applicable to a particular storm is taken as that at the beginning of the first day of rain. Thus a value of 1.8 in would be used for the storm of the ninth and tenth in Fig. 8-4.

8-5 Storm Analysis

In any statistical correlation, it is extremely important that the basic data be as consistent and reliable as possible. The consistency tests for precipitation data discussed in Sec. 3-10 should be applied whenever the normal annual precipitation varies appreciably over the catchment. The streamflow records should be carefully reviewed in each case (Sec. 4-16) and adjustments made if necessary.

Methods of storm analysis should be rigorous and objective. Only that storm rainfall which produced the runoff being considered should be included. Small showers occurring after the hydrograph had started to recede should not be included if they had little effect upon the amount of runoff. Similarly, showers occurring before the main storm should be excluded from the storm rainfall and included in the antecedent-precipitation index. Long, complex storms should be separated into as many short storm periods as possible by hydrograph analysis.

Runoff also depends upon rainfall intensity, but for basins of 250 km² (100 mi²) or more, an average intensity as reflected by amount and duration is usually adequate. In this case duration can be estimated with sufficient accuracy from 6-hr rainfall data. An objective rule is preferable, such as "the sum in hours of those 6-hr periods with more than 5 mm (0.2 in) of rain plus one-half the intervening periods with less than 5 mm (0.2 in)." Although experimental infiltration data indicate rates commonly in excess of 2.5 mm/hr (0.1 in/hr), relations such as Fig. 8-5 consistently show the effect of duration on storm runoff to be of the order of 0.25 mm/hr (0.01 in/hr). The difference is largely caused by intercorrelations and the inclusion of interflow with surface runoff.

8-6 Multivariate Relations for Total Storm Runoff

If storm characteristics and basin conditions are to be represented adequately in a runoff relation, a number of independent variables must be included. The relationship is not an additive one, and the usual multivariate linear correlation is not satisfactory. The coaxial graphical method† of correlation was first shown to be particularly useful for this work [28]. Betson et al. [29] subsequently demonstrated an analytical correlation technique.

---

† Graphical correlation methods were presented in Appendix A of previous editions. Increasing availability of computers has resulted in restricted use of graphical techniques, and the detailed presentation of such methods is now considered unjustified.

To illustrate the coaxial method assume that a relation for estimating storm runoff is desired, using antecedent precipitation, date (or week number), and rainfall amount and duration as variables. Values of these variables are compiled for 50 or more storms. With the exception of rainfall amount, the variables should be more closely related to the fraction of rainfall which does not run off than to the runoff volume. It is therefore convenient to calculate an auxiliary variable equal to the storm rainfall minus the storm runoff. This variable is called the *recharge* in the subsequent discussion. Once a satisfactory relation for estimating recharge is completed, it is a simple matter to revise the precipitation quadrant (chart *C*) so that the final answer is in terms of runoff since rainfall minus recharge should equal runoff. Equations (8-10) and (8-11) show that the correlation can be made to yield runoff, even though the season quadrant is unchanged from that based on recharge.

A three-variable relation is developed first (Fig. 8-5, chart *A*) by (1) plotting antecedent precipitation versus recharge, (2) labeling the points with week

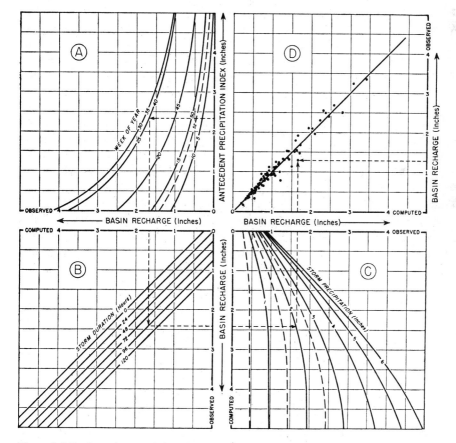

Figure 8-5 Basin-recharge relation for Monocacy River at Jug Bridge, Maryland. (*U.S. National Weather Service.*)

number, and (3) fitting a smooth family of curves representing the various weeks. Chart B is placed with its horizontal scale matching that of chart A to facilitate plotting. Points are plotted in chart B with observed recharge as ordinate and recharge computed from chart A as abscissa, and these points are labeled with duration. A family of smooth curves is drawn to represent the effect of duration on recharge. Charts A and B together are a graphical relation for estimating recharge from antecedent index, week, and duration. Storm precipitation is then introduced (chart C) by (1) plotting recharge computed from charts A and B against observed recharge, (2) labeling the points with rainfall amount, and (3) fitting a family of curves. Charts A, B, and C constitute the *first approximation* to the desired relation. Chart D indicates the overall accuracy of the derived charts.

Since the variables are intercorrelated and the first charts are developed independently of factors subsequently introduced, revision of the charts may improve the overall relation. The process is one of successive approximations. To check the week curves, the other curve families are assumed to be correct and the adjusted abscissa for a point in chart A is determined by entering charts B and C in reverse order with observed recharge, rainfall amount, and duration. The ordinate for the adjusted point is the observed antecedent-precipitation index. In other words, the week curves should be revised to fit this adjusted point if the relation is to yield a computed recharge equal to the observed recharge. The second (and subsequent) approximations for duration and rainfall are made in the same manner. In each case the points are plotted by entering the chart sequence from both ends with observed values to determine the adjusted coordinates.

Although the method presented in the previous paragraphs is general and can be used as described, certain modifications simplify the procedure and require fewer approximations. Since storm rainfall is extremely important, the first plotting of chart A may show little correlation and the construction of the curves will be difficult. However, an important advantage in having the rainfall parameter in the last chart is that the possibility of computing runoff in excess of rainfall or of computing negative values of runoff is eliminated. Moreover, the arrangement of Fig. 8-5 results in the determination of a unified index of moisture conditions in the first chart, which is a decided advantage in river forecasting. If the plotting of chart A is limited to storms having rainfall within a specified class interval (2 to 4 in, for example), the construction of the curves is simplified, provided that there are sufficient data. Only limited data are required since the general curvature and convergence are always as shown in the example. The relations are quite similar throughout a geographic region, and charts A and B for one basin may be used as a first approximation for another basin in the area.

One analytical technique [29] uses the equations

$$I_{ps} = c + (a + dI_s)e^{-bl} \tag{8-10}$$

$$Q = (P^n + I_{ps}^n)^{1/n} - I_{ps} \tag{8-11}$$

where I_{ps} is a runoff index approximating the first quadrant of a coaxial plot, I_s is a fixed function of week number ranging between $+1$ and -1, I is the antecedent-precipitation index, e is the base of napierian logarithms, P is storm rainfall, Q is direct runoff, and a, b, c, d, and n are statistically derived coefficients. With only five constants the functions are quickly derived on a computer but are more constrained in form than the graphical solution. In comparative tests on catchments in the Tennessee River basin, the analytic method gave results generally slightly better than the graphical method. McCallister [30] describes another method of using the computer to develop a runoff relation based on the variables used in the coaxial method.

Since it is impossible to segregate the water passing a gaging station according to the portion of the basin in which it fell, statistical runoff relations must be based on basin averages of the variables. Unfortunately, a relation based on storms of uniform areal distribution will yield runoff values which are too low when applied to storms with extremely uneven distributions. This can be demonstrated by computing the runoff for 4, 6, and 8 in of rainfall, assuming all other factors remain fixed. While 6 is the average of 4 and 8, the average runoff from the 4- and 8-in rainfalls is not equal to that from a 6-in rain. An uneven distribution of antecedent precipitation produces similar results. Runoff relations based on uniform areal conditions can be used to compute the runoff in the vicinity of each rainfall station, and the average of these runoff values will, in general, more nearly approach the observed runoff from the basin when either the storm or the antecedent precipitation is quite variable.

8-7 Relations for Incremental Storm Runoff

In order to determine increments of runoff throughout a storm for application of a unit hydrograph, Fig. 8-5 may be used with accumulated rainfall up to the end of each period and the successive values of runoff subtracted to obtain increments. When applied to small catchments, however, there is a marked tendency to underestimate the peak flows, because the relation does not properly account for the time distribution of rainfall. Some of the errors for larger basins are certainly caused by the same factor.

The principal problem in developing an incremental runoff relation lies in our inability to determine short-period increments of runoff by hydrograph analysis. A process of successive iterations has been proposed [31], and another approach [32] modifies the relation for total storm runoff by introducing a second (short-term) antecedent precipitation index as illustrated in Fig. 8-6. Since duration is constant for an incremental relation (6-hr, for example), this factor can be eliminated by making suitable adjustments in one of the other quadrants. In comparing Fig. 8-6 with Fig. 8-5, it will be seen that the duration quadrant has been replaced by one for the short-term (retention) index, and that the precipitation quadrant has been converted to one yielding runoff instead of recharge. It has been found that the function for the retention quadrant can be assumed as

$$I_{psr} = I_{ps}(B)^{I_r} \qquad (8\text{-}12)$$

where B is a constant less than unity, I_{ps} is the runoff index from the season quadrant, and I_{psr} is the integrated index reflecting also the retention index I_r.

Two coefficients must be evaluated for the retention index quadrant—the recession factor for computing I_r [corresponding to k in Eq. (8-8)] and B in Eq. (8-12). This is accomplished by a trial-and-error procedure. Using assumed values of B, incremental runoff is computed and summed for each storm event

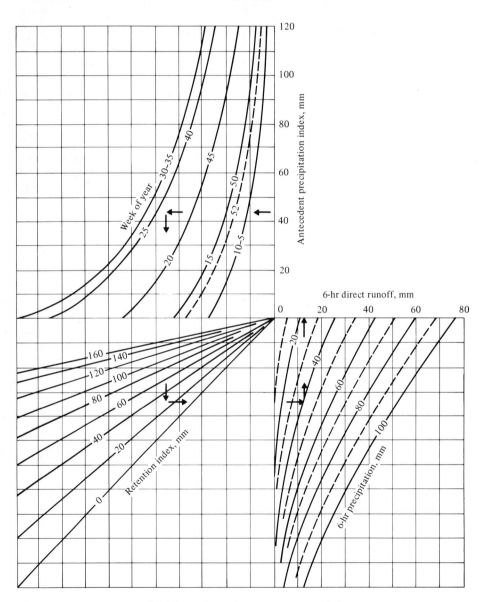

Figure 8-6 Incremental runoff relation using a short-term retention index.

under study. Comparing computed and observed total storm runoff for several assumed values of B leads to a solution for any selected recession factor. Values of the short-term recession factor (daily) are the order of 0.4 to 0.5 and B ranges from 0.6 to 0.8 (0.98 to 0.99 with precipitation in millimeters). The t-hr recession factor is equal to the daily factor to the $t/24$ power.

8-8 Infiltration Approach to Runoff Estimates

The infiltration approach assumes that the surface runoff from a given storm is equal to that portion of the rainfall which is not disposed of through (1) interception and depression storage, (2) evaporation during the storm, and (3) infiltration. If items 1 and 2 are invariable or insignificant or can be assigned reasonable values, one need be concerned only with rainfall, infiltration, and runoff. In the simplest case, where the supply rate i_s is at or in excess of the infiltration capacity, surface runoff is equivalent to the storm rainfall less surface retention and the area under the capacity curve.

The procedure appears to be simple and to offer a solution to the estimation of short period increments of runoff. Experience has shown otherwise. If the rainfall intensity is always above the infiltration-capacity curve (Fig. 8-1) the problem is merely one of defining the infiltration curve which is a function of the antecedent moisture conditions. If rainfall intensities fluctuate above and below the infiltration curve, the matter is confused, since the curve inherently assumes that the infiltration capacity decreases because a fixed amount of water was added to the soil moisture during an interval. If $i_s < f_p$, the increment of soil moisture is less than assumed and the drop in the infiltration curve correspondingly less.

The time-intensity pattern of rainfall is rarely uniform over the catchment, and the applicable infiltration-capacity curve varies from point to point depending on soils, vegetation, and antecedent moisture. Finally, the infiltration approach ignores other storm-flow generation mechanisms (Sec. 8-2) which, in addition to groundwater accretion, must be determined in some other way. For these and other reasons the infiltration approach never proved satisfactory as a tool for hydrograph prediction.

8-9 Infiltration Indexes

Difficulties with the theoretical approach to infiltration led to the use of infiltration indexes [33]. The simplest of these is the Φ *index,* defined as that rate of rainfall above which the rainfall volume equals the runoff volume (Fig. 8-7). The W *index* is the average infiltration rate during the time rainfall intensity exceeds the capacity rate; i.e.,

$$W = \frac{F}{t} = \frac{1}{t}(P - Q_S - S) \tag{8-13}$$

where F is total infiltration, t is time during which rainfall intensity exceeds infiltration capacity, P is total precipitation corresponding to t, Q_S is surface

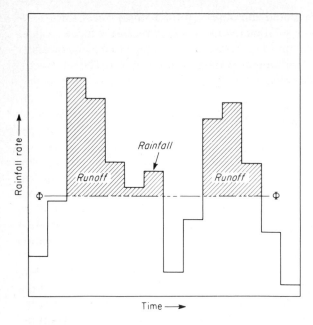

Figure 8-7 Schematic diagram showing the meaning of the Φ index.

runoff, and S is the effective surface retention. The W index is essentially equal to the Φ index minus the average rate of retention by interception and depression storage. While the segregation of infiltration and retention would seem to be a refinement, the task of estimating the retention rate is such that combining it with infiltration is probably equally satisfactory.

With very wet conditions, when the infiltration capacity is at a minimum and the retention rate is very low, the values of W and Φ are almost equal. Under these conditions, the W index becomes the W_{min} index by definition. This index is used principally in studies of maximum flood potential.

The derivation of infiltration indexes is simple, and the approach is cloaked in an aura of logic through the name *infiltration index*. Actually the indexes are no more than average loss rates, and their magnitude is highly dependent on antecedent conditions, so that they are in no way superior to the multivariate relations (Secs. 8-6 and 8-7). The Φ index has been used in some unit-hydrograph studies to define the time pattern of excess rainfall. In such cases the actual runoff volume is known, and there is no problem in calculating Φ. Since it is known that the actual infiltration is not constant, however, the runoff pattern derived with the Φ index cannot be correct.

ESTIMATING SNOWMELT RUNOFF

The storage and melting of snow plays an important role in the hydrology of some areas. In such areas, reliable predictions of the rate of melt and release of liquid water from a snowpack are requisite to the efficient design and operation of water resources projects and the issuance of river forecasts and warnings. Energy-balance snowpack models have recently been developed which also simulate temperature, albedo, density, water equivalent, and other snowcover characteristics during both accumulation and ablation [34–36]. The heat transfer within the pack must be modeled to determine the changes in energy storage as required in the energy-balance equation [34], and it is usually necessary to simulate other unobserved variables in applying conceptual models (Chap. 12). Modeling is simplified if interest is restricted to periods of melt when the pack is *ripe* (saturated and isothermal at 0°C).

8-10 Physics of Snowmelt

Snowmelt and evaporation (including sublimation) are both thermodynamic processes, and both are amenable to the energy-balance approach [Eq. (5-2)]. In applying the energy balance to a snowpack, the rate of melt and release of liquid water are of primary concern. Heat exchange with the soil is more important when treating snowmelt than in the case of lake evaporation, but the exchange at the snow-air interface dominates the melt process.

For each centimeter of water melted in a snowpack at 0°C (32°F), heat must be supplied in the amount of 80 cal/cm² (203 cal/cm² per inch of melt). Meltwater may refreeze with diurnal cooling, or it may drain from the pack, thus contributing to soil moisture, groundwater recharge, or streamflow.

Snowmelt computations are simplified by the fact that the surface is always at the freezing point during melt, and at the same time they are made more complex by variations in albedo, the percentage of solar and diffuse radiation reflected by the surface (Sec. 2-2). The energy for snowmelt is derived from (1) net radiation, (2) conduction and convective transfer of sensible heat from the overlying air, (3) condensation of water vapor from the overlying air, (4) conduction from the underlying soil, and (5) heat supplied by incident rainfall.

Components of radiation exchange are discussed in Sec. 5-3 in connection with evaporation. Although only 5 to 10 percent of incident shortwave radiation is reflected by a free-water surface, 80 to 90 percent is reflected by a clean, dry snow surface. As snow ages, its albedo drops to 50 percent or less because of changes in crystalline structure, density, and amount of dirt on the surface. The shortwave radiation absorbed by the pack penetrates to varying depths depending largely upon snow density. About one-half of the radiation entering a homogeneous pack is absorbed in the upper 1.7 cm (0.7 in) with density of 0.1 and, with density of 0.7, one-half will penetrate about 7 cm or 2.8 in [37].

Snow radiates essentially as a blackbody, and the outgoing longwave radiation at 0°C or 32°F during a 24-hr period is equivalent to about 84 mm (3.3 in) of negative melt. The net loss by longwave radiation is materially less than this amount since the atmosphere reradiates a portion back to the earth. Net longwave radiation depends primarily on air and snow temperatures, atmospheric vapor pressure, and extent and type of cloud cover. With clear skies net loss by longwave radiation is equal to about 20 mm (0.8 in) of melt per day when dewpoint and air and snow temperatures are near freezing. Numerous empirical formulas have been derived to estimate this factor [38, 39] and net all-wave radiation, but they are not particularly satisfactory. Radiometer observations of net radiation could provide the needed data if an adequate network of stations were established. While an adequate network is perhaps feasible for flat terrain and vegetation which does not penetrate the snow surface, representative measurements under forest cover and in rugged terrain are difficult to achieve.

Heat exchange between a snowpack and the atmosphere is also effected by conduction, convection, condensation, and evaporation [34, 39, 40]. Although it is readily shown that conduction in still air is very small, convective exchange can be an important factor [41]. As in the case of evaporation, extensive research has been reported [42] on the theoretical aspects of turbulent transport between the atmosphere and an underlying snow surface. If the dewpoint of the air is above freezing, condensation on the snow occurs, with consequent release of latent heat. If the dewpoint is less than the snow-surface temperature, the vapor pressure of the snow is greater than that of the air and evaporation occurs. The rate of transfer of sensible heat (convection) is proportional to the temperature difference between the air T_a and the snow T_0, while vapor transport is proportional to the vapor-pressure difference $e_a - e_0$. The transfer rates by both processes are proportional to wind velocity v. Since the latent heat of vaporization is about 7.5 times the latent heat of fusion, condensation of unit depth of water vapor on the snow surface produces 8.5 units of liquid water, including condensate. The two processes can be described by similar equations for melt:

$$M_h = k_h(T_a - T_0)v \tag{8-14}$$

$$M_e = k_e(e_a - e_0)v \tag{8-15}$$

where the exchange coefficients k_h and k_e are dependent on the units used, levels of observation above the surface, and other factors. During periods of melt, $T_0 = 0°C = 32°F$ and $e_0 = 6.11$ mb $= 0.18$ in Hg.

Temperatures beneath the soil surface do not vary as much or as rapidly as surface temperatures, and there is usually an increase of temperature with depth during the winter and early spring. This results in a transfer of heat from the soil to the snowpack above. While the heat transfer is small on a daily basis, it may be equivalent to several inches of melt during an entire season—sufficient to keep the soil saturated and to permit a prompt response of streamflow when melting from other causes takes place.

Raindrop temperatures correspond closely to the surface wet-bulb temperature. As the drops enter a snowpack, their temperature is reduced to 0°C (32°F) and an equivalent amount of heat is imparted to the snow. The melt (millimeters) from rain is given by

$$M_r = \frac{PT_w}{80} \tag{8-16}$$

where P is the rainfall in millimeters, T_w is the wet-bulb temperature in degrees Celsius, and 80 is the latent heat of fusion in calories per gram. The heat available in 10 mm of rain at 10°C will melt about 1.2 mm of water from the snow. There has been a tendency to overemphasize the importance of rainfall melt because warm rains are accompanied by high humidity and temperature and often by moderate to strong winds. Equation (8-16) becomes

$$M_r = \frac{P(T_w - 32)}{144} \qquad (8\text{-}16a)$$

if M_r and P are in inches and T_w is in degrees Fahrenheit.

From a theoretical point of view, it would appear that snowmelt computations could be made by applying the energy-budget approach, and this is basically true for snowmelt at a point [34]. On a natural basin, however, the hydrologist is confronted with numerous complications, among the more obvious being the effects of variations in elevation, slope, aspect, and forest cover on radiation exchange, temperature, wind, and other key factors in snowmelt. The heat required to melt snow depends on its thermal quality (Sec. 3-16), which may also vary within a basin. Finally, since the hydrologist is concerned with the runoff from snowmelt, the retention of liquid water within the snowpack must be taken into account. Gerdel [43] found that ripe snow can retain about 5 percent liquid water by weight but, if all or part of the pack is "dry," retention becomes more difficult to estimate [44]. For these reasons and because of the scarcity of pertinent data, the theoretical approach is generally inapplicable. Even so, computer simulation (Chap. 12) is usually based on a simplified energy budget to the extent that data are available.

8-11 Estimating Snowmelt Rates and Consequent Runoff

Air temperature is the single most reliable index to snowmelt. It so completely reflects radiation, wind, and humidity that residual errors are usually not materially correlated with these factors. Since snowmelt does not occur with temperatures appreciably below freezing, the temperature data are commonly converted to degree-days or degree-hours above some base, usually the freezing point (see Sec. 2-11). A day with a mean temperature of 10°C represents 10 degree-days above 0°C (or 18 degree-days, 50° minus 32°, with computations in Fahrenheit). If the minimum temperature for the day were not below freezing, there would be 10×24 or 240 degree-hours (Celsius). On the other hand, there are more than 240 degree-hours on a day with a mean temperature of 10°C and a minimum temperature below freezing.

A variety of different relationships have been suggested for forecasting snowmelt. Most commonly, however, a *degree-day factor* or the ratio of snowmelt to concurrent degree-days is utilized. If the actual rate of snowmelt were known, the degree-day factor might well be substantially constant. Actually, the rate of runoff must be used in lieu of rate of melt, and a plot of accumulated snowmelt runoff versus accumulated degree-days tends toward an ogive shape.

Early in a period of melt some heat may be required to bring the snow temperature up to the freezing point. In addition, some of the initial meltwater will be stored in the snow or soil so that the degree-day factor will appear to be low. As time passes, the degree-day factor will increase as an increasing fraction of each day's meltwater reaches the stream gage. Near the end of the melting period the snow cover may become thin and patchy, and the apparent melt per degree-day will again decrease unless the runoff is calculated for the actual area of snow surface. Values of the degree-day factor generally range from near zero to 9 mm per Celsius degree-day (0.2 in/F°-day), with values between 2 and 5 mm per Celsius degree-day (0.05 and 0.1 in/F°-day) being most common.

In many basins of high relief, winter conditions are typified by semipermanent snow cover in the headwaters while the lower portions of the basins have only intermittent snow cover. The lower limit of snow cover (*snow line*) is a dynamic feature, moving up and down the slope. In practice, the snow line is often considered as a contour of elevation, a reasonably realistic assumption following a storm. As melting proceeds, however, snow cover recedes more rapidly on southerly and barren slopes, and the snow line can be defined only as an average elevation of the lower limit of snow. Since surface air temperature is an inverse function of elevation, the rate of snowmelt decreases with elevation. If the freezing isotherm is below the snow line, there is no melting within the basin. Thus, temperature at an index station must be considered in conjunction with the extent of snow cover in estimating snowmelt. Mean daily surface air temperature in mountain regions drops about 0.5 to 1 Celsius degree per 100 m increase in elevation (3 to 5 Fahrenheit degrees per 1000 ft) (Sec. 2-13). Given the temperature at an index station, an area-elevation curve for the basin, and the average snow-line elevation, one can compute the area subject to melting on the basis of an assumed rate of change of temperature with elevation. The summation of the portion of the basin lying within each elevation zone (above the snow line) multiplied by the applicable degree-day value for the elevation gives a weighted value of degree-days over the melting zone [45].

SEASONAL AND ANNUAL RUNOFF RELATIONS

The hydrologist is frequently required to estimate monthly, seasonal, or annual volumes of runoff. Such estimates may be needed for operational purposes or to provide a data base for evaluating reservoir storage requirements. A more reliable basis for such estimates is the calculation of short-term runoff (daily or storm period) with relations such as those discussed in Sec. 8-6. However, if it is necessary to estimate flows for many years, such computations would be extremely tedious. Simulation methods (Chap. 12) offer the most promising combination of reliability and ease of calculation [46]. In relatively humid regions, where precipitation is reasonably uniformly distributed through the year, and in cases where the runoff is largely from the melting of a seasonal snowpack, variations in antecedent conditions and distinctions between direct and groundwater runoff may be less important and simpler relationships may prove adequate. Pending improvement of long-range precipitation forecasts, prediction of seasonal or annual runoff is feasible only where precipitation accumulates as snow during the winter and melts during the subsequent spring and early summer. Precipitation during the forecast period is usually required as input for the runoff relations used in preparing water-supply forecasts. Predicted precipitation can be used to meet this requirement, or statistical values can be used to provide probability forecasts [47, 48].

8-12 Precipitation-Runoff Relations

A simple plotting of annual precipitation versus annual runoff will often display good correlation (Fig. 8-8), particularly in areas where the major portion of the

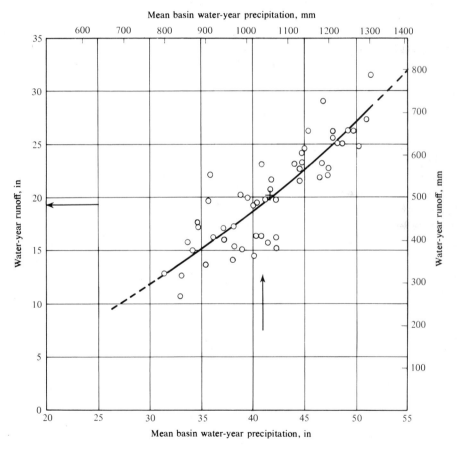

Figure 8-8 Relation between annual precipitation and runoff for the Merrimack River above Lawrence, Mass. (11,550 km² or 4460 mi²). Correlation coefficient = 0.87 and standard error of estimate = 58 mm (2.3 in).

precipitation falls in the winter months. Numerous refinements can be introduced if a moderate increase in reliability justifies the additional effort required in analysis [49]. First, there is the question of consistent data throughout the period of record under study. Observed runoff should be adjusted for storage changes in reservoirs and, in some cases, for increased evaporation losses from the reservoirs. Adjustment may also be needed for consumptive loss from irrigation or other water uses. Finally, changes in observational techniques, discharge section, gage location, and other factors have contributed to discharge records which in some cases show definite time trends unexplained by precipitation. Testing precipitation and runoff data by double-mass analysis (Chaps. 3 and 4) may disclose inconsistencies which can be eliminated through appropriate adjustments.

Figure 8-8 is based on averages of precipitation observed at 30 stations in and near the basin. If records are available for only a few stations, application of station weights may improve results. Summer precipitation and a portion of that occurring in the fall and spring are in the form of rain and fall on bare ground, and much of this precipitation is lost to soil moisture and evapotranspiration. Winter snows accumulate as a snowpack, and evapotranspiration losses are relatively small. The effectiveness of precipitation in producing streamflow thus depends on when it occurs, and the application of monthly weights may improve the reliability of the relation.

The ratio of mean basin precipitation to that observed at specific points depends on storm type, which may in turn vary with season and from year to year. The interdependence of precipitation pattern and meteorological causes can influence optimum station and monthly weights, and may even explain apparent time trends in the relation between precipitation and streamflow [50].

Station and monthly weights can be derived through least-squares analysis. The high intercorrelation of precipitation observed at stations within a limited area may result in regression coefficients (weights) which overemphasize the relative value of the better stations, and if so, the results should be tempered toward equal weighting. Derived monthly weights may tend to be erratic, particularly if based on a short record, but they can be adjusted to provide a smooth transition throughout the year.

There is often a substantial lag between precipitation and the subsequent discharge of that portion which recharges the groundwater. Where groundwater is an appreciable part of the total flow, introduction of precipitation during the previous year as an additional parameter may improve the reliability of the relation. Streamflow during the previous year and flow during one or more winter months [51] have also been used as indexes of groundwater carry-over.

8-13 Use of Snow Surveys

The application of snow-survey data to the preparation of water-supply forecasts is appealing because of the rather simple relation envisioned. If the seasonal flow results primarily from melting of a mountain snowpack, measurements of the water in the snowpack before melt begins should indicate the volume of runoff to be expected. For many years snow surveys were made near the end of the snow-accumulation season, but now they are usually made monthly or more often during the accumulation season. Daily observations telemetered from snow pillows are also used. Because of drifting, variations in

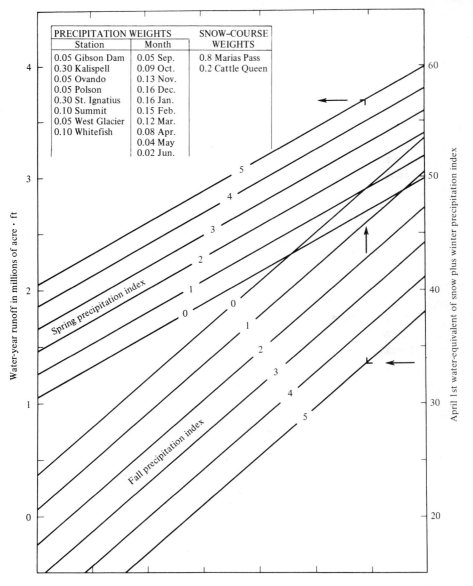

Figure 8-9 Annual runoff relation for South Fork, Flathead River, above Hungry Horse Reservoir, Montana. (*U.S. National Weather Service.*)

winter melt from point to point, etc., snow cover is usually more variable than precipitation within a given area. It is not economically feasible to take a sufficient number of samples to determine directly the volume of water stored in the pack, and the surveys (Sec. 3-16) constitute nothing more than an index. Accordingly, samples must be taken at the same points and in a consistent manner each year. Care must be taken to avoid changes in exposure brought about by forest fires, growth of timber and underbrush, etc.

Although there is good correlation between snow-survey data and seasonal runoff, it is now recognized that reliable water-supply forecasts cannot be made from snow surveys alone [52, 53]. Runoff subsequent to the surveys is also dependent upon (1) groundwater storage, (2) antecedent soil-moisture deficiency, and (3) precipitation during the runoff period. It has been found [54, 55] that snow-survey data can best be treated as an independent measure of winter precipitation in a multiple correlation (Fig. 8-9), or as a check on simulated snow cover [56].

The areal extent of snow cover has been found to be indicative of seasonal snowmelt runoff in some basins. Where data are inadequate for conventionally based streamflow forecasts, it appears that predictions based on satellite-derived snow-cover data may prove useful [57].

PROBLEMS

8-1 Derive a mean rainfall-versus-runoff curve from the data shown in the table on page 259 for the Ramapo River at Pompton Lakes, New Jersey (drainage area = 160 mi^2). Compute the average error of the relation and the bias for the tabulated summer (May to October) and winter (November to April) storms.

8-2 Derive an average loss rate (recharge divided by duration) for each of the storms in the table on page 259. Using the average of these loss rates, compute the runoff for each storm and the average error of estimate and bias of the estimates.

8-3 Given the initial conditions depicted by arrows in the season and retention quadrants of Fig. 8-6, determine 6-hr runoff increments for three successive periods during which rainfall was 30, 40, and 20 mm (assume daily recession factors of 0.90 and 0.50, respectively, for updating the antecedent precipitation and retention indexes). What are the equivalent 6-hr recession factors? How much more runoff would be expected under the same initial conditions if the 90 mm of rain had fallen in one 6-hr period?

8-4 Compute the Φ and W indexes for the storm depicted in Fig. 8-1.

8-5 Find the contribution to recharge and runoff (combined) for a day on which net radiation = 100 cal/cm^2; convective transfer to the snowpack = 50 cal/cm^2; condensation on the snowpack = 0.5 mm depth; rainfall (at 8°C) = 10 mm; and water equivalent of the snowpack is initially 40 mm. Assume the snowpack to be "ripe."

8-6 Develop the equation for air temperature throughout the day as a function of the daily mean and range, assuming a sine curve with minimum at 0600 hr and maximum at 1800 hr.

8-7 Given maximum and minimum daily temperatures of 55 and 35°F, respectively, and assuming temperature to follow a sine curve, compute the degree-days and degree-hours for a 24-hr period using a base of 32°F. Repeat the computations with temperatures of 45 and 25°F. Comment on the results.

Data for problems 8-1, 8-2, and 8-12.

Storm no.	Date	Ante-cedent-precip-itation index	Week of year	Storm dura-tion, hr	Storm precip-itation, in	Storm runoff, in	Basin recharge, in
1	3/15/40	0.90	11	12	2.03	1.23	0.80
2	4/22/40	2.20	16	18	1.95	1.05	0.90
3	5/17/40	0.58	20	9	1.50	0.46	1.04
4	9/2/40	2.70	35	12	1.81	0.33	1.48
5	10/3/40	1.00	40	12	1.07	0.10	0.97
6	12/17/40	0.75	51	12	1.13	0.33	0.80
7	2/8/41	0.50	6	15	1.95	0.99	0.96
8	4/7/41	0.29	14	20	1.30	0.53	0.77
9	12/25/41	1.20	52	15	1.60	0.53	1.07
10	2/1/42	0.65	5	9	1.06	0.30	0.76
11	2/18/42	1.03	7	12	0.79	0.22	0.57
12	3/4/42	0.45	9	9	1.83	0.89	0.94
13	8/10/42	1.26	32	15	2.87	0.49	2.38
14	8/18/42	3.29	33	6	2.81	1.06	1.75
15	9/28/42	1.00	39	18	3.58	1.01	2.57
16	2/7/43	1.06	6	18	0.67	0.17	0.50
17	6/2/43	1.07	22	9	1.47	0.30	1.17
18	11/10/43	2.05	45	21	2.33	0.92	1.41
19	3/8/44	0.45	10	12	0.95	0.50	0.45
20	4/25/44	1.57	17	21	1.94	1.10	0.84
21	1/2/45	1.25	1	6	1.30	0.74	0.56
22	7/23/45	5.45	30	12	3.55	2.39	1.16
23	8/7/45	2.56	32	9	1.61	0.42	1.19
24	11/30/45	1.42	48	24	2.25	0.92	1.33
25	5/28/46	1.44	22	15	2.90	1.44	1.46
26	9/25/46	1.03	39	12	2.27	0.43	1.84
27	6/9/47	1.27	23	10	1.54	0.31	1.23
28	5/14/48	1.11	20	27	2.27	0.71	1.56
29	5/30/48	1.56	22	6	1.18	0.34	0.84
30	7/14/48	1.00	28	9	1.83	0.22	1.61
31	12/31/48	0.70	52	36	4.67	2.77	1.90
32	11/26/50	1.12	48	15	3.30	1.12	2.18
33	3/31/51	1.00	13	36	5.25	3.51	1.74
34	10/8/51	0.40	41	18	1.85	0.11	1.74
35	3/12/52	0.81	11	12	2.83	2.09	0.74
36	4/6/52	0.89	14	18	3.10	1.55	1.55
37	6/2/52	2.17	22	15	4.10	2.07	2.03
38	8/17/52	2.66	33	15	3.08	0.83	2.25
39	9/2/52	1.20	35	6	3.94	1.11	2.83
40	12/4/52	2.11	49	12	1.50	0.71	0.79
41	3/12/53	1.01	11	36	3.50	2.15	1.35
42	3/23/53	2.33	12	12	2.00	1.08	0.92

8-8 If the daily range in temperature is 14 Celsius degrees, what combination of maximum and minimum temperatures will result in the largest degree-day error (assume sine curve to compute hourly temperatures)? What is the magnitude of the error?

8-9 Assuming the temperature-index station to be at an elevation of 1000 ft, the variation of temperature with elevation to be 4 Fahrenheit degrees per 1000 ft, and the following area-elevation characteristics:

Elevation, ft	Percent of area	Elevation, ft	Percent of area
1000	0	5000	80
2000	5	6000	90
3000	30	7000	96
4000	60	8000	100

compute weighted degree-days for index-station temperatures of 60 and 80°F and snow-line elevations of 3000 and 5000 ft for each temperature.

8-10 For a basin selected by the instructor, compile the necessary data and derive a precipitation-runoff relation of the type shown in Fig. 8-8.

8-11 Assuming $S_i = 0.1$ in and $E = 0.002$ in/hr, what would be the accumulated interception loss at the end of 4 and 20 hr for the storm depicted in Fig. 8-1? Supposing $S_d = 0.35$ in, what would be the depression storage at the same times?

8-12 Using the data for Prob. 8-1, develop a coaxial relation similar to Fig. 8-5. Use the two left-hand quadrants of Fig. 8-5 as a first approximation. What is the average error of this relation?

REFERENCES

1. Linsley, R. K., M. A. Kohler, and J. L. H. Paulhus: "Applied Hydrology," pp. 263–268, McGraw-Hill, New York, 1949.
2. Rowe, P. B., and T. M. Hendrix, Interception of Rain and Snow by Second-Growth Ponderosa Pine, *Trans. Am. Geophys. Union,* vol. 32, no. 6, pp. 903–908, December 1951.
3. Merriam, R. A.: A Note on the Interception Loss Equation, *J. Geophys. Res.,* vol. 65, no. 11, pp. 3850–3851, November 1960.
4. Corbett, E. S., and R. P. Crouse: Rainfall Interception by Annual Grass and Chaparral . . . Losses Compared, *U.S. For. Serv. Res. Pap.* PSW-48, 1968.
5. Johnston, R. S.: Rainfall Interception in a Dense Utah Aspen Clone, *U.S. For. Serv. Res. Note* INT-143, July 1971.
6. Trimble, G. R., Jr., and S. Weitzman: Effect of a Hardwood Forest Canopy on Rainfall Intensities, *Trans. Am. Geophys. Union,* vol. 35, no. 2, pp. 226–234, April 1954.
7. Horton, R. E.: Rainfall Interception, *Mon. Weather Rev.,* vol. 47, no. 9, pp. 603–623, September 1919.
8. ———: The Role of Infiltration in the Hydrologic Cycle, *Trans. Am. Geophys. Union,* vol. 14, pp. 446–460, 1933.
9. Mosley, M. P.: Streamflow Generation in a Forested Watershed, New Zealand, *Water Resour. Res.,* vol. 15, no. 4, pp. 795–806, August 1979.
10. Philip, J. R.: An Infiltration Equation with Physical Significance, *Soil Sci.,* vol. 77, p. 153, 1954.

11. Collis-George, N.: Infiltration for Simple Soil Systems, *Water Resour. Res.,* vol. 13, no. 2, pp. 395–403, April 1977.
12. Dunin, F. X.: Infiltration, Its Simulation for Field Conditions, chap. 8 in J. C. Rodda (ed.): "Facts of Hydrology," pp. 199–227, Wiley, London, 1976.
13. Free, G. R., G. M. Browning, and G. W. Musgrave: Relative Infiltration and Related Physical Characteristics of Certain Soils, *U.S. Dept. Agric. Tech. Bull.* 729, 1940.
14. Zingg, A. W.: The Determination of Infiltration Rates on Small Agricultural Watersheds, *Trans. Am. Geophys. Union,* vol. 24, pt. 2, pp. 476–480, 1943.
15. Horton, R. E.: Determination of Infiltration Capacity for Large Drainage Basins, *Trans. Am. Geophys. Union,* vol. 18, pt. 2, pp. 371–385, 1937.
16. Sherman, L. K.: Comparison of F-Curves Derived by the Methods of Sharp and Holtan and of Sherman and Mayer, *Trans. Am. Geophys. Union,* vol. 24, pt. 2, pp. 465–467, 1943.
17. Musgrave, G. W., and H. N. Holtan: Infiltration, sec. 12 in V. T. Chow (ed.): "Handbook of Applied Hydrology," McGraw-Hill, New York, 1964.
18. Betson, R. P.: What Is Watershed Runoff?, *J. Geophys. Res.,* vol. 69, no. 8, pp. 1541–1552, April 1964.
19. Crawford, N. H., and R. K. Linsley: Digital Simulation in Hydrology: Stanford Watershed Model IV, *Stanford Univ., Dept. Civ. Eng. Tech. Rep.* 39, 1966.
20. Kirkby, M. J., and R. J. Chorley: Throughflow, Overland Flow, and Erosion, *Bull. Int. Assoc. Sci. Hydrology,* vol. 12, no. 1, pp. 5–21, 1967.
21. Chorley, R. J.: The Hillslope Hydrological Cycle, chap. 1 in M. J. Kirkby (ed.): "Hillslope Hydrology," pp. 13–16, Wiley, Chichester, 1978.
22. Freeze, R. A.: Role of Subsurface Flow in Generating Surface Runoff: 2 Upstream Source Areas, *Water Resour. Res.,* vol. 8, no. 5, pp. 1272–1283, October 1972.
23. Hewlett, J. D., and A. R. Hibbert: Factors Affecting the Response of Small Watersheds to Precipitation in Humid Areas, in W. E. Sopper and H. W. Lull (ed.): "Forest Hydrology," pp. 275–290, Pergamon, New York, 1967.
24. Dunne, T., and R. D. Black: Partial Area Contributions to Storm Runoff in a Small New England Watershed, *Water Resour. Res.,* vol. 6, no. 5, pp. 1296–1311, October 1970.
25. Hoyt, W. G.: An Outline of the Runoff Cycle, *Penn. State Coll. Sch. Eng. Tech. Bull.* 27, pp. 57–67, 1942.
26. Linsley, R. K., and W. C. Ackermann: Method of Predicting the Runoff from Rainfall, *Trans. ASCE,* vol. 107, pp. 825–846, 1942.
27. Kohler, M. A.: Meteorological Aspects of Evaporation Phenomena, *Gen. Assemb. Int. Assoc. Sci. Hydrol., Toronto,* vol. 3, pp. 421–436, 1957.
28. Kohler, M. A., and R. K. Linsley: Predicting the Runoff from Storm Rainfall, *U.S. Weather Bur. Res. Pap.* 34, 1951.
29. Betson, R. P., R. L. Tucker, and F. M. Haller: Using Analytical Methods to Develop a Surface-Runoff Model, *Water Resour. Res.,* vol. 5, no. 1, pp. 103–111, February, 1969.
30. McCallister, J. P.: Role of Digital Computers in Hydrologic Forecasting and Analysis, *Int. Assoc. Sci. Hydrol. Publ.* 63, pp. 68–77, 1963.
31. Miller, J. F., and J. L. H. Paulhus: Rainfall-Runoff Relation for Small Basins, *Trans. Am. Geophys. Union,* vol. 38, no. 2, pp. 216–218, April 1957.
32. Sittner, W. T., C. E. Schauss, and J. C. Monro: Continuous Hydrograph Synthesis with an API-Type Hydrological Model, *Water Resour. Res.,* vol. 5, no. 5, pp. 1007–1022, October 1969.
33. Cook, H. L.: The Infiltration Approach to the Calculation of Surface Runoff, *Trans. Am. Geophys. Union,* vol. 27, no. 5 pp. 726–747, October 1946.
34. Anderson, E. A.: A Point Energy and Mass Balance of a Snow Cover, *NOAA Tech. Rep. NWS* 19, February 1976.
35. Outcalt, S. I., C. Goodwin, G. Weller, and J. Brown: A Digital Computer Simulation of the Annual Snow and Soil Thermal Regimes at Barrow, Alaska, *Cold Regions Res. Dev. Lab. Res. Rep.* 331, Hanover, N.H., 1975.

36. Eggleston, K. O., E. K. Israelsen, and J. P. Riley: Hydrid Computer Simulation of the Accumulation and Melt Processes in a Snowpack, Utah Water Res. Lab. PRWG 65-1, Uath State University, Logan, 1971.
37. Thomas, C. W.: On the Transfer of Visible Radiation through Sea Ice and Snow, *J. Glaciology*, vol. 4, no. 34, pp. 481–484, 1963.
38. Anderson, E. R.: Energy-Budget Studies, Water-Loss Investigations, vol. 1, Lake Hefner Studies, *U.S. Geol. Surv. Prof. Pap.* 269 (reprint of *U.S. Geol. Surv. Circ.* 229), pp. 71–119, 1954.
39. "Snow Hydrology, Summary Report of the Snow Investigations," pp. 145–166, North Pacific Division, U.S. Corps of Engineers, Portland, Oreg., June 1956.
40. Wilson, W. T.: An Outline of the Thermodynamics of Snowmelt, *Trans. Am. Geophys. Union*, vol. 22, pt. 1, pp. 182–195, 1941.
41. Light, P.: Analysis of High Rates of Snow Melting, *Trans. Am. Geophys. Union*, vol. 22, pt. 1, pp. 195–205, 1941.
42. See [34], pp. 8–22.
43. Gerdel, R. W.: Dynamics of Liquid Water in Deep Snow Packs, *Trans. Am. Geophys. Union*, vol. 26, pt. 1, pp. 83–90, 1945.
44. Colbeck, S. C.: An Analysis of Water Flow in Dry Snow, *Water Resour. Res.*, vol. 12, no. 3, pp. 523–527, June 1976.
45. Linsley, R. K.: A Simple Procedure for the Day-to-Day Forecasting of Runoff from Snowmelt, *Trans. Am. Geophys. Union*, vol. 24, pt. 3, pp. 62–67, 1943.
46. Kuehl, D. W.: Volume Forecasts Using the SSARR Model in a Zone Mode, *Proc. West. Snow Conf.*, pp. 38–47, 1979.
47. Twedt, T. M., J. C. Schaake, and E. L. Peck: National Weather Service Extended Streamflow Prediction, *Proc. West. Snow Conf.*, pp. 52–57, 1977.
48. Twedt, T. M., R. J. C. Burnash, and L. R. Ferral: Extended Streamflow Prediction during the California Drought, *Proc. West. Snow Conf.*, pp. 92–96, 1978.
49. Kohler, M. A., and R. K. Linsley: Recent Developments in Water Supply Forecasting, *Trans. Am. Geophys. Union*, vol. 30, no. 3, pp. 427–436, June 1949.
50. Peck, E. L.: The Little Used Third Dimension, *Proc. West. Snow Conf.*, pp. 33–40, 1964.
51. Peck, E. L.: Low Winter Streamflow as an Index to the Short- and Long-Term Carryover from Previous Years, *Proc. West. Snow Conf.*, pp. 41–48, 1954.
52. U.S. Soil Conservation Service, Snow Survey and Water Supply Forecasting, "SCS National Engineering Handbook," sec. 22, April 1972.
53. "Review of Procedures for Forecasting Inflow to Hungry Horse Reservoir, Montana," Water Management Subcommittee, Columbia Basin Interagency Committee, June 1953.
54. Hannaford, J.: Multiple-graphical Correlation for Water Supply Forecasting, *Proc. West. Snow Conf.*, pp. 26–32, 1956.
55. Kohler, M. A.: Water-Supply Forecasting Developments, *Proc. West. Snow Conf.*, pp. 62–68, 1957.
56. See [46], p. 39.
57. Rango, A., V. V. Salomonson, and J. L. Foster: Seasonal Streamflow Estimation in the Himalayan Region Employing Meteorological Satellite Snow Cover Observations, *Water Resour. Res.*, vol. 13, no. 1, pp. 109–112, February 1977.

BIBLIOGRAPHY

Gartska, W. U.: Snow and Snow Survey, sec. 10 in V. T. Chow (ed.): "Handbook of Applied Hydrology," McGraw-Hill, New York, 1964.
Nordenson, T. J., and M. M. Richards: River Forecasting, sec. 25-IV in V. T. Chow (ed.): "Handbook of Applied Hydrology," McGraw-Hill, New York, 1964.

Ogrosky, H. O., and V. Mockus: Hydrology of Agricultural Lands, sec. 21 in V. T. Chow (ed.): "Handbook of Applied Hydrology," McGraw-Hill, New York, 1964.

U.S. Soil Conservation Service: Snow Surveying and Water Supply Forecasting, sec. 22 in "SCS National Engineering Handbook," April 1972.

Viessman, W., Jr., J. W. Knapp, G. L. Lewis, and T. E. Harbaugh: "Introduction to Hydrology," 2d ed., pp. 67–81, 343–398, 451–478, IEP, New York, 1977.

World Meteorological Organization: "Guide to Hydrological Practices," 3d ed., Geneva, pp. 5.37–5.47, 5.89–5.97, WMO no. 168, 1974.

HYDROLOGIC ROUTING

As discharge in a channel increases, stage also increases and with it the volume of water in temporary storage in the channel. During the falling portion of a flood an equal volume of water must be released from storage. As a result, a flood wave moving down a channel appears to have its time base lengthened and (if volume remains constant) its crest lowered. The flood wave is said to be *attenuated*. Wave movement in natural channels traditionally has been treated in design and prediction by applying *hydrologic routing* procedures [1]. Such procedures solve the continuity equation (or *storage equation*) for an extended reach of the river, usually bounded by selected gaged points. The *hydraulic routing* methods discussed in Chap. 10 deal directly with the hydraulic characteristics of the channel and, in some cases, also take dynamic effects into account. *Streamflow routing* is a general term applied to methods used to predict unsteady flow in streams.

Given the flow at an upstream point, routing can be used to compute the flow at a downstream point. The principles of routing apply also to computation of the effect of a reservoir on the shape of a flood wave. Hydraulic storage occurs not only in channels and reservoirs but also as water flowing over the ground surface. Hence, storage is effective at the very inception of the flood wave, and routing techniques may be used to compute the hydrograph which will result from a specified pattern of rainfall excess.

9-1 Wave Movement

One of the simplest wave forms is the *monoclinal rising wave* in a uniform channel. Such a wave consists of an initial steady flow, a period of uniformly

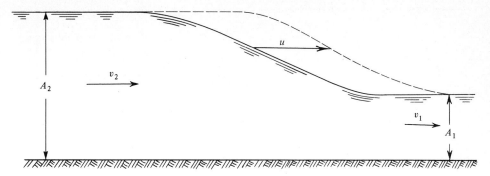

Figure 9-1 Definition sketch for analysis of a monoclinal rising wave.

increasing flow, and a continuing steady flow at the higher rate. Figure 9-1 shows the initial wave profile and the profile after an elapsed time of 1 s. From the law of continuity and assuming negligible effects from any change in wave shape, the difference between inflow and outflow must equal the change in storage within the reach:

$$u\,(A_2 - A_1) = A_2 v_2 - A_1 v_1 \tag{9-1}$$

where u and v are velocities of the wave and water, respectively, and A is the cross-sectional area of the channel. Solving Eq. (9-1) for the wave velocity and substituting the discharge q for Av gives

$$u = \frac{A_1 v_1 - A_2 v_2}{A_1 - A_2} = \frac{q_1 - q_2}{A_1 - A_2} \tag{9-2}$$

The velocity of a monoclinal wave is thus a function of the area-discharge relation for the stream (Fig. 9-2). Since velocity usually increases with stage, area-discharge curves are usually concave upward. The slopes of the secants OA and OB represent the water velocities at sections 1 and 2, respectively ($v_1 = q_1/A_1 = \tan\theta_1$), while the slope of the secant AB represents the wave velocity [Eq. (9-2)]. It may be concluded that (1) wave velocity is greater than the water velocity in most channels; (2) for a given peak flow, the wave having the highest initial flow will travel fastest; and (3) for a wave of very small height

$$u = \frac{dq}{dA} = \frac{1}{B}\frac{dq}{dy} \tag{9-3}$$

where B is the channel top width. Equation (9-3) is known as *Seddon's law* after the man who first demonstrated its validity on the Mississippi River [2]. Theoretical aspects of the law were independently derived by Kleitz (1858) and others [3], but Seddon was unaware of these works.

From the Chézy formula for flow in a wide, open channel (assuming depth equal to hydraulic radius),

$$v = Cy^{1/2}s^{1/2} \tag{9-4}$$

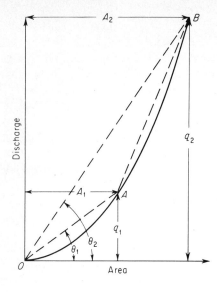

Figure 9-2 Typical area-discharge relation for a stream, and its influence on wave velocity.

and

$$q = Av = vBy = CBy^{3/2}s^{1/2} \qquad (9\text{-}5)$$

where s is water-surface slope. Differentiating gives

$$\frac{dq}{dy} = \tfrac{3}{2}CBy^{1/2}s^{1/2} = \tfrac{3}{2}Bv \qquad (9\text{-}6)$$

Substituting Eq. (9-6) into Eq. (9-3),

$$u = \tfrac{3}{2}v \qquad (9\text{-}7)$$

The derived ratio between water and wave velocities depends on channel shape and the flow formula used, and involves the implicit assumption that discharge is a single-valued function of area. Values shown in Table 9-1 may be used as rough estimates of flood-wave velocity in natural channels. Waves resulting from abrupt changes in flow like those which occur with the opening of a flood gate require consideration of other equations governing wave movement (Chap. 10).

Table 9-1 Theoretical ratio between wave and water velocity for regular sections

Shape	Manning	Chézy
Triangle	1.33	1.25
Wide rectangle	1.67	1.50
Wide parabola	1.44	1.33

9-2 Waves in Natural Channels

Controlled experiments [4, 5] in flumes of regular cross section have confirmed the equations developed in Sec. 9-1. Reasonable checks have also been obtained in natural streams where the effect of local inflow is negligible, as in Seddon's work on the lower Mississippi [2] and Wilkinson's [6] study of waves downstream from TVA dams.

Simple mathematical treatment of flood waves is necessarily limited to uniform channels with fairly regular cross section. The hydrologist must deal with nonuniform channels of complex section with nonuniform slope and varying roughness. Most flood waves are generated by nonuniform lateral inflow along all the channels of the stream system. Thus natural flood waves are considerably more complex than the simplified cases which yield to mathematical analysis, but theoretical treatment is particularly useful in studies of surges in canals, impulse waves in still water (including seiches and tides), and waves released from dams. Improvements and widespread availability of electronic computers have led to renewed interest in hydraulic routing methods (Chap. 10).

Natural flood waves are generally intermediate between pure translation and pondage, which occurs in a broad reservoir or lake. Figure 9-3 shows an example of a flood wave moving with nearly pure translation, i.e., little change in shape. Figure 9-4 illustrates the great modifications which can occur when a flood wave moves through a reservoir in which discharge is a function of the quantity of water in storage. Momentum forces predominate in pure translatory waves, and such waves have relatively short time bases compared with the dimensions of the system in which they move. Most natural flood waves move under friction control and have time bases considerably exceeding the dimensions of the stream system.

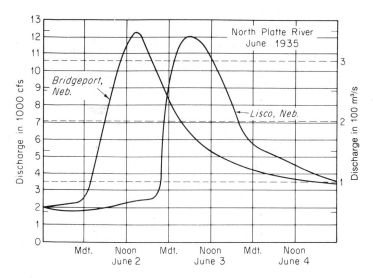

Figure 9-3 Example of near translatory wave movement, North Platte River between Bridgeport and Lisco, Neb.

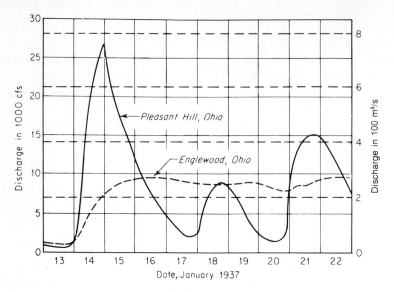

Figure 9-4 Reduction of discharge by storage in the Englewood retarding basin, Stillwater River, Ohio.

9-3 The Storage Equation

The continuity equation may be expressed as

$$I - O = \frac{dS}{dt} \tag{9-8}$$

or

$$\Delta S = S_2 - S_1 = \int_{t_1}^{t_2} I \, dt - \int_{t_1}^{t_2} O \, dt \tag{9-9}$$

where I is inflow rate, O is outflow rate, S is storage (all for a specific reach of a stream), and t is time. To provide a form more convenient for hydrologic routing, it is commonly assumed that the average of the flows at times t_1 and t_2, the beginning and end of the *routing period*,† equals the average flow during the period:

$$\frac{I_1 + I_2}{2} t - \frac{O_1 + O_2}{2} t = S_2 - S_1 \tag{9-10}$$

Most storage-routing methods are based on Eq. (9-10). It is assumed that I_1, I_2, O_1, and S_1 are known and O_2 and S_2 must be determined. Since there are two unknowns, a second relation between storage and flow is needed to complete a solution. The major difficulties in storage routing are involved in this latter relation.

---

† The time difference between t_1 and t_2 is hereafter designated t, the routing period.

The assumption that $(I_1 + I_2)/2 = \bar{I}$ implies that the hydrograph is a straight line during the routing period t. Thus the controlling factor in selecting the routing period is that it be sufficiently short to ensure that this assumption is not seriously violated. The routing period should never be greater than the time of travel through the reach, for if it were, it would be possible for the wave crest to pass completely through the reach during a routing period. If the routing period is shorter than is really necessary, the work of routing is increased since the same computations are required for each routing period. Generally a routing period between one-half and one-third of the time of travel will work quite well. Since hydrologic routing is a solution of the continuity equation, the computed outflow volume for a flood must equal the inflow volume adjusted for any residual change in storage. If the volumes do not check, a serious error in the procedure exists. Minor computational errors usually are compensated quite rapidly. If outflow in an interval is overestimated, the storage at the end of the interval will be too low and the outflow in the next interval will be somewhat too low. Such errors rarely lead to instability in the solution.

9-4 Determination of Storage

Before a relation between storage and flow can be established, it is necessary to determine the volume of water in the stream at various times. The obvious method for finding storage is to compute volumes in the channel from cross sections by using the prismoidal formula. The water surface is usually assumed to be level between cross sections. Total storage in the reach for any given flow conditions is the sum of the storage increments between successive cross sections. For the summation, the elevation in any subreach is the elevation indicated by a backwater curve for the midpoint of the subreach (Fig. 9-5). This method requires extensive surveys to provide adequate cross sections and computation of water-surface profiles for many unsteady, nonuniform flow situations in order to represent the range of conditions expected. The method is difficult and relatively costly and is used where no alternative is possible. It

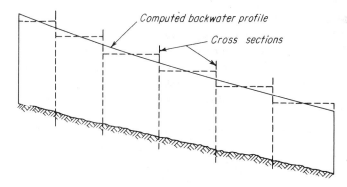

Figure 9-5 Computation of reach storage from channel cross sections.

would be used, for example, to compute storage in a reach in which channel alteration or levee construction is planned, since conditions after construction would be quite unlike those existing before construction.

Storage-elevation curves for reservoirs are usually computed by planimetering the area enclosed within successive contours on a topographic map. The measured area multiplied by the contour interval gives the increment of volume from the midpoint of one contour interval to the midpoint of the next higher interval. A level water surface is usually assumed, a condition which is satisfied in most reservoirs. In reservoirs with relatively small cross-sectional area, the water surface may be far from level when large flows occur (Fig. 9-6). Under such conditions a computation similar to that described above for natural channels must be used.

The common method of determining storage in a reach of natural channel is to use Eq. (9-10) with observed flows. Figure 9-7 shows the inflow and outflow hydrographs for a reach of river. When inflow exceeds outflow, ΔS is positive, and when outflow exceeds inflow, ΔS is negative. Since routing involves only ΔS, absolute storage volumes are not necessary and the point of zero storage can be taken arbitrarily. Thus, the storage at any time is the sum of the positive and negative storage increments subsequent to the selected zero point. The computation is illustrated in the figure.

9-5 Treatment of Local Inflow

One of the most annoying problems in flood routing is the treatment of the *local inflow* which enters the reach between the inflow and outflow stations. If the

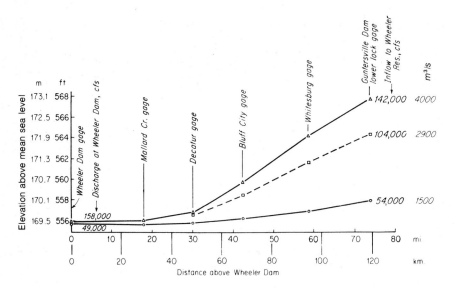

Figure 9-6 Profiles of the water surface in Wheeler Reservoir on the Tennessee River. (*Data from TVA.*)

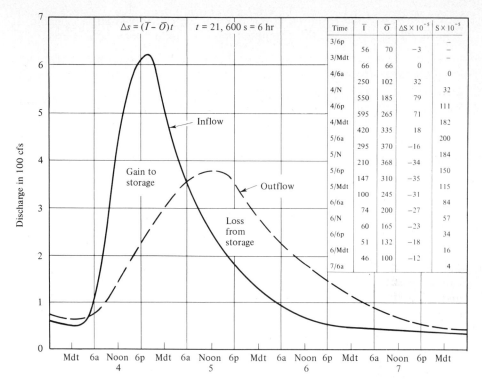

Figure in upper region shows:

$$\Delta s = (\bar{I} - \bar{O})t \qquad t = 21,600\ s = 6\ hr$$

Time	$\bar{I}$	$\bar{O}$	$\Delta S \times 10^{-5}$	$S \times 10^{-5}$
3/6p				—
	56	70	−3	
3/Mdt				—
	66	66	0	
4/6a				0
	250	102	32	
4/N				32
	550	185	79	
4/6p				111
	595	265	71	
4/Mdt				182
	420	335	18	
5/6a				200
	295	370	−16	
5/N				184
	210	368	−34	
5/6p				150
	147	310	−35	
5/Mdt				115
	100	245	−31	
6/6a				84
	74	200	−27	
6/N				57
	60	165	−23	
6/6p				34
	51	132	−18	
6/Mdt				16
	46	100	−12	
7/6a				4

Labels in figure: Inflow; Outflow; Gain to storage; Loss from storage. Y-axis: Discharge in 100 cfs. X-axis: Mdt 6a Noon 6p Mdt (days 4, 5, 6, 7).

Figure 9-7 Calculation of channel storage from inflow and outflow hydrographs.

local inflow enters mainly near the upstream end of the reach, it is usually added to the main-stream inflow to obtain total inflow. At a major junction, inflow stations (for which the flows are added) should be upstream from back-water effects. If local inflow occurs primarily near the downstream end of the reach, it may be *subtracted* from the outflow before storage is computed. In this case, the main-stream flow is routed through the reach and the local inflow added after routing is complete. Between these two extremes lie many pos-sibilities of combining various percentages of the local with the main-stream inflow before routing and adding the remainder to the outflow after routing. If the local inflow is relatively small compared with the main-stream inflow, any reasonable treatment consistently applied should give satisfactory results. If local inflow is large, consideration should be given to reducing the length of the reach.

The total volume of ungaged local inflow can be found by subtracting measured outflow from measured inflow for a period beginning and ending at about the same low flow, that is, $\Delta S = 0$. The time distribution of the ungaged local inflow is usually assumed to agree with the observed flows in a small tributary similar in size and character to the typical streams of the ungaged area. This procedure throws all errors of flow measurement into the ungaged local inflow, and the resulting flows may not appear altogether reasonable. If influent seepage is large, the computed ungaged inflow may be negative.

9-6 Reservoir Routing

A reservoir in which the discharge is a function of water-surface elevation offers the simplest of all routing situations. Such a reservoir may have ungated sluiceways and/or an uncontrolled spillway. Reservoirs having sluiceway or spillway gates may be treated as simple reservoirs if the gates remain at fixed openings. The known data on the reservoir are the elevation-storage curve and the elevation-discharge curve (Fig. 9-8). Equation (9-10) can be transformed [7] into

$$I_1 + I_2 + \left(\frac{2S_1}{t} - O_1\right) = \frac{2S_2}{t} + O_2 \qquad (9\text{-}11)$$

Solution of Eq. (9-11) requires a routing curve showing $2S/t + O$ versus O (Fig. 9-8). All terms on the left-hand side of the equation are known, and a value of $2S_2/t + O_2$ can be computed. The corresponding value of O_2 can be determined

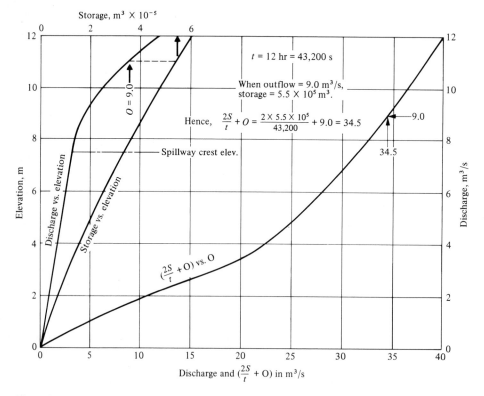

Figure 9-8 Routing curves for a typical reservoir.

Table 9-2 Routing with the $2S/t + O$ curve of Fig. 9-8
Data available at start of routing shown in boldface and all tabular values are in cubic meters per second

Date	Hour	I	$\dfrac{2S}{t} - O$	$\dfrac{2S}{t} + O$	O
1	Noon	**2.0**	5.6	9.0	1.7
	Midnight	**5.2**	8.2	12.8	2.3
2	Noon	**10.1**	14.9	23.5	4.3
	Midnight	**12.2**	16.2	37.2	10.5
3	Noon	**8.5**	16.3	36.9	10.3
	Midnight	**4.7**	16.3	29.5	6.6
4	Noon	**2.3**		23.3	4.2

from the routing curve. The computation is then repeated for succeeding routing periods. Table 9-2 illustrates a typical solution. It should be noted that $2S/t - O$ is easily computed as $(2S/t + O) - 2O$.

Routing in a reservoir with gated outlets depends on the method of operation. A general equation is obtained by modifying Eq. (9-10) to

$$\frac{I_1 + I_2}{2} t - \frac{O_1 + O_2}{2} t - \bar{O}_R t = S_2 - S_1 \qquad (9\text{-}12)$$

where O is uncontrolled outflow and O_R is regulated outflow. If O is zero, Eq. (9-12) becomes

$$\bar{I}t - \bar{O}_R t + S_1 = S_2 \qquad (9\text{-}13)$$

which can be readily solved for S_2 and reservoir elevation. If O is not zero, the routing equation becomes

$$I_1 + I_2 - 2\bar{O}_R + \left(\frac{2S_1}{t} - O_1\right) = \left(\frac{2S_2}{t} + O_2\right) \qquad (9\text{-}14)$$

The solution of Eq. (9-14) is identical with that for Eq. (9-11), except for the inclusion of O_R.

If the gates are set at fixed openings so that the discharge is a function of head, the solution requires a family of $2S/t + O$ curves for various gate openings. The routing method is again the same as that for Eq. (9-11) except that the curve appropriate to the existing gate opening is used each time.

9-7 Routing in River Channels

Routing in natural river channels is complicated by the fact that storage is not a function of outflow alone. This is illustrated when the storage computed in Fig. 9-7 is plotted against simultaneous outflow. The resulting curve is usually a wide loop indicating greater storage for a given outflow during rising stages than during falling (Fig. 9-9). The storage beneath a line parallel to the streambed is

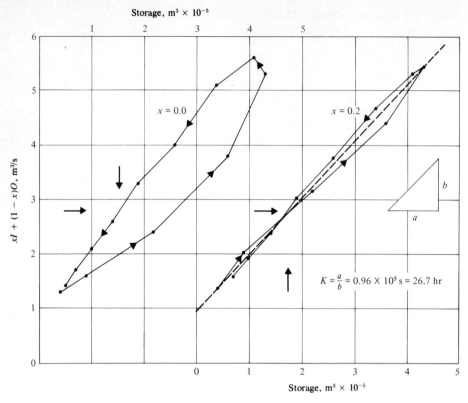

Figure 9-9 Determination of the Muskingum storage constants [Eq. (9-16)].

called *prism storage*; between this line and the actual profile, *wedge storage* (Fig. 9-10). During rising stages a considerable volume of wedge storage may exist before any large increase in outflow occurs. During falling stages, inflow drops more rapidly than outflow, and the wedge-storage volume becomes negative. Routing in streams requires a storage relationship which adequately represents the wedge storage. This is usually done by including inflow as a parameter in the storage equation.

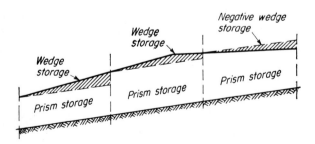

Figure 9-10 Some possible water-surface profiles during the passage of a flood wave.

9-8 Channel Routing: Analytical Methods

One expression for storage in a reach of a stream is

$$S = \frac{b}{a} \left[xI^{m/n} + (1 - x)O^{m/n} \right] \qquad (9\text{-}15)$$

where a and n are constants from the mean stage-discharge relation for the reach, $q = ag^n$, and b and m are constants in the mean stage-storage relation for the reach, $S = bg^m$. In a uniform rectangular channel, storage would vary with the first power of stage ($m = 1$) and discharge would vary as the $\frac{5}{3}$ power (Manning formula). In a natural channel with overbank floodplains the exponent n may approach or become less than unity. The constant x expresses the relative importance of inflow and outflow in determining storage. For a simple reservoir, $x = 0$ (inflow has no effect), while if inflow and outflow were equally effective, x would be 0.5. For most streams, x is between 0 and 0.3 with a mean value near 0.2.

The Muskingum method [8] assumes that $m/n = 1$ and lets $b/a = K$. Equation (9-15) then becomes

$$S = K[xI + (1 - x)O] \qquad (9\text{-}16)$$

The constant K, known as the *storage constant*, is the ratio of storage to discharge and has the dimension of time. It is approximately equal to the travel time through the reach and in the absence of better data is sometimes estimated in this way. If flow data on previous floods are available, K and x are determined by plotting S versus $xI + (1 - x)O$ for various values of x (Fig. 9-9). The best value of x is that which causes the data to plot most nearly as a single-valued curve. The Muskingum method assumes that this curve is a straight line with slope K. The units of K depend on the units of flow and storage.

If Eq. (9-16) is substituted for S in Eq. (9-10) and like terms are collected, the resulting equation reduces to

$$O_2 = c_0 I_2 + c_1 I_1 + c_2 O_1 \qquad (9\text{-}17)$$

where

$$c_0 = -\frac{Kx - 0.5t}{K - Kx + 0.5t} \qquad (9\text{-}17a)$$

$$c_1 = \frac{Kx + 0.5t}{K - Kx + 0.5t} \qquad (9\text{-}17b)$$

$$c_2 = \frac{K - Kx - 0.5t}{K - Kx + 0.5t} \qquad (9\text{-}17c)$$

Combining Eqs. (9-17a) to (9-17c) gives

$$c_0 + c_1 + c_2 = 1 \qquad (9\text{-}17d)$$

In these equations t is the routing period in the same time units as K. With K, x, and t established, values c_0, c_1, and c_2 can be computed. All three coefficients must be positive in order for Eq. (9-17) to give valid results (Sec. 10-5). The routing operation is simply a solution of Eq. (9-17) with the O_2 of one routing period becoming the O_1 of the succeeding period. Table 9-3 illustrates a typical computation.

Since most routing procedures involve computation of cumulative storage, the outflow at anytime can be determined only by routing from the last known value of outflow. An expression for O_4 can be written from Eq. (9-17) as

$$O_4 = c_0 I_4 + c_1 I_3 + c_2 c_0 I_3 + c_1 c_2 I_2 + c_2^2 c_0 I_2 + c_2^2 c_1 I_1 + c_2^3 O_1 \qquad (9\text{-}18)$$

Since c_2 is less than unity, c_2^3 will usually be negligible, and combining coefficients gives

$$O_4 = a I_4 + b I_3 + c I_2 + d I_1 \qquad (9\text{-}19)$$

Equation (9-19) provides a means of computing outflow at any time if the preceding inflows are known.

Since x is dependent upon the relative importance of wedge storage, it also depends on the length of the reach. The distance between gaging stations is usually such that x is significantly greater than zero. The time delay from channel storage can be simulated, however, by successive routings [9, 10] through a number of increments of reservoir-type storage $(x = O)$. This technique may be visualized as dividing the reach into a number of unit lengths for which wedge storage is small relative to the prismatic storage. The optimum number of unit reaches and corresponding value of K are usually determined by trial and error.

The Muskingum method assumes that K is constant at all flows. While this assumption is generally adequate, in some cases the storage-flow relation is nonlinear and an alternative method must be found. One approach is to assume that K is a function of inflow [9, 11]. It is apparent that Eq. (9-11) can be modified for channel routing by deriving a curve relating $2S/t + aI + O$ and $aI + O$.

9-9 Channel Routing: Graphical Methods

A variety of graphical methods for solving the routing equation have been suggested. The Muskingum storage equation, when $x = 0$, can be expressed as

$$\frac{dS}{dt} = K \frac{dO}{dt} \qquad (9\text{-}20)$$

Combining Eqs. (9-8) and (9-20) gives

$$\frac{I - O}{K} = \frac{dO}{dt} \qquad (9\text{-}21)$$

Equation (9-21) is the basis for a very simple graphical routing method [12]. Given an inflow hydrograph and the initial segment of the outflow hydrograph, the latter can be extended by placing a straightedge as indicated in Fig. 9-11. No routing period is involved, and it will be noted that K need not be constant but

Table 9-3 Application of the Muskingum method
Based on $K = 11$ hr, $t = 6$ hr, $x = 0.13$; hence, $c_0 = 0.124$, $c_1 = 0.353$, and $c_2 = 0.523$; values known at the beginning of routing are shown in boldface

Date	Hour	I	$c_0 I_2$	$c_1 I_1$	$c_2 O_1$	O
1	6 a.m.	**10**	...	...	...	**10**
	Noon	**30**	3.7	3.5	5.2	12.4
	6 p.m.	**68**	8.4	10.6	6.5	25.5
	Midnight	**50**	6.2	24.0	13.3	43.5
2	6 a.m.	**40**	5.0	17.7	22.7	45.4
	Noon	**31**	3.8	14.1	23.7	41.6
	6 p.m.	**23**	2.9	10.9	21.8	35.6

can be expressed as a function of O. The procedure is therefore suitable for routing through an ungated reservoir for which a K curve (dS/dO versus O) can be constructed.

It is not necessary to determine K by the procedure described in Sec. 9-8. Instead, K may be found by reversing the routing procedure described above. A straight line conforming to the slope of the outflow hydrograph at any time t is projected to a discharge value equal to the inflow at that time. The time difference between the inflow and this projection is K. When the value of K for a number of selected points on the rising and falling limbs of a series of historical events has been determined, the relation between K and O can readily be derived.

The graphical procedure described above assumes pure reservoir action ($x = 0$), and the peak of the outflow hydrograph must fall on the receding limb

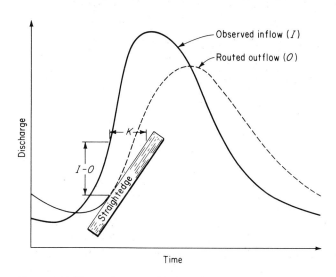

Figure 9-11 Illustrating a graphical routing method.

of the inflow hydrograph [Eq. (9-21)]. A graphical construction introducing the equivalent of the Muskingum x could be derived, but a simpler solution is available. The factor x may be viewed as a measure of the translatory component of the wave motion. Figure 9-12 shows that, with a constant K, the translation of the hydrograph increases as x increases. Thus, the effect of increasing x can be introduced by lagging the inflow hydrograph and decreasing K. If the lag T_L is constant, it is immaterial whether the inflow is lagged and then routed or the routed flow is lagged. A completely flexible procedure would utilize both variable K and T_L as functions of flow [13]. Since, with no translation, the outflow peak would fall on the inflow recession, a measure of T_L is the time difference between the outflow peak and the occurrence of an equal flow during the recession of inflow (Fig. 9-13). Values of T_L can be determined from the hydrographs of several historic floods and plotted as a T_L-versus-I curve (Fig. 9-14). Using the historic data, inflow is lagged according to the T_L-versus-I curve, and a K-versus-O curve is constructed from the lagged inflow and observed outflow as described earlier. Routing is accomplished by first lagging the inflow hydrograph (Fig. 9-14) and then routing with a straightedge as illustrated in Fig. 9-11. Templates can be devised to achieve both operations simultaneously [13], and the process is readily computerized [14].

9-10 Deriving Basin Outflow by Routing

The shape of the hydrograph from a basin depends on the travel time through the basin and on the shape and storage characteristics of the basin. When excess rainfall (runoff) is considered to be inflow and the hydrograph to be

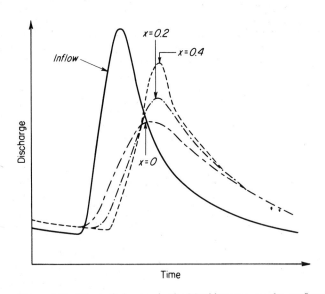

Figure 9-12 Effect of changes in the Muskingum x on the outflow hydrograph.

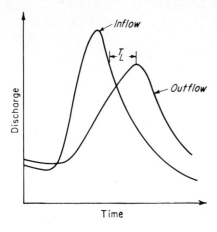

Figure 9-13 Determination of the lag term T_L.

outflow, the problem is analogous to storage routing [15, 16]. The unit hydrograph itself is basically a set of average routing coefficients.

The nature of the problem suggests the use of lag-and-route methods (Sec. 9-9). Inflow can be lagged by dividing the basin into zones by isochrones of travel time from the outlet. The area between isochrones is then measured, and a time-area diagram (Fig. 9-15) is plotted. This diagram may be viewed as inflow

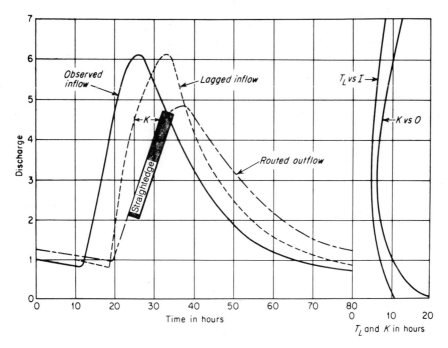

Figure 9-14 Graphical flood routing with a variable lag and storage factor.

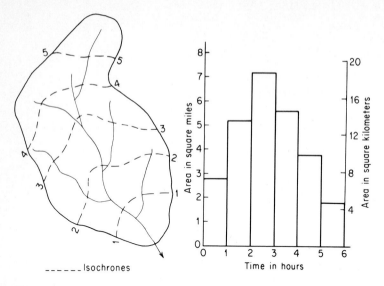

Figure 9-15 Derivation of the time-area diagram for a basin.

to a hypothetical reservoir with storage characteristics equivalent to those of the basin and located at the basin outlet. Thus routing the time-area diagram by the Muskingum method (Sec. 9-8) with $x = 0$ (or some other suitable method) yields the outflow hydrograph after adjustment for units. Because of the method of constructing the time-area diagram, such a hydrograph would be the result of an instantaneous rainfall (duration 0 hr) [17], and it is called an instantaneous unit hydrograph (Secs. 7-7 and 7-10).

The technique outlined above need not be limited to deriving unit hydrographs. For a storm of duration equal to the interval between isochrones, the average runoff can be estimated for each time zone and expressed in cubic meters per second.† The resulting time-runoff diagram is then routed through storage to give the actual outflow hydrograph. If rain lasts for several time periods, the time-runoff diagrams are lagged and superimposed (Fig. 9-16) and the summation is routed. The method accounts for time-intensity variations and areal distribution of rainfall, two factors which the unit hydrograph cannot readily consider. For this reason the routing approach can be applied to much larger basins than the simple unit-hydrograph approach can. This approach is commonly used in hydrologic simulation (Sec. 12-2).

The routing of excess rainfall over the basin, often referred to as the *isochrone* method, is not as readily applied as one might think. There is no simple, rigorous means of deriving the time-area diagram: it is usually assumed that travel time is proportional to channel distance from each point to the outflow station, possibly taking variations in slope into account and noting that inflow

---

† $q = 2.78AQ/t \text{ m}^3/\text{s}$, where A is in square kilometers, Q is in centimeters, and t is in hours [$q = (645.3AQ)/t \text{ ft}^3/\text{s}$ with A in square miles and Q in inches].

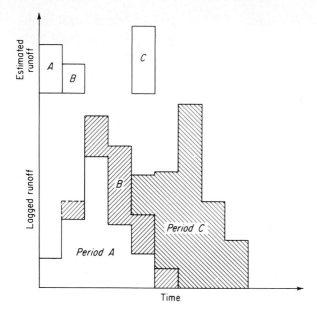

Figure 9-16 Time-runoff diagram for a long storm.

must equal outflow at the time of the outflow peak. The time-area curve so derived is only a first approximation, which may require adjustment to yield an optimum fit when coupled with the best value of K. In other words, a trial-and-error approach is used in which various values are tried until a combination that gives a good fit to historical floods is found. Although laborious, this procedure is satisfactory for gaged streams but obviously unsuited to ungaged basins.

An estimate of K may be obtained from data on the recession of flow for a basin. From Eqs. (7-3) and (9-16), assuming $x = 0$,

$$K = \frac{1}{-\ln K_r} \tag{9-22}$$

where K_r is the direct runoff recession constant for the stream.

Clark [17] suggested that K (in hours) be estimated from

$$K = \frac{cL}{\sqrt{s}} \tag{9-23}$$

where L is the length of the main stream in kilometers, s is the mean channel slope, and c varies from about 0.5 to 1.4 (0.8 to 2.2 with L in miles). Linsley, in a discussion of Clark's paper, suggested the formula

$$K = \frac{bL\sqrt{A}}{\sqrt{s}} \tag{9-24}$$

where A is the drainage area in square kilometers and b varies from about 0.01 to 0.03 (0.04 to 0.08 with L in miles and A in square miles) for the basins tested.

The isochrone method is commonly used in models for hydrologic simulation (Sec. 12-3). Kinematic routing is sometimes used in preference to the

Muskingum method in the simulation of basin outflow, but channel slope, roughness, and cross-sectional area are then required for many points throughout the basin (Sec. 10-5).

9-11 Gage Relations

A discussion of wave travel and routing would be incomplete without brief mention of a simple empirical solution which is often quite successful. *Gage relations* are graphs correlating an observed stage or discharge at one or more upstream stations with the resulting stage or discharge at a downstream station. Gage relations are most effective at crest (Fig. 9-17). If such a relation is to be reliable, the quantity of local inflow between the stations in each flood must bear a reasonably fixed relation to the reach inflow at the upstream station. Since such a proportional relation is unlikely, gage relations are most effective when the local inflow is relatively small compared with the main-stream inflow. It is also necessary that the peak of the local inflow bear a fixed time relation to the peak of the main-stream inflow. If a slight difference in time of occurrence can cause a considerable difference in the resulting outflow (Fig. 9-18), gage relations will not be successful. Thus gage relations are most useful on large streams where local inflow is small with respect to main-stream flow and rates of change of flow are relatively low.

More complex gage relations can be constructed to account for variable local inflow, and it is also possible to derive charts for routing in terms of stage [18]. Stage relations and stage routing are useful when dealing with streams for which discharge data are not available. It should be emphasized that any change in the channel, either natural or artificial, may result in changes in the stage-discharge and stage-storage relationships for the reach.

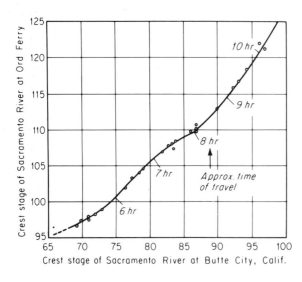

Figure 9-17 A simple crest-stage relation. (*U.S. National Weather Service.*)

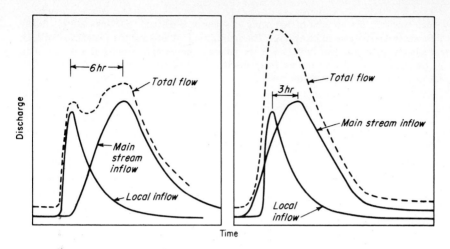

Figure 9-18 Effect of timing of flood peaks on reach outflow.

PROBLEMS

9-1 If the channel width for the stream depicted in Fig. 4-11 is 30 ft at a stage of 4 ft, what would be the velocity of translatory monoclinal wave of small height?

9-2 Find the ratio between wave velocity and water velocity for a semicircular channel when $y = r$; when $y = 0.2r$. Use the Chézy formula.

9-3 Given the stage-discharge relation defined by Eq. (4-3) and the values $a = 0.055$, $b = 2.50$, and $k = 10.25$, compute the velocity of a monoclinal wave of small height, assuming a channel width of 20 m and a stage of 1.3 m.

9-4 If the channel width for the stream depicted in Fig. 4-11 is 10 m at a stage of 1.0 m, what would be the velocity of a monoclinal wave of small height?

9-5 Given the hydrographs tabulated below (cubic meters per second), find the 12-hourly storage values for the reach in cubic meters. Ignore local inflow.

Date	Hour†	Inflow	Outflow	Date	Hour†	Inflow	Outflow
1	M	36	58	7	N	196	341
2	N	43	46		M	153	272
	M	121	42	8	N	124	218
3	N	346	61		M	101	180
	M	575	149	9	N	84	150
4	N	717	326		M	71	124
	M	741	536	10	N	60	104
5	N	612	674		M	52	86
	M	440	681	11	N	46	73
6	N	328	560		M	41	62
	M	251	437	12	N	37	52

† M = midnight; N = noon.

9-6 A small reservoir has an area of 300 acres at spillway level, and the banks are essentially vertical for several feet above spillway level. The spillway is 15 ft long and has a coefficient of 3.75. Taking the inflow hydrograph of Prob. 9-5 as the inflow to the reservoir (cubic feet per second), compute the maximum pool level and maximum discharge to be expected if the reservoir is initially at the spillway level at midnight on the first.

9-7 A small reservoir has an area of 4000 hectares at spillway level and the banks are essentially vertical for several meters above spillway level. The spillway is 50 m long and has a coefficient of 2.0. Taking the outflow hydrograph of Prob. 9-5 as the inflow to the reservoir in cubic meters per second, compute the maximum discharge and pool level to be expected if the reservoir is at spillway level at midnight on the first.

9-8 Tabulated below are the elevation-storage and elevation-discharge data for a small reservoir. Taking the inflow hydrograph of Prob. 9-5 as the reservoir inflow (cubic feet per second), and assuming the pool elevation to be 875 at midnight on the first, find the maximum pool elevation and peak outflow rate.

Elevation	Storage, acre · ft	Discharge, ft³/s	Elevation	Storage, acre · ft	Dishcarge ft³/s
862	0	0	882	1220	100
865	40	0	884	1630	230
870	200	0	886	2270	394
875	500	0	888	3150	600
880	1000	0			

9-9 Take the data of Prob. 9-8 as elevation in meters, storage in 10^5 m³, and flows in cubic meters per second, and determine the maximum pool level and outflow rate.

9-10 Find the Muskingum K and x for the flood of Prob. 9-5 (flows in cubic meters per second). Determine also the routing coefficients for $t = 12$ hr and $t = 24$ hr. Which routing period would you recommend? Why?

9-11 Taking the outflow hydrograph of Prob. 9-5 as the inflow (cubic feet per second) to a reach with $K = 27$ hr and $x = 0.2$, find the peak outflow using the Muskingum method of routing.

9-12 Write the routing equation for the case when storage is a function of $aI + bO$.

9-13 Using the graphical method of Sec. 9-9, find the K curve for the flood of Prob. 9-5 assuming that T_L is constant (note that flow units are immaterial).

9-14 Using the outflow hydrograph of Prob. 9-5 as the inflow to a reach for which the lag $T_L = 6$ hr and $K = 18$ hr, find the peak outflow and time of peak by the graphical method of Sec. 9-9.

REFERENCES

1. Chow, V. T.: "Open Channel Hydraulics," p. 586, McGraw-Hill, New York, 1959.
2. Seddon, J.: River Hydraulics, *Trans. ASCE,* vol. 43, pp. 217–229, 1900.
3. Lighthill, M. J., and G. B. Whitham: I. Flood Movement in Long Rivers, *Proc. R. Sci.,* ser. A, vol. 229, pp. 281–316, 1955.
4. Horton, R. E.: Channel Waves Subject Chiefly to Momentum Control, *Perm. Int. Assoc. Navig. Congr. Bull.* 27, 1939.
5. Moots, E. E.: A Study in Flood Waves, *Univ. Iowa Stud. Eng. Bull.* 14, 1938.

6. Wilkinson, J. H.: Translatory Waves in Natural Channels, *Trans. ASCE,* vol. 110, pp. 1203–1236, 1945.
7. Goodrich, R. D.: Rapid Calculation of Reservoir Discharge, *Civ. Eng.,* vol. 1, pp. 417–418, 1931.
8. McCarthy, G. T.: The Unit Hydrograph and Flood Routing, presented at *Conf. North Atl. Div., U.S. Corps Eng., June 1938*; see also "Engineering Construction: Flood Control," pp. 147–156, The Engineer School, Ft. Belvoir, Va., 1940.
9. Rockwood, D. M.: Application of Streamflow Synthesis and Reservoir Regulation—"SSARR"—Program to the Lower Mekong River, *Int. Assoc. Sci. Hydrol. Publ.* 80, pp. 329–334, 1968.
10. World Meteorological Organization: "Guide to Hydrological Practices," WMO no. 168, 3d ed., p. 5.64, Geneva, 1974.
11. Prasad, R.: A Nonlinear Hydrologic System Response Model, *J. Hydraul. Div. ASCE,* vol. 93, pp. 201–222, July 1967.
12. Wilson, W. T.: A Graphical Flood-Routing Method, *Trans. Am. Geophys. Union,* vol. 21, pt. 3, pp. 893–898, 1941.
13. Kohler, M. A.: Mechanical Analogs Aid Graphical Flood Routing, *J. Hydraul. Div. ASCE,* vol. 84, pp. 1585.1–1585.14, April 1958.
14. Hydrologic Research Laboratory, National Weather Service River Forecast System Forecast Procedures, *NOAA Tech. Mem.* NWS HYDRO-14, December 1972.
15. Popov, E. G.: "Hydrological Forecasts," pp. 126–128, Gidrometeoizdat, Moscow, 1957.
16. See [10], pp. 5.60–5.61.
17. Clark, C. O.: Storage and the Unit Hydrograph, *Trans. ASCE,* vol. 110, pp. 1419–1488, 1945.
18. Kohler, M. A.: A Forecasting Technique for Routing and Combining Flow in Terms of Stage, *Trans. Am. Geophys. Union,* vol. 25, pt. 6, pp. 1030–1035, 1944.

BIBLIOGRAPHY

Chow, V. T.: "Open Channel Hydraulics," pp. 586–621, McGraw-Hill, New York, 1959.
Eagleson, P. S.: "Dynamic Hydrology," pp. 325–368, McGraw-Hill, New York, 1970.
Keulegan, G. H.: Wave Motion, chap. 11 in H. Rouse (ed.): "Engineering Hydraulics," pp. 711–768, Wiley, New York, 1950.
Klemeš, V.: Applications of Hydrology to Water Resources Management, *Opera. Hydrol. Rep.,* no. 4, World Meteorological Organization, Geneva, 1973.
Lawler, E. A.: Flood Routing, sec. 25-II in V. T. Chow (ed.): "Handbook of Applied Hydrology," pp. 25.34–25.59, McGraw-Hill, New York, 1964.
Linsley, R. K., M. A. Kohler, and J. L. H. Paulhus: "Applied Hydrology," pp. 465–541, McGraw-Hill, New York, 1949.
Němec, J.: "Engineering Hydrology," pp. 274–281, McGraw-Hill, London, 1972.
Thomas, H. A.: The Hydraulics of Flood Movement in Rivers, *Carnegie Inst. Tech. Eng. Bull.,* 1935.
Viessman, W., Jr., J. W. Knapp, G. L. Lewis, and W. E. Harbaugh: "Introduction to Hydrology," 2d ed., pp. 232–244, 380–384, IEP, New York, 1977.

TEN

HYDRAULIC ROUTING†

The routing methods discussed in Chap. 9 are adequate for a considerable majority of the problems encountered in hydrology. There are special cases where hydrologic routing methods may not suffice. Streams where there may be a reversal of flow, complex channel systems as in a delta, and situations with variable backwater are among these special cases. Routing of unusual events such as flood waves caused by failure of a dam or by the probable maximum flood (Chap. 13) may also require special treatment. Ungaged streams for which determination of storage as described in Chap. 9 is impracticable can sometimes be treated with a method which requires only information on the geometry of the channel. *Hydraulic routing* based on a solution of the energy or momentum equations is an alternate to the hydrologic methods. Solution of these equations is a relatively complex task and generally requires the use of a computer.

10-1 Governing Equations

Hydraulic routing rests on three assumptions: (1) water density is constant; (2) stream length affected by the flood wave is many times greater than the depth of flow; and (3) flow is essentially one-dimensional. Waves satisfying these assumptions are called *shallow-water* or *translatory* waves. Since vertical accelerations are negligible, pressure distribution in the wave is hydrostatic.

---

† The original draft of this chapter was written by Dr. Delbert D. Franz, Vice-president, Linsley, Kraeger Associates, Ltd.

There are a variety of forms for the shallow-water equations, none of which is suitable for all purposes [1–3]. The general version is the *integral form,* written below for an element of channel $x_2 - x_1$ (Fig. 10-1):

$$\int_{x_1}^{x_2} [A(x, t_2) - A(x, t_1)] \, dx = \int_{t_1}^{t_2} [Q(x_1, t) + I(t) - Q(x_2, t)] \, dt \quad (10\text{-}1)$$

$$\underbrace{\int_{x_1}^{x_2} [Q(x, t_2) - Q(x, t_1)] \, dx}_{\text{momentum change}} = g \int_{t_1}^{t_2} [\underbrace{Qv(x_1, t)}_{\text{momentum in}} - \underbrace{Qv(x_2, t)}_{\text{momentum out}}] \, dt$$

$$+ g \int_{t_1}^{t_2} [\underbrace{J(x_1, t)}_{\substack{\text{pressure} \\ \text{force on} \\ \text{upstream} \\ \text{face}}} - \underbrace{J(x_2, t)}_{\substack{\text{pressure} \\ \text{force on} \\ \text{down-} \\ \text{stream} \\ \text{face}}} + \underbrace{\int_{x_1}^{x_2} J_x^y \, dx}_{\substack{\text{downstream} \\ \text{pressure} \\ \text{force} \\ \text{on sides}}}] \, dt$$

$$+ g \int_{t_1}^{t_2} [\underbrace{\int_{x_1}^{x_2} A s_0 \, dx}_{\substack{\text{downstream} \\ \text{component} \\ \text{of weight}}} - \underbrace{\int_{x_1}^{x_2} A s_f \, dx}_{\substack{\text{friction force} \\ \text{opposing} \\ \text{motion}}}] \, dt \quad (10\text{-}2)$$

where

$$J = \int_0^y (y - z) B(x, z) \, dz \quad (10\text{-}3)$$

$$J_x^y = \int_0^y (y - z) \frac{\partial B}{\partial x} (x, z) \, dz \quad (10\text{-}4)$$

$$s_f = \frac{Q|Q|}{K^2} \quad (10\text{-}5)$$

$$K = \frac{1.49}{n} AR^{2/3} \quad (10\text{-}6)$$

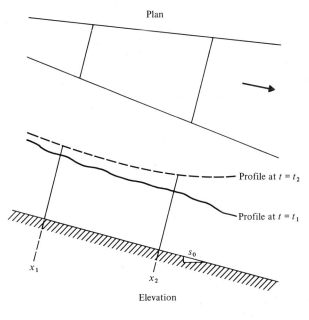

Plan

Profile at $t = t_2$

Profile at $t = t_1$

x_1

x_2

s_0

Elevation

Figure 10-1 An element of a stream channel.

In these equations $A(x, t)$ is the cross-sectional area of flow as a function of distance x and time t, y is depth, J is the first moment of A about the water surface, J_x^y is the rate of change of J with respect to distance at constant depth, s_0 is the channel bottom slope, s_f is the friction slope, K is conveyance, v is the average velocity (Q/A), B is the top width of flow, R is the hydraulic radius, n is the Manning roughness coefficient, g is the acceleration of gravity, and z is a dummy integration variable. The lateral inflows I are assumed to enter the element normal to the mean flow direction. Hence, they do not contribute momentum in the direction of flow.

Equation (10-1) states the principle of the conservation of volume (mass). Equation (10-2) provides for conservation of momentum; i.e., the change of momentum in the interval $t_2 - t_1$ is given by the integral of the inflow of momentum less the outflow of momentum plus the integral of the net downstream force acting on the water.

All other forms of the shallow-water equations are derived from Eqs. (10-1) and (10-2). Taking limits of these equations as the time and distance limits become small yields two partial differential equations known as the *conservation form*:

$$\frac{\partial A}{\partial t} + \frac{\partial Q}{\partial x} = q \qquad (10\text{-}7)$$

$$\frac{\partial Q}{\partial t} + g\frac{\partial J}{\partial x} + \frac{\partial}{\partial x}\left(\frac{Q^2}{A}\right) = gA(s_0 - s_f) + gJ_x^y \qquad (10\text{-}8)$$

where q is given by

$$I(t) = \int_{x_1}^{x_2} q(x, t)\, dx \qquad (10\text{-}9)$$

Expanding terms in the conservation form and simplifying the momentum equation using appropriate substitution from the continuity equation yields the *St. Venant* or *divergent form*:

$$A\frac{\partial v}{\partial x} + vB\frac{\partial y}{\partial x} + B\frac{\partial y}{\partial t} + vA_x^y = q \qquad (10\text{-}10)$$

$$\frac{\partial v}{\partial t} + v\frac{\partial v}{\partial x} + g\frac{\partial y}{\partial x} = g(s_0 - s_f) - \frac{vq}{A} \qquad (10\text{-}11)$$

where

$$A_x^y = \int_0^y \frac{\partial B}{\partial x}(x, z)\, dz \qquad (10\text{-}12)$$

Here A_x^y is the rate of change of area with distance at constant depth. The *characteristic form* derived [4, 5] by transformation of the St. Venant form is

$$\left[\frac{\partial v}{\partial t} + \frac{dx}{dt}\frac{\partial v}{\partial x}\right] \pm \frac{g}{c}\left[\frac{\partial y}{\partial t} + \frac{dx}{dt}\frac{\partial y}{\partial x}\right] = g(s_0 - s_f) - \frac{gvq}{A} \pm \frac{c}{A}(vA_x^y - q)$$

$$(10\text{-}13)$$

$$\frac{dx}{dt} = v \pm c = u \tag{10-14}$$

where $c = \sqrt{gA/B}$ = celerity of a very small wave. *Celerity* is defined as the velocity of the wave relative to the water on the low side of the wave.

These forms of the governing equations show a progression from the physical to the mathematical. The integral and conservation forms are closely tied to the physical features of the flow; i.e., each term is associated with a specific flux, volume, or force. The mathematical manipulations used to derive the other forms obscure these relationships. On the other hand the characteristic form gives a convenient visualization of simple wave phenomena not possible with the other forms. The major assumptions made in deriving the shallow-water equations are: (1) hydrostatic pressure distribution, (2) one-dimensional flow, (3) fixed channel geometry, (4) small channel slope, (5) uniform velocity distribution, and (6) friction losses in unsteady flow which can be approximated by the losses in steady uniform flow [6, 7].

10-2 Dynamic Wave Velocity

The monoclinal wave discussed in Sec. 9-1 is one limiting form for a shallow-water wave. Another limiting form is the *abrupt wave,* formed, for example, when a sudden change in a sluice gate opening occurs (Fig. 10-2). The depth change takes place over a distance only 6 to 10 times greater than the larger depth. An observer moving along the bank of the channel at wave velocity u will see a steady profile as for the monoclinal wave. However, a single-valued relation between flow rate and area is not assumed, so that an equation for conservation of momentum must be used to solve for wave velocity. The momentum of the water contained in the wave profile between sections 1 and 2 (Fig. 10-2) is constant because the velocity of the wave and the velocities at 1 and 2 are constant. Therefore, when the effects of boundary friction and bottom slope are ignored, conservation of momentum requires

$$(v_2 - u)A_2\rho v_2 - (v_1 - u)A_1\rho v_1 + \rho g(J_2 - J_1) = 0 \tag{10-15}$$

where ρ is the water density. Conservation of water volume requires

$$u = \frac{Q_2 - Q_1}{A_2 - A_1} = \frac{A_2 v_2 - A_1 v_1}{A_2 - A_1} \tag{10-16}$$

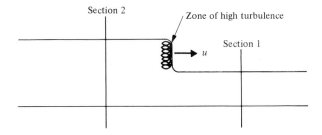

Section 2

Zone of high turbulence

Section 1

u

Figure 10-2 An abrupt wave.

by the same arguments used in Sec. 9-1. The terms $(v_2 - u)A_2$ and $(v_1 - u)A_1$ represent the flow of water into and out of the wave profile, respectively, as viewed by an observer moving with the wave. The terms ρv_2 and ρv_1 give the momentum (mass times velocity) per unit volume entering and leaving the profile. Hence, the first two terms of Eq. (10-15) give the inflow and outflow of momentum, respectively. Substitution of the expression for v_2 from Eq. (10-16) into Eq. (10-15) and solving for u yields

$$u = v_1 \pm \sqrt{\frac{g(J_2 - J_1)}{A_1(1 - A_1/A_2)}} \qquad (10\text{-}17)$$

The radical term is the celerity of the abrupt wave. Abrupt translatory waves occur as tidal bores in many estuaries, as surges in power canals and tailraces, and occasionally as flood waves caused by intense small-area storms.

Equation (10-17) simplifies to

$$u = v_1 \pm \sqrt{gA/B} \qquad (10\text{-}18)$$

when the height of the abrupt wave is vanishingly small. Equations (10-17) and (10-18) apply to a general cross section. For a rectangular channel Eq. (10-18) becomes

$$u = v_1 \pm \sqrt{gy} \qquad (10\text{-}19)$$

where y is the depth of water. Waves with speed given by Eqs. (10-17) through (10-19) are called *dynamic waves* because the inertia of the water is significant in determining their motion.

10-3 Numerical Techniques

The governing equations represent continuous variation in time and space of the flow and water depth in the channel. Analytical solutions, if they existed, would give these values for any point within the domain of the routing problem. The numerical methods which must be used are limited to results at discrete points in time and space. These points are conveniently displayed on what is called the xt plane (Fig. 10-3). The points at which flow and depth are sought are at the grid intersections.

The xt plane is convenient for visualizing flow phenomena because all variables are functions of x and/or t. Variables such as depth, flow, velocity, and celerity are dependent variables, while x and t are independent variables. The variation of a dependent variable in time and space is a three-dimensional surface constructed on the xt plane. For simplicity, only the paths of the significant features of the flow are drawn on the plane. Figure 10-3a shows the path of an abrupt wave. The inverse slope of the path line is the velocity of propagation of the wave.

Two unknowns exist at each point in the xt plane, usually depth and flowrate. Other combinations such as area and flowrate, depth and velocity, and elevation and flowrate can be used. Knowledge of one pair allows computation

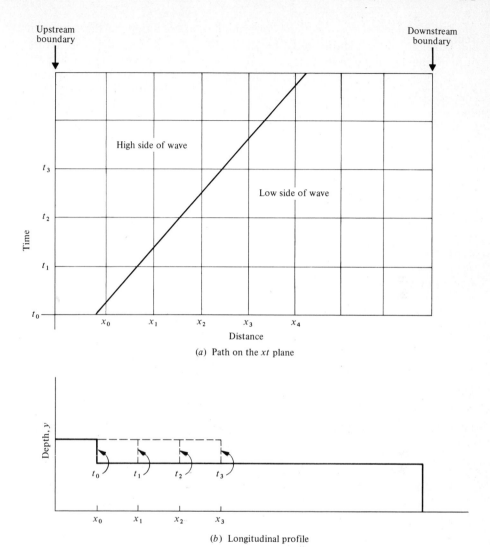

(a) Path on the xt plane

(b) Longitudinal profile

Figure 10-3 An abrupt wave in a prismatic channel, (a) path on the xt plane, (b) longitudinal profile.

of all other pairs. There are $n + 1$ points along the channel for n elements and there are $2n + 2$ unknowns to be determined for each discrete time in the xt plane. The numerical equations relate dependent variables at two times. All dependent variables are assumed known at any time and the equations are solved for the unknowns at the next time.

The size and shape of the channel are derived from cross sections taken normal to the average flow direction. The area, top width, and conveyance are computed from the cross sections. The bottom slope of the channel is usually taken as the average slope computed from the difference in bottom elevations and the distance between sections. Information on conditions at the upstream and downstream boundaries must be given. The number, location, and nature of the boundary conditions must be consistent with the governing equations.

Several concepts from numerical analysis become important in streamflow routing. The concepts are properties of the scheme used to represent the continuous governing equations [8]. The assumptions and the numerical methods used to define the discrete equations are called the numerical scheme or the difference scheme.

Stability of a scheme relates to the growth of round-off errors in the computations. If round-off errors introduced in the computations become so large that the solution is destroyed, the scheme is *unstable*. Error growth for a stable scheme is bounded and remains small relative to the solution. Some schemes are *conditionally stable;* i.e., they are stable for a range of time and distance steps. *Unconditionally stable* schemes are stable for all finite time and distance steps. Stability can be conveniently evaluated only for linearized equations. In this context stability refers to the numerical stability of the computations and does not relate to the physical stability of the solution of the governing equations.

A scheme is *consistent* if, in the limit, the discrete equations become the continuous equations as time and distance steps approach zero. A scheme is *convergent* if the solution of the discrete equations approaches the solution of the continuous equations as the time and distance steps approach zero. An unstable scheme cannot be convergent. A consistent scheme is not always stable and therefore may not be convergent.

The error of a solution for finite time and distance steps may or may not be small for a convergent scheme. Various accuracy ratios can be estimated between solutions of highly simplified continuous equations and the discrete equations [9], but the accuracy of the solution is ultimately dependent on the sensitivity of the solution to changes in time or distance steps and on the approximations used in developing the discrete equations.

A scheme is called *conservative* if it mimics the conservation properties of the governing equations. A conservative scheme is not necessarily the most accurate scheme but errors in conservation must be limited for an accurate solution. This property of the scheme must be evaluated with the approximations used in the scheme in mind. The evaluation of the volume of inflow and outflow of water, for example, must be done using a rule of approximate integration consistent with the rules used in the scheme. Some schemes are inherently conservative while others are nonconservative and cannot be made conservative. Some external check on the size of the error in conservation must be made for these schemes. Such a check is wise in any case to assist in detecting major blunders in the implementation of the scheme.

The numerical dissipation of a scheme relates the attenuation of flood waves in the discrete solution to the attenuation of flood waves in the continuous solution. The scheme is said to be *dissipative* if the attenuation of a wave in the discrete solution is greater than the attenuation of a wave of similar length in the continuous solution. The practical evaluation of this property is limited to simplified linear governing equations for which exact solutions can be developed. Typically, the discrete and continuous solutions for simplified equations are developed using Fourier series. A Fourier series represents a wave using the sum of sine and cosine terms with varying wavelength and wave amplitude. The attenuation in amplitude for each term of the series occurring over one time step can be evaluated for both the discrete and the continuous solutions. Normally the continuous solution of the simplified governing equations shows no attenuation. The numerical scheme is called *nondissipative* if it also shows no attenuation. Schemes which attenuate one or more wavelengths are called dissipative. Finally a scheme which amplifies the amplitude is unstable. Quite naturally the methods used to define the numerical dissipation properties of a scheme are also used to define the stability of a scheme [8, 9].

10-4 Routing with Complete Equations

The equations presented in Sec. 10-1 are commonly called the complete equations, or full equations, although they still involve assumptions and simplifications of the physical processes.

The formulation and numerical solution of the discrete equations must be guided by the behavior of the waves described by the governing equations. Imagine that a small change in flow occurs at time t_0 at point L along the channel (Fig. 10-4). Two infinitesimal dynamic waves are created by this disturbance, one moving upstream and the other downstream (assuming subcritical flow).

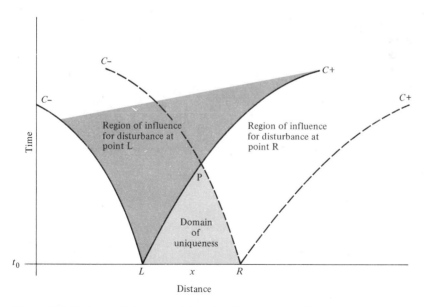

Figure 10-4 Regions of influence and domain of uniqueness on the xt plane.

The velocity of these waves is given by Eq. (10-14). Water velocity and wave celerity are not constant, and therefore their trajectories in the xt plane are curved (trajectories C+ and C−, Fig. 10-4). At any time points along the channel to the right of the C+ trajectory and to the left of the C− trajectory will be unaffected by the change in flow which occurred at point L at time t_0. Points between the C+ and C− trajectories will be affected by the change. The region of the xt plane between the two trajectories is called the *region of influence* of point L at time t_0.

If a similar change also occurs at point R, it also creates a region of influence in the xt plane (Fig. 10-4). A disturbance at any point between L and R will influence values at point P. Disturbances to the left of L or right of R will have no influence on values at P if the trajectories of the same family (C+, C−) do not intersect. Such an intersection would imply two values of depth at a point, which can occur only if there is a discontinuity at that point. Such a discontinuity exists if an abrupt wave forms. The velocity of this wave is no longer given by Eq. (10-18) but by Eq. (10-17). In this case the trajectories of a family can meet and follow a single trajectory for the abrupt wave. Excluding the case of a discontinuity, the line segment LR defines the interval of dependence and the area enclosed by the trajectories is the *domain of uniqueness* (Fig. 10-4).

In subcritical flow (Fig. 10-5a) water velocity is less than celerity and the C− trajectory has a negative slope while the C+ trajectory has a positive slope.

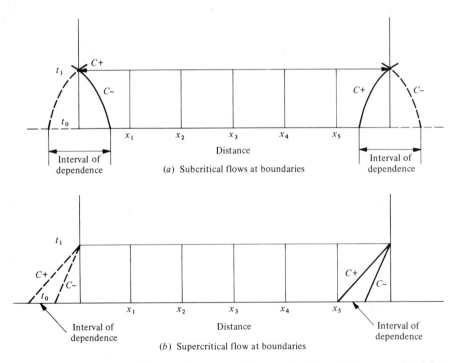

Figure 10-5 Boundary-condition requirements, (a) in subcritical flow, (b) in supercritical flow.

The interval of dependence of the boundary points is then not wholly within the solution domain. Boundary conditions must be specified to permit a solution at the boundary points. With supercritical flow (Fig. 10-5b), velocity exceeds celerity and the trajectories have slopes of the same sign. The interval of dependence of the downstream point is wholly within the solution domain, and no boundary condition is needed. If one is specified, the solution will be unstable. Conversely, two boundary conditions must be supplied at the upstream boundary where the interval of dependence falls outside the solution domain.

Method of characteristics. An observer who moves along the C+ trajectory in the xt plane at the speed of a very small abrupt wave will encounter only waves moving from right to left along the C− trajectories crossed. A similar observer on the C− trajectory will encounter only waves moving from left to right on the C+ trajectories. The observer could sum all the changes encountered along the C+ or C− trajectory because these changes are isolated from those being propagated along the observer's path. This discussion has taken the view of small discrete changes, which in the limit becoming infinitesimally small but increase in number without bound. Hence, every point in the xt plane has a unique pair of trajectories passing through it. If summation is replaced by integration, the discrete changes merge into a continuous variation of depth and flow rate. The characteristic form of the governing equations, Eqs. (10-13) and (10-14), describes this limit. If the channel is prismatic and there is no lateral inflow, Eq. (10-13) can be written as

$$\frac{dv}{dt} \pm \frac{g}{c}\frac{dy}{dt} = g(s_0 - s_f) \tag{10-20}$$

The bracketed terms in Eq. (10-13) have been written as total differentials, and these differentials must be taken in the direction given by Eq. (10-14) defining the trajectories of the disturbances. These trajectories are called *characteristics*. A more detailed treatment [4, 10] shows that the partial derivatives of the solution may suffer discontinuities across a characteristic even though the solution is continuous. Thus the characteristics may be viewed as paths taken by potential discontinuities in the derivatives.

Integration along the characteristics gives the following relationships for the points shown in Fig. 10-6:

$$v_P - v_S = \int_{y_S}^{y_P} \frac{g}{c}\, dy + \int_{t_S}^{t_P} g(s_0 - s_f)\, dt \tag{10-21}$$

$$x_P - x_S = \int_{t_S}^{t_P} (v + c)\, dt \tag{10-22}$$

$$v_P - v_F = \int_{y_F}^{y_P} \frac{g}{c}\, dy + \int_{t_F}^{t_P} g(s_0 - s_f)\, dt \tag{10-23}$$

$$x_P - x_F = \int_{t_F}^{t_P} (v - c)\, dt \tag{10-24}$$

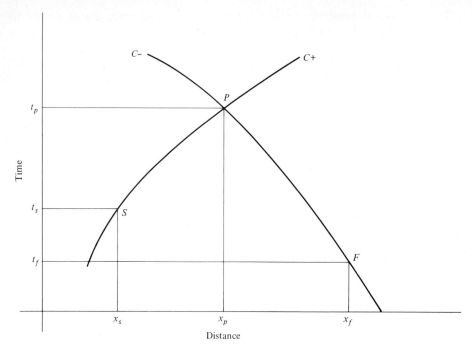

Figure 10-6 Points for integration in method of characteristics.

Given the unknowns at points S and F, simultaneous solution of Eqs. (10-21) through (10-24) defines the four unknowns at point P. The discrete equations may be derived from the continuous equations by using the trapezoidal rule to approximate the integrals in Eqs. (10-21) through (10-24) [11]. The *characteristics grid* method of solution uses the equations to define a grid of points on the xt plane as the solution is developed (Fig. 10-7a). This grid must be built from the initial and boundary conditions. The numbers along selected grid lines indicate the $(C+, C-)$ pairs which could be used to start the solution and show the order of computations.

Figure 10-7b shows a *specified-intervals* grid. In this method of solution, points are selected in a regular pattern and the characteristics lines through the points are projected backward in time to intersect the time line at which conditions are known [12]. The points of intersection are found as part of the solution by using an interpolation rule along line LMR to locate points S and F. This method is more convenient than the characteristics grid but the interpolations decrease the accuracy. The time step is also limited by the Courant condition discussed later [Eq. (10-25)]. Both methods mimic the continuous solution closely in that they approximately follow its behavior.

The methods do not work well on nonprismatic channels [13]. Neither scheme is conservative and conservation errors must be closely monitored. The

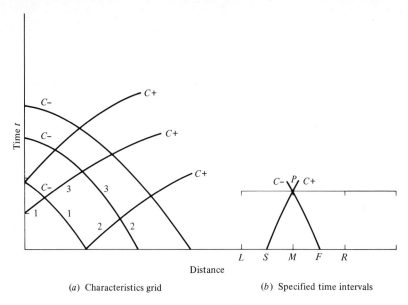

(a) Characteristics grid   (b) Specified time intervals

Figure 10-7 Example pattern for method of characteristics.

method of characteristics is useful in special cases and provides insight into the behavior of solutions.

Explicit and implicit schemes. An explicit scheme permits solution of the equations node by node using information within one or two distance steps from the unknown node. Implicit schemes are viewed as requiring a simultaneous solution for all unknown values. A classification scheme based on the required order of the computation is more closely related to the physical behavior of the solution. This classification [8] defines an explicit scheme as one for which the order of computation of the unknowns at the nodes may be arbitrary. Conversely, an implicit scheme is one for which the order of computations is fixed, including schemes requiring simultaneous solution of the equations.

Figure 10-8 shows a pattern of points on the xt plane for an explicit scheme. Values at point P can be determined from information along LMR. Data from $L'M'R'$ are insufficient to determine the values at point P' because $L'M'R'$ does not contain the interval of dependence of point P'. An application to implicit schemes is shown in Fig. 10-9. The case of subcritical flow requires a simultaneous solution since only one condition is known at the boundaries and no element has sufficient equations to solve for the four unknowns. The domain of uniqueness contains all the unknown points no matter what time step is used. The simultaneous solution is not valid for supercritical flow, because both flow and depth are known at the upstream boundary. A possible scheme for this case is shown in Fig. 10-9b. The scheme is implicit

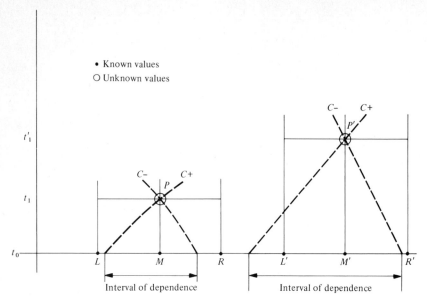

Figure 10-8 Interval of dependence and domain of uniqueness applied to explicit schemes.

because the computations must proceed from the upstream boundary to the downstream boundary, but the unknown point is always within the domain of uniqueness.

The time-step limitation implied for explicit schemes (assuming that the characteristics are straight lines) is

$$\Delta t \leq \frac{\Delta x}{u} \qquad (10\text{-}25)$$

where u is the wave velocity. This is the *Courant condition,* and it requires that the time step be less than the time required for a small disturbance to traverse Δx. The term $u\,\Delta t/\Delta x$, called the *Courant number,* must be less than unity for explicit schemes. The Courant condition does not apply to implicit schemes but the time step may be limited by stability. These arguments are based on the way information about flow changes is propagated. The time step may be limited by such things as the propagation of round-off errors even though information propagation arguments suggest an unlimited time step. Continued violation of time-step limits will eventually result in an invalid solution, if not total break-down, of the computations.

All discrete forms of the complete equations given in this chapter are de-rived from the integral form of the governing equations because it is conceptu-ally simple. The same schemes have been applied to a variety of differential forms of the governing equations, including many not described in this text. No one combination of scheme and form is universally better than all others.

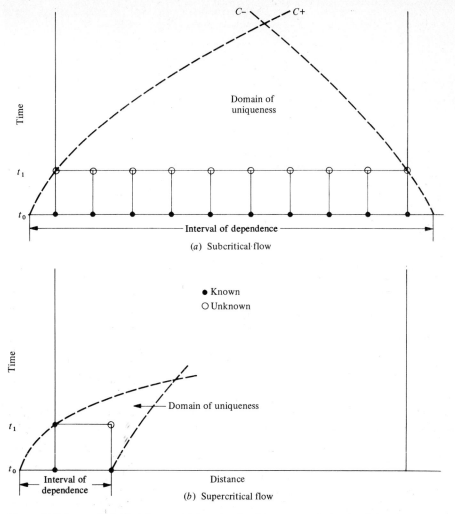

$C-$ $C+$

Domain of
uniqueness

t_1

t_0

Interval of dependence

(a) Subcritical flow

● Known
○ Unknown

Domain of uniqueness

t_1

t_0

Interval of
dependence

Distance

(b) Supercritical flow

Figure 10-9 Interval of dependence and domain of uniqueness applied to implicit schemes.

A relatively simple explicit form derived from Eqs. (10-1) and (10-2) uses the following approximations (Fig. 10-10):

Midpoint rule:

$$\int_{x_1}^{x_2} f(x,\ t)\ dt = f\left(\frac{x_1 + x_2}{2},\ t\right)(x_2 - x_1) \qquad (10\text{-}26)$$

Rectangular, right-hand rule:

$$\int_{t_1}^{t_2} f(x,\ t)\ dt = f(x,\ t_1)(t_2 - t_1) \qquad (10\text{-}27)$$

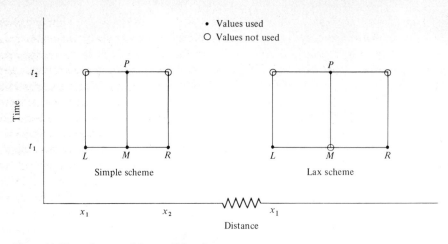

Figure 10-10 xt plane used for explicit schemes.

$$\int_{t_1}^{t_2} \int_{x_1}^{x_2} f(x,\, t) \; dx \; dt = f\left(\frac{x_1 + x_2}{2},\, t_1\right)(t_2 - t_1)(x_2 - x_1) \qquad (10\text{-}28)$$

where $f(x,\, t)$ is any function in the governing equations. These approximations yield

$$2\Delta x A_P - 2\Delta x A_M = \Delta t Q_L - \Delta t Q_R \qquad (10\text{-}29)$$

$$2\Delta x Q_P - 2\Delta x Q_M = \Delta t F_L - \Delta t F_R + \Delta x \, \Delta t G_M \qquad (10\text{-}30)$$

where $F = Qv + J$ and $G = gA(s_0 - s_f)$. Equations (10-29) and (10-30) assume a prismatic channel and no lateral inflow. When solved for the unknowns, these equations yield

$$A_P = A_M + \frac{\Delta t}{2\Delta x} (Q_L - Q_R) \qquad (10\text{-}31)$$

$$Q_P = Q_M + \frac{\Delta t}{2\Delta x} (F_L - F_R + G_M \, \Delta x) \qquad (10\text{-}32)$$

Unfortunately this simple scheme is unstable when friction forces are ignored [4]. The friction term has a weak stabilizing influence but only for limited conditions and times. Thus, the scheme is unusable. Replacing terms at point M with the average of the corresponding terms at L and R makes the scheme conditionally stable with the time step limited by the Courant condition. This scheme, called the *diffusing* or *Lax scheme,* is

$$A_P = \tfrac{1}{2}(A_L + A_R) + \frac{\Delta t}{2\Delta x} (Q_L - Q_R) \qquad (10\text{-}33)$$

$$Q_P = \tfrac{1}{2}(Q_L + Q_R) + \frac{\Delta t}{2\Delta x}\left[F_L - F_R + \frac{\Delta x}{2}(G_L + G_R)\right] \qquad (10\text{-}34)$$

The time step is also constrained by

$$t < \frac{2}{\left| \dfrac{\partial G}{\partial Q} \right|} \tag{10-35}$$

which becomes limiting in rough channels at shallow depths.

This scheme is not consistent with the governing equations since the limit of the discrete equations contains terms dependent on Δx and Δt which are not present in the governing equations [14]. These terms cause the scheme to be dissipative, which is useful in representing flows containing an abrupt wave because the dissipative terms represent approximately the energy loss across the flow discontinuity [14, 15].

Equations (10-33) and (10-34) may be applied at all nodes except those on the boundary. Values at the boundary nodes can be calculated using the boundary conditions and an equation approximating the relationship along the characteristic line through the boundary point [14].

The diffusing scheme, although simple, must be used with caution. A steady-state profile for this scheme is a straight line and, hence, nearly steady flows will not be well represented. The scheme has not worked well in nonprismatic channels [13]. A weighted average between the diffusing scheme and the unstable explicit scheme has been used for modeling dam-break flows with some success [16], with terms added to the discrete equations to improve the modeling of flow in nonprismatic channels. Applications of this scheme should be limited to rapidly varying, discontinuous flows in essentially prismatic channels.

The *Preissman weighted four-point scheme* is an implicit scheme widely used for various forms of the governing equations [2, 17]. It is derived here using the generalized two-point integration rule for the time integrals and the trapezoidal rule for the distance integrals in Eqs. (10-1) and (10-2). The generalized two-point rule is

$$\int_{t_1}^{t_2} f(t) \, dt = [\theta f_2 + (1 - \theta)f_1](t_2 - t_1) \tag{10-36}$$

where f_1 and f_2 are integrand values at t_1 and t_2, and θ is an integrating weight, $0 \le \theta \le 1$. Applying these rules yields

$$\tfrac{1}{2}\Delta x(A_{1,2} + A_{2,2}) - \tfrac{1}{2}\Delta x(A_{1,1} + A_{2,1}) = \Delta t[\theta Q_{1,2} + (1 - \theta)Q_{1,1}]$$
$$- \Delta t[\theta Q_{2,2} + (1 - \theta)Q_{2,1}] + \Delta t[(1 - \theta)I_1 + \theta I_2] \tag{10-37}$$

as the approximate equation for conservation of water volume. Here $Q_{i,j}$ is the flow at point x_i at time t_j in the xt plane, $A_{i,j}$ is the corresponding area, Δx is the length of an element, Δt is the time step, I_j is the lateral inflow at time t_j, and θ is an integrating weight for integrals with respect to time.

The approximation of the momentum equation is simplified by noting that

$$J(x_1, t) - J(x_2, t) + \int_{x_1}^{x_2} J_x \, dx \approx \frac{A(x_1, t) + A(x_2, t)}{2} [y(x_1, t) - y(x_2, t)]$$

(10-38)

with an error of the same order as for the trapezoidal rule [18]. Letting $G = gA(s_0 - s_f)$, the discrete form of Eq. (10-2) for the weighted four-point scheme is

$$[\tfrac{1}{2}(Q_{2,2} + Q_{1,2}) - \tfrac{1}{2}(Q_{2,1} + Q_{1,1})] \, \Delta x$$
$$= g\{\theta Q_{1,2}v_{1,2} + (1 - \theta)Q_{1,1}v_{1,1} - [\theta Q_{2,2}v_{2,2} + (1 - \theta)Q_{2,1}v_{2,1}]\} \, \Delta t$$
$$+ g\left[\theta \frac{A_{1,2} + A_{2,2}}{2}(y_{1,2} - y_{2,2}) + (1 - \theta)\frac{A_{1,1} + A_{2,1}}{2}(y_{1,1} - y_{2,1})\right] \Delta t$$
$$+ [\tfrac{1}{2}\theta(G_{1,2} + G_{2,2}) + \tfrac{1}{2}(1 - \theta)(G_{1,1} + G_{2,1})]\Delta x \, \Delta t$$

(10-39)

Each stream element yields two discrete equations, one each for continuity and momentum. The $2n$ equations for n elements are not sufficient to uniquely define $(2n + 2)$ unknowns. Two additional equations conforming to the boundary conditions must be supplied. These conditions will be one of the following: (1) flow as a function of time; (2) depth as a function of time; (3) a relationship between depth and flow; and (4) flow and depth as a function of time (supercritical flow). The physical situation and the state of the flow determine which condition is required. Often the flow is known as a function of time at the upstream boundary, and a stage-discharge relation is known or assumed at the downstream boundary. In this case the following system of equations results:

$$Q_1 - Q(t) \qquad\qquad = 0$$
$$F_1(Q_1, A_1, Q_2, A_2) \qquad = 0$$
$$G_1(Q_1, A_1, Q_2, A_2) \qquad = 0$$
$$\vdots \qquad\qquad\qquad \vdots$$

(10-40)

$$F_n(Q_n, A_n, Q_{n+1}, A_{n+1}) \; = 0$$
$$G_n(Q_n, A_n, Q_{n+1}, A_{n+1}) \; = 0$$
$$Q_{n+1} - f(A_{n+1}) \qquad\quad = 0$$

where $Q(t)$ is flow at the upstream boundary, $f(A)$ is the relation between flow and area at the downstream boundary, $F_j(\;)$ is the continuity equation for the jth element, A_i is the unknown area at node i at t_2, Q_i is the unknown flow at node i at t_2; and n is the number of elements. Single subscripts are used to simplify notation, and only unknown values are shown as arguments of functions. The momentum equation and the relation between flow and area at the downstream boundary are nonlinear equations. Choice of A and Q as dependent variables renders the continuity equation linear, but other dependent variables would make the continuity equation nonlinear.

A general approach to the solution of a nonlinear system is to replace the nonlinear equations with linear equations which approximate the nonlinear equations over some interval. The solution of the linear equations is used as the solution of the nonlinear system or as the first estimate of this solution. To that end, Eq. (10-40) may be expanded in a Taylor series through linear terms to yield a linear approximation for each equation as follows:

$$
\begin{aligned}
\Delta Q_1 &&&&&&= 0 \\
F_1(Q_1^0,A_1^0,Q_2^0,A_2^0) &+ B_1\,\Delta Q_1 &+ C_1\,\Delta A_1 &+ D_1\,\Delta Q_2 &+ E_1\Delta A_2 &&= 0 \\
G_1(Q_1^0,A_1^0,Q_2^0,A_2^0) &+ K_1\Delta Q_1 &+ L_1\,\Delta A_1 &+ M_1\,\Delta Q_2 &+ N_1\,\Delta A_2 &&= 0 \\
\vdots && \vdots & \vdots & \vdots & \vdots & \vdots \\
F_n(Q_n^0,A_n^0,Q_{n+1}^0,A_{n+1}^0) &+ B_n\,\Delta Q_n &+ C_n\,\Delta A_n &+ D_n\,\Delta Q_{n+1} &+ E_n\,\Delta A_{n+1} &&= 0 \\
G_n(Q_n^0,A_n^0,Q_{n+1}^0,A_{n+1}^0) &+ K_n\Delta Q_n &+ L_n\,\Delta A_n &+ M_n\,\Delta Q_{n+1} &+ N_n\,\Delta A_{n+1} &&= 0 \\
\Delta Q_{n+1} - f'(A_{n+1}^0)\,\partial A_{n+1} &&&&&&= 0
\end{aligned}
$$

$$(10\text{-}41)$$

where the superscript 0 denotes the reference values about which the Taylor series are developed: $\Delta Q_i = Q_i - Q_i^0$; $\Delta A_i = A_i - A_i^0$; $B_i = \Delta F_i/\Delta Q_i$; . . . ; $E_i = \Delta F_i/\Delta A_{i+1}$; $K_i = \Delta G_i/\partial Q_i$; . . . ; $N_i = \Delta G_i/\Delta A_{i+1}$; and $i = 1, \ldots, n$. Equation system (10-41) may be represented more compactly by the matrix form given for a two-element representation of a stream

$$
\begin{bmatrix}
1 & 0 & 0 & 0 & 0 & 0 \\
B_1 & C_1 & D_1 & E_1 & 0 & 0 \\
K_1 & L_1 & M_1 & N_1 & 0 & 0 \\
0 & 0 & B_2 & C_2 & D_2 & E_2 \\
0 & 0 & K_2 & L_2 & M_2 & N_2 \\
0 & 0 & 0 & 0 & 1 & -f'
\end{bmatrix}
\begin{bmatrix}
\Delta Q_1 \\ \Delta A_1 \\ \Delta Q_2 \\ \Delta A_2 \\ \Delta Q_3 \\ \Delta A_3
\end{bmatrix}
=
\begin{bmatrix}
0 \\ -F_1 \\ -G_1 \\ -F_2 \\ -G_2 \\ 0
\end{bmatrix}
\qquad (10\text{-}42)
$$

Coefficient matrix   vector of unknowns   right-hand vector

The solution of this system yields corrections to the reference values of the Taylor series expansion. The corrected reference values are used to recompute the coefficient matrix and the right-hand vector to develop the next set of corrections. The corrections will become small if the process converges. This approach is the Newton-Raphson method applied to an equation system [19]. The correction process converges rapidly if the reference values used for the expansions are close to the solution. Frequently the reference values are taken as the known values at t_1. The Newton-Raphson method may have severe problems if a good set of initial reference values is not available.

A useful alternative approach [2] is to linearize the equations and compute only one correction, using methods other than the Taylor series. In any case the time and distance steps must be controlled to obtain an accurate solution. Convergence of the Newton-Raphson method after many corrections indicates that the time and/or distance steps are too large. An exact solution of the discrete equations is only an approximate solution of the continuous equations.

Simultaneous solution of the linear equation system can be implemented efficiently because the structure of the coefficient matrix is not changed as more elements are added. There are no more than two (sometimes only one) nonzero entries for the upstream and downstream boundary conditions. There are no more than four nonzero entries for each element equation, and the two equations for each element are centered on the main diagonal of the matrix. The computational effort for solution of such a *banded matrix* is proportional to the number of equations. Use of the band structure in the solution greatly reduces effort. Solutions with several hundred equations are still economical [4, 20]. Figure 10-9a shows that a simultaneous solution is required for subcritical flow but a sequential solution beginning at the upstream boundary is required for supercritical flow [8, 21]. The discrete equations remain unchanged and only the details of the solution process are altered.

The weighted four-point scheme is unconditionally stable for $\theta \geq 0.5$, but the time and distance steps are limited by the accuracy of the assumed linear variation of functions between node points. The use of $\theta > 0.5$, although reducing integration accuracy, may damp parasitic oscillations which often appear in streamflow routing with the complete equations [8, 21]. These oscillations have wavelengths about $2\Delta x$ and may grow sufficiently to destroy the solution. The oscillations are generated by the numerical approximations inherent in the discrete equations [15]. Use of $\theta > 0.5$ introduces dissipation which damps the shorter wavelengths most strongly. The weighted four-point scheme is widely used [21–25].

10-5 Kinematic Routing

The complete equations may not be needed to describe some unsteady flows adequately. Simplifications can be made if some forces influencing the flow are small relative to the other forces. The major forces affecting unsteady flow are pressure, gravity, and friction. Inertia of the water influences the response of the flow to these forces. The dominant forces in many natural streams are gravity and friction. The kinematic flow approximation results if all factors other than friction and gravity are ignored in the motion equation; i.e.,

$$s_0 = s_f = \frac{Q^2}{K^2} \tag{10-43}$$

A method which uses Eqs. (10-1) and (10-43) or variants thereof is a *kinematic method*. Certain kinematic wave problems can be solved exactly for simple flows such as overland flow on a plane [26, 27]. More general cases must be solved by numerical methods.

A convenient form for the continuity equation is

$$[(1 - \phi)A_{1,2} + \phi A_{2,2}] \Delta x - [(1 - \phi)A_{1,1} + \phi A_{2,1}] \Delta x =$$
$$[(1 - \theta)Q_{1,1} + \theta Q_{1,2}] \Delta t - [(1 - \theta)Q_{2,1} + \theta Q_{2,2}] \Delta t$$
$$+ [(1 - \theta)I_1 + \theta I_2] \Delta t \tag{10-44}$$

This is the same as Eq. (10-37) except that the generalized two-point integration rule is used for both distance and time integrals with ϕ as the distance integrating weight. Equation (10-43) is the second equation required to solve for flow and depth at the downstream end of the element. Computations begin at the upstream element and move downstream sequentially through the remaining elements. Only one boundary condition is possible—the stage or inflow into the most upstream element.

The kinematic motion equation can be represented as $Q = f(A)$ where the flow is viewed as dependent on area instead of depth. Solving this equation simultaneously with Eq. (10-44) gives flow and area at the downstream end of an element at the end of a time step (x_2, t_2). Restrictions on the time step may be necessary to ensure stability, validity, or accuracy of the solution.

Such restrictions can, in general, be derived only for linear equations. Thus $Q = f(A)$ is linearized to

$$Q - Q_0 = f'(A)(A - A_0) = u(A - A_0) \tag{10-45}$$

where values with subscript 0 are any convenient reference values, and u is the kinematic wave velocity (Sec. 9-1). Using Eq. (10-45) to replace area terms with flow terms in Eq. (10-44) and solving for the unknown flow gives

$$Q_{2,2} = Q_{1,1} \left[\frac{(1 - \phi)K + (1 - \theta)\,\Delta t}{\phi K + \theta\,\Delta t} \right] + Q_{1,2} \left[\frac{-(1 - \phi)K + \theta\,\Delta t}{\phi K + \theta\,\Delta t} \right]$$
$$+ Q_{2,1} \left[\frac{\phi K - (1 - \theta)\,\Delta t}{\phi K + \theta\,\Delta t} \right] + \left[\frac{(1 - \theta)I_1 + \theta I_2}{\phi K + \theta\,\Delta t} \right] \tag{10-46}$$

where $K = \Delta x/u$ = time for a small kinematic wave to traverse the element. Equation (10-46) is similar to the equations for Muskingum routing (Sec. 9-8). Cunge [28] found that the Muskingum weighting coefficient x must satisfy $0 \le x \le \frac{1}{2}$ for the method to be stable. Thus, the weighting coefficient for distance integrals, ϕ in Eq. (10-46), must satisfy $\frac{1}{2} \le \phi \le 1$ for stability. This is not the only necessary constraint. Without further restrictions Eq. (10-46) can predict that outflow decreases when inflow increases, and the outflow may over- or undershoot a newly established steady flow. Such meaningless solutions are avoided if each of the bracketed terms on flowrate (Eq. 10-46) is required to be ≥ 0. This leads to the following restriction on time and distance increments:

$$\frac{1 - \phi}{\theta} K \le \Delta t \le \frac{\phi}{1 - \theta} K \tag{10-47}$$

The values of the weighting coefficients that yield the most accurate estimates of the integrals in Eq. (10-1), $\theta = \phi = \frac{1}{2}$, require that $\Delta t = K$. This equality cannot be maintained because the kinematic wave velocity is a function of stream stage.

Selecting the most accurate approximation for each term in the governing equations does not necessarily lead to a workable algebraic solution of these equations. The choice of $\theta = \phi = 1$ allows an unrestricted choice of time steps, but this choice relates only to stability and validity. Accuracy requires that the range of acceptable time steps be restricted. Note that $\theta = \phi = 1$ implies that the volume of water in an element is the product of downstream area and element length and that the volume of inflow or outflow during an interval is the product of the time step and the inflow or outflow rate at the end of the interval. Time and distance steps should be selected with these approximations in mind.

Schaake [29] presented a method based on Eq. (10-44) with $\theta = \phi = 1$ but eliminating the flowrate terms instead of the area terms. The resulting equation is

$$A_{2,2} = \frac{\Delta t}{K + \Delta t} A_{1,2} + \frac{K}{K + \Delta t} A_{2,1} + \frac{I_2 \Delta t}{u(K + \Delta t)} \qquad (10\text{-}48)$$

The kinematic wave speed u in Eq. (10-48) is estimated by

$$u = f'[\tfrac{1}{2}(A_{1,2} + A_{2,1})] \qquad (10\text{-}49)$$

for each time step and element. The flowrate corresponding to $A_{2,2}$ given by Eq. (10-48) should be calculated from

$$Q_{2,2} = Q_{1,2} + u(A_{2,2} - A_{1,2}) \qquad (10\text{-}50)$$

if the scheme is to be conservative. Computation of the flowrate from $Q_{2,2} = f(A_{2,2})$ introduces an error in volume because Eq. (10-48) results from Eq. (10-44) by eliminating $Q_{2,2}$ using Eq. (10-50). Equations (10-50) and (10-48) form a conservative scheme but the area-discharge equation, $Q = f(A)$, is not satisfied. A possible resolution is to use Eqs. (10-48) and (10-50) and then check for the error in the area-discharge relation. The error would be evaluated as

$$\frac{|Q_{2,2} - f(A_{2,2})|}{f(A_{2,2})} \leq \varepsilon \qquad (10\text{-}51)$$

where ε is the maximum acceptable relative error in the discharge-area relationship. Precise agreement can be obtained with nonlinear solution techniques but may not be warranted in view of the approximations in kinematic assumptions and in Eq. (10-44). Violations of the tolerance [Eq. (10-51)] should be corrected by reducing the distance or time steps. A nonlinear solution of Eqs. (10-44) and $Q = f(A)$ with $\theta = \phi = 1$ has been developed [30].

An alternative formulation which conveniently incorporates diversions [31] uses Eq. (10-44) with $\phi = 1$ and rearranged to show $Q_{2,2}$ as a linear function of $A_{2,2}$

$$Q_{2,2} = -\frac{\Delta x}{\theta \, \Delta t} A_{2,2} + \left[\frac{1 - \theta}{\theta} (Q_{1,1} - Q_{2,1}) + Q_{1,2} \right] + \frac{\Delta x}{\theta \, \Delta t} A_{2,1} \qquad (10\text{-}52)$$

Equation (10-52) is plotted as a straight line in Fig. 10-11. The area-discharge relationship is approximated by a series of connected straight lines. The intersection of the two lines is the solution. The error tolerance of the area-discharge function is established by the number and location of the approximating line segments. Diversions are treated as additive to the discharge curve (Fig. 10-11).

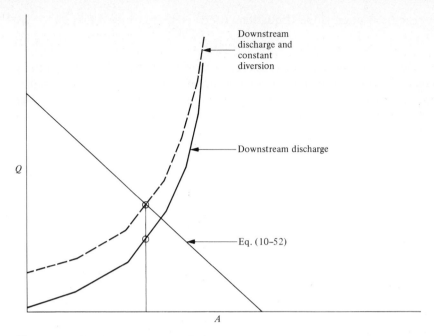

Figure 10-11 Graphical kinematic routing.

Area and total discharge are given by the intersection of the total area-discharge curve and the line representing Eq. (10-52).

The intercept given by Eq. (10-52) at $A_{2,2} = 0$ may be negative if $\theta < 1$, and there is no solution to the equation. If the inflow to the element is zero, the element has emptied before the end of the interval, and $Q_{2,2} = 0$. Otherwise the time step is too large and should be reduced.

Care must be exercised in kinematic routing to avoid unrealistic results. Rapid decreases in channel slope or flow capacity, or an increase in roughness may increase the neglected terms sufficiently to invalidate the kinematic assumption. The zero-inertia assumption (Sec. 10-6) may be valid because the inertia terms are generally smaller than the water-surface slope terms [32]. Kinematic routing is frequently used as the overland flow and streamflow routing component of catchment models. A recent urban runoff model [33] used the Schaake modification for kinematic routing. Additional applications of kinematic routing using a variety of solution schemes are available [34–36].

10-6 Zero-Inertia Routing

A less radical simplification of the complete equations results if only the inertial terms are neglected in the momentum equation. The equation becomes

$$\frac{\partial y}{\partial x} = s_0 - s_f \tag{10-53}$$

where s_f is the friction slope. Equation (10-53) is approximated [37]

$$y_{2,2} - y_{1,2} = s_0 \, \Delta x - \tfrac{1}{2}(S_{2,2} - S_{1,2}) \, \Delta x \qquad (10\text{-}54)$$

where $S = s_f = $ the friction slope. The equations cannot be solved sequentially as in kinematic routing but must be solved simultaneously. A solution similar to that for the complete equations [Eqs. (10-37) through (10-42)] may be used. The zero-inertia approximation has been applied to border irrigation applications [37] and routing in streams [38, 39].

PROBLEMS

10-1 Write two forms of momentum flux equivalent to Qv and show why all three forms represent momentum flux.

10-2 Derive the St. Venant form of the governing equations from the conservation form.

10-3 Why does the term J_x^y appear in Eq. (10-8)? Hint: Set $Q = 0$ and examine the rest of the terms.

10-4 Derive Eq. (10-18) from Eq. (10-17).

10-5 Show that the first moment of area about the water surface, J, may also be evaluated as

$$J = \int_0^y A(x,z) \, dz$$

Integration of Eq. (10-3) by parts or geometrical construction should prove useful.

10-6 A reservoir of rectangular cross section is 52 ft deep at the dam and 20 ft deep 15,000 ft upstream. Compute the time for an infinitesimally small, shallow wave initiated at the dam to reach the point where the water is 20 ft deep. Assume channel slope is constant and flowrate is zero.

10-7 What boundary conditions would be required for the complete equations in the case of a channel which discharges into a lake?

10-8 Show the trajectory of the wave in Prob. 10-6 on the xt plane.

10-9 Suppose that another infinitesimally shallow wave is initiated where the depth is 20 ft at the same time as the wave in Prob. 10-6. Compute the point on the xt plane where the two waves will meet. Draw the trajectories and show the domain of uniqueness on the xt plane.

10-10 Does the diffusing scheme conserve water? What about the weighted four-point scheme?

10-11 Show the pattern of zero and nonzero elements in the coefficient matrix of the weighted four-point scheme for a stream represented by five elements with the flow given at the upstream boundary and with the downstream boundary subject to tidal fluctuation.

10-12 What values of θ and ϕ [Eq. (10-46)] yield the Muskingum method? How might Eq. (10-46) be used to improve on the assumption of constant K in Muskingum routing?

10-13 A uniform rectangular channel 3 m wide with $n = 0.015$ and slope of 0.0004 is flowing at a depth of 1.6 m. A sudden gate opening increases the depth to 1.9 m. What is the celerity of the resulting abrupt wave?

REFERENCES

1. Liggett, J. A.: "Basic Equations of Unsteady Flow," in K. Mahmood and V. Yevjevich, (eds.): "Unsteady Flow in Open Channels," pp. 29–62, Water Resources Publications, 1975.
2. Liggett, J. A., and J. A. Cunge: Numerical Methods of Solution of the Unsteady Flow Equations, ibid., pp. 89–182.

3. Abbott, M. B.: "Computational Hydraulics," pp. 1–55, Pitman, London, 1980.
4. Strelkoff, T.: Numerical Solution of the St. Venant Equations, *J. Hydraul. Div. ASCE,* vol. 96, pp. 223–252, January 1970.
5. See [3], pp. 56–116.
6. Yen, B. C.: Open Channel Flow Equations Revisited, *J. Eng. Mech. Div. ASCE,* vol. 99, pp. 979–1009, October 1973.
7. Strelkoff, T.: One-Dimensional Equations of Open-Channel Flow, *J. Hydraul. Div. ASCE,* vol. 95, pp. 861–876, May 1969.
8. See [3], pp. 117–219.
9. Basco, D. R.: Introduction to Numerical Methods, Verification of Mathematical and Physical Models in Hydraulic Engineering, *Proc. 26th Ann. Hydraul. Div. Specialty Conf. ASCE,* pp. 280–302, 1978.
10. See [3], pp. 56–116.
11. Liggett, J. A., and D. A. Woolhiser: Difference Solutions of the Shallow Water Equations, *J. Eng. Mech. Div. ASCE,* vol. 93, pp. 39–71, April 1967.
12. Ames, W. F.: "Numerical Methods for Partial Differential Equations," pp. 209–212, Barnes and Noble, New York, 1969.
13. Franz, D. D.: Dam Break Flood Analysis: Problems, Pitfalls, and Partial Solutions, *Proc. Dam-Break Flood Model Workshop,* October 1977, U.S. National Tech. Inf. Serv. NTIS PB 275 437, pp. 354–383, U.S. Water Resources Council, Washington, D.C.
14. Terzidis, G., and T. Strelkoff: Computation of Open-Channel Surges and Shocks, *J. Hydraul. Div. ASCE,* vol. 96, pp. 2581–2610, December 1970.
15. See [3], pp. 220–250.
16. Rajar, R.: Mathematical Simulation of Dam-Break Flows, *J. Hydraul. Div. ASCE,* vol. 104, pp. 1011–1026, July 1978 (errata, March 1979; discussion, August 1979; closure, March 1980).
17. Fread, D. L.: Flood Routing in Meandering Rivers with Flood Plains, *Rivers '76,* vol. 1, *Symp. Inland Waterways for Navigation, Flood Control, and Water Diversions, ASCE,* pp. 16–35, 1976.
18. Cunge, J. A.: Discussion of M. Amein and H. Chu, Implicit Numerical Modeling of Unsteady Flows, *J. Hydraul. Div. ASCE,* vol. 102, pp. 120–124, January 1976.
19. Dahlquist, G., A. Bjorck, and N. Anderson (translator): "Numerical Methods," pp. 249–254, Prentice-Hall, Englewood Cliffs, N.J., 1974.
20. Fread, D. L.: Discussion of M. Amein and C. S. Fang: Implicit Flood Routing in Natural Channels, *J. Hydraul. Div. ASCE,* vol. 97, pp. 1155–1159, July 1971.
21. Fread, D. L.: The Development and Testing of a Dam-Break Flood Forecasting Model, *Proc. Dam-Break Flood Routing Model Workshop,* October 1977, U.S. National Tech. Info. Serv. NTIS PB 275 437, U.S. Water Resources Council, Washington, D.C.
22. Fread, D. L.: NWS Operational Dynamic Wave Model, *Proc. 25th Ann. Hydraul. Div. Specialty Conf. ASCE,* pp. 455–464, 1978.
23. Chaudry, Y. M., and D. N. Contractor: Application of the Implicit Method to Surges in Open Channels, *Water Resour. Res.,* vol. 9, pp. 1605–1612, December 1973.
24. Amein, M., and H. Chu: Implicit Numerical Modelling of Unsteady Flows, *J. Hydraul. Div. ASCE,* vol. 101, pp. 717–731, June 1975.
25. Fread, D. L.: Technique for Implicit Dynamic Routing in Rivers with Tributaries, *Water Resour. Res.,* vol. 9, pp. 918–926, August 1973.
26. Henderson, F. M.: "Open Channel Flow," pp. 394–398, Macmillan, New York, 1966.
27. Overton, D. E., and M. E. Meadows: "Stormwater Modelling," pp. 58–86, Academic Press, New York, 1976.
28. Cunge, J. A.: On the Subject of a Flood Propagation Method (Muskingum Method), *J. Hydraul. Res.,* vol. 7, no. 2, pp. 205–230, 1969.
29. Schaake, J. C., Jr.: Deterministic Urban Runoff Model, *Treatise on Urban Water Systems,* pp. 357–383, Colorado State University, July 1971.
30. Li, R. M., D. B. Simons, and M. A. Stevens: Nonlinear Kinematic Wave Approximation for Water Routing, *Water Resour. Res.,* vol. 11, pp. 245–252, April 1975.

31. Johanson, R. C., J. C. Imhoff, and H. H. Davis: User's Manual for Hydrologic Simulation Program—Fortran (HSPF), Environmental Protection Agency Report, EPA-600/9-80-015, April 1980.
32. See [26], pp. 374–381.
33. Dawdy, D. R., J. C. Schaake, Jr., and W. M. Alley: User's Guide for Distributed Routing Rainfall-Runoff Model, *U.S. Geol. Surv. Water-Resources Invest.* 78–90, September 1978.
34. Devries, J. J., and R. C. MacArthur: Introduction and Application of Kinematic Wave Routing Techniques Using HEC-1, *Hydrol. Eng. Center Training Doc.* 10, U.S. Army Corps of Engineers, May 1979.
35. Harley, B. M., F. E. Perkins, and P. S. Eagleson, A Modular Distributed Model of Catchment Dynamics, *Mass. Inst. Tech. Ralph M. Parsons Lab. for Water Resour. and Hydrodynamics Rep.* 133, December 1970.
36. Terstriep, M. L., and J. B. Stall: The Illinois Urban Drainage Simulator, ILLUDAS, *Ill. State Water Surv. Bull.* 58, 1974.
37. Strelkoff, T., and N. D. Katapodes: Border Irrigation Hydraulics with Zero Inertia, *J. Irrig. Drain. Div. ASCE,* vol. 103, pp. 325–342, September 1977.
38. Brakensiek, D. L., A. L. Heath, and G. H. Comer: Numerical Techniques for Small Watershed Flood Routing, *U.S. Agric. Res. Serv. ARS* 41:113, February 1966.
39. Harder, J. A., and L. V. Armacost: Wave Propagation in Rivers, *Univ. Calif. Hydraul. Eng. Lab. HEL* 8-1, June 1966.

BIBLIOGRAPHY

Abbott, M. B.: "An Introduction to the Method of Characteristics," Thames and Hudson, London; and Elsevier, New York, 1966.
———: Continuous Flows, Discontinuous Flows, and Numerical Analysis, *J. Hydraul. Res.,* vol. 12, no. 4, pp. 417–467, 1974.
Cunge, J. A., F. M. Holly, Jr., and A. Verwey: "Practical Aspects of Computational River Hydraulics," Pitman, London, 1980.
Eagleson, P. S.: "Dynamic Hydrology," pp. 325–368, McGraw-Hill, New York. 1970.
Fread, D. L.: Effects of Time Step Size in Implicit Dynamic Routing, *Water Resour. Bull. AWRA,* vol. 9, no. 2, pp. 338–351, 1973.
———: Calibration Technique for 1-D Unsteady Flow Models, *J. Hydraul. Div. ASCE,* vol. 104, pp. 1027–1044, July 1978.
Mahmood, K., V. Yevjevich, and W. A. Miller, Jr. (eds.): "Unsteady Flow in Open Channels," 3 vols., Water Resources Publications, Fort Collins, Colo., 1975.
Ponce, V. M., H. Indlekofer, and D. B. Simons: Convergence of Four-Point Implicit Water Wave Models, *J. Hydraul. Div. ASCE,* vol. 104, pp. 947–958, July 1978.
Thomas, H. A.: The Hydraulics of Flood Movement in Rivers, *Carnegie Inst. Tech. Eng. Bull.,* 1935.
Whitaker, S.: "Introduction to Fluid Mechanics," pp. 211–284, Prentice-Hall, Englewood Cliffs, N.J., 1968.
Wood, E. F., B. M. Harley, and F. E. Perkins: Operational Characteristics of a Numerical Solution for the Simulation of Open Channel Flow, *Mass. Inst. Tech. Ralph M. Parsons Lab. for Water Resour. and Hydrodynamics Rep.* 150, 1972.
———, ———, and ———: Transient Flow Routing in Channel Networks, *Water Resour. Res.,* vol. 11, no. 3, pp. 423–430, June 1975.
Yen, B. C.: Methodologies for Flow Prediction in Urban Storm Drainage Systems, *Univ. Ill. Water Resour. Center Res. Rep.* 72, 1973.
Yevjevich, V., and A. H. Barnes: Flood Routing through Storm Drains, pt. IV, Numerical Computer Methods of Solution, *Colo. State Univ. Hydrol. Pap.* 43, 1970.

ELEVEN

EROSION, SEDIMENTATION, AND THE RIVER BASIN

Previous chapters discussed ways in which the form of a river basin influences its hydrologic behavior. At the same time, the rate and volume of runoff has much to do with the shaping of the landscape into its visible form. This feedback suggests that there should be some important relationships between basin form and hydrologic performance, but quantitative definition of these relationships has largely eluded researchers.

The erosion and transport of sediment by water is a key process in shaping a river basin, and has important economic and environmental consequences. Predicting rates of erosion and transport of sediment and its deposition on the floodplains or in reservoirs, lakes, and estuaries is one of the tasks of the hydrologist. This chapter discusses methods of estimating erosion and deposition for engineering purposes. It also presents an introduction to *geomorphology*—the study of the formation of the landscape—insofar as it is influenced by water.

11-1 Physical Descriptors of Catchment Form

If basin form and hydrologic characteristics are to be related, the basin form must also be represented by quantitative descriptors. Many such measures have been proposed. Only limited progress has been made in relating the physical and hydrologic features for a number of reasons. The definition of descriptors is necessarily arbitrary, and satisfactory definitions have not yet been established. Descriptors must be measured from maps or aerial photographs, and differing scales and cartographic standards may yield values which are not truly equivalent. A few descriptors which seem to have special relevance in hydrology are discussed in the following paragraphs.

Stream order. Horton [1] suggested a classification of *stream order* as a measure of the amount of branching within a basin. A first-order stream is a small, unbranched tributary (Fig. 11-1). A second-order stream has only first-order tributaries. A third-order stream has only first- and second-order tributaries. The order of a particular drainage basin is determined by the order of the principal stream.

Order is extremely sensitive to the map scale used. A careful study of aerial photographs will often show three or four orders of streams (mostly ephemeral rills and channels) not indicated on a $1:24,000$ scale topographic map. The $1:24,000$ scale map will show one or two orders more than a $1:62,500$ scale map. Even standard maps are not consistent in delineation of streams. Thus if order is to be used as a comparative parameter, it must be carefully defined. For some uses it may be desirable to make an adjustment to estimates of order on the basis of a detailed field survey of a few small tributary basins.

Horton also introduced the *bifurcation ratio* to describe the ratio of the number of streams of any order to the number in the next lowest order. The bifurcation ratios within a basin tend to be about the same magnitude. Generally, bifurcation ratios are found to be between 2 and 4 with a mean value near 3.5. This observation led to the *law of stream numbers*

$$N_u = r_b^{k-u} \tag{11-1}$$

where N_u is the number of streams of order u, r_b is the bifurcation ratio, and k is the order of the main stream. Similarly, Horton suggested the law of stream lengths.

$$\bar{L}_u = \bar{L}_1 r_e^{u-1} \tag{11-2}$$

where $\bar{L}$ is the average length of streams of order u and r_e is the length ratio. An equivalent equation also applies to the average area $\bar{A}$ of basins of order u

$$\bar{A}_u = \bar{A}_1 r_a^{u-1} \tag{11-3}$$

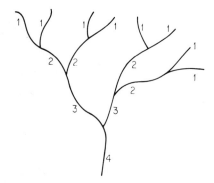

Figure 11-1 Definition sketch for stream order.

Equations (11-1) to (11-3) indicate a geometric progression of number, length, and area. Graphically, they suggest a linear plot of order number versus the logarithms of number, length, or area. The relationships have been confirmed under a wide range of conditions. The equations can be used by measuring N, L, and A for the two highest orders in the basin and then estimating these values for all lower orders.

Some geomorphologists favor reversing the numbering system so that the main stream is order one and tributaries of increasingly higher order. Both approaches are arbitrary, but Horton's system seems to be the most widely used procedure.

Drainage density. The total length of streams within a catchment divided by the drainage area defines the *drainage density,* the length of channels per unit area. A high drainage density reflects a highly dissected basin, which should respond relatively rapidly to a rainfall input, while low drainage density reflects a poorly drained basin with slow hydrologic responses. Observed values of drainage density range from as low as 3 in parts of the Appalachian area of the United States to 400 or more in Badlands National Monument, South Dakota. Low drainage densities are observed where soil materials are resistant to erosion or very permeable and where the relief is small. High values may be expected where soils are easily eroded or relatively impermeable, slopes are steep, and vegetal cover is scanty.

Length of overland flow. The average length of overland flow $\bar{L}_o$ may be approximated by

$$\bar{L}_o = \frac{1}{2D} \tag{11-4}$$

where D is the drainage density. This approximation ignores the effects of ground and channel slope, which make the actual overland path longer than the estimate. The error is probably of little significance. Horton suggested that the denominator be multiplied by $\sqrt{1 - s_c/s_g}$, where s_c and s_g are the average channel and ground slopes, respectively. This modification reduces the approximation inherent in Eq. (11-4).

Area relations. Data for a number of the larger rivers of the world [2] seem to conform to the equation

$$L = 1.27A^{0.6} \tag{11-5}$$

where L is main-channel length (km) and A is drainage area (km²). Data by Hack [3] show a similar relation. The exponent generally seems to be between 0.6 and 0.7, suggesting that a basin tends to elongate as it grows larger. The coefficient in Eq. (11-5) becomes 1.4 with dimensions in miles.

Basin shape. The shape of a catchment affects the streamflow hydrograph and peak-flow rates. Numerous efforts to develop a factor which describes basin shape by a single numerical index have been reported. Basins tend to the form of a pear-shaped ovoid, but geologic controls result in many substantial deviations from this shape. Horton [4] suggested the dimensionless form factor R_f as an index of shape,

$$R_f = \frac{A}{L_b^2} \qquad (11\text{-}6)$$

where A is basin area and L_b is the length of the basin measured from outlet to divide near the head of the longest stream along a straight line. This index or its reciprocal has been used as an indicator of shape in some synthetic unit hydrograph procedures.

Equation (11-6) implies no special assumption of basin shape. For a circle $R_f = \pi/4 = 0.79$; for a square with the outlet at the midpoint of one side $R_f = 1$; and for a square with the outlet at one corner $R_f = 0.5$. Values for other geometrical shapes are easily calculated. Numerous writers have suggested the use of a circle [5] or lemniscate [6] as a reference shape. The resulting shape factors reduce to Horton's R_f multiplied by a constant.

11-2 Descriptors of Catchment Relief

The topography or relief of a basin may have more influence on its hydrologic response than catchment shape, and numerous descriptors of relief have been advanced by various writers. Problems encountered in relating catchment relief and hydrologic response are similar to those discussed in Sec. 11-1. Some of the more useful descriptors are discussed in this section.

Channel slope. The slope of a channel affects velocity of flow and must play a role in hydrograph shape. Typical channel profiles (Fig. 11-2) are concave upward. In addition all but the very smallest basins contain several channels, each with its own profile. Thus the definition of average channel slope of a basin is difficult. Commonly, only the main stream is considered in describing the channel slope of a catchment. A simple and widely used measure of channel slope is the slope of a line (AB, Fig. 11-2) drawn such that the area under it equals the area under the main-channel profile. Taylor and Schwarz [7] calcu-

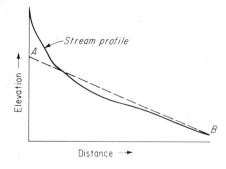

Figure 11-2 One method of defining mean channel slope.

lated the slope of a uniform channel having the same length and time of flow as the main channel. Since the velocity is proportional to square root of slope, the procedure used by Taylor and Schwarz is equivalent to weighting channel segments by the square root of their slope, which gives relatively less weight to the steep upstream reaches of the stream. Thus, if the channel were divided into n equal segments, each of slope s_i, a simple index of slope would be

$$R_s = \left(\frac{\sum_{i=1}^{i=n} \sqrt{s_i}}{n} \right)^2 \qquad (11\text{-}7)$$

Land slope. The slope of the ground surface is a factor in the overland-flow process and hence a parameter of hydrologic interest, especially on small basins where the overland-flow process may be a dominant factor in determining hydrograph shape. Because of variation in land-surface slope in the usual catchment, a method of defining an average or index value is required.

The distribution of land-surface slope can be determined by establishing a grid or a set of randomly located points over a map of the catchment. The slope of a short segment of line normal to the contours is determined at each grid intersection or random point. The mean, median, and variance of the resulting distribution can be calculated. The accuracy of the results depends on the adequacy of the map used.

Area-elevation data. When one or more of the factors of interest in a hydrologic study vary with elevation, it is useful to know how the catchment area is distributed with elevation. An area-elevation (or *hypsometric*) curve can be constructed by planimetering the area between contours on a topographic map and plotting the cumulative area above (or below) a given elevation versus that elevation (Fig. 11-3). In some cases it is convenient to use percentage of area instead of actual area, particularly if a comparison between basins is desired. If a grid with about 100 or more intersections is superimposed on a catchment map and the number of grid intersections in each elevation range is noted, an area-elevation curve can be constructed which will be about as accurate as one derived by planimetry but requires less effort.

The curve of Fig. 11-3 is typical of area-elevation curves of geologically mature catchments. Very small basins, however, may show quite different characteristics. Snowmelt computations in mountainous areas must usually be made for elevation zones because both snow depth and temperature are functions of elevation. Precipitation in mountainous areas may sometimes be weighted by elevation in calculations of basin average precipitation.

Aspect. The aspect of a slope is the direction toward which the slope faces. Because of the influence of insolation on snowmelt, aspect may be of interest in dealing with snow. Precipitation amounts are often influenced by the aspect of a slope relative to the direction of the wind. Customarily, aspect is used as a

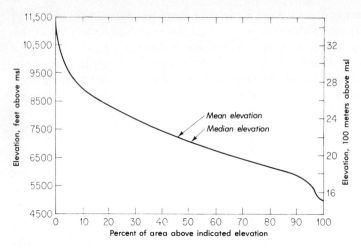

Figure 11-3 Area-elevation curve for the Teton River catchment above St. Anthony, Idaho.

characteristic of a particular point or at most of a specific hillside. The distribution of aspect may be determined in manner similar to that described for land slope. A grid or set of random points is superimposed over a map of the catchment, and the bearing of a line normal to the contours at each intersection is recorded. From these data a polar diagram can be plotted with aspect shown as angle from north and percentage area as distance from the origin. Typically, such plots appear nearly circular, indicating a uniform distribution of aspect. However, very small basins may show a substantial departure from uniformity.

11-3 Hydraulic Geometry

Hydraulic geometry describes the character of the channels of a basin: the variation of mean depth, top width, and velocity at a particular cross section and between cross sections. These relations apply to alluvial channels, where the cross section is readily adapted to the flows which occur, but are less reliable where rock outcrops control the channel characteristics.

The basic equations of hydraulic geometry are [8]

$$B = aq^b \tag{11-8}$$

$$D = cq^f \tag{11-9}$$

$$v = kq^m \tag{11-10}$$

where q is discharge, B is top width, D is mean depth, v is mean velocity, and a, b, c, f, k, and m are numerical constants. Since $q = BDv$, it follows that $ack = 1$ and $b + f + m = 1$. The equations appear as straight lines on log-log plots, the exponents representing the slope of the lines and the coefficients the intercept when $q = 1$. Plotting data for a fairly large number of streams indicates marked regional conformity in the values of the exponents, and agreement between regions is close enough to suggest the possibility of some degree of universality to the values.

Table 11-1 presents values of the exponents b, f, and m determined by various investigators. At-station values describe the variation of B, D, and v with q at a given cross section, while the between-station values reflect the change in the values of B, D, and v as q increases in the downstream direction along a stream. Figure 11-4 illustrates the interrelation between the two sets of equations. In Fig. 11-4 the solid lines represent the between-station relations and the dashed lines the at-station relations with slopes plotted at approximately the mean values as given in Table 11-1. Note that Eq. (11-9) is the equation of a rating curve and that the exponent f remains approximately the same at a station and between stations. The width, however, tends to increase more rapidly in a downstream direction than it does at a station, while velocity increases rapidly with q at a station but remains relatively constant in the downstream direction.

To be meaningful, the between-station relations must be defined for flows of the same frequency. Most of the data have been derived for a flowrate equal to the mean annual discharge, which represents a flow at about one-third bankfull depth and is equaled or exceeded about 25 percent of the time (about 90 days/year). However, the bank-full flow with a return period of about 1.5 years and the 10-year flow have also been used.

Much additional work needs to be done to establish the quantitative relations of hydraulic geometry and to explain fully the reasons underlying the relations. In part, they are related to the sediment-transport characteristics of the stream [12]. However, it is clear that for most streams a reasonable relationship can be derived and used, for example, to supplement cross-section data as a basis for routing flow from the catchment.

Table 11-1 Values of the exponents in Eqs. (11-8) to (11-10)

	At station			Between stations		
	b	f	m	b	f	m
Average, midwestern states [8]	0.26	0.40	0.34	0.5	0.4	0.1
Ephemeral streams in semi-arid United States [9]	0.29	0.36	0.34	0.5	0.3	0.2
Average, 158 United States stations [2]	0.12	0.45	0.43			
10 stations on the Rhine River [2]	0.13	0.41	0.43			
Appalachian streams [10]				0.55	0.36	0.09
Kaskaskia River, Ill. [11]				0.51	0.39	0.14

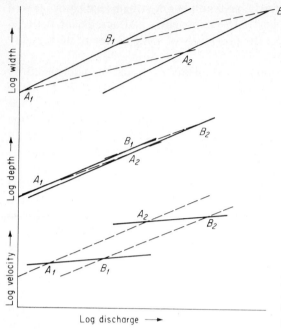

Figure 11-4 Interrelationships in hydraulic geometry between stations A and B and between high and low flows at each station.

11-4 Stream Patterns

When viewed in plan, stream channels may be described as meandering, braided, or straight. A *meandering stream* flows in large, more or less symmetrical loops, or bends (Fig. 11-5). The median length of meandering streams appears to be about 1.5 times the valley length [13]; i.e., the *sinuosity* averages about 1.5. The wavelength of meanders ranges from 7 to 11 times the channel width, and the radius of curvature of the bend usually ranges between 2 and 3 times the channel width. The amplitude of the meanders or width of the meander belt varies considerably and seems to be controlled more by the characteristics of the bank material than by other factors. Amplitude usually ranges from 10 to 20 times channel width.

A *braided* stream (Fig. 11-6) consists of many intertwined channels (*anabranches*) separated by islands. Braided streams tend to be very wide and relatively shallow with coarse bed material. No formal statements about the geometry of braided streams are possible. Few long straight channels exist in nature, but many lack sufficient curvature to be called a meandering stream. A *straight* stream is commonly defined as one with a sinuosity of less than 1.25.

The important question for the engineering hydrologist is to explain why a channel adopts one of the patterns described above. Braided channels are usually found in reaches where the banks are easily erosible—sandy material with little vegetal protection. Bed material is relatively coarse and of heterogeneous particle sizes. The slope of the braided reach is greater than that of

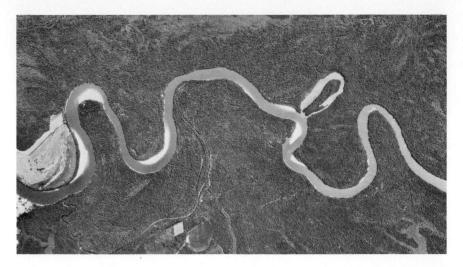

Figure 11-5 A meandering river with a cutoff meander loop, or oxbow lake. (*U.S. National Ocean Survey.*)

adjacent unbraided reaches. Hydraulically, the braided reach is less efficient than the unbraided reach. The total width of branches in a braided reach may be 1.5 to 2 times that of an undivided channel, and the depth of flow is correspondingly less. Braiding is thus a way of dissipating energy when stream slope steepens. Velocity increases that would otherwise lead to erosion are thus avoided.

An initially straight channel, either in a laboratory flume or in the field, will usually develop meanders as water flows through it if the bank material is

Figure 11-6 A braided stream. (*U.S. Geological Survey.*)

erosible. A meandering channel may be 1.5 to 2 times as long as a nonmeandering channel. Its slope is correspondingly reduced, but head losses are increased both because of the longer channel length and because of the bend losses. Without these losses, velocities would be higher, with corresponding tendency to downcut the channel. Many meandering streams cannot downcut because they discharge into a water body with fixed elevation. If downcutting cannot occur, some other device is required to dissipate the available energy.

Thus both braiding and meandering can be explained as means of energy dissipation. Braiding will occur when bed material is coarse and heterogeneous and banks easily erosible. Meandering is likely to occur on flatter slopes where the material is finer and the banks somewhat more cohesive.

In either case the stream is in a kind of equilibrium—equilibrium in the sense that it will maintain its grade but obviously not in the sense that there will be no channel changes. In the braided channel there is continual shifting and changing between individual anabranches, and meanders undergo a more or less continuous process of erosion in the concave bend and deposition at the subsequent point bar, so that the meanders seem to be constantly moving downstream. Any attempt by man to change the natural pattern of a stream requires careful planning and usually costly revetment work to prevent erosion of banks and return to the original pattern.

11-5 Floodplains

The floodplain of a river is the valley floor adjacent to the incised channel, which may be inundated during high water. The river tends to swing back and forth across the valley bottom, reworking the floodplain deposits and eroding first one valley side and then the other. Floodplains are built up primarily from deposition of sediment in the river channel and deposition of fine sediments on the floodplain when flooded. Additionally, organic materials may accumulate in cutoff meander loops (*oxbow lakes*). Often a natural levee will form along the banks of the incised channel caused by the deposit of coarse sediment as the water from the stream invades the floodplain. Sediment deposition in the channel plus natural levees on the bank can lead to a situation in which the stream flows at a higher elevation than its floodplain. This condition develops quite frequently in streams flowing across an alluvial cone.

As suggested earlier, floodplains tend to be flooded at fairly low recurrence intervals. Leopold et al. [14] report numerous evaluations of the flow magnitude required to overflow the floodplain. Return periods generally range between 1 and 2 years, and a general statement that floodplains of the eastern and central United States are inundated by floodwaters in 2 out of 3 years is quite reasonable. The universality of this finding is questionable, however. Nixon [15] made a similar analysis of British streams and found that flooding occurred on the average about twice each year. There is some difficulty in defining precisely what the floodplain is and a problem in defining precisely the bankfull stage. Thus it is not clear if Nixon's data indicate conditions substantially different from those in the United States. In any case it is clear that the floodplain is subject to frequent flooding, and hence its use for buildings and other purposes should be carefully regulated. Transverse slope of a floodplain is usually quite small, and it is often difficult to detect natural levees by visual inspection. Consequently in studies for which the floodplain characteristics are important, one must have either adequately detailed maps or special field surveys which satisfactorily define the information needed.

11-6 The Erosion Process

Soil can be *eroded,* i.e., moved from its current location, by the action of wind, water, gravity (landslides), and human activity. Water erosion may be viewed as starting with the *detachment* of soil particles by the impact of raindrops [16]. The kinetic energy of the drops can splash soil particles into the air. On level ground the particles are redistributed more or less uniformly in all directions, but on a slope there is a net transport downslope (Fig. 11-7). If overland flow is occurring, some falling particles will be entrained in the flowing water and moved even farther downslope before settling to the soil surface. Overland flow is predominantly laminar and cannot detach soil particles from the soil mass, but it can move loose particles already on the soil surface. The splash and overland-flow processes are responsible for *sheet erosion,* the relatively uniform degradation of the soil surface. Sheet erosion is difficult to detect except as the soil surface is lowered below old soil marks on fence posts, tree roots are exposed, or small pillars of soil capped by stones remain.

Raindrops vary in diameter d from 0.5 to 6 mm (0.02 to 0.25 in) and terminal velocity v varies with diameter (Table 3-1) from about 2 to 9 m/s (7 to 30 ft/s). Since kinetic energy is proportional to d^3v^2, the erosive power of the largest drops may be 10,000 times that of the smaller. This conforms with the observation that a few intense storms account for most of the erosion. The effect is augmented by the fact that overland flow is more likely to occur during intense rains.

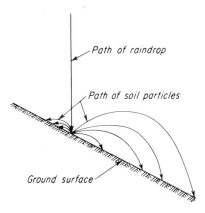

Path of raindrop

Path of soil particles

Ground surface

Figure 11-7 Downhill transport of soil particles by splash.

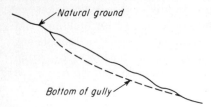

Figure 11-8 Profile of a typical gully.

At some point on the slope sufficient overland flow may accumulate to cause a small rivulet. If turbulence in the flow is strong enough to dislodge particles from the bed and banks of the channel, *gully erosion* may occur. As the gully deepens, its profile is steepest near its head (Fig. 11-8). Erosion is most rapid in this region, and the gully tends to grow headward.

Mass movement of soil, as either the slow downward *creep* of the soil mass or the rapid collapse of a slope (*landslide*), is an important mechanism delivering soil to the streams in steep canyons with unstable side slopes. Landslides occurring as the result of earthquakes or the saturation of the slopes during heavy rains may create temporary dams whose subsequent overtopping and erosion may create critical flood waves downstream.

11-7 Factors Controlling Erosion

The most important factors controlling erosion are rainfall regime (Sec. 11-6), vegetal cover, soil type, and land slope. Because of the important role of raindrop impact, vegetation provides significant protection against erosion by absorbing the energy of the falling drops and generally reducing the drop sizes which reach the ground. Vegetation may also provide mechanical protection to the soil against gully erosion. In addition, a good vegetal cover generally improves infiltration capacity through the addition of organic matter to the soil. Higher infiltration capacity means less overland flow and consequently less erosion.

A cohesive soil will resist splash erosion more readily than loose soils. Generally, splash erosion increases with an increasing fraction of sand in the soil because of the loss of cohesion. Splash erosion decreases with an increasing percentage of water-stable aggregates. A soil whose individual grains do not tend to form aggregates will erode more readily than one in which aggregates are plentiful.

Rates of erosion are greater on steep slopes than on flat slopes. The steeper the slope the more effective splash erosion is in moving soil downslope. Overland-flow velocities are also greater on steep slopes, and mass movements are more likely to occur in steep terrain. Length of slope is also important. The shorter the slope length, the sooner the eroded material reaches the stream, but this is offset by the fact that overland-flow discharge and velocity increase with length of slope.

Land use is also an important factor in fixing the rate of erosion. Poor cropping practices or careless construction of roads may greatly accelerate erosion. Removal of vegetation by fire or lumbering may also increase the erosion hazard. Proper soil conservation practices may greatly reduce erosion losses. The universal soil loss equation [17] attempts to combine all these factors, but it is difficult to express the rainfall regime in a single index number, and field determination of soil erosibility is not yet generally available. Hence, this equation and others of a similar nature are, at best, approximate.

11-8 Suspended-Sediment Transport

Sediment moves in the stream as *suspended sediment* in the flowing water and as *bed load,* which slides or rolls along the channel bottom. A third term, *saltation,* is used to describe the movement of particles which seem to bounce along the bed. The processes are not independent, for material which appears as bed load at one section may be in suspension at another. Another useful distinction is between *bed-material load,* represented by those particles of grain size normally found in the stream bed, and *wash load,* made up of particles smaller than are usually found in the bed. Wash load consists of the fine material washed into the stream during rainfall and which normally travels through the system without redepositing.

The settling velocity of suspended particles in still water is approximated by Stokes' law:

$$v_s = \frac{2(\rho_g - \rho)gr^2}{9\mu} \tag{11-11}$$

where ρ_g and ρ are densities of the particle and the liquid, respectively, r is the radius of the particle, and μ is the absolute viscosity of the water. Generally considered applicable to particles from 0.0002 to 0.2 mm in diameter, the equation assumes that viscosity offers the only resistance to settling, that the particles are rigid and spherical, and that their fall is not impeded by other particles.

In turbulent flow the gravitational settling of particles is counteracted by upward transport in turbulent eddies. Since the concentration of suspended material is greatest near the bottom of the stream, upward-moving eddies carry more sediment than downward-moving eddies. The system is in equilibrium if gravity movement and turbulent transport are in balance and the amount of suspended material remains constant.

The general two-dimensional nonequilibrium equation for suspended-sediment transport is

$$v\,\frac{\partial c_s}{\partial x} = v_s\,\frac{\partial c_s}{\partial y} + \frac{\partial \varepsilon_x}{\partial x}\,\frac{\partial c_s}{\partial x} + \frac{\partial \varepsilon_y}{\partial y}\,\frac{\partial c_s}{\partial y} + \varepsilon_x\,\frac{\partial^2 c_s}{\partial x^2} + \varepsilon_y\,\frac{\partial^2 c_s}{\partial y^2} \tag{11-12}$$

where c_s is the sediment concentration for a particular size of particle, v_s is settling velocity, ε is a mixing coefficient, and x and y are longitudinal and vertical dimensions, respectively. No solution of this equation has been achieved without so many simplifying assumptions that the solution is of little value in dealing with natural streams. The literature is replete with analytic and experimental studies of suspended-load transport. Most effort has been directed toward deriving a function which would describe the vertical variation of sediment concentration in the stream. Such a function combined with a vertical distribution of flow velocity would permit calculation of suspended-load transport. Alternatively, given sediment samples at one or two depths, the total sediment load could be calculated in a manner analogous to current-meter gaging of flow. Several functions have been derived which conform reasonably well with observed sediment-concentration variations in the vertical [18]. However, such functions can apply only to a limited particle-size range and must be summed over the total range of particle size. When these functions are applied to the range of particle sizes in the bed material, a fair approximation to the transport of suspended bed material seems possible, but a large and variable wash-load component precludes a reliable computation of total suspended load. Hence, methods of suspended-load measurement independent of a knowledge of the sediment-concentration gradient were developed (Sec. 11-10).

11-9 Bed-Material Transport

For many years analysis of bed-load transport has been based on the classical equation of du Boys [19]

$$G_i = Y \frac{\tau_0}{w} (\tau_0 - \tau_c) \tag{11-13}$$

where G_i is the rate of bed-load transport per unit width of stream, Y is an empirical coefficient depending on the size and shape of the sediment particles, w is the specific weight of water, τ_0 is the shear at the streambed, and τ_c is the magnitude of shear at which transport begins. Numerous variations on the original du Boys formula have been proposed [18], all using the concept of a critical tractive force to initiate motion. This approach ignores the modern concepts of turbulence and the boundary layer as they affect entrainment of bed particles. The successful application of Eq. (11-13) lies in the proper selection of the coefficient Y. Most available values are determinations from studies with small flumes. Table 11-2 summarizes values given by Straub [20].

Accuracy of instruments for bed-load measurement is so uncertain (Sec. 11-10) that field comparison of bed-load formulas is difficult. The validity of bed-load formulas is therefore quite indefinite. Recent work on the bed-load problem has utilized the concepts of turbulent flow and statistical variation of fluid forces at a point. A widely used approach is that of Einstein [21], who defines the *intensity* of bed-load transport as

$$\Phi = \frac{G_i}{w} \sqrt{\frac{\rho}{\rho_s - \rho} \frac{1}{gd^3}} \tag{11-14}$$

Table 11-2 Factors in Eq. (11-13) for bed-load movement [20]

Particle diameter, mm	Y		τ_c	
	ft⁶/lb²·s	m⁶/kg²·s	lb/ft²	kg/m²
$\frac{1}{8}$	0.81	0.0032	0.016	0.078
$\frac{1}{4}$	0.48	0.0019	0.017	0.083
$\frac{1}{2}$	0.29	0.0011	0.022	0.107
1	0.17	0.0007	0.032	0.156
2	0.10	0.0004	0.051	0.249
4	0.06	0.0002	0.090	0.439

and the *flow intensity* as

$$\psi = \frac{\rho_s - \rho}{\rho} \frac{d}{sR} \tag{11-15}$$

where w is the specific weight of water, ρ is the density of water, ρ_s is the density of the bed material, d is the grain diameter, s is the channel slope, and R is the hydraulic radius. Einstein divided the total hydraulic radius in portions depending on grain roughness R' and on bedforms, R''. In practice R is frequently taken as the total hydraulic radius of the channel or the mean depth. An empirical relation between Φ and ψ (Fig. 11-9) permits the solution of Eqs.

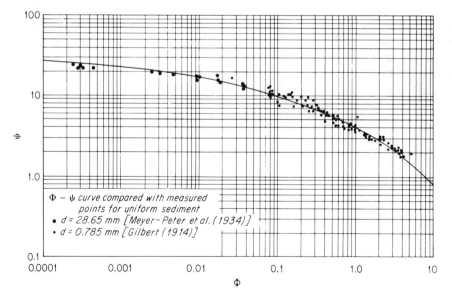

$\Phi - \psi$ curve compared with measured
points for uniform sediment
• $d = 28.65$ mm [Meyer-Peter et al. (1934)]
• $d = 0.785$ mm [Gilbert (1914)]

Figure 11-9 Plot of Einstein's function Φ versus ψ. (*From* [21].)

(11-14) and (11-15) for G_i. The method applies to a specific narrow range of particle sizes and must be repeated to cover the entire range of the bed-material sizes. However, if the range of particle size is small, Einstein suggests a single solution using d_{35}, the particle size for which 35 percent of the bed material is finer. Graf [18] illustrates a detailed computation with the Einstein procedure. This procedure has also been programmed for computer solution.

11-10 Sediment Measurement

Early suspended-sediment observations were made with open bottles or complex grab samplers, which failed to provide adequate data for a number of reasons. A good sampler must cause minimum disturbance of streamflow, avoid errors from short-period fluctuations in sediment concentration, and give results which can be related to velocity measurements. These requirements seem to be met in a series of samplers [22] designed at the Iowa Hydraulic Laboratory under the sponsorship of several federal agencies. The samplers (Fig. 11-10) consist of a streamlined shield enclosing a glass bottle as a sample container. A vent permits escape of air as water enters the bottle and controls the inlet velocity so that it is approximately equal to the local stream velocity. Nozzle tips of various sizes are available to control the rate at which the bottle fills. The large models have the bottle fully enclosed and are fitted with tail vanes to keep the sampler headed into the current when cable-supported. The bottle can be replaced with a collapsible plastic bag which permits a longer sampling time [23].

The sampler is lowered through the stream at constant vertical speed until the bottom is reached and is then raised to the surface at constant speed. The result is an integrated sample with the relative quantity collected at any depth in

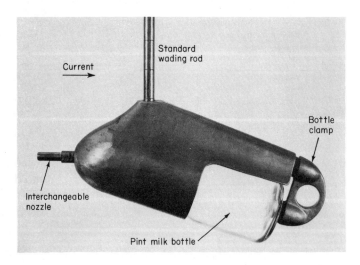

Figure 11-10 The DH-48 depth-integrating hand sampler for use on small streams.

proportion to the velocity (or discharge) at that depth. The duration of the traverse is determined by the time required to nearly fill the sample bottle and can be computed from the filling-rate curves for the particular nozzle when the stream velocity is known. A number of traverses are made at intervals across the section to determine the total suspended-sediment load for the section. Thus there is no problem of whether a point sample is representative of sediment load in the section. Point samplers are used only where it is impossible to use the depth-integrating type because of great depths or high velocity or for studies of sediment distribution in streams. Because of the shape of the sampler, the nozzle cannot be lowered to the streambed, and consequently a few inches of the depth near the bed is not sampled. This may represent a large error [24] in shallow streams.

The collected samples are filtered and the sediment dried. The ratio of dry weight of sediment to total weight of the sample is the sediment concentration, usually expressed in parts per million or milligrams per liter. Other analyses which may be performed include determination of grain-size distribution, fall velocity, and, occasionally, heavy-mineral or chemical analysis. The latter tests may be useful in tracing the original source of the sediment.

Bed-load samplers must rest on the streambed and trap the moving bed load without disturbing the flow. Most portable samplers consist of boxes or bags of wire mesh with supporting frame and a tail vane to keep the entrance pointed into the current. Permeable fabric bags are used when the bed material consists of fine particles which would pass through the wire mesh. The difficulty of designing a device which does not cause some acceleration of the flow is obvious.

On small streams a permanent bed-load trap may consist of a grated opening in the streambed into which the bed material may fall. The trapped material is later excavated and measured. Turbulence-producing weirs have also been designed which throw the bed load into suspension locally so that it can be sampled with a suspended-sediment sampler. Comparison of the samples thus obtained with those from a section upstream of the weir indicates the quantity of bed load. This method is suitable only when the bed material is relatively fine.

Because of the considerable variability of sediment load in streams, a continuous monitoring device would be very useful, and a number of such devices have been proposed. Pumping samplers which draw water from a fixed point in the stream at preset time intervals and store the samples in bottles for later analysis have been tested, as have samplers which pass the water through a cell where the attenuation of a light beam by the suspended sediment is measured with a photocell [25]. Photocell probes which are immersed in the stream [26, 27] and an x-ray probe [28] have also been tried. A significant limitation of all such devices is that they sample only one point in the cross section. An intensive investigation with conventional samplers must precede their installation to determine a representative location within the stream. At best this will be an approximation since the most representative location will vary with stage and possibly other factors. Thus the absolute accuracy of such monitoring devices is unknown. Nevertheless, they do provide a means of obtaining a record of the variation of sediment load with time and, if used in conjunction with a regular program of sampling, should provide better information than is obtained from occasional samples alone.

11-11 Sediment-Rating Curves

Sediment measurements, like current-meter measurements, give only occasional samples of the sediment discharge. A *sediment-rating curve* relating suspended-sediment discharge and water discharge (Fig. 11-11) is commonly used to estimate sediment load on days when no measurements are available. The figure clearly shows that such relations are approximate. A given flowrate may result from melting snow or rains of differing intensity, and a different sediment load would result in each case. Areal distribution of runoff may be a factor if different portions of the basin are more prolific sediment sources than others. Sediment-rating curves should be used with caution and where possible applied to small and relatively homogeneous basins. However, when they are used to estimate mean annual sediment yield, the errors in the sediment rating will tend to compensate and the resulting answer should be reasonably satisfactory if a sufficiently long record is used.

11-12 Sediment Yield of a Catchment

The average annual sediment production from a catchment is dependent on many factors such as climate, soil type, land use, topography, and the presence of reservoirs. Adequate data for a complete analysis of all factors are difficult to obtain. Langbein and Schumm [29] used data from a number of basins to construct the curve of Fig. 11-12, which relates average annual sediment production per unit area to mean annual precipitation. Maximum production rates occur at about 300 mm (12 in) of mean annual precipitation because such areas usually have little protective vegetal cover. With heavier rainfalls, vegetal cover reduces the erosion, and with lesser rainfalls the erosion also decreases. Where mass movement of soil is important, the maximum may be shifted upward.

Fleming [30] utilized data from over 250 catchments about the world to derive relations [Eq. (11-16) and Table 11-3] for mean annual suspended load Q_s in tons as a function of mean annual discharge in cubic feet per second for various vegetal covers:

$$Q_s = aQ^n \qquad (11\text{-}16)$$

Errors of ± 50 percent may be expected from these relations.

For catchments without sediment records the relations of Table 11-3 may be viewed as offering an order-of-magnitude estimate of sediment yield. If

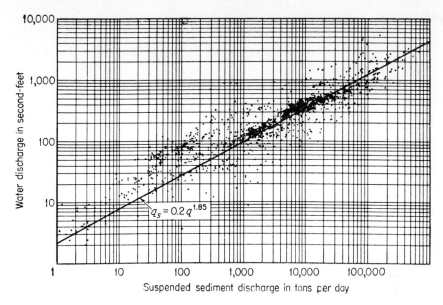

Figure 11-11 Sediment-rating curve for the Powder River at Arvada, Wyoming. (*From* [26].)

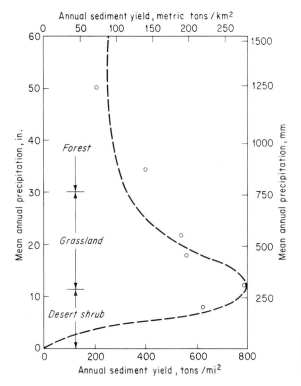

Figure 11-12 Sediment yield as a function of mean annual precipitation. (*From* [29].)

Table 11-3 Values of a and n in Eq. (11-16) for various cover types

Vegetal cover	n	For Q_s in tons, Q in ft³/s	For Q_s in metric tons, Q in m³/s
Mixed broadleaf and coniferous	1.02	120	4,000
Coniferous forest and tall grassland	0.82	3,500	59,000
Short grassland and scrub	0.65	19,000	177,000
Desert and scrub	0.72	38,000	446,000

Source: Adapted from [30].

possible, such estimates should be compared with sediment data on similar basins in the same region.

11-13 Sediment Simulation

Negev [31] developed a model for suspended-load transport and tested it with good results. In this model the amount of soil splash R is made a function of the hourly precipitation amount i

$$R = K_1 i^j \tag{11-17}$$

The transport of splash residue S is a function of the residue in storage on the ground surface R_s and the overland flow rate q_0

$$S = K_2 R_s q_0{}^k \tag{11-18}$$

and the sediment washed from impervious areas E is

$$E = K_3 R \tag{11-19}$$

The total wash load then becomes

$$W = R + S + E \tag{11-20}$$

Unless overland flow occurs, the only source of wash load is from impervious areas and is very small.

Gully erosion G is also related to overland flow

$$G = K_4 q_0{}^m \tag{11-21}$$

Negev divided the sediment from gully erosion into two portions. The first B has particle size substantially the same as the bed-material load in the stream

$$B = (1 - K_5)G \tag{11-22}$$

and the second portion, which he called *interload I*, represents material finer than about 95 percent of the bed-material load:

$$I = G - B = K_5 G \tag{11-23}$$

The total suspended-load transport q_s is then

$$q_s = K_6 I_s q^n + K_7 q^r \qquad (11\text{-}24)$$

where q is the mean daily streamflow and I_s is the quantity of interload in storage in the streambed calculated by maintaining a running balance of the input from erosion and the outflow of suspended interload material.

The procedure involves a number of coefficients which must be found by calibration (Chap. 12). The exponent j can be assumed as 3.0, and k and m as 2.5. Negev assumed $n = r =$ the slope of a sediment-rating curve for the station. The coefficient K_5 can be estimated from particle-size data. The overland flowrate is calculated with a flow-simulation model (Chap. 12).

By generating a long sediment record, a better estimate of the mean annual sediment yield may be obtained than from a short observed record because of the great variability in sediment transport from year to year. A limitation on the simulation approach is that it requires calibration against at least a short but continuous (daily) sediment record. Occasional, infrequent sediment samples do not provide a good basis for calibration. A program of daily sediment samples or continuous monitoring is important in improving our ability to deal with sediment problems. The Negev model did not attempt to estimate bed-load transport because of the lack of bed-load data available for development of algorithms.

11-14 Reservoir Sedimentation

The rate at which the capacity of a reservoir is reduced by sedimentation depends on (1) the quantity of sediment inflow, (2) the percentage of this inflow trapped in the reservoir, and (3) the density of the deposited sediment. The quantity of sediment inflow may be estimated by any of the methods discussed in Secs. 11-11 to 11-13, or if data are available, by reference to mean-annual-yield data per unit area from similar basins in the region. Table 11-4 presents some selected values of sediment yield derived from reservoir surveys. These

Table 11-4 Selected data on rates of sediment production

Location	Drainage area		Annual sediment production	
	mi²	km²	Tons/mi²	Metric tons/km²
Bayview Reservoir, Ala.	72	186	1,769	620
San Carlos Reservoir, Ariz.	12,900	33,411	389	136
Morena Reservoir, Calif.	112	290	3,340	1,170
Black Canyon Reservoir, Idaho	2,540	6,579	172	60
Pittsfield Reservoir, Ill.	1.8	4.7	3,090	1,082
Mission Lake, Kan.	11.4	29.5	2,705	947
High Point Reservoir, N.C.	63	163	544	191
Tygart Reservoir, W.Va.	1,182	3,061	51	18

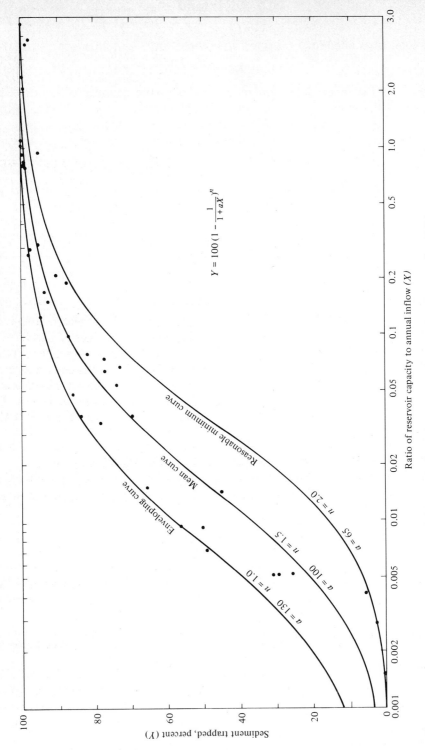

$$Y = 100 \left(1 - \frac{1}{1 + aX}\right)^n$$

Figure 11-13 Trap efficiency of normal ponded reservoirs as a function of capacity-inflow ratio. (*After* [33].)

Table 11-5 Constants in Eq. (11-25) for the specific weight of sediment in lb/ft^3 [34]
For kg/m^3 multiply by 16.1

Reservoir condition	Sand		Silt		Clay	
	w_1	K	w_1	K	w_1	K
Sediment always submerged	93	0	65	5.7	30	16.0
Moderate reservoir drawdown	93	0	74	2.7	46	10.7
Considerable reservoir drawdown	93	0	79	1.0	60	6.0
Reservoir normally empty	93	0	82	0.0	78	0.0

data are generally obtained by surveying the reservoir with sounding lines or echo-sounding equipment and are published periodically [32].

By comparing sediment accumulation in reservoirs with estimates of sediment inflow based on measured sediment transport, Brune [33] derived a relationship between reservoir *trap efficiency,* the percent of incoming sediment retained in the reservoir, and the ratio of reservoir capacity to mean annual water inflow. Trap efficiency must increase with the residence time of the sediment-laden water in the reservoir. Since most sediment enters a reservoir during periods of high inflow, much of it will be discharged over the spillway if the capacity inflow ratio is small. If the ratio is large, there is much less spillage and much of this will be water which has been in storage for an extended period. Figure 11-13, which is modified from Brune [33] can be used to estimate the fraction of the sediment inflow which is trapped. As the reservoir is filled with sediment, the trap efficiency will decrease so that it may be necessary to make the computation for several time intervals with appropriate adjustment of the trap efficiency. The curves apply to total load, and an increment of bed load should be added to suspended-load data.

Coarse sediment is usually deposited in a delta at the head of the reservoir. Fine sediment which remains in suspension can be carried through to the dam and may eventually settle in the deepest part of the pool. If the inflowing water is denser than the surface water in the reservoir, it will plunge below the surface as a *density current* or *turbidity current* and carry its load to the dam even though sediment may not be visible at the reservoir surface.

The volume† occupied by the sediment in the reservoir will depend on the specific weight of the deposited material. The specific weight varies with the kind of sediment and the age of the deposits. Older sediments have more time to consolidate and are under a superimposed load from the more recent deposits. Lane and Koelzer [34] found that dry specific weight w_t at time t can be defined by

$$w_t = w_1 + K \log t \qquad (11\text{-}25)$$

---

† Gross volume of sediment is generally used in computing rate of filling of reservoirs, but part of the water in the pore spaces of the sediment may be recoverable.

Table 11-6 Range of specific weights of reservoir sediments after 50-year accumulation [35]

	Permanently submerged		Aerated	
Material	lb/ft³	kg/m³	lb/ft³	kg/m³
Clay	40–60	640–960	60–80	960–1280
Silt	55–75	880–1200	75–85	1200–1360
Sand	85–100	1360–1600	85–100	1360–1600
Poorly sorted sand and gravel	95–130	1520–2080	95–130	1520–2080

where w_1 is the initial specific weight and K is a consolidation coefficient (Table 11-5). If a mixture of materials is present in the sediment, a weighted average specific weight should be calculated. Equation (11-25) applies to each annual accumulation of sediment, and the average weight of the total sediment accumulation at time t must be found by integrating from year 1 to year t. Table 11-6 presents average specific weights after 50 years used by the U.S. Soil Conservation Service for general design purposes.

Illustrative example 11-1 A reservoir has a capacity of 6×10^6 m³ and a drainage area of 200 km². Streamflow averages 350 mm of runoff per year and sediment production is estimated at 1100 metric tons per km². Using the mean curve of Fig. 11-13 and assuming an average in-place density of 1500 kg/m³, how long will it take to reduce the reservoir capacity to 1×10^6 m³?

$$\text{Annual inflow} = 0.35 \times 200 \times 10^6 = 70 \times 10^6 \text{ m}^3$$

$$\text{Annual sediment inflow} = \frac{1100 \times 1000 \times 200}{1500} = 133,000 \text{ m}^3$$

We will calculate the time it takes to fill successive increments of one million cubic meters of storage.

Average capacity	C/I ratio	Trap efficiency	Time to fill, years
5.5	0.079	0.84	9.0
4.5	0.064	0.80	9.4
3.5	0.050	0.76	9.9
2.5	0.036	0.69	10.9
1.5	0.021	0.56	13.4
		Total	52.6

Note that if the mean value of trap efficiency had been used and a single increment of 5×10^6 m³ considered, the time would have been

$$\frac{5 \times 10^6}{133,000 \times 0.76} = 49.5 \text{ years}$$

The difference in the two answers is undoubtedly less than the uncertainty in the input data.

Illustrative example 11-2 A sediment consists of 5 percent sand, 43 percent silt, and 52 percent clay. What is the in-place density of this material after 30 years assuming the sediment is always submerged? From Eq. (11-25), the first line of Table 11-5, and the sediment composition data:

$$w = 0.05 \times 93 + 0.43(65 + 5.7 \log 30) + 0.52(30 + 16 \log 30)$$
$$w = 0.05 \times 93 + 0.43(65 + 5.7 \times 1.477) + 0.52(30 + 16 \times 1.47)$$
$$w = \quad 4.65 \quad + \quad 31.57 \quad + \quad 27.89$$
$$w = \quad 64.11 \quad \text{lb/ft}^3$$

PROBLEMS

11-1 Find the equation of a sediment rating passing through the points $q = 10$, $q_s = 4$ and $q = 1000$, $q_s = 8000$ as in Fig. 11-11.

11-2 (a) Find the equation applicable to the graph of Fig. 11-11 with sediment load expressed in acre · feet per day and cubic meters per day. Assume an in-place density of 100 lb/ft³.

(b) The drainage area of the Powder River at Arvada is 6050 mi². Write the equation in terms of average depth of erosion in inches.

(c) What is the approximate extreme departure (error) of the plotted points from the line in percent?

11-3 A proposed reservoir has a capacity of 3000 acre · ft and a tributary area of 50 mi². If the annual streamflow averages 5 in of runoff and the sediment production is 0.69 acre · ft/mi², what is the probable life of the reservoir before its capacity is reduced to 500 acre · ft? Use the mean curve from Fig. 11-13. Repeat the computations using the two envelope curves.

11-4 A reservoir has a capacity of 50,000 acre · ft, and the annual inflow averages 78,000 acre · ft. The estimated sediment production of the area is 950 tons/mi², and the drainage area is 1120 mi². Sediment samples indicate that the grain-size distribution is 24 percent sand, 33 percent silt, 43 percent clay. When will 80 percent of the reservoir capacity be filled with sediment? Use line 3 of Table 11-5.

11-5 A reservoir has a capacity of 4×10^6 m³ and a drainage area of 250 km². The annual streamflow averages 375 mm of runoff, and the sediment production is 1250 metric tons per square kilometer. The sediment has an average in-place specific weight of 1500 kg/m³. Using the mean curve of Fig. 11-13, find the time required to reduce the reservoir capacity to 1×10^6 m³.

11-6 A reservoir has a capacity of 50×10^6 m³ and an annual inflow of 75×10^6 m³. Sediment production of the watershed is 275 metric tons per square kilometer. Area of the catchment is 1000 km². The grain-size distribution of the sediment is 8 percent sand, 41 percent silt, and 51 percent clay. The sediment is expected to be always submerged. How long will it take for the reservoir to become 80 percent filled with sediment?

11-7 Secure a topographic map (preferably 1 : 24,000 scale) and outline the catchment of a small stream on it. From the map determine the stream order, bifurcation ratio, length ratio, and area ratio. Does the stream appear to conform to Eqs. (11-1) to (11-3)? If your classmates used other streams in the same area, is there any regional consistency in the ratios?

11-8 If it is possible to obtain a 1 : 62,500 or other small-scale map of the same area you used in Prob. 11-7, repeat the analysis. What differences in values occur when the map scale changes?

11-9 For the catchment of Prob. 11-7 (or similar) determine the drainage density and length of overland flow.

11-10 Collect data on area and channel length for at least 10 streams. Do these data appear to conform to Eq. (11-5)? If not, can you suggest why they do not conform?

11-11 For the catchment of Prob. 11-7 (or other basin) determine average channel slope as given by line AB, Fig. 11-2, and by Eq. (11-7). How well do you think these estimates actually represent the true slope of the stream?

11-12 Construct an area-elevation curve for the catchment you studied in Prob. 11-7 (or other basin).

11-13 From a map or field survey of a meandering stream calculate the sinuosity, wavelength of meanders, radius of curvature, and amplitude of meanders. How do these values check with the average values given in Sec. 11-4?

11-14 Test Eqs. (11-8) through (11-10) on field data for a stream in your vicinity. Do the values of b, f, and m agree with those in Table 11-1? Does there seem to be regional consistency in your area?

REFERENCES

1. Horton, R. E.: Erosional Development of Streams, *Geol. Soc. Am. Bull.,* vol. 56, pp. 281–283, 1945.
2. Leopold, L. B., M. G. Wolman, and J. P. Miller: "Fluvial Processes in Geomorphology," p. 145, Freeman, San Francisco, 1964.
3. Hack, J. T.: Studies of Longitudinal Stream Profiles in Virginia and Maryland, *U.S. Geol. Surv. Prof. Pap.* 294-B, pp. 45–97, 1957.
4. Horton, R. E.: Drainage Basin Characteristics, *Trans. Am. Geophys. Union,* vol. 13, pp. 350–361, 1932.
5. Schumm, S. A.: Evolution of Drainage Systems and Slopes in Badlands at Perth Amboy, New Jersey, *Geol. Soc. Am. Bull.* vol. 67, pp. 597–646, 1956.
6. Chorley, R. J., D. E. G. Malm, and H. A. Pogorzelski: A New Standard for Estimating Drainage Basin Shape, *Am. J. Sci.,* vol. 255, pp. 138–141, 1957.
7. Taylor, A. B., and H. E. Schwarz: Unit-Hydrograph Lag and Peak Flow Related to Drainage Basin Characteristics, *Trans. Am. Geophys. Union,* vol. 33, pp. 235–246, April 1952.
8. Leopold, L. B., and T. G. Maddock: The Hydraulic Geometry of Stream Channels and Some Physiographic Implications, *U.S. Geol. Surv. Prof. Pap.* 252, 1953.
9. Leopold, L. B.: Downstream Change of Velocity of Rivers, *Am. J. Sci.,* vol. 251, pp. 606–624, 1953.
10. Bush, L. M., Jr.: Drainage Basins, Channels and Flow Characteristics of Selected Streams in Central Pennsylvania, *U.S. Geol. Surv. Prof. Pap.* 282-F, pp. 145–181, 1961.
11. Stall, J. B., and Y. S. Fok: Hydraulic Geometry of Illinois Streams, *Univ. Ill. Water Resour. Cent. Res. Rep.* 15, July 1968.
12. See [2], pp. 248–291.
13. See [2], p. 296.
14. See [2], pp. 319–322.
15. Nixon, M.: A Study of the Bankfull Discharges of Rivers in England and Wales, *Proc. Inst. Civ. Eng.,* pp. 157–174, 1959.
16. Ellison, W. D.: Studies of Raindrop Erosion, *Agric. Eng.,* vol. 25, pp. 131–136, 181, 182, 1944.

17. Wischmeier, W. H., and D. D. Smith: Predicting Rainfall Erosion Losses East of the Rocky Mountains, *Agricultural Handbook 282,* U.S. Department of Agriculture, 1965.
18. Graf, W. H.: "The Hydraulics of Sediment Transport," McGraw-Hill, New York, 1971.
19. du Boys, P.: Le Rhône et les rivières à lit affouillable, *Ann. Ponts Chaussées,* ser. 5, vol. 18, pp. 141–145, 1879.
20. Straub, L. G.: H.R. Doc. no. 238, 73d Cong., 2d Sess., p. 1135, 1935.
21. Einstein, H. A.: The Bed-Load Function for Sediment Transportation in Open Channel Flows, *U.S. Dept. Agric. Soil Conserv. Serv. Tech. Bull.* 1026, 1950.
22. U.S. Interagency Committee on Water Resources: A Study of Methods Used in Measurement and Analysis of Sediment Loads in Streams, *Rep.* 14, Minneapolis, Minn., 1963.
23. Stevends, H. H., G. A. Lentz, and D. W. Hubbell: Collapsible-Bag, Suspended-Sediment Sampler, *J. Hydraul. Div. ASCE,* vol. 106, pp. 611–616, April 1980.
24. Colby, B. R.: Relationship of Unmeasured Discharge to Mean Velocity, *Trans. Am. Geophys. Union,* vol. 38, pp. 708–717, October 1957.
25. Thorpe, G. B.: A New Suspended Solids Recorder, *Ind. Electron.,* vol. 2, pp. 415–418, September 1964.
26. Fleming, G.: Suspended Solids Monitoring: A Comparison between Three Instruments, *Water Eng.,* vol. 73, pp. 377–382, September 1969.
27. Jackson, W. H.: An Investigation into Silt in Suspension in the River Humber, *Doc. Harbour Auth.,* vol. 45, August 1964.
28. Murphree, C. E., G. C. Bolton, J. R. McHenry, and D. A. Parsons: Field Test of an X-Ray Sediment Concentration Gage, *J. Hydraul. Div. ASCE,* vol. 94, pp. 515–528, March 1968.
29. Langbein, W. B., and S. A. Schumm: Yield of Sediment in Relation to Mean Annual Precipitation, *Trans. Am. Geophys. Union,* vol. 39, pp. 1076–1084, December 1958.
30. Fleming, G.: Design Curves for Suspended Load Estimation, *Proc. Inst. Civ. Eng.,* vol. 43, pp. 1–9, 1969.
31. Negev, M.: A Sediment Model on a Digital Computer, *Stanford Univ. Dept. Civ. Eng. Tech. Rep.* 76, March 1967.
32. U.S. Department of Agriculture: Summary of Reservoir Sediment Deposition Surveys Made in the United States through 1970, *Misc. Publ.* 1266, July 1973.
33. Brune, G. M.: Trap Efficiency of Reservoirs, *Trans. Am. Geophys. Union,* vol. 34, pp. 407–418, June 1953.
34. Lane, E. W., and V. A. Koelzer: Density of Sediments Deposited in Reservoirs, *U.S. Corps Eng. St. Paul Dist. Rep.* 9, 1943.
35. Gottschalk, L. C.: Sedimentation, sec. 17 in V. T. Chow (ed.): "Handbook of Applied Hydrology," McGraw-Hill, New York, 1964.

BIBLIOGRAPHY

Birot, P.: "The Cycle of Erosion in Different Climates," University of California Press, Berkeley, 1968.
Blench, T.: "Regime Behaviour of Rivers and Canals," Butterworth's, London, 1957.
Bogardi, J.: "Vizgolkasok Horddalekszallitasa," in Hungarian (Sediment Transport in Alluvial Streams), Akademiai Kiado, Budapest, 1970. (Translated into English by International Courses in Hydrology at Budapest.)
King, C. A. M.: "Landforms and Geomorphology: Concepts and History," Benchmark Papers in Geology, vol. 28, Academic Press, New York, 1976.
Leliavsky, S.: "Introduction to Fluvial Hydraulics," Constable, London, 1955.
Morisawa, M.: "Streams: Their Dynamics and Morphology," McGraw-Hill, New York, 1968.
Painter, R. B.: Sediment, chap. 6 in J. C. Rodda (ed.): "Facets of Hydrology," Wiley, London, 1976.

Raudkivi, A. J.: "Loose Boundary Hydraulics," Pergamon Press, Oxford, 1967.

Schumm, S. A.: "River Morphology," Benchmark Papers in Geology, vol. 2, Academic Press, New York, 1972.

———: "Slope Morphology," Benchmark Papers in Geology, vol. 6, Academic Press, New York, 1973.

———: "Drainage Basin Morphology," Benchmark Papers in Geology, vol. 41, Academic Press, New York, 1977.

Strahler, A. N.: Dynamic Basis for Geomorphology, *Geol. Soc. Am. Bull.*, pp. 923–928, 1952.

———: Quantitative Geomorphology of Drainage Basins and Channel Networks, sec. 4-II in V. T. Chow (ed.): "Handbook of Applied Hydrology," McGraw-Hill, New York, 1964.

TWELVE

DETERMINISTIC HYDROLOGIC MODELS

In the mathematical sense the word *model* describes a system of assumptions, equations, and procedures intended to describe the performance of a prototype system. The methodology of hydrology described in the preceding chapters includes many such mathematical models. Most of them are relatively simple, but the discussion of the hydrologic system shows that it is complex and not easily described by a simple model. Since the advent of the digital computer the term *hydrologic model* has come to mean a relatively complex mathematical description of the hydrologic cycle designed for solution on a computer. Some models attempt to describe the actual physical processes of the hydrologic cycle so as to simulate† actual hydrologic events such as the transformation of a series of rainfall inputs to the resulting streamflow hydrograph. These models have been described as *deterministic, physically based,* and *conceptual.* Another class of hydrologic models seeks to reproduce the statistical behavior of a hydrologic time series without regard to an actual event. These *stochastic models* are the subject of Chap. 14.

Models, both deterministic and stochastic, have contributed much to hydrology. They have given us a better understanding of the hydrologic processes, are useful research tools, and can be used to develop solutions to engineering problems with a detail and accuracy which could not be achieved by more conventional paper and pencil analysis.

---

† Simulation: The representation of a system by a device (as a computer) that imitates the behavior of the system. ("Webster's 7th New Collegiate Dictionary," 1971.)

12-1 Types of Deterministic Models

If a model precisely imitated the processes of the hydrologic system, it might be supposed that only one such model would be needed. For many reasons such precise imitation of the processes of hydrology has not been attained, and many models have been developed for different purposes and with different philosophical concepts.

Models intended primarily for research are usually designed to simulate part of the hydrologic cycle at a point [1, 2]. These *research models* utilize formal mathematical expressions for the algorithms, usually differential equations. Such models are useful for exploring the sensitivity of the hydrologic process to specific site characteristics or to watch the variation of the process over short time periods.

Conceptual models embody a series of functions which are considered to describe the catchment processes involved. The algorithms are usually simplified by use of empirical relations in order to speed the solution and to adapt the model to cope with the point-to-point variations in the hydrologic process within the catchment [3–11]. Most deterministic models for engineering application fall within the conceptual category.

Lumped models treat a whole catchment, or a significant portion of it, as if it were homogeneous in character and subject to uniform rainfall. *Distributed models* divide the catchment into a large number of small subareas, simulate each separately, and combine them to obtain catchment response. In principle, the distributed model deals with the variations over a catchment more logically than the lumped model. However, unless the input rainfall and the catchment characteristics are known with comparable detail, the solution may be no better than that of a lumped model. Computations in the detail of the distributed model are usually too time-consuming for engineering applications. However, the distributed model is useful for research on the consequences of heterogeneity in a catchment.

Continuous models are capable of generating outflow hydrographs over long periods of time (several years). Of course, such models use finite time periods in the computation and the computed flows are discrete points in time. *Event models* are designed to simulate a single event such as the hydrograph of a single storm. A continuous model must include algorithms which maintain a continuous water balance for the catchment so that the conditions antecedent to each storm event are known. In the event model, initial conditions must be specified as part of the input. Most event models were developed specifically to design urban drainage systems and other small projects [12]. The probability of a computed peak flow is sometimes assumed equal to that of the input rainfall (Sec. 13-17). Continuous models are useful for simulation of long flow records for use in design, evaluating the impact of changes in a catchment on streamflow, and forecasting of streamflow (Chap. 16).

General models are designed to be used on any catchment, with catchment characteristics represented by input parameters. *Catchment specific models* are prepared for a single catchment and have appropriate catchment characteristics "built in." Specific models are often constrained to the range of conditions actually represented in the data base on which they are built, while general models are adapted to a much wider range of conditions.

Models are rarely static. They undergo frequent modification by their developer or by a subsequent user. Consequently it is difficult to fit a specific model into a class, and only the most general type of classification will be used in this chapter. Readers interested in a particular model should secure an operations manual and/or program documentation and determine whether the model will serve their specific purpose.

12-2 Structure of a Conceptual Model

A number of conceptual models have been devised for the simulation of discharge, water quality, sediment load, etc. Several of these models are discussed briefly in Sec. 12-4. The Stanford Watershed Model [3, 13], one of the earliest, is described in more detail here to illustrate how such a model is formulated.

A flow diagram of the Stanford Watershed Model (SWM) is shown in Fig. 12-1. The SWM is a general, continuous, lumped, conceptual model. For simulating runoff from rainfall, the input data are hourly rainfall and daily potential evapotranspiration (Chap. 5). Additional data are required if snowmelt is to be simulated (Sec. 12-5).

In application, the catchment to be modeled may be divided into subareas called *segments*. Each segment is assumed to be homogeneous with respect to soils, vegetation, topography, precipitation, and other factors which influence runoff. Each segment is represented by a set of parameters defining the characteristics of the area. Each segment is simulated separately and the results from the several segments combined into an output for the catchment as a whole. Increasing the number of segments allows a more detailed description of the basin but at the same time increases the running time in the computer in direct proportion to the number of segments. Moreover, with multiple segments, determination of some parameters becomes somewhat more difficult. For practical reasons, the number of segments is usually kept small. Often the number of segments is made equal to the number of rain gages to be used in the simulation, since rainfall differences are often the most important factor in determining the catchment response. At the same time this scheme allows some representation of other areal variations within the catchment. For simplicity the following discussion treats a "basin" consisting of a single segment.

If a realistic water-balance accounting is to result, the point observations of rainfall must represent the areal average over the segment. Hence, the first step in the simulation is to multiply station rainfall by a constant K1 equal to the ratio of basin normal annual rainfall to station normal annual rainfall. *Interception* loss (Sec. 8-1) is simulated by an interception-storage capacity (usually between 0 and 5 mm or 0 and 0.2 in). All rainfall is assumed to enter interception storage until it is full. Interception storage is depleted by evaporation at the

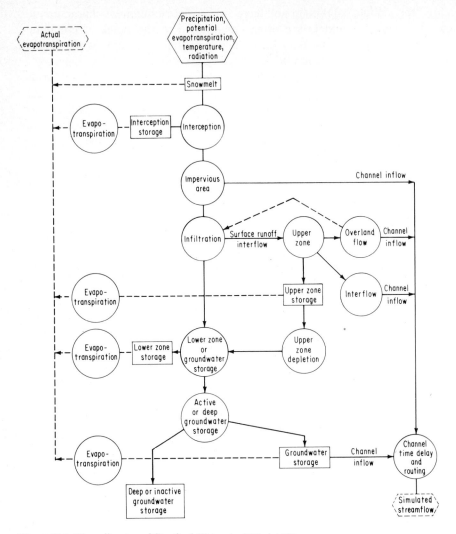

Figure 12-1 Flow diagram of Stanford Watershed Model IV.

potential rate until reduced to zero. Evaporation during a storm increases interception loss (Sec. 8-1). Impervious area in the catchment is simulated by diverting a constant percentage of each hourly increment of rain to the stream as *impervious-area runoff*. The percentage is equal to the percentage of impervious area directly connected to the streams in the catchment. The channel surfaces of the stream system, lakes, and swamps are also treated as impervious area. For most rural basins impervious area is small (1 or 2 percent), but for urban areas this factor is usually more than 20 percent.

It is assumed that *infiltration capacity* (Sec. 8-2) at any time varies over the segment. For lack of better information this variation is assumed to be linear

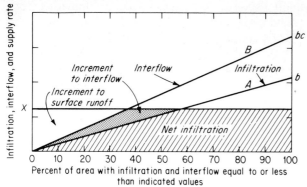

(Fig. 12-2). The position of this line is varied by varying the value of b, as a function of the ratio of moisture in the *lower-zone storage* LZS to the nominal capacity of this zone LZSN (Fig. 12-3). The *net infiltration* to the lower zone is the hatched trapezoid of Fig. 12-2 and is a function of the current soil-moisture ratio (LZS/LZSN) and the *supply rate* X (precipitation minus interception).

Interflow (Sec. 7-1) is calculated by a similar process. Line B of Fig. 12-2 divides the rainfall excess triangle into two portions, surface runoff and interflow. The position of line B is fixed by multiplying b by a factor c, which is greater than 1 and is also a function of LZS/LZSN. The fraction of interflow increases as soil moisture increases (Fig. 12-3). Interflow runoff is placed in interflow storage, from which a fixed fraction of the storage is released to the stream in each time interval. This produces an exponential decline in the rate of interflow in the stream. The fraction is set by an interflow recession constant KI determined from observed flow data (Sec. 7-2).

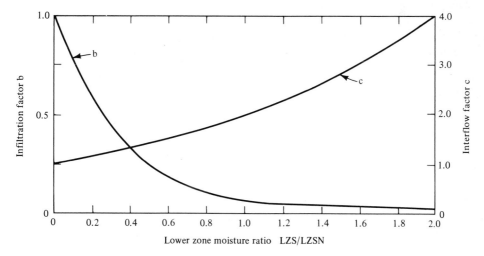

Figure 12-3 Variation of the infiltration factor b and interflow factor c as a function of *LZS/LZSN* when INFILTRATION and INTERFLOW = 1.0.

A portion of surface runoff becomes *upper-zone storage* simulating both depression storage (Sec. 8-1) and upper-soil storage, and the balance becomes overland flow. The fraction ΔUZS entering upper-zone storage is a function of the ratio UZS/UZSN (Fig. 12-4). The balance of the surface runoff enters the overland-flow process. An empirical function derived from experimental data relates outflow from overland flow to the volume of water in overland-flow detention. A simple continuity computation calculates the detention volume from

$$V_2 = V_1 + \Delta V - q_0 \tag{12-1}$$

where q_0 is outflow at the beginning of the interval, ΔV is the increment to overland-flow detention, and V_1 and V_2 are the detention volumes at the beginning and end of the interval. There is an obvious approximation in using q_0 for $\bar{q}$, but if the interval is short, the error is small and an iterative solution is not warranted. Any water left in detention at the end of the interval is returned to the rainfall supply to infiltration, simulating delayed infiltration from overland flow. For small watersheds, the time delay in overland flow may be important, but for all catchments the delayed infiltration is an important factor in the runoff process.

A portion of the upper-zone storage may pass on to the lower zone whenever UZS/UZSN is greater than LZS/LZSN (upper-zone depletion). This water and the net infiltration are divided between *lower-zone soil-moisture storage* and *groundwater storage* on the basis of a function of LZS/LZSN (Fig. 12-5). More water goes to soil moisture when LZS/LZSN < 1 and more to groundwater when the ratio exceeds unity. A fixed percentage of the water entering groundwater may be diverted to deep groundwater which does not contribute to streamflow. The remainder goes to groundwater storage from which water enters the stream in accord with an exponential recession (Sec. 7-2).

Water in upper-zone and interception storage is assumed to evaporate at the potential rate. Water in lower-zone storage is depleted by evapotranspira-

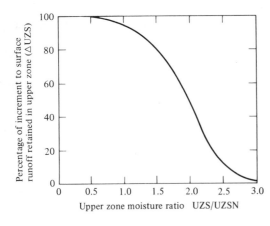

Figure 12-4 Fraction of rainfall excess retained in upper-zone storage.

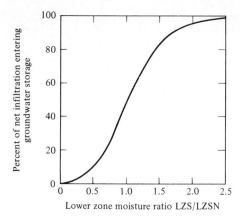

Figure 12-5 Division of infiltrated water between lower-zone moisture storage and groundwater.

tion at a rate which is a function of the ratio LZS/LZSN. When the ratio is high, evapotranspiration is near the potential rate and decreases as the ratio decreases. If shallow groundwater exists, evaporation from it is permitted.

The runoff quantities calculated in the water-balance computation must be routed through the channels of the basin to calculate the outflow hydrograph. This can be accomplished by an application of lag and route techniques. For this purpose a time-area diagram converted to flow units and Muskingum type routing are convenient (Chap. 9). If several segments are used, the time-area diagram can reflect the differing runoff in each segment. Kinematic-wave routing is also a useful method. Kinematic routing requires that channel cross sections, slopes, and roughness be defined. Compensating for this, however, the kinematic routing permits calculation of flow and stages at the end of each reach. Further, the routing is defined by the physical characteristics of the channel with natural nonlinearities and discontinuities represented far better than is possible with an empirical routing procedure.

12-3 Parameters and Calibration

Table 12-1 lists the 15 parameters that are set to adapt the general simulation program to a specific catchment. The parameters A, LL, and SS can be determined from maps (Chap. 11); NN must be estimated from knowledge of the catchment, but its exact magnitude is not very critical; and K3 can be estimated from aerial photography or by field inspection. A mean-annual-precipitation map for the catchment is required to estimate K1. The two groundwater parameters K24L and K24EL are often zero. When appropriate, they must be estimated from a priori knowledge of the groundwater situation in the watershed. If no information exists regarding possible deep groundwater storage, the value of K24L can sometimes be determined in the calibration process. The recession parameters KK24 and KI can be estimated from a short streamflow record by the methods discussed in Sec. 7-2. If no record exists for the basin, estimates of

Table 12-1 Parameters of the Stanford Watershed Model for rainfall runoff

K1	Ratio of normal basin rainfall to normal station rainfall
EPXM	Interception storage
A	Percent of impervious area
INFILTRATION	Average infiltration rate
INTERFLOW	Interflow function
UZSN	Upper-zone nominal capacity
LL	Average length of overland flow
SS	Average slope of overland flow
NN	Manning's n for overland flow
LZSN	Lower-zone nominal capacity
K24L	Portion of groundwater assigned to deep storage
K3	Evaporation-loss index (percent of area with deep-rooted vegetation)
K24EL	Percent of area with shallow groundwater subject to direct evapotranspiration
KK24	Groundwater recession constant
KI	Interflow recession constant

these values from similar basins nearby may be used. EPXM is relatively unimportant and is usually assigned a value between 2 and 5 mm (0.1 and 0.2 in), depending on vegetation density.

The remaining four parameters, INFILTRATION, INTERFLOW, UZSN, and LZSN, are the most important ones from the hydrologic viewpoint. IN-FILTRATION and INTERFLOW are relative numbers which position the two lines on Fig. 12-3. A value of INFILTRATION other than 1.0 will raise or lower the curve for b in Fig. 12-3 and shift the level of infiltration accordingly. Similarly a value of INTERFLOW > 1 will cause a larger fraction of the runoff to appear as interflow. Evapotranspiration increases with larger values of UZSN and LZSN until the loss is limited by the available water or the potential evapotranspiration. These storages are defined as "nominal" values rather than absolute capacities. Actual amounts of water in storage can increase above these nominal values, but as they do so the functions of Figs. 12-4 and 12-5 limit the input to the storage so that a limiting capacity is approached asymptotically. LZSN is intended as an index to the transient storage in the soil. Essentially permanent soil moisture is of no concern in the simulation process. A first estimate of LZSN (mm) can be taken as

$$\text{LZSN} = \begin{cases} 100 + 0.25P & \text{seasonal precipitation} \\ 100 + 0.125P & \text{precipitation distributed} \\ & \text{throughout the year} \end{cases} \quad (12\text{-}2)$$

where P is mean annual precipitation in millimeters. With P in inches, the constant 100 in Eq. (12-2) should be changed to 4. As a first estimate, UZSN may be taken as one-tenth of LZSN.

The values of the parameters can be estimated and the model operated without verification against observed data. Since it is unlikely that the best parameters are obtained from an initial estimate, a process of calibration against observed data is recommended for establishing INFILTRATION, INTERFLOW, UZSN, and LZSN. In the process it may be discovered that changes in K1, K24L, or K24EL are indicated.

In the calibration process initial values are assumed for all parameters and a period of streamflow is simulated and compared with historic flows. If necessary, parameters are changed and the comparison repeated until a satisfactory "fit" of the observed and simulated data is attained. The period of comparison should be as long as the data permit, preferably 5 years or more. It is possible to obtain what seems to be a good calibration on a short sample, only to discover later that it is not a reliable calibration. The goodness of "fit" obtained in calibration is highly dependent on the adequacy and quality of the data. With good data, runoff volumes should agree within a few percent and flood peaks should be randomly distributed about the observed data with no significant bias. If the catchment under study is ungaged, calibration against a similar catchment in the area is advisable. In this case the parameters may be modified before actual use if differences between the catchments warrant.

Automatic parameter-determination routines may be incorporated into a single-segment model [14, 15]. Observed flows are input as data, and an objective function (commonly the sum of squared differences between observed and simulated mean daily flow) is calculated. A parameter is adjusted by some predetermined amount, and the simulation is repeated. By successive iterations it is possible to approach an optimum set of parameters (minimum objective function). The process is facilitated by programming a systematic search procedure. For a multisegment basin, automatic adjustment of parameters is practicable only if all segments are assumed to have the same values for LZSN, UZSN, INFILTRATION, and INTERFLOW. If all segments have different parameters, the number of iterations increases exponentially as the number of segments and the computation time becomes too great.

Manual adjustment is greatly enhanced by direct display of output hydrographs on a cathode-ray tube [16]. With a time-sharing computer system, direct interaction between the hydrologist and the simulation program permits rapid adjustment of parameters.

While calibration is normally limited to determining the four parameters UZSN, LZSN, INFILTRATION, and INTERFLOW, it sometimes becomes evident during calibration that other parameters require adjustment. This is especially true of K1 (precipitation adjustment), K24L (percolation to deep groundwater), and K24EL (evapotranspiration from groundwater). The necessity for adjustment of these parameters is usually evident from an inability to obtain a water balance. If groundwater flows are consistently high but reducing INFILTRATION would yield too much surface runoff, it may be appropriate to invoke K24L. If total water balance is in error, an adjustment in K1 may be indicated. A thorough understanding of the simulation algorithm and the effects of the individual parameters is essential for good calibration.

Errors or lack of representivity in the input data and errors in the stream-

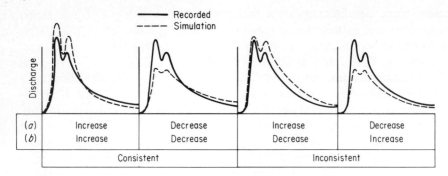

Figure 12-6 Some consistent and inconsistent indications for the adjustment of INFILTRATION based on (*a*) surface-runoff volume and (*b*) groundwater flow.

flow data used for calibration are a significant problem. If parameters are adjusted to bring erroneous cases to a fit, the parameters may be seriously biased. This is a special problem with automatic optimization if a squared-difference criterion is used, since some of the largest differences between observed and simulated flow may result from data errors or inadequacies. Adjustment of parameters controlling volume of runoff is fairly straightforward through comparison of simulated and observed values of monthly flow volume. Adjustment of parameters influencing hydrograph shape is more difficult because differences between observed and simulated hydrographs may arise from errors in timing, estimated volume of storm runoff, or the initial assumption of shape parameters. In some cases the comparison will indicate ambiguous situations (Fig. 12-6), which cannot be corrected by parameter adjustment.

The problem of fitting is not unique to simulation but is present with all hydrologic techniques. Because simulation is capable of more precise results, and because it is tested against several years of observed data, there is a tendency to be more concerned with accuracy of calibration and the selection of optimal parameters than with conventional techniques. Inadequate input data are probably the most serious constraint on simulation. No matter how good the model is, it cannot overcome the limitations of poor data. Scattered random errors in the data are usually not serious, but data containing a significant bias will yield biased answers. The closeness of fit in calibration is controlled by the input data, and there is a tendency to revise parameters and rerun the calibration many times in an attempt to get a better fit than is possible with the data. Johanson [17] investigated the limits of accuracy in calibration in relation to the adequacy of the input data (Sec. 3-6).

12-4 Other Deterministic Models

Many models and modeling systems have been suggested and programmed for computer solution. Some are combinations of conventional methods adapted for computer solution. The widely used program HEC-1 [18] uses a simple loss rate formulation and unit hydrograph to reconstruct floods from rainfall data. The Storm Water Management Model (SWMM) developed for the U.S. Environmental Protection Agency [19] offers several choices for simple rainfall-runoff estimates and utilizes kinematic routing (Chap. 10) to form the hydrograph. SWMM was designed for application to urban storm-drain systems and includes algorithms for simulating some quality parameters. STORM [20] was also designed to simulate urban storm flow. A model developed by the British Road Research Laboratory assumed that all runoff was from impervious areas and utilized Muskingum routing to generate the hydrograph. This model was subsequently modified and a simple rainfall-runoff estimation procedure added to create ILLUDAS [21] used for storm drainage design. This group of models are event models intended to take as input a design rainfall pattern and to produce design flows assumed to have the same return period as the rainfall (Sec. 13-17).

One can visualize a catchment as a system of reservoirs (either linear or nonlinear), and this is the basis of a model (Tank 1) by Sugawara [22]. A tank with side outlet representing runoff, bottom outlet representing infiltration, and a storage component (Fig. 12-7) can represent the runoff process in a catchment or a portion thereof. Several such tanks in parallel can represent a large catchment. Without the bottom outlet or storage, the tank can simulate channel routing. Both types of tanks can be combined as required to simulate a stream system. A model by the British Water Resources Board, called DISPRIN [23], one used by the Institute of Meteorology and Hydrology of Romania [24] called IMH2-SVP, and one suggested by Dawdy and O'Donnell [25] operate on a basically similar concept.

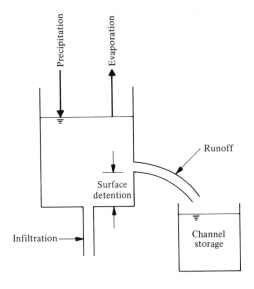

Figure 12-7 A simple tank model.

The Streamflow Simulation and Reservoir Regulation Model (SSARR) was originally a system using Muskingum routing to simulate the flow of the Columbia River system [9] but now includes a rainfall-runoff relation and snowmelt routines to simulate the runoff from the land [26]. Sittner et al. [7] describe a model using an API-type relationship (Sec. 8-6) and Muskingum routing. The Australian Bureau of Meteorology CBM model [27] is similar.

It is possible to use an analog computer (or hybrid) for modeling [28, 29]. Such computers are less readily available than the digital computer, and there seems to have been relatively little operational use made of such models.

The SWM has undergone many modifications. Efforts at the University of Texas [5] and Kentucky [14] explored various modifications of the basic algorithms. The National Weather Service Model (NWSH) is patterned after the SWM but modified to meet the specific needs of forecasting [11]. The Sacramento model uses a moisture accounting procedure similar to SWM and a unit hydrograph to form the outflow hydrograph [30]. Numerous daily models utilize a simplified moisture accounting scheme [4].

The USDAHL model developed by the U.S. Agricultural Research Service differs considerably from most models [31]. The catchment is divided into three zones—uplands, hillsides, and bottom land. Overland flow from the upper zone is assumed to cascade over the lower zones en route to the channel and is subject to infiltration during the flow. Infiltration is a function of available storage and crop conditions. The model is intended for application to agricultural problems and is probably best suited for relatively small catchments.

HSPF, developed for the U.S. Environmental Protection Agency [32], is a package of programs based on a modified Stanford Watershed Model. The package includes routines for simulation of water quality and washoff of agricultural chemicals and other pollutants as well as for simulation of streamflow.

12-5 Snowmelt Simulation

For many catchments a program module to deal with snowmelt is required. Computed values of melt can be introduced into the simulation model with the rainfall input, and the runoff process can be simulated as for rainfall alone. The snowmelt routine must maintain a heat balance for the snow and a water balance for the snow on the ground. Anderson and Crawford [10] developed one of the first snowmelt routines as an adjunct of the SWM. Anderson subsequently developed a more complex point melt model [2]. The latter model is a useful research tool, but this section will concentrate on estimating snowmelt on a catchment [33].

Because of the effect of elevation on temperature the basin must be divided into elevation zones if the range of elevation is large. Since temperature differences between 3 and 5 Celsius degrees can occur over an elevation difference of 500 m, this is about the largest interval that should be used, and 250 m (800 ft) would be better. With maps in feet, an interval of 1000 ft is common.

Temperature in each elevation zone must be estimated from the nearest temperature station by use of an appropriate lapse rate. Hence, the elevation difference between the station and the median elevation of each zone that it represents is an input parameter. The lapse rate is usually fixed at an average value of 1.8 Celsius degrees per 250 m (4 Fahrenheit degrees per 1000 ft), but it could be made variable depending on whether precipitation is occurring. If there are two or more temperature stations, a lapse rate can be calculated for each observation.

Records of water equivalent of snowfall are generally unavailable, and the program must distinquish between snowfall and rain and maintain a running account of depth, density, and water equivalent of snow on the ground. The decision between rain and snow is usually based on the current temperature in the segment, with rain assumed to occur at temperatures above 0°C (32°F) and snow at temperatures below that level. If snow is assumed, an initial density must be assigned by specifying an average density of new snow at some reference temperature and a variation with temperature about that value.

If there is new snow, the water equivalent of snow on the ground is computed as the sum of the water equivalent at the end of the previous interval and the precipitation amount during the current interval. A snow-correction factor which is >1 allows for adjusting reported precipitation to account for the underregistration resulting from the effect of wind on the gage catch (Sec. 3-5). Whenever the program determines that the precipitation during an interval is snow, the reported catch is multiplied by the snow-correction factor.

Snow depth at any time is computed from the water equivalent of the snowpack and the density. If the density ρ is less than ρ_{max}, depth is decreased each hour by use of Eq. (12-3):

$$D_2 = D_1[1 - 0.00002D_1(\rho_{max} - \rho)] \qquad (12\text{-}3)$$

where D is snow depth and the subscripts 1 and 2 refer to the beginning and end of an interval, respectively. The new density is then calculated from the water equivalent and depth. The process simulates the consolidation of the snowpack under its own weight. As a result of melt or rainfall on the snowpack the liquid-water content of the snowpack can be allowed to reach a maximum expressed as a fraction of the water equivalent of the pack, usually from 0.01 to 0.05 (Sec. 3-16).

It may be assumed that a segment is completely snow-covered if the current water equivalent exceeds the largest value of water equivalent previously attained during the season. The fraction of area snow-covered can be taken as the ratio of the current water equivalent to the previous maximum value.

Several options are available for simulating the melting process. If only temperature data are available, a simple degree-day function may be used (Sec. 8-11). With wind and dewpoint data, Eqs. (8-13) and (8-14) can be employed, supplemented by Eq. (8-15) when rain occurs. Finally, if radiation measurements are available, an energy-balance computation can be programmed with albedo expressed as a function of snow density and back radiation calculated from snow temperature. This requires an estimate of the cooling of the snow below 0°C by back radiation.

A major problem in the simulation of snow accumulation and melt is the inadequacy of field data in mountain areas. Unless reasonably representative data are available, the simple degree-day function may yield results as good as more elaborate approaches. In calibrating the snowmelt routines, snow-course data may be helpful in verifying the reliability of the snow-accumulation algorithms.

12-6 Reliability of Hydrologic Simulation

It has always been difficult to estimate the accuracy to be expected when a hydrologic procedure is applied to a specific investigation. If the procedure is used to simulate flows where an existing record is available, it is reasonably simple to compare observed and simulated flows and determine the accuracy of the simulation. If the test period is sufficiently long, some generalizations about the accuracy of record extensions may be drawn. Commonly, however, hydrologic procedures are applied to catchments with at best a short record and in many cases no record at all. Any statements concerning the accuracy of these estimates must rest on the judgment of the user relative to the comparability of the particular study catchment to those in which tests have been made on actual data.

The accuracy of simulation depends on three factors—accuracy of input data, effectiveness of parameter evaluation, and inherent errors in the model. If the input data are in some way biased, the output is likely to be biased. It is particularly important that the streamflow data used for calibration be unbiased. If a bias does exist, the results of the study will be in error, and the hydrologist probably will not be aware of this error. Random errors in input data are less troublesome. However, an error in the precipitation or streamflow of a major flood may lead to an error in parameter evaluation in an attempt to fit this special case. This problem may be particularly acute when automatic parameter determinations are used.

Parameter evaluation should aim at a set of values which gives unbiased answers. Overall volume estimates should have near zero errors, and base flows should be closely simulated, with the simulated flows both above and below the observed flows. A small but persistent bias in base flow will mean an equal volume error in storm flow but with opposite sign. Simulated peaks should be randomly distributed above and below the observed peaks, and the observed and simulated peaks should exhibit similar variance. If calibration is thorough and the calibration period sufficiently long to give some statistical validity to the results, the extended flow record should be reasonably satisfactory. Generally the user of a simulation will have some feeling for the reliability. A good "fit" in calibration is quite obvious and a poor "fit" is equally visible. *If parameters are estimated without benefit of calibration, major errors may result.*

The World Meteorological Organization sponsored a comparative test of 10 simulation models on 6 standard data sets [24]. Six years of calibration data were provided, but only 2 years of test data were used. The accuracy and adequacy of the data sets were, in some cases, questionable. Nevertheless, the conclusions from the comparison are significant. The models were grouped into three categories. *Explicit moisture accounting models* included the National Weather Service Model, Sacramento Model and SSARR model. *Implicit moisture accounting models* included two tank-type models. *Index models* included two models using an API-type index to runoff, and a linear systems model [34]. In the latter model an instantaneous unit hydrograph is applied directly to the rainfall in a "black box" approach.

It was concluded that all models performed more or less equally well on humid basins, but that the explicit accounting models showed a clear superiority in semiarid and arid catchments. With poor-quality data the simpler models showed relatively better results than the more complex models.

In the humid regions, particularly where rainfall is largely from frontal systems and there is significant orographic control, simulation results are good, probably because there is less uncertainty regarding rainfall amounts. In humid regions with rainfall more or less uniformly distributed throughout the year, soil moisture amounts do not vary widely from storm to storm and hence have less influence on runoff than other factors such as total rainfall. Thus the added complexity of moisture accounting is less useful. In semiarid and arid catchments with a large portion of the rainfall occurring in thunderstorms, there are much larger variations in soil moisture and the accuracy of areal rainfall estimates decreases. As the quality of the data decreases, the more complex models encounter difficulty because their algorithms require a moisture balance, which, because of poor data, may be difficult to achieve with reasonable parameters. The simpler models with fewer constraints may be able to adapt to the poor data, but it will not be known whether the answers are better simply because they more closely fit data which may be in error.

In selecting a model, a prospective user should be concerned with the data handling. Simple procedures for handling data will facilitate the simulation process and give the hydrologist more time to think about the hydrology of the problem. Clear and adequate operating instructions and program documentation are also important. The program should be reasonably logical from the hydrologic viewpoint—a program with illogical algorithms is not likely to be uniformly effective. The program should include built-in checks on the model performance. For example, an overall water balance can be computed monthly to confirm that continuity is being satisfied as the program operates.

PROBLEMS

12-1 In the infiltration-interflow function (Fig. 12-2) if b is 0.9 in/hr, c is 1.3, and $X = 0.1$ in/hr, what surface runoff and interflow will occur? Suppose $X = 0.5$ in/hr. What is the result?

12-2 If UZSN = 1 in, LZSN = 10 in, and UZS/UZSN = 0.7, what part of the surface runoff of Prob. 12-1 will be retained in surface storage? What is the increment to overland flow? Repeat with UZS = 2.0 in.

12-3 If LZSN = 10 in and LZS = 7 in, what part of the net infiltration of Prob. 12-1 will be added to groundwater storage? Repeat with LZS = 18 in.

12-4 If a simulation model is available, calibrate it for a small stream in your area. Discuss the results in terms of such topics as adequacy of input data, interaction between model parameters, etc.

12-5 Using the simulation model, explore its sensitivity to input data and parameters. For example, compare output when the potential evapotranspiration is increased 20 percent. What is the effect of a 10 percent change in precipitation? How sensitive is the model to changes in parameters? To what extent do you think your results are specific for the conditions you have assumed? For example, would the changes resulting from an assumed change in evapotranspiration be different if the precipitation were different?

12-6 Construct a simple tank model for some stream in your area which is gaged. Calibrate it and test the calibration on a period of record not used in the calibration. Make some sensitivity tests of the parameters. Would a more complex model give better results?

REFERENCES

1. Freeze, R. A.: Mathematical Models of Hillslope Hydrology, sec. 6 in M. J. Kirkby (ed.): "Hillslope Hydrology," Wiley, New York, 1978.
2. Anderson, E. A.: A Point Energy and Mass Balance Model of a Snow Cover, *NOAA Tech. Rep. NWS* 19, February 1976.
3. Crawford, N. H., and R. K. Linsley: Digital Simulation in Hydrology: Stanford Watershed Model IV, *Stanford Univ., Dept. Civ. Eng. Tech. Rep.* 39, 1966.
4. Boughton, W. C.: A Mathematical Catchment Model for Estimating Runoff, *J. Hydrol. (New Zealand)*, vol. 7, pp. 75–100, 1968.
5. Claborn, B. J., and W. Moore: Numerical Simulation in Watershed Hydrology, *Univ. Tex. Hydraul. Eng. Lab. Tech. Rep. HYD* 14-7001, 1970.
6. Holtan, H. N., and N. C. Lopez: An Engineering-Oriented Model of Watershed Hydrology, *Proc. Int. Seminar Hydrol. Prof.*, vol. 3, pp. 572–605, 1969.
7. Sittner, W. T., C. E. Schauss, and J. C. Monro: Continuous Hydrograph Synthesis with an API-Type Hydrologic Model, *Water Resour. Res.*, vol. 5, pp. 1007–1022, October 1969.
8. Sugawara, M.: On the Analysis of the Runoff Structure about Several Japanese Rivers, *Jap. J. Geophys.*, vol. 2, 1961.
9. Rockwood, D. M.: Columbia Basin Streamflow Routing by Computer, *J. Waterw. Div. ASCE*, vol. 84, p. 1874, December 1958.
10. Anderson, E. A., and N. H. Crawford: The Synthesis of Continuous Snowmelt Runoff Hydrographs on a Digital Computer, *Stanford Univ. Dept. Civ. Eng. Tech. Rep.* 36, 1964.
11. National Weather Service River Forecast System Forecast Procedures, *NOAA Tech. Mem. NWS Hydro*-14, December 1972.
12. Chen, C. W., and R. P. Shubinski: Computer Simulation of Urban Storm Water Runoff, *J. Hydraul. Div. ASCE*, vol. 97, pp. 289–301, February 1971.
13. Linsley, R. K., and N. H. Crawford: Computation of a Synthetic Streamflow Record on a Digital Computer, *Int. Assoc. Sci. Hydrol. Publ.* 51, pp. 526–538, 1960.
14. Liou, E. Y.: OPSET: Program for Computerized Selection of Watershed Parameter Values for the Stanford Watershed Model, *Univ. Ky. Water Resour. Inst. Res. Rep.* 34, 1970.
15. Monro, J. C.: Direct Search Optimization in Mathematical Modeling and a Watershed Model Application, *U.S. Natl. Weather Serv., NOAA Tech. Mem. NWS HYDRO*-12, April 1971.
16. Franz, D. D., and R. K. Linsley: An Interactive Time Series Plotting System for Classroom Instruction in Hydrology, *Stanford Univ. Dept. Civ. Eng. Tech. Rep.* 143, February 1971.
17. Johanson, R. C.: Precipitation Network Requirements for Streamflow Estimation, *Stanford Univ. Dept. Civ. Eng. Tech. Rep.* 147, August 1971.

18. HEC-1, Flood Hydrograph Package, Hydrologic Engineering Center, Corps of Engineers, U.S. Army, January 1973.
19. Metcalf and Eddy, Inc., University of Florida, and Water Resources Engineers, Inc.: Storm Water Management Model, vols. 1–4, *U.S. Environmental Protection Agency Rep.* 110224DOC, 1971.
20. Urban Storm Water Runoff: STORM User's Manual, U.S. Army Corps of Engineers, Hydrologic Engineering Center, Davis, Calif., 1976.
21. Terstriep, M. L., and J. B. Stall: The Illinois Urban Drainage Area Simulator, *Bull.* 58, *Illinois State Water Survey,* 1974.
22. Sugawara, M.: The Flood Forecasting by a Series Storage Type Model, *Int. Symp., Floods and Their Computation,* Leningrad, 1967.
23. Jamieson D. G., and J. C. Wilkinson: Operating Multi-Purpose Reservoirs for Water Supply and Flood Alleviation, *Water Resour. Res.,* vol. 8, pp. 899–903, 1972.
24. World Meteorological Organization: Intercomparison of Conceptual Models Used in Operational Hydrological Forecasting, *Opera. Hydrol. Rep.* 7, Annex III, p. 47, Geneva, 1975.
25. Dawdy, D. R., and T. O'Donnell: Mathematical Models of Catchment Behavior, *J. Hydraul. Div., ASCE,* vol. 91, pp. 123–137, 1965.
26. Schermerhorn, V. P., and D. W. Kuehl: Operational Streamflow Forecasting with the SSARR Model, *Int. Assoc. Sci. Hydrol. Publ.* 80, pp. 317–328, 1968.
27. Heatherwick, G.: Development of Flood Forecasting System for the Latrobe River at Yalloun, *World Meteorol. Org. Tech. Note* 92, pp. 110–132, 1969.
28. Narayana, V. V. D., J. P. Riley, and E. K. Israelsen: Analog Computer Simulation of the Runoff Characteristics of an Urban Watershed, Utah State Univ. Water Res. Lab., 1969.
29. Ida, Y., T. Kiya, and K. Sasaki: Flood Forecast and Flood Control by Computer, *Proc.* 13*th Cong. Int. Assoc. Hydraul. Res.,* vol. 1, September 1969.
30. Burnash, R. J. C., R. L. Ferrel, and R. A. McGuire: A Generalized Streamflow Simulation System, Conceptual Modeling for Digital Computers, U.S. National Weather Service, Sacramento, Calif., 1973.
31. U.S. Agricultural Research Service: USDAHL-74, Revised Model of Watershed Hydrology, *Tech. Bull.* 158, 1975.
32. Johanson, R. C., J. C. Imhoff, and H. Davis: Users Manual for Hydrologic Simulation Program—Fortran (HSPF), Environmental Protection Agency Rep. EPA-600/80-015, April 1980.
33. Anderson, E. A.: National Weather Service River Forecast System—Snow Accumulation and Ablation Model, *NOAA Tech. Mem. NWS HYDRO-17,* November 1973.
34. Natale, L., and E. Todini: Black Box Identification of a Linear Flood Wave Propagation Model, *Proc. Int. Assoc. Hydraulic Res.,* 1973.

BIBLIOGRAPHY

Aitken, A. D.: Assessing Systematic Errors in Rainfall-Runoff Models, *J. Hydrol.,* vol. 20, pp. 121–137, October 1973.
Armstrong, B. L.: Derivation of Initial Soil Moisture Accounting Parameters from Soil Properties for the National Weather Service River Forecast System, *NOAA Tech. Mem. NWS HYDRO* 37, 1978.
Dawdy, D. R., and J. M. Bergmann: Effect of Rainfall Variability on Streamflow Simulation, *Water Resour. Res.,* vol. 5, pp. 958–966, 1969.
Fleming, G.: "Deterministic Simulation in Hydrology," American Elsevier, New York, 1974.
Huff, D. D.: Simulation of the Hydrologic Transport of Radioactive Aerosols, Ph.D. thesis, Comm. Hydrology, Stanford University, 1967.
Ibbitt, R. P., and T. O'Donnell: Fitting Methods for Conceptual Catchment Models, *J. Hydraul. Div. ASCE,* vol. 97, pp. 1331–1342, September 1971.

James, L. D.: An Evaluation of Relationships between Streamflow Patterns and Watershed Characteristics through the Use of OPSET: A Self-Calibrating Version of the Stanford Watershed Model, *Univ. Ky. Water Resour. Inst. Res. Rep.* 36, 1970.

Johnston, P. R., and D. H. Pilgrim: A Study of Parameter Optimization for a Rainfall-Runoff Model, *Water Res. Lab. Rep.* 131, University of New South Wales, February 1973.

Overton, D. E., and M. E. Meadows: "Stormwater Modeling," Academic Press, New York, 1976.

Sittner, W. T.: WMO Project on Intercomparison of Conceptual Models Used in Hydrological Forecasting, *Hydro. Sci. Bull.,* vol. 21, pp. 203–213, 1976.

——— and K. M. Krouse: Improvement of Hydrologic Simulation by Utilizing Observed Discharge as an Indirect Input (Computed Hydrograph Adjustment Technique—CHAT), *NOAA Tech. Mem. NWS HYDRO*-38, 1979.

Watt, W. E., and C. I. R. Kidd: QUURM—A Realistic Urban Runoff Model, *J. Hydrol.,* vol. 27, pp. 225–235, 1975.

THIRTEEN

PROBABILITY IN HYDROLOGY: A BASIS FOR PLANNING

Water resources systems must be planned for future events for which no exact time of occurrence can be forecast. Hence, the hydrologist must give a statement of the probability† that streamflows (or other hydrologic factors) will equal or exceed (or be less than) a specified value. These probabilities are important to the economic and social evaluation of a project [1]. In most cases absolute control of floods or droughts is impossible. Planning to control a flood of a specific probability recognizes that a project will be overtaxed occasionally and damages will be incurred. However, repair of the damages should be less costly in the long run than building initially to protect against the worst possible event. The planning goal is not to eliminate all floods but to reduce the frequency of flooding and, hence, the resulting damages. If the socioeconomic analysis is to be correct, the probabilities of flooding must be estimated accurately. For major projects, the failure of which seriously threatens human life, a more extreme event, the probable maximum flood, has become the standard for designing the spillway.

This chapter deals with techniques for defining probability from a given set of data and with special methods employed for determining the spillway-design flood for major dams. Methods for probability analysis using synthetically generated data are discussed in Chap. 14.

---

† Probability: a mathematical basis for prediction, which, for an exhaustive set of outcomes, is the ratio of the outcomes that will produce a given event to the total number of possible outcomes (paraphrased from "Webster's 7th New Collegiate Dictionary," 1971). Frequency: the number of cases in a class when events are classified according to differences in one or more attributes.

FLOOD PROBABILITY

The following section discusses basic concepts of probability analysis with specific reference to flood peaks. In general these methods are also applicable to other hydrologic parameters with minor differences discussed in subsequent sections.

13-1 Selection of Data

If probability analysis is to provide reliable answers, it must start with a data series that is relevant, adequate, and accurate. *Relevance* implies that the data must deal with the problem. Most flood studies are concerned with peak flows, and the data series will consist of selected observed peaks. However, if the problem is duration of flooding, e.g., for what periods of time a highway adjacent to a stream is likely to be flooded, the data series should represent duration of flows in excess of some critical value. If the problem is one of interior drainage of a leveed area, the data required may consist of those flood volumes occurring when the main river is too high to permit gravity drainage.

Adequacy refers primarily to length of record, but sparsity of data-collecting stations is often a problem. The observed record is merely a sample of the total population of floods that have occurred and may occur again. If the sample is too small, the probabilities derived cannot be expected to be reliable. Available streamflow records are too short to provide an answer to the question: How long must a record be to define flood probabilities within acceptable tolerances? Table 13-1 gives some estimates derived from synthetic data series.

Table 13-1 suggests that extrapolation of frequency estimates beyond probability of 0.01 is extremely risky with the data series generally available [3, 4]. Studies by Ott [5] and Nasseri [6] indicate that 80 percent of the estimates of the 100-year flood based on 20 years of record will be too high and that 45 percent of the overestimates will exceed 30 percent. If the available record is short, an attempt should be made to extend the record rather than to extrapolate from the limited sample.

Accuracy refers primarily to the problem of homogeneity. Most flow records are satisfactory in terms of intrinsic accuracy, and if they are not, there is

Table 13-1 Length of record in years required to estimate floods of various probabilities with 95 percent confidence [2]

Design probability	Acceptable error	
	10%	25%
0.1	90	18
0.02	110	39
0.01	115	48

little that can be done with them. If the reported flows are unreliable, they are not a satisfactory basis for frequency analysis. Even though reported flows are accurate, they may be unsuitable for probability analysis if changes in the catchment have caused a change in the hydrologic characteristics, i.e., if the record is not internally homogeneous. Dams, levees, diversions, urbanization, and other land-use changes may introduce inconsistencies (Sec. 4-16). Such records should be adjusted before use to current conditions or to natural conditions.

If the analysis is concerned with probabilities less than 0.5, a data series of *annual floods,* the largest event in each year, is the best choice. For more frequent events, the *partial duration series* (or simply *partial series*) is better (Sec. 13-10).

13-2 Plotting Positions

Probability analysis seeks to define the flood flow with a probability p of being equaled or exceeded in any year. *Return period*† T_r is often used in lieu of probability to describe a design flood. Return period and probability are reciprocals, i.e.,

$$p = \frac{1}{T_r} \tag{13-1}$$

To plot a series of peak flows as a cumulative frequency curve it is necessary to decide on a probability or return period to associate with each peak. There are various formulas for defining this value, known as the *plotting position* [7, 8]. The one most commonly used in the United States is Weibull's [9]:

$$p = \frac{m}{n + 1} \qquad \text{or} \qquad T_r = \frac{n + 1}{m} \tag{13-2}$$

where n is the number of years of record and m is the rank of the event in order of magnitude, the largest event having $m = 1$. The U.K. National Environmental Research Council [10] recommends Gringorten's formula [11, 12]

$$p = \frac{m - 0.44}{n + 0.12} \qquad \text{or} \qquad T_r = \frac{n + 0.12}{m - 0.44} \tag{13-3}$$

Equation (13-2) assigns a return period to the largest event in a data series equal to $n + 1$ years. Table 13-2 presents the theoretical distribution of the return period for floods having specified average return periods. Note that the probability that the return period will be less than the average value is greater than 0.5. Over a long period of time 25 percent of the intervals between floods ≥ 100-year flood will be less than 29 years, while an equal number will be greater than 139 years.

The Gringorten plotting position [Eq. (13-3)] gives longer return periods for

---

† *Return period* and *recurrence interval* are used interchangeably to mean the average interval in years within which a given event will be equaled or exceeded.

Table 13-2 Theoretical distribution of the return period

Average return period $\bar{T}_r$	Actual return period T_r exceeded various percentages of the time						
	1%	5%	25%	50%	75%	95%	99%
2	8	5	3	1	0	0	0
5	22	14	7	3	1	0	0
10	45	28	14	7	3	0	0
30	137	89	42	21	8	2	0
100	459	300	139	69	29	5	1
1,000	4,620	3,000	1,400	693	288	51	10
10,000	46,200	30,000	14,000	6,932	2,880	513	100

the higher floods in a series. This recognizes that the true return period of the higher floods is probably longer than the value computed with Eq. (13-2).

The probability j that the actual probability of the mth of n floods is less than p_0 can be obtained from [13]

$$ j = \frac{n!}{m!(n-m)!} \, m \int_1^{1-p_0} p^{m-1}(1-p)^{n-m} \, dp \qquad (13\text{-}4) $$

which can be used to compute Table 13-3. There are approximately 2 chances out of 3 that the true return period of the largest event in a data series is greater

Table 13-3 Average return periods for various levels of probability

Rank from top m	Number of years of record n	Probability				
		0.01	0.25	0.50	0.75	0.99
1	2	1.11	2.00	3.41	7.46	200
	5	1.66	4.13	7.73	17.9	498
	10	2.71	7.73	14.9	35.3	996
	20	4.86	14.9	29.4	70.0	1990
	60	13.5	43.8	87.0	209.0	5970
2	3	1.06	1.48	2.00	3.06	17.0
	6	1.42	2.57	3.78	6.20	37.4
	11	2.13	4.41	6.76	11.4	71.1
	21	3.61	8.12	12.7	21.8	138
	61	9.62	23.0	36.6	63.4	408
3	4	1.05	1.32	1.63	2.19	7.10
	7	1.31	2.06	2.75	3.95	14.1
	12	1.86	3.32	4.62	6.86	25.6
	22	3.03	5.86	8.35	12.6	48.6
	62	7.76	16.1	23.3	35.8	140
4	5	1.03	1.24	1.46	1.83	4.50
	8	1.25	1.80	2.27	3.04	8.26
	13	1.70	2.77	3.63	5.02	14.4
	23	2.67	4.72	6.36	8.98	26.6
	63	6.63	12.5	17.2	24.8	75.2

than the period of record. The table shows the uncertainty in the plotting position assigned to the highest event in a series, but for $m \geq 4$ the range of uncertainty is reasonably narrow. If frequency analysis is intended to gain information on floods with return periods less than $n/5$, one may plot flood magnitude against plotting position and fit a curve by eye. For longer return periods it is better to fit a theoretical distribution (Sec. 13-3) to the data. Note that tests for the best theoretical distribution are greatly influenced by the assumed plotting positions.

13-3 Theoretical Distributions of Floods

Statistical distributions are usually demonstrated by use of samples numbering in the thousands. No such samples are available for streamflow and it is not possible to state with certainty that a specific distribution applies to flood peaks. Numerous distributions have been suggested on the basis of their ability to "fit" the plotted [13–15] data from streams (Sec. 13-2). Despite much effort, tests [16] suggest that there is no best distribution for floods. Intuitively at least, there is no reason to expect that a single distribution will apply to all streams worldwide. The log-Pearson Type III has been adopted [17] for use by United States federal agencies for flood analysis. The first asymptotic distribution of extreme values (EV1), commonly called the *Gumbel Type I distribution,* has been widely used and is recommended in the United Kingdom. These two distributions are described in the following sections.

Chow [18] has shown that most frequency functions can be generalized to

$$X = \bar{X} + K \sigma_X \qquad (13-5)$$

where X is a flood of specified probability, $\bar{X}$ is the mean of the flood series, σ_X is the standard deviation of the series, and K, a frequency factor defined by a specific distribution, is a function of the probability level of X.

13-4 Log-Pearson Type III Distribution

The recommended procedure for use of the log-Pearson distribution is to convert the data series to logarithms and compute:†

Mean: $$\overline{\log X} = \frac{\Sigma \log X}{n} \qquad (13-6)$$

---

† Alternatively,

$$\sigma_{\log x} = \sqrt{\frac{\Sigma (\log x)^2 - (\Sigma \log x)^2/n}{n - 1}}$$

$$G = \frac{n^2 \Sigma (\log x)^3 - 3n \Sigma \log x \Sigma (\log x)^2 + 2(\Sigma \log x)^3}{n(n - 1)(n - 2)(\sigma_{\log x})^3}$$

where $\log x = \log X - \overline{\log X}$.

Standard deviation: $\qquad \sigma_{\log X} = \sqrt{\dfrac{\Sigma (\log X - \overline{\log X})^2}{n - 1}}$ $\qquad$ (13-7)

Skew coefficient: $\qquad G = \dfrac{n \, \Sigma \, (\log X - \overline{\log X})^3}{(n - 1)(n - 2)(\sigma_{\log x})^3}$ $\qquad$ (13-8)

The value of X for any probability level is computed from Eq. (13-5) modified

$$\log X = \overline{\log X} + K \sigma_{\log x} \qquad (13\text{-}9)$$

where K is taken from Table 13-4. The cumulative frequency distribution will plot as a straight line on log-normal paper when the skew coefficient $G = 0$.

The Type III distribution is one of a family of distributions suggested by Pearson [19]. There are no theoretical arguments for the application of this distribution to hydrologic data. It is a skew distribution bounded on the left and therefore of the general shape of most hydrologic distributions. When skew is zero, the Pearson Type III is identical to the log-normal distribution which was once widely used in hydrology. With a third parameter, skew, the distribution can be "fitted" to most data sets. Reliable estimates of skew require very large samples, however. In lieu of using Eq. (13-8) to compute skew, regional average values are often used [17]; although this may not be more reliable it does lead to more consistency between values for various streams in the region.

The probability density function for Type III is

$$p(X) = p_0 \left(1 - \frac{X}{a}\right)^c e^{-cX/2} \qquad (13\text{-}10)$$

where $\qquad\qquad\qquad c = \dfrac{4}{\beta} - 1$ $\qquad\qquad$ (13-11)

$$a = \frac{c}{2} \frac{\mu_3}{\mu_2} \qquad (13\text{-}12)$$

$$p_0 = \frac{n}{a} \frac{c^{c+1}}{e^c \Gamma(c + 1)} \qquad (13\text{-}13)$$

$$\beta = \frac{\mu_3^2}{\mu_2^3} \qquad (13\text{-}14)$$

where μ_2 is the variance, μ_3 is the third moment about the mean $= \sigma^6 G$, Γ is the gamma function, and e is the base of napierian logarithms.

13-5 Extreme-Value Type I Distribution

Fisher and Tippett [20] found that the distribution of the maximum (or minimum) values selected from n samples approached a limiting form as the size of the samples increased. When the initial distributions within the samples are exponential, the Type I distribution results. This distribution is given by

$$p = 1 - e^{-e^{-y}} \qquad (13\text{-}15)$$

Table 13-4 K Values for the log-Pearson Type III distribution

	Recurrence interval, years							
	1.0101	1.2500	2	5	10	25	50	100
Skew coefficient G	Percent chance							
	99	80	50	20	10	4	2	1
3.0	−0.667	−0.636	−0.396	0.420	1.180	2.278	3.152	4.051
2.8	−0.714	−0.666	−0.384	0.460	1.210	2.275	3.114	3.973
2.6	−0.769	−0.696	−0.368	0.499	1.238	2.267	3.071	3.889
2.4	−0.832	−0.725	−0.351	0.537	1.262	2.256	3.023	3.800
2.2	−0.905	−0.752	−0.330	0.574	1.284	2.240	2.970	3.705
2.0	−0.990	−0.777	−0.307	0.609	1.302	2.219	2.912	3.605
1.8	−1.087	−0.799	−0.282	0.643	1.318	2.193	2.848	3.499
1.6	−1.197	−0.817	−0.254	0.675	1.329	2.163	2.780	3.388
1.4	−1.318	−0.832	−0.225	0.705	1.337	2.128	2.706	3.271
1.2	−1.449	−0.844	−0.195	0.732	1.340	2.087	2.626	3.149
1.0	−1.588	−0.852	−0.164	0.758	1.340	2.043	2.542	3.022
0.8	−1.733	−0.856	−0.132	0.780	1.336	1.993	2.453	2.891
0.6	−1.880	−0.857	−0.099	0.800	1.328	1.939	2.359	2.755
0.4	−2.029	−0.855	−0.066	0.816	1.317	1.880	2.261	2.615
0.2	−2.178	−0.850	−0.033	0.830	1.301	1.818	2.159	2.472
0	−2.326	−0.842	0.	0.842	1.282	1.751	2.054	2.326
−0.2	−2.472	−0.830	0.033	0.850	1.258	1.680	1.945	2.178
−0.4	−2.615	−0.816	0.066	0.855	1.231	1.606	1.834	2.029
−0.6	−2.755	−0.800	0.099	0.857	1.200	1.528	1.720	1.880
−0.8	−2.891	−0.780	0.132	0.856	1.166	1.448	1.606	1.733
−1.0	−3.022	−0.758	0.164	0.852	1.128	1.366	1.492	1.588
−1.2	−3.149	−0.732	0.195	0.844	1.086	1.282	1.379	1.449
−1.4	−3.271	−0.705	0.225	0.832	1.041	1.198	1.270	1.318
−1.6	−3.388	−0.675	0.254	0.817	0.994	1.116	1.166	1.197
−1.8	−3.499	−0.643	0.282	0.799	0.945	1.035	1.069	1.087
−2.0	−3.605	−0.609	0.307	0.777	0.895	0.959	0.980	0.990
−2.2	−3.705	−0.574	0.330	0.752	0.844	0.888	0.900	0.905
−2.4	−3.800	−0.537	0.351	0.725	0.795	0.823	0.830	0.832
−2.6	−3.889	−0.499	0.368	0.696	0.747	0.764	0.768	0.769
−2.8	−3.973	−0.460	0.384	0.666	0.702	0.712	0.714	0.714
−3.0	−4.051	−0.420	0.396	0.636	0.660	0.666	0.666	0.667

Source: Adapted from [10].

where p is the probability of a given flow being equaled or exceeded, e is the base of napierian logarithms, and y, the reduced variate, is a function of probability (Table 13-5). Then

$$X = \bar{X} + (0.7797y - 0.45)\sigma_X \qquad (13\text{-}16)$$

where $\bar{X}$ is the mean of the data series and σ_X is its standard deviation. This equation is equivalent to Eq. (13-5) with K equal to the term in parentheses.

Table 13-5 Values of K [Eq. (13-5)] for the extreme-value (Type 1) distribution

Return period, years	Probability	Reduced variate y	K
1.58	0.63	0.000	−0.450
2.00	0.50	0.367	−0.164
2.33	0.43	0.579	0.001
5	0.20	1.500	0.719
10	0.10	2.250	1.30
20	0.05	2.970	1.87
50	0.02	3.902	2.59
100	0.01	4.600	3.14
200	0.005	5.296	3.68
400	0.0025	6.000	4.23

Table 13-5 gives values of K for various return periods. When using the Grin-gorten plotting position [Eq. (13-3)], no correction for record length is considered necessary. Two or more computed values of X define a straight line on extreme-value probability paper. (See Figs. 13-1 and 13-2.)†
From Eq. (13-15)

$$y = -\ln[-\ln(1 - p)] \qquad (13\text{-}17)$$

Equation (13-3) is a very good approximation to the unbiased mean value of probability associated with a given point assuming the extreme-value distribution [12].

The solution by Eq. (13-16) utilizes the method of moments. The method of maximum likelihood is generally superior but more difficult to solve. Maximizing log likelihood requires the determination of α and β in

$$y_X = \alpha(X - \beta) \qquad (13\text{-}18)$$

where

$$\beta = \frac{1}{\alpha} \ln\left(\frac{n}{\sum\limits_1^n e^{-\alpha X_m}}\right) \qquad (13\text{-}19)$$

and

$$F(\alpha) = 0 = \sum_1^n X_m e^{-\alpha X_m} - \left(\bar{X} - \frac{1}{\alpha}\right)\sum_1^n e^{-\alpha X_m} \qquad (13\text{-}20)$$

† Extrapolation of a straight line is easier than extrapolation of a curve. Hence, frequency analysis is simplified by using special plotting papers scaled so that a specific distribution will plot as a straight line. On such probability paper, the ordinate is normally flowrate, and the abscissa is probability or return period. For some distributions the logarithm of flow is the ordinate. The abscissa scale is warped to achieve the straight-line plot.

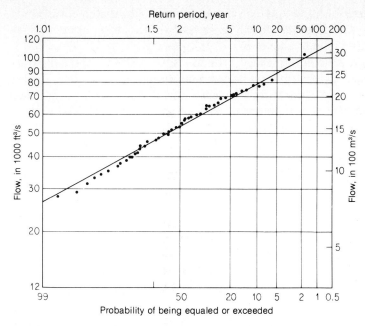

Figure 13-1 Flood-frequency curve for the Clearwater River at Kamiah, Idaho. The curve is plotted on log-normal probability paper and fitted by the log-Pearson Type III function. Points plotted with Eq. (13-2).

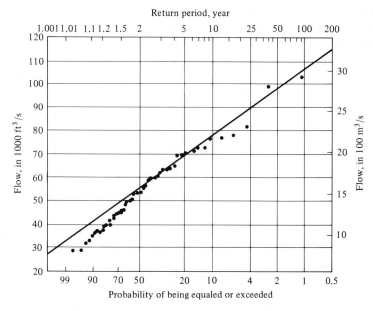

Figure 13-2 Flood-frequency curve for the Clearwater River at Kamiah, Idaho, plotted on extreme-value paper and fitted with the Type I (Gumbel) distribution. Points plotted with Eq. (13-3).

where X_m is the mode of X and y_x is the value of the reduced variate y corresponding to a specific X. Equation (13-20) can be solved by a Taylor series approximation starting with an initial value of $\alpha = 1.2825/\sigma_x$.

Gumbel [21] was the first to suggest the use of the extreme-value distribution for floods, and the distribution is commonly referred to as the Gumbel distribution. Gumbel's argument for the use of this distribution was that each year of record constituted a sample with $n = 365$ and the annual flood was the maximum value from the sample. Hence, it could be assumed that the flood case conformed to the conditions specified by Fisher and Tippett.

Illustrative example 13-1 Using the data of Table 13-6, find the magnitude of the 10- and 100-year floods with the log-Pearson Type III and Gumbel distributions. For Gumbel,

$$\bar{q} = 54{,}971 \text{ ft}^3/\text{s} \qquad \sigma_q = 16{,}483 \text{ ft}^3/\text{s}$$

From Table 13-5, $K_{10} = 1.30$ and $K_{100} = 3.14$

$$q_{10} = 54{,}971 + 1.30 \times 16{,}483 = 76{,}400 \text{ ft}^3/\text{s}$$

$$q_{100} = 54{,}971 + 3.14 \times 16{,}483 = 107{,}000 \text{ ft}^3/\text{s}$$

For log-Pearson Type III from Table 13-6

$$\overline{\log q} = 4.7212 \qquad \sigma_{\log q} = 0.1296 \qquad G = 0.0434$$

Table 13-6 Flood-probability analysis for the Clearwater River at Kamiah, Idaho
Drainage area = 4850 mi² (12,560 km²)

Year	Month	Discharge, ft³/s	Order m	Plotting† position, years
1911	June	39,500	45	1.24
1912	May	61,900	19	2.95
1913	May	76,600	5	11.20 (12.1)
1914	May	42,200	42	1.33
1915	May	28,200	55	1.02
1916	June	56,000	25	2.24
1917	June	70,500	10	5.60
1918	May	52,800	28	2.00
1919	May	52,000	31	1.81
1920	May	43,600	41	1.37
1921	May	69,700	12	4.67
1922	June	62,400	18	3.11
1923	May	49,600	32	1.75
1924	May	58,900	22	2.55
1925	May	59,800	20	2.80
1926	April	35,900	50	1.12

† Calculated with Weibull equation (13-2). Positions calculated with Eq. (13-3) shown in parentheses.

Table 13-6 (*Continued*)

Year	Month	Discharge, ft³/s	Order m	Plotting† position, years
1927	June	68,600	13	4.31
1928	May	72,100	7	8.00 (8.4)
1929	May	52,700	29	1.93
1930	April	31,000	53	1.06
1931	May	40,800	43	1.30
1932	May	72,100	8	7.00
1933	June	81,400	3	18.67 (21.5)
1934	April	45,900	37	1.51
1935	May	44,000	40	1.40
1936	May	63,200	16	3.50
1937	May	34,300	51	1.10
1938	April	63,400	15	3.73
1939	May	46,000	36	1.56
1940	May	37,100	47	1.19
1941	May	28,900	54	1.04
1942	May	37,100	48	1.17
1943	May	52,200	30	1.87
1944	May	34,200	52	1.08
1945	May	44,400	38	1.47
1946	May	36,600	49	1.14
1947	May	69,900	11	5.09
1948	May	99,000	2	28.00 (35.3)
1949	May	76,200	6	9.33
1950	June	62,600	17	3.29
1951	May	44,200	39	1.44
1952	April	49,200	34	1.65
1953	June	53,100	27	2.07
1954	May	58,800	23	2.43
1955	June	64,100	14	4.00
1956	May	77,800	4	14.00 (15.5)
1957	May	71,200	9	6.22
1958	May	59,600	21	2.67
1959	June	55,100	26	2.15
1960	May	49,600	33	1.70
1961	May	58,600	24	2.33
1962	April	39,700	44	1.27
1963	May	38,200	46	1.22
1964	June	103,000	1	56.00 (98.4)
1965	May	47,900	35	1.60

Σ	3,023,400	
Mean	54,971	
σ_q	16,483	
$\overline{\log q}$	4.7212	
$\sigma_{\log q}$	0.1296	
G	0.0434	

Source: Data from *U.S. Geological Survey Water-Supply Papers.* This station was discontinued in 1965.

From Table 13-4, $K_{10} = 1.286$ and $K_{100} = 2.358$.

$$\log q_{10} = 4.7212 + 0.1296 \times 1.286 = 4.8879$$

$$q_{10} = \text{antilog } 4.8879 = 77,200 \text{ ft}^3/\text{s}$$

$$\log q_{100} = 4.7212 + 0.1296 \times 2.358 = 5.0268$$

$$q_{100} = \text{antilog } 5.0268 = 106,000 \text{ ft}^3/\text{s}$$

All answers should be rounded to three significant figures.

13-6 Selection of Design Frequency

Analyses described in preceding sections define average probability level or average return period. This information is useful for numerous problems. For example, in calculating average annual flood damages, the damage caused by a particular flow should be multiplied by its average probability of occurrence to find its contribution to the average annual damage. The design of any structure with essentially unlimited life would be based on average probabilities. Design of highway culverts to pass flows of annual probability p implies that, on the average, pN culverts will be overtopped each year, where N is the total number of culverts.

There are situations when one is concerned with the probability of a flood occurring during a specified interval of future time. For example, what flood probabilities exist during the construction period of a dam? The probability J_k that a flood with average probability of occurrence p will be exceeded exactly k times during an N-year period is given by [13] the binomial distribution

$$J_k = \frac{N!}{k!(N-k)!} (1-p)^{N-k} p^k \tag{13-21}$$

The probability of one *or more* exceedances in N years is found by taking $k = 0$ and noting that the probability of exceedance is one minus the probability of nonexceedance,

$$J_{1 \text{ or more}} = 1 - (1-p)^N \tag{13-22}$$

For example, there is a 10 percent chance that the 460-year event will be equaled or exceeded during the next 50 years (Table 13-7). *Note that it is assumed that the annual probability of occurrence is known accurately.* Since this is unlikely, some further uncertainty remains in the estimates. Thus it is important to obtain as accurate an estimate of p as possible. The procedure outlined cannot be looked upon as a certain safety factor. If p is underestimated appreciably, this procedure may lead to gross overdesign, a strong possibility if the record is short (Sec. 13-1).

13-7 Regional Flood Frequency

Analysis of historic records is of little value in flood-frequency studies for ungaged catchments or catchments with only a few years of record. One proce-

Table 13-7 Return period required for specified risk of occurrence within project life

Permissible risk of failure	Expected life of project, years				
	1	10	25	50	100
0.01	100	910	2440	5260	9100
0.10	10	95	238	460	940
0.25	4	35	87	175	345
0.50	2	15	37	72	145
0.75	1.3	8	18	37	72
0.99	1.01	2.7	6	11	22

dure designed to overcome this problem is regional-flood-frequency analysis [22]. Records from a number of streams are assembled, and a standard period is selected. Some annual peaks may be estimated so that all records are complete for the standard period. Flow data for each station are made nondimensional by dividing by the mean annual flood $\bar{q}$ for the station. Median ratios for each order number are then plotted (omitting estimated values), and a regional curve fitted (Fig. 13-3). Statistical tests may be used to confirm the homogeneity of the data. An independent relation defining the mean annual flood in terms of basin characteristics is then required. Often this is a simple relation between $\bar{q}$ and drainage area (Fig. 13-4), although other parameters such as basin slope, mean elevation, percentage of area in lakes, and mean annual precipitation over the catchment may be employed.

A probability curve for an ungaged basin is then constructed by first determining the mean annual flood and then converting the ratio scale of the regional-frequency curve to flow by multiplying the ratios by the estimated $\bar{q}$. The weakness of the regional-frequency approach lies in the assumption that all streams will show the same variance of $q/\bar{q}$ (slope of the cumulative-frequency

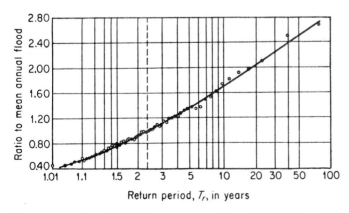

Figure 13-3 Regional-flood-frequency curve for selected stations in the Youghiogheny and Kiskiminetas River basins, Pennsylvania and Maryland. (*U.S. Geological Survey.*)

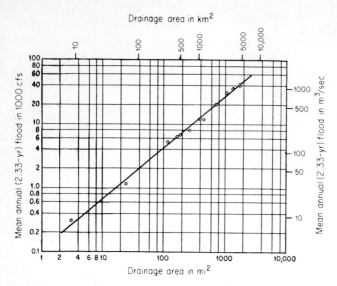

Figure 13-4 Variation of mean annual flood with drainage area, Youghiogheny and Kiskiminetas River basins, Pennsylvania and Maryland (*U.S. Geological Survey.*)

function). Statistical tests may indicate that several streams *could* have the same function, but they can never prove the functions are identical. Differences between frequency curves of several streams may be real differences arising from different regimes. They may, on the other hand, be apparent differences arising from the variability of short samples.

If the regional approach is used, considerable care should be taken to select streams as nearly similar in hydrologic characteristics as possible. They should have similar vegetal cover, land use, topographic conditions, and geologic characteristics, and the range of drainage area should not be excessive. They should also have similar rainfall and evapotranspiration regimes. Combining data from a number of catchments may help to reduce the uncertainty arising from short records if the records are sufficiently independent. However, if maximum events at most stations derive from the same storms and tend to plot as outliers from the station-frequency curve, the composite curve will have the same outliers.

13-8 Frequency Analysis from Synthetic Data

Since the major problems of frequency analysis result from short records, it is clearly desirable to use the longest possible records. Often rainfall data will be available for a longer period than streamflow data, and techniques discussed in Chaps. 7 to 10 and Chap. 12 may be used to estimate flood peaks prior to the beginning of a streamflow record. These same techniques can be used to develop data for ungaged catchments. Because there will be some error in the estimated peaks, the effective record length is not equal to the total of observed and estimated records, but an improvement in the frequency analysis can result if the estimates are made with sufficient care. It is important that there be no bias in the estimates of peaks as this will bias the frequency analysis. Random errors will tend to cancel.

Procedures for estimating peak flows may be derived, if necessary, from 2 or 3 years of observed data, although a longer sample is to be preferred. Hence, records which are quite useless for frequency analysis may be extended. Simulation techniques (Chap. 12) are especially useful for record extension.

It is not necessary that rainfall stations used for estimating peak flows be in the catchment under study. The only requirement is that the stations used experience the same precipitation-frequency regime as the study basin [23]. This may permit use of precipitation data from as far as 200 km (120 mi). If remote data are used, calculated peaks may not necessarily reproduce historic peaks on the catchment, but if the data used represent the same climatic regime, the resulting frequency curve will approximate the true curve for the study basin.

Stochastically generated rainfall data (Chap. 14) may be used [5, 24–26] as a basis for simulating long streamflow records (Chap. 12). Figure 13-5 illustrates a frequency curve so derived. If stochastic data are used, exact correspondence with historic events cannot be expected. Hence, a short extension of an observed record by stochastic methods is not advisable. Instead, a long stochastic record should be utilized. This approach allows one to escape the constraints of the short streamflow records now available. Ott [5] and Nasseri [6] used this

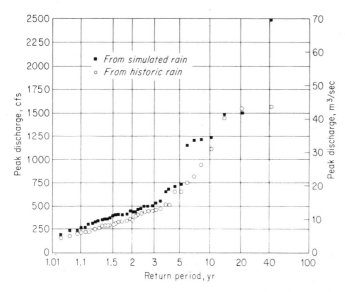

Figure 13-5 Comparison of flood-frequency curves for Silver Creek near San Jose, Calif., based on historic data and on flows simulated using stochastically generated hourly rainfall. (*From* [26].)

technique to develop 500-year flood samples for study of frequency-analysis methods. The accuracy of the result depends on both the stochastic procedure and the deterministic simulation being unbiased. Since there is a decided bias in the use of short flow records, the use of stochastically based data offers the prospect of improved results and an opportunity to investigate the question of the most appropriate distribution to be applied in hydrologic probability analysis. Ott found the log-Pearson III distribution to give the best fit to data from North Carolina but the Gumbel distribution best fit the California stream which he studied.

13-9 Conditional Probability

Hydrologic problems often involve occurrence of two or more simultaneous events. These situations require special treatment. If the events are independent, the probability of their joint occurrence is the product of their individual probabilities, $p_1 p_2$. If the events are completely dependent, i.e., one cannot happen without the other, the probability of their joint occurrence is equal to the probability of either event occurring. Few hydrologic events are either independent or completely dependent.

In the design of internal drainage for an area protected by levees, it may be possible to discharge accumulated water by gravity when the stage in the main river is below some threshold value h. The question might be: What is the probability distribution of flood flows from the internal drainage when the main-river stage exceeds h? The answer may be obtained by selecting only tributary flood peaks when the main-river stage is h or higher for frequency analysis. The series will probably be different from the annual flood series for the tributary. In some years there may be no occurrences if the main stream did not exceed the critical stage. If the occurrence is relatively rare, there may be few events for analysis. Extension of the record is indicated to provide a larger data base.

Some knowledge may be desired of the interrelation between high stages in the main river and floods on the tributary rather than a simple threshold analysis. In this case, the problem can be treated as one involving multiple thresholds, and the results may be expressed as a matrix (Table 13-8) showing the number of tributary floods in various magnitude classes for selected ranges of main-stream stage or flow. Alternatively, a frequency curve can be drawn for each class interval of main-stream stage or above each threshold level. In each case, n equals the record length in calculating plotting position. The data constitute several partial series and are not subject to analysis by rigorous statistical methods. The relatively limited sample in each row is the major difficulty.

Some problems can be handled by combining the several variables into a single variable which can be analyzed by conventional methods. For example, if the probability of average rainfall over a catchment is required, the observed rainfalls should be averaged and the averages treated as a data series for analysis. Similarly, one might add simultaneous flows of two or more streams to

Table 13-8 Conditional probability of flood flows on a tributary being equaled or exceeded when the main stream is at various levels
Probabilities are shown in parentheses

Main-stream stage, ft	Tributary peak flow, ft³/s						Total cases in row
	0–99	100–199	200–299	300–399	400–499	500–599	
0–5	2 (0.25)	1 (0.15)		1 (0.10)	1 (0.05)		5 (0.25)
5–8	1 (0.20)	1 (0.15)	2 (0.10)				4 (0.20)
8–11		1 (0.20)	1 (0.15)	1 (0.10)	1 (0.05)		4 (0.20)
11–14	1 (0.25)		1 (0.20)	2 (0.15)	1 (0.05)		5 (0.25)
14–17			1 (0.10)			1 (0.05)	2 (0.10)
Total cases in column	4 (0.20)	3 (0.15)	5 (0.25)	4 (0.20)	3 (0.15)	1 (0.05)	20 (1.00)

determine the probability of combined discharges into a reservoir. Unless the events are perfectly correlated, it is incorrect to add the 10-year flows to obtain the 10-year combined flow. This procedure will almost always result in an overestimate of flow for any return period.

13-10 Frequent Events

When the design problem requires consideration of events with return periods less than 5 years, a partial flood series is preferable to the annual series. Use of

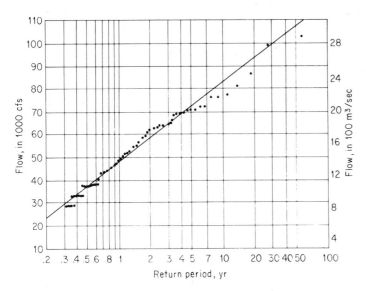

Figure 13-6 Flood-frequency curve for the Clearwater River at Kamiah, Idaho, based on partial duration flow data. (Weibull plotting position.)

Table 13-9 Corresponding return periods in years for annual and partial series [27]

Partial series	0.5	1.0	1.45	2.0	5.0	10	50	100
Annual series	1.16	1.58	2.00	2.54	5.52	10.5	50.5	100.5

such short return periods might be appropriate in urban drainage, where damage is small. The partial series is made up of all floods above some selected base value. The base is usually chosen so that two or three events are included for each year. The partial series can then indicate the probability of events being equaled or exceeded 2 or more times per year (Fig. 13-6). Minor peaks associated with a larger flood are usually not included in the series. While the decision must be largely arbitrary, it should take into account the purpose of the study. If it is felt that each peak constitutes an independent event of interest for the study, then all should be counted. Table 13-9 compares corresponding return periods for the annual and partial series as derived on theoretical grounds [27]. Return period for any flowrate is approximately 0.5 year less for the partial series than for the annual series. Because the partial series is arbitrarily selected, it cannot be expected to fit a standard distribution.

PROBABILITY OF RUNOFF VOLUME

13-11 Distributions

Monthly and annual volumes of runoff seem to conform to the normal or lognormal distributions, although the gamma distribution is sometimes used. Frequency analysis of runoff volume requires fewer data than for flood peaks. No systematic analysis has been undertaken to define the required record length for a stable volume frequency analysis, but 30 to 50 years would seem reasonably adequate for most purposes. Arid areas, where many years or months have zero or near zero runoff, require longer records.

13-12 Drought

Hydrologic drought may be defined as a period during which streamflows are inadequate to supply established uses under a given water-management system. No more specific definition seems possible, since each situation must be analyzed separately. Commonly, hydrologists are involved in two kinds of drought studies. They may be concerned with low flows which would be limiting for a water supply served by diversion without storage and are also critical in water-pollution problems. They may also be concerned with extended periods of low

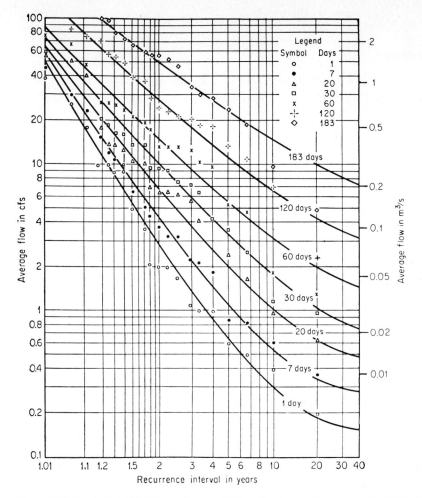

Figure 13-7 Frequency of minimum flows for Yellow Creek near Hammondsville, Ohio, 1915–1935. (*From W. P. Cross and E. E. Webber, Ohio Stream-Flow Characteristics, Ohio Dept. Nat. Resour. Bull. 13, pt. 2, table 1, December 1950.*)

flow as they may affect the yield of a storage reservoir. This latter problem is best treated by stochastic analysis (Chap. 14) since occurrences of drought conditions in recorded streamflow are generally too few for frequency analysis.

For low-flow analysis, it is necessary to select the time period of concern (1, 5, 10 days, etc.) and to abstract the annual minimum flows for this interval [28]. The data are usually plotted using Eq. (13-2), and a curve is fitted by eye. Figure 13-7 illustrates such a plot for several time intervals on log-normal probability paper. Gumbel suggested that a log–extreme-value distribution would be appropriate. This can be treated as a Type I distribution using logarithms of flow or plotted on extreme-value paper with a logarithmic scale of flow.

PRECIPITATION PROBABILITY

13-13 Distributions

The preceding discussion on flood probability applies generally to precipitation. Annual maximum hourly or daily amounts ordinarily conform to a Gumbel Type I, log-Pearson, log-normal, or gamma distribution. In humid areas, where the mean is high, monthly, seasonal, or annual totals will approximate a normal distribution. In drier areas a skew distribution such as the log-Pearson, log-normal, gamma, square-root, or cube-root normal may give a better fit. Various comparisons [29–31] have been made of different distributions, but the results are generally inconclusive. Maximum annual water equivalent of snow on the ground is best fitted with a log-normal distribution [32].

13-14 Generalization of Rainfall-Frequency Data

Precipitation records are often too short to permit reliable frequency analysis except for relatively short return periods, say 10 years or less. In such cases reliable rainfall-frequency values for longer return periods can be obtained by applying ratios of rainfall magnitudes for different return periods, say 100- to 10-year rainfall, computed from long-record stations with precipitation regimes similar to those of the short-record stations for which the data are required. The need for such generalization is greatest for short-duration rainfall since recording-gage records are generally much shorter than those of nonrecording gages. Also, recording-gage networks are relatively sparse, and there is a greater need for using records from as many gages as possible.

Several empirical relationships have been developed [33–36] for estimating rainfall-frequency values when precipitation data are generally inadequate for frequency analysis. Rainfall-frequency values for stations with adequate data are correlated with such readily available climatological data as mean annual precipitation, mean annual number of rainy days, and mean annual number of thunderstorm days. These relationships, which are usually developed and presented in graphic form, are then used to estimate rainfall-frequency values for localities with inadequate data. Such relations have been used to estimate rainfall-frequency values for durations from 20 min to 24 hr for return periods from 2 to 100 years.

13-15 Adjustment of Fixed-Interval Precipitation Amounts

Analysis of short-duration rainfall is ordinarily performed on data published as clock-hour or observational-day amounts. Since it is unlikely that an intense rainfall will occur entirely within an interval between fixed observation times, data of this type give an underestimate of true maximum amounts for the specified durations. Rainfall-frequency data based on hourly or daily amounts should be increased by 13 percent to approximate true values for 60 min or 24 hr

Table 13-10 Ratio of true to fixed-interval maximum precipitation

Number of observational increments	1	2	3–4	5–8	9–24
Ratio	1.13	1.04	1.03	1.02	1.01

Source: J. F. Miller. "Two- to Ten-day Precipitation for Return Periods of 2 to 100 Years in the Contiguous United States," *U.S. Natl. Wea. Serv. Tech. Rep.* 49, p. 2, 1964.

(Table 13-10). The adjustment decreases as the number of observations on which the maximum value is based increases. For example, 24-hr maximum amounts based on a maximum of four consecutive 6-hr rainfall increments require, on the average, an increase of about 3 percent. The same adjustment would apply to 96-hr rainfall-frequency values computed from observational-day data.

13-16 Rainfall-Frequency Maps

Except in mountainous terrain, rainfall intensity-frequency variations over short distances may be expected to be small. Thus, it is practical to prepare maps of rainfall amounts for various frequencies and durations. Figure 13-8 is one of a series [36] of such maps which cover durations from 30 min to 24 hr and return periods from 1 to 100 years. Similar maps are available [34] also for Alaska, Hawaii, and the West Indies and for durations up to 10 days. These maps provide a good basis for estimating intensity-frequency values except in mountainous areas, where the scale is too small for reliable interpolation. Larger-scale frequency maps of 6- and 24-hr precipitation have been prepared [37] on a state-by-state basis for the mountainous western United States. These maps are based on both observed precipitation and on precipitation estimated from multiple correlations with slope, elevation, distance to moisture source, and distance and elevation of intervening barriers.

All the above precipitation-frequency maps depict point-precipitation values only. These values are often assumed to apply up to 10 mi^2 (25 km^2). Areal average precipitation must be less than the maximum point value in the area. Figure 13-9 indicates the average reduction of point precipitation of various durations for size of area. It should be noted, however, that depth-area-duration relations of the type depicted vary with storm type and intensity and may exhibit regional differences [38].

13-17 Design Storm

Because of difficulties in estimating flood frequency for ungaged catchments or where records are short, the concept of the design storm has developed. A rainfall time-intensity pattern is selected and the runoff hydrograph or peak flow is calculated by techniques discussed in Chaps. 7 and 8 or, in many cases, by an empirical formula. The technique is sometimes extended to large areas, and an

Figure 13-8 One-hour rainfall (inches) to be expected on the average once in 10 years. (Map scale precludes reliable depiction for mountainous region west of the 105th meridian. Detailed maps for that region are published in reference [37]). (*U.S. National Weather Service.*)

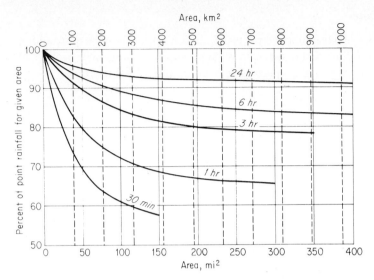

Figure 13-9 Reduction of point rainfall-frequency values for application to drainage areas up to 400 mi² (1040 km²). (*U.S. National Weather Service.*)

areal pattern of rainfall is specified for the design storm. It is commonly assumed that the probability of the derived flood flow is the same as that of the design rainfall. This assumption is rarely right and often grossly in error. A storm includes a time-intensity pattern, an areal pattern, and the total rainfall. It is impossible to assign a frequency to such a complex event. Usually only the total rainfall is considered. Since the time-intensity and areal patterns affect runoff volume and flood peak, storms with the same rainfall volume seldom produce the same flood peak. In addition, a storm occurs in a sequence of events which fix antecedent conditions and, in turn, runoff volume and hydrograph shape.

A single design storm, even if the frequency is known accurately, is inadequate for the economic analysis which should be made for flood mitigation, storm drainage, culvert design, etc. The preferred procedure is to synthesize the longest possible flood series and derive a frequency relation from the synthetic data. The design-storm approach is seldom warranted except when there has been a policy decision that the structure should be designed for the probable maximum event (Sec. 13-19).

PROBABILITY OF OTHER METEOROLOGICAL PHENOMENA

The hydrologist sometimes finds it necessary to consider the probability of occurrence of meteorological events other than precipitation. The appropriate frequency distributions for various meteorological phenomena are discussed briefly below.

13-18 Distributions

Just as in the case of streamflow and precipitation, there is no well-founded theoretical basis for choosing distributions representing various meteorological phenomena [39]. Decisions are made largely on empirical evidence.

Wind speed. No satisfactory distribution has been identified for the probability distribution of instantaneous, hourly, or daily mean values of wind speed. Such data are often bimodal with maxima at or near zero and at some higher value. The extreme-value Type 1 distribution usually fits the annual maximum wind speeds.

Wind direction. Wind direction data commonly show two (or more) maxima depending on the specific location of the station with respect to the general circulation and on local effects which may influence wind direction, e.g., land- and sea-breeze phenomena. Wind direction distributions are commonly represented graphically as a wind rose, showing the percent of time the wind blows from each of 8 or 16 compass points. No formal distribution is applicable in most cases.

Maximum and minimum temperature. Annual summer maxima and winter minima conform reasonably well to the normal distribution. Summer maxima sometimes show positive skew and winter minima may show negative skew.

Number of occurrences per month or year. Data such as the number of days with thunderstorms, frost, or rain in excess of some fixed amount usually fit the Poisson distribution when the probability of the event is relatively small. If the probability of occurrence is between about 0.2 and 0.8, the binomial distribution [Eq. (13-16)] is more appropriate. An example of the latter case might be the number of days per year with rain in a humid area.

Amount of clouds. The frequency distribution of cloud amounts is U-shaped since values of zero cloudiness and total cloudiness are more frequent than intermediate values. There is no suitable formal distribution.

Atmospheric pressure. The distribution of atmospheric pressure by seasons is nearly normal, sometimes with a slight negative skew.

Humidity. Since saturation vapor pressure increases logarithmically with temperature, its distribution is positively skewed. The distribution of relative humidity depends on the local climate, season, and time of day. The sum of two normal distributions with different means and fitted by inspection will sometimes lead to a reasonable fit.

PROBABLE MAXIMUM EVENTS

In some situations where substantial risk of loss of life exists, it may be appropriate to design a facility against what appears to be the worst possible conditions. No probability can be realistically attached to such an event, and no meaningful economic evaluation is feasible. It must simply rest on a policy decision that maximum protection be provided. The probable maximum flood is accepted as the standard for design of spillways on dams where failure could lead to catastrophic loss of life. It might be appropriate also to locate water-treatment plants, wastewater-treatment facilities, and other essential public utilities above the level of the probable maximum flood (PMF). The PMF is large and almost always beyond the possibility of control with conventional flood-mitigation works. If it were to occur, flooding would be extensive and damage would be heavy. Consideration of the PMF in design serves only to eliminate the possibility of the addition of a sudden dam failure or the loss of a potable-water supply to already serious flood conditions.

13-19 Hydrometeorological Studies

Determination of the PMF begins with the determination of the probable maximum precipitation (PMP). There is a reasonable basis on which to analyze the basic factors of major floods, i.e., storm rainfall and snowmelt: maximize them to their upper physical limits consistent with accepted meteorological knowledge; and then reassemble them into more critical but meteorologically and hydrologically acceptable combinations or chronological sequences.

PMP is usually derived [40–42] by (1) taking the results of depth-area-duration analyses (Sec. 3-12) of precipitation in major storms that have or could have occurred in the area of interest, (2) adjusting them for maximum moisture charge and rate of moisture inflow, and (3) enveloping the adjusted values for all storms to obtain the depth-area-duration curves of PMP. The use of storms outside the area of interest, called *storm transposition,* involves modification for differences in factors affecting rainfall, e.g., elevation, latitude, and distance from moisture source. Changes in shape and orientation of isohyetal patterns are also considered. Storm patterns are not transposed in mountainous regions because it is not possible to adjust rainfall accurately for orographic influences.

Storm models have also been used [41–44] to estimate PMP. A model using wind and moisture as parameters is hypothesized, tested against major observed storms, and then used to estimate PMP by introducing into the model maximum values of the parameters.

PMP can also be estimated by statistical procedures. In one approach [41, 42, 45, 46], PMP for a point and given duration is expressed as the mean of the annual series plus K standard deviations, as in Eq. (13-4). The empirically derived coefficient K varies with rainfall duration and inversely with the mean of the series and ranges between 5 and 30. Various adjustments are made to the mean and standard deviation for outliers and length of record, and depth-area curves similar to those of Fig. 13-9 are used to adjust the estimated point PMP values for size of area. The approach should not be interpreted to imply that a specific probability can be assigned to the PMP.

In many areas snowmelt is an important, and sometimes predominant, factor in major floods. In such cases the PMF requires determination [40, 47, 48] of optimum snow cover, maximum melting rates, and the PMP consistent with the optimum snowmelt conditions. This requires that the seasonal variation of the PMP be determined so that the magnitudes of possible floods, with and without snowmelt, can be estimated for various times of the year and compared.

Generalized charts of PMP have been prepared for the United States and a few other places. Figure 13-10 shows enveloping values of 6-hr PMP for 1000 mi^2 (2600 km^2) for central and eastern United States [49]. Estimates for areas from 10 to 20,000 mi^2 (26 to 52,000 km^2) and for durations from 6 to 72 hr can be obtained from similar charts. These PMP estimates have been compared [50] with the greatest observed storm rainfalls in the area shown in Fig. 13-10, and the ratios were found to be generally between 1.2 and 2.0. Ratios for 6- and 24-hr 10-mi^2 (26 km^2) PMP versus the greatest observed rainfalls (for the same durations and area) west of the Continental Divide were also about 1.2 to 2.0. Comparison of 6- and 24-hr amounts over 10 mi^2 with corresponding 100-year rainfalls showed ratios averaging between 4 and 6 for most of the United States but with some values as low as 2 and as high as 8 west of the Divide. Some of the extreme ratios may result from uncertainty as to the magnitude of the 100-year rainfalls in the desert areas of the west.

13-20 Probable Maximum Flood

The PMF must be estimated by transforming the PMP using hydrologic methods. This conversion may be accomplished with rainfall-runoff relations and unit hydrographs (Sec. 16-9), or simulation techniques can be used. The PMF is normally several times larger than the largest flood of record on a catchment. In the process of extrapolating from events of common occurrence, it is important that the estimate be neither too low nor too high. The temptation is to assume a very low loss rate. In addition it is not uncommon to assume that the unit hydrograph peak be increased by 25 to 50 percent to allow for the faster movement of the flood through the stream system. Such assumptions should be carefully considered. An unusually low loss rate implies that the basin is either largely impermeable or that it would experience heavy rains antecedent to the PMP. In high rainfall areas with a distinct rainfall season it is common for high soil moistures to persist through much of the wet season, but in arid regions most storms occur on relatively dry ground.

The question of adjustment of the unit hydrograph is basically an assumption that the unit hydrograph is nonlinear. Since we have few records of extraordinarily large floods, empirical data on which to judge the unit hydrograph

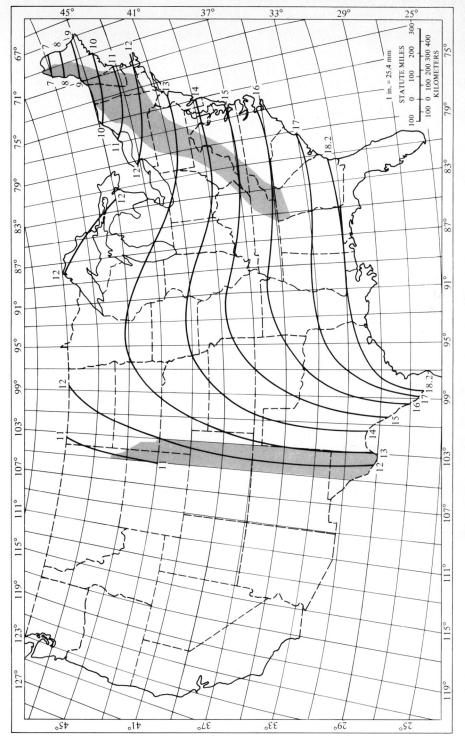

Figure 13-10 Probable maximum 6-hr precipitation for 1000 mi² (2600 km²). Stippling marks areas of reduced reliability because of complex orographic effects. (*U.S. National Weather Service.*)

383

linearity concept are not conclusive. However, judgment suggests that any non-linearity is related to the characteristics of the channel system. Channels with large floodplains may result in unit hydrographs with lesser peaks from larger floods. Streams in pronounced canyons may experience velocity increases in major floods. Such velocity increases are likely to cause relatively higher peaks unless the synchronization of peaks from various tributaries is altered. Use of kinematic routing to derive the outflow hydrograph (Sec. 9-10) could be more reliable than the assumption of an arbitrary change in a unit hydrograph. Simulation with a carefully calibrated model is likely to be more reliable than the arbitrary selection of a loss rate.

PROBLEMS

13-1 Using the data given in Table 13-9 and the frequency curve of Fig. 13-1 construct a partial-duration frequency curve. Plot the frequency curve of Fig. 13-6 on the same chart for comparative purposes. Are the two curves of the same functional form? Are the differences between the two curves significant in view of possible sampling errors?

13-2 Assuming the curve of Fig. 13-2 to represent the true average return period, for what discharge would you design to provide 50 percent assurance that failure would not occur within the next 20 years?

13-3 Using the data in Table 13-6, compute the frequency distributions for the five 10-year periods beginning with 1911, 1921, 1931, 1941, and 1951. Plot the five curves so derived on a single sheet of extreme-probability paper for comparative purposes. What is the extreme error at the 50-year return period, assuming the curve of Fig. 13-2 to yield the true average return period?

13-4 Obtain annual and partial-duration flood data for a selected gaging station and plot the two series on extreme-value paper and semilogarithmic paper, respectively. Fit a curve to each by eye, and also compute the frequency curve for annual data. How does the curve fitted by eye compare with the Gumbel curve? What is the percentage error at return periods of 10 and 100 years? How do the annual and partial-duration return periods compare relative to those given in Table 13-9?

13-5 Obtain excessive-precipitation data for a first-order weather station and derive intensity-frequency curves for durations of 5, 15, 30, 60, and 120 min. Compare your results for the 60-min duration with Fig. 13-8.

13-6 Compile an annual series of minimum flow for a selected gaging station and plot on extreme-value paper. Fit curves to the plotted data by eye and then by calculation. Do the data appear to follow the theory of extreme values? Does plotting log q help?

13-7 Obtain a report giving estimates of PMP for your area. Construct depth-area-duration curves from the data given. How do these amounts compare with observed rainfall? Estimate the PMF for some basin in your area. What is the ratio of the PMF peak to the maximum observed peak on this basin?

REFERENCES

1. James, L. D., and R. R. Lee: "Economics of Water Resources Planning," McGraw-Hill, New York, 1971.
2. Benson, M. A.: Characteristics of Frequency Curves Based on a Theoretical 1000-Year Record, *U.S. Geol. Surv. Water-Supply Pap.* 1543-A, pp. 51–74, 1952.

3. Glos, E., and R. Krause: Estimating the Accuracy of Statistical Flood Values by Means of Long-Term Discharge Records and Historical Data, *Int. Assoc. Sci. Hydrol. Publ.* 84, pp. 144–151, 1967.
4. Melentijevich, M.: Estimation of Flood Flows using Mathematical Statistics, *Int. Assoc. Sci. Hydrol. Publ.* 84, pp. 164–172, 1967.
5. Ott, R. F.: Streamflow Frequency Using Stochastically Generated Hourly Rainfall, *Stanford Univ. Dept. Civ. Eng. Tech Rep.* 151, December 1971.
6. Nasseri, I.: Regional Flow Frequency Analysis Using Multi-Station Stochastic and Deterministic Models, *Stanford Univ. Dept. Civ. Eng. Tech. Rep.* 210, 1976.
7. Benson, M. A.: Plotting Positions and Economics of Engineering Planning, *J. Hydraul. Div. ASCE,* vol. 88, pp. 57–71, November 1962.
8. Benson, M. A.: Evolution of Methods for Evaluating the Occurrence of Floods, *U.S. Geol. Surv. Water-Supply Pap.* 1580-A, 1962.
9. Weibull, W.: A Statistical Theory of the Strength of Materials, *Ing. Vetenskapsakad. Handl. (Stockh.),* vol. 151, p. 15, 1939.
10. Flood Studies Report, vol. 1, p. 1.3.2, Natural Environmental Research Council, London, 1975.
11. Gringorten, I. I.: A Plotting Rule for Extreme Probability Paper, *J. Geophys. Res.,* vol. 68, pp. 813–814, 1963.
12. Cunnane, C.: Unbiased Plotting Positions—A Review, *J. Hydrol.* vol. 37, pp. 205–222, 1978.
13. Thomas, H. A., Jr.: Frequency of Minor Floods, *J. Boston Soc. Civ. Eng.,* vol. 35, pp. 425–442, 1948.
14. Methods of Flow Frequency Analysis, *Subcomm. Hydrol., Inter-Agency Comm. Water Resour., Washington, Notes Hydrol. Act. Bull.* 13, April 1966.
15. Réméniéras, G.: Statistical Methods of Flood Frequency Analysis, in "Assessment of the Magnitude and Frequency of Flood Flows," *Water Resour. Ser.* 30, pp. 50–108, United Nations/World Meteorological Organization, New York, 1967.
16. Benson, M. A.: Uniform Flood-Frequency Estimating Methods for Federal Agencies, *Water Resour. Res.,* vol. 4, pp. 891–898, October 1968.
17. Guidelines for Determining Flood Flow Frequency, *Bull.* 17, Hydrology Committee, Water Resources Council, Washington, rev. June 1977.
18. Chow, V. T.: A General Formula for Hydrologic Frequency Analysis, *Trans. Am. Geophys. Union,* vol. 32, pp. 231–237, April, 1951.
19. Pearson, K.: "Tables for Statisticians and Biometricians," 3d ed., Cambridge University Press, London, 1930.
20. Fisher, R. A., and L. H. C. Tippett: Limiting Forms of the Frequency Distributions of the Smallest and Largest Member of a Sample, *Proc. Camb. Phil. Soc.,* vol. 24, pp. 180–190, 1928.
21. Gumbel, E. J.: "Statistics of Extremes," Columbia University Press, New York, 1958.
22. Dalrymple, T.: Regional Flood Frequency, *High. Res. Board Res. Rep.* 11-B, pp. 4–20, 1950. The U.S. Geological Survey has summarized flood data and presented regional frequency methods for the United States in *Water-Supply Pap.* 1671–1689, 1964.
23. Paulhus, J. L. H., and J. F. Miller: Flood Frequencies Derived from Rainfall Data, *J. Hydraul. Div. ASCE,* vol. 83, pap. 1451, December 1957.
24. Valencia, D., and John C. Schaake: A Disaggregation Model for Time Series Analysis and Synthesis, *Mass. Inst. Tech., Ralph M. Parsons Lab. Rep.* 149, June 1972.
25. Franz, D. D.: Hourly Rainfall Synthesis for a Network of Stations, *Stanford Univ. Dept. Civ. Eng. Tech. Rep.* 126, 1970.
26. Pattison, A.: Synthesis of Hourly Rainfall Data, *Water Resour. Res.,* vol. 1, pp. 489–498, 1965.
27. Langbein, W. B.: Annual Floods and Partial Duration Series, *Trans. Am. Geophys. Union,* vol. 30, pp. 879–881, December 1949.
28. Stall, J. B., and J. C. Neill: Partial Duration Series for Low-Flow Analysis, *J. Geophys. Res.,* vol. 66, pp. 4219–4225, December 1961.

29. Hershfield, D. M.: An Empirical Comparison of the Predictive Value of Three Extreme-Value Procedures, *J. Geophys. Res.,* vol. 67, pp. 1535–1542, April 1962.
30. Huff, F. A., and J. C. Neill: Comparison of Several Methods for Rainfall Frequency Analysis, *J. Geophys. Res.,* vol. 64, pp. 541–548, May 1959.
31. Markovic, R. D.: Probability Functions of Best Fit to Distributions of Annual Precipitation and Runoff, *Colo. State Univ. Hydrol. Pap.* 8, August 1965.
32. Thom, H. C. S.: Distribution of Maximum Annual Water Equivalent of Snow on the Ground, *Mon. Weather Rev,* vol. 94, pp. 265–271, April 1966.
33. Hershfield, D. M., L. L. Weiss, and W. T. Wilson: Synthesis of Rainfall Intensity-Frequency Regimes, *Proc. ASCE,* vol. 81, sep. no. 744, July 1955.
34. Hershfield, D. M.: Rainfall Frequency Atlas for the United States, *U.S. Weather Bur. Tech. Pap.* 40, May 1961.
35. Bell, F. C.: Generalized Rainfall-Duration-Frequency Relationships, *J. Hydraul. Div. ASCE,* vol. 95, pp. 311–327, January 1969.
36. See United States Data Sources at end of Chap. 3.
37. Miller, J. F., R. H. Frederick, and R. J. Tracey: Precipitation-Frequency Atlas for the Conterminous Western United States (by States), *NOAA Atlas,* 2, 11 vols., 1973.
38. Myers, V. A., and R. M. Zehr: A Methodology for Point-to-Area Rainfall Frequency Ratios, *NOAA Tech. Rep. NWS,* 24, February 1980.
39. Thom, H. C. S.: Some Methods of Climatological Analysis, *World Meteorol. Org., Tech. Note* 81, 1966.
40. World Meteorological Organization: Estimation of Maximum Floods, WMO no. 233, *Tech Pap.* 126, *Tech. Note* 98, pp. 1–116, Geneva, 1969.
41. World Meteorological Organization: Manual for Estimation of Probable Maximum Precipitation, *Opera. Hydrol. Rep.* 1, WMO no. 332, Geneva, 1973.
42. Wiesner, C. J.: "Hydrometeorology," pp. 147–225, Chapman & Hall, London, 1970.
43. Generalized Estimates of Probable Maximum Precipitation and Rainfall: Frequency Data for Puerto Rico and Virgin Islands, *U.S. Weather Bur. Tech. Pap.* 42, 1961.
44. Interim Report: Probable Maximum Precipitation in California, *U.S. Weather Bur. Hydrometeorol. Rep.* 36, October 1961.
45. Hershfield, D. M.: Method for Estimating Probable Maximum Rainfall, *J. Am. Waterw. Assoc.,* vol. 57, pp. 965–972, August 1965.
46. McKay, G. A.: Statistical Estimates of Precipitation Extremes for the Prairie Provinces, PFRA Engineering Branch, Canada Dept. of Agriculture, April 1965.
47. Bruce, J. P., and R. H. Clark: "Introduction to Hydrometeorology," pp. 235–239, Pergamon, New York, 1966 (rev. 1980).
48. Runoff from Snowmelt, Engineering and Design Manual EM 1110-2-1406, U.S. Corps of Engineers, Washington, January 1960.
49. Probable Maximum Precipitation Estimates, United States East of the 105th Meridian, U.S. National Weather Service, *Hydrometeorol. Rep.* 51, June 1978.
50. Riedel, J. T., and L. C. Schreiner: Comparison of Generalized Estimates of Probable Maximum Precipitation with Greatest Observed Rainfalls, *NOAA Tech. Rep. NWS* 25, March 1980.

BIBLIOGRAPHY

Alekseev, G. A.: Objective Statistical Methods of Computation and Generalization of the Parameters of Maximum Rainfall Runoff, *Proc. Symp. Floods and Their Computation,* IASH-UNESCO-WMO, 1969.
Chow, V. T.: Frequency Analysis, sec. 8-I in V. T. Chow (ed.): "Handbook of Hydrology," McGraw-Hill, New York, 1964.
Gilman, C. S.: Rainfall, sec. 9 in V. T. Chow (ed.): "Handbook of Hydrology," McGraw-Hill, New York, 1964.

Gringorten, I. I.: Fitting Meteorological Extremes by Various Distributions, *Q. J. R. Meteorol. Soc.,* vol. 88, pp. 170–176, 1962.

Gumbel, E. J.: On the Plotting of Flood Discharge, *Trans. Am. Geophys. Union,* pt. II, pp. 699–716, 1943.

Hazen, A.: "Flood Flows," Wiley, New York, 1930.

Jenkinson, A. F.: The Frequency Distribution of the Annual Maximum (or Minimum) Values of Meteorological Events, *Q. J. R. Meteorol. Soc.,* vol. 81, pp. 158–171, 1955.

Kimball, B. F.: On the Choice of Plotting Positions on Probability Paper, *J. Am. Stat. Assoc.,* vol. 55, pp. 546–560, 1960.

Linsley, R. K., and J. B. Franzini: "Water Resources Engineering," 3d ed., McGraw-Hill, New York, 1979.

National Bureau of Standards: Probability Tables for the Analysis of Extreme-Value Data, *Appl. Math. Ser.,* no. 22, 1953.

Viessman, W. Jr., J. W. Knapp, G. L. Lewis, and T. E. Harbaugh: "Introduction to Hydrology," 2d ed., pp. 157–229, IEP, New York, 1977.

World Meteorological Organization: Guide to Hydrological Practices, 3d ed., WMO no. 168, Geneva, 1974.

UNITED STATES DATA SOURCES

Generalized maps of point-precipitation–frequency data for durations up to 10 days and return periods up to 100 years are available for the entire United States and a few foreign areas in various issues of the *U.S. Weather Bureau Technical Paper* and *NOAA Atlas* series. Area-reduction curves are provided for reducing point values to areas up to 400 mi² (1000 km²). Estimates of probable maximum precipitation for various catchments in the United States and in a few foreign countries are available in the U.S. National Weather Service's *Hydrometeorological Report* series. Generalized estimates for the entire United States can be obtained from the *Hydrometeorological Report* series and various issues of the *U.S. Weather Bureau Technical Paper* series. See also Chaps. 3 and 4 for precipitation and streamflow data.

FOURTEEN

STOCHASTIC HYDROLOGY†

"The science of conjecture, or the stochastic science, is defined as the art of estimating as best one can, the probability of things so that in our judging and acting we may choose to follow the best, the safest, the surest, or the most soul-searching way [1]."

In statistics the word stochastic is synonymous with random, but in hydrology it has been used in a special way to refer to a time series which is partially random. Stochastic hydrology fills the gap between deterministic methods (Chaps. 8 and 12) and probabilistic hydrology (Chap. 13). In deterministic hydrology the time variability is assumed to be explained totally by other variables as processed through an appropriate model. In probabilistic hydrology we are not concerned with time sequence but only with the probability, or chance, that an event will be equaled or exceeded. In stochastic hydrology, the time sequence is all-important.

A simple example of a stochastic process is that of drawing colored balls from an urn. The important feature is the actual order in which the balls are withdrawn from the urn. Valuable information is contained in the sequence of withdrawals: red, red, black, green, black, white, green, green, etc. Average probability statements, by contrast, are concerned only with the relative numbers of different colored balls withdrawn. Stochastic representation preserves chance properties associated with sequencing of events.

† The original draft of this chapter was written by Professor Stephen J. Burges, University of Washington.

A typical hydrologic time series is the quantitative description of stream-flow or precipitation history at a point. There is a finite amount of information contained in any hydrologic time series; the most complete description is the continuous observed time record. The same record can be described in terms of mechanisms (mathematical relationships) with various degrees of precision. It is possible to generate (by mathematical functions) time series that differ from the observed time series but retain many of its properties. Each generated sequence is so constructed that events have the same probability of occurrence as in the observed sequence.

Stochastic hydrology is meaningful only in a design or operational decision-making sense. The designer usually wishes to see how the particular facility will perform for representative future hydrologic inputs. Designers do not know what future flows or precipitation events will occur, but they can assume that these events will have the same stochastic properties as the observed historical record. At present it is not possible to evaluate or project climatic change from available hydrologic and meteorological records [2]. Most analysts, therefore, assume the future will be statistically similar to the past. This assumption forms the principal basis of stochastic hydrology, namely, generation of equiprobable input sequences, each sequence having similar statistical properties. Each input sequence yields a sequence of outputs from the system under investigation. A stochastic analysis using many input sequences thus yields a probability distribution of system response.

Stochastic methods were first introduced into hydrology to cope with the problem of reservoir design. The required capacity of a reservoir depends on the sequence of flows, especially a sequence of low flows. If a reservoir operates on an annual cycle, i.e., it fills and is partially or wholly emptied each year, it may be possible to assess its *reliability,* the probability that it will deliver its intended yield each year, on the basis of an analysis of the historic flow record if this record is sufficiently long. If the reservoir operates with carryover storage, i.e., it provides storage to meet required withdrawals over a period of several dry years, the historic record is unlikely to provide adequate information on reliability. Records are generally too short to define the probability of a series of subnormal years. Stochastic methods provide a means of estimating the probability of sequences of dry years during any specified period of future time. Even if the historic record suggests that a reservoir will operate on an annual cycle, the possibility of two or more dry years in sequence exists and thus stochastic analysis should be part of the hydrologic study for all reservoirs dependent on natural stream inflows.

Stochastic methods in combination with deterministic methods also offer the prospect of improving estimates of flood frequency (Chap. 13) since this task also depends on long records for reliable estimates.

Attempts to solve the problem of short record length by various statistical means were probably first initiated by Hazen [3], who suggested combining records from several stations into a single long record. Sudler [4] wrote historic flows on cards and dealt the cards randomly without replacement from the deck to construct a synthetic record of 1000 years. This procedure produces a variety of sequences of historic flows for storage analysis. The advent of the computer has made more sophisticated techniques [5] possible, and these have superseded the early (and erroneous) procedures.

14-1 The Stochastic Generator—Single-Site Applications

Basic to stochastic analysis is the assumption that the process is stationary, i.e., that the statistical properties of the process do not vary with time. Thus the properties of the historic record can be used to derive long synthetic sequences. These sequences must display statistical properties similar to those of the historic sequence.

Some properties of hydrologic time series can be investigated in a time domain through correlogram analysis. In some situations it is convenient to work in a frequency domain, using spectral analysis to identify the principal harmonics contained in the series; but the short length of hydrologic time series limits the usefulness of these methods. Correlogram and spectral analysis enable some deterministic trends (usually climate) to be revealed [6]. When "trends" have been detected they should be subtracted from the original series, and the residual series examined. Of usual interest are the probability distributions of the elements of the residual series. For example, if a monthly time scale is the basic unit used in analysis, the probability distributions of the flow volumes (or residual flow volumes) for each month are of interest.

Basically a time series can be modeled mathematically as the combination of a deterministic and residual random component. A stochastic generation equation can be very simple, preserving only the mean, variance, and lag-1 serial correlation. More sophisticated generators [7, 8] attempt to preserve low-frequency (as well as high-frequency) fluctuations in the time series; the simpler generators [9] are restricted to preserving high-frequency fluctuations.

Although many properties are needed to describe a historic sequence fully, stochastic analysis need include only those important for the physical system under study. In fact, this is paramount in any type of mathematical simulation, and it reflects the important coupling of inputs, system demands, and system operation. It is therefore important to identify the most suitable generation scheme for the problem at hand.

The same basic approach to stochastic generation applies to streamflow, precipitation, evapotranspiration, and other hydrometeorologic factors. There are two basic approaches. The first is an aggregation method where the model is built for a monthly or seasonal time increment and annual values result from summing the increments. The second approach is a disaggregation method in which data are generated (e.g., on an annual scale) and disaggregated to smaller time units. At present the most satisfactory stochastic models are for monthly and yearly time increments. Models using daily or weekly increments have been given little attention and pose some special problems. Incorporation of catchment physics into these models rather than use of purely stochastic methods is likely to be most satisfactory.

The properties that a flow model is to replicate dictate its form. If one examines the series of annual flow data, there may be periods when high flows follow high flows and low flows follow low flows; periods when the first part of the record seems wetter than the second; and other such features. These phenomena require that some sort of persistence be built into the model. Persistence structures range from high frequency to low frequency (long periods of low or high flow) to mixtures of both. In addition to persistence it is necessary to model the marginal probability density function from which each annual observation might have been drawn. Usually these data are positively skewed. In some situations when the annual *coefficient of variation* (standard deviation divided by the mean) is less than about 0.5, the data are frequently close to normally distributed.

With record lengths usually available it is virtually impossible to identify any true long-term persistence. A lumped measure of aperiodic low frequencies is the Hurst coefficient [10–13], which can be computed when records are moderately long, say about 50 years. While studying long-term storage requirements on the Nile River, Hurst found that

$$R_n = \sigma_n \left(\frac{n}{2}\right)^h \tag{14-1}$$

where σ_n is the standard deviation, n the length of the series, and the range R_n is the difference between the greatest cumulative excess above the mean and the greatest cumulative deficiency below the mean flow. The exponent h is the *Hurst coefficient.* For a Markov† process $h = 0.5$. Hurst tested some 800 time series (streamflow, varves, tree rings, precipitation, and temperature) and found $0.5 < h < 1$, with a mean value of 0.73 and a standard deviation of 0.09. One implication of $h > 0.5$ is that there is long-term persistence in natural time series and that the Markov process is not a valid model. Assuming the hydrologic regime to be quasi-stationary the range of persistence falls between a simple autoregressive structure [6] and one that preserves the Hurst phenomenon.

There is much controversy concerning the Hurst phenomenon, and it is questionable if the calculated values of h have any meaning. It is useful to think of h as an approximate index to drought phenomena. Annual models that can replicate h have the capability of replicating historically observed droughts as well as other events of greater and lesser severity. No causality for h has been demonstrated. By using h as a model parameter, the median value of the synthetic droughts can be adjusted to approximate the historic droughts. We do not know if the Hurst phenomenon exists, but if it does, the impact on hydrologic design would be large [14–16].

---

† A *Markov process* is one in which any event is dependent only on the preceding event.

Annual autoregressive model. The simplest annual autoregressive model is an AR(1)† or lag-1 Markov model. This model regresses flow in time period i against flow in period $i - 1$. Using Q to represent annual flow volume [5]:

$$Q_i = \bar{Q} + \rho(Q_{i-1} - \bar{Q}) + t_i\sigma\sqrt{1 - \rho^2} \tag{14-2}$$

where t is a random variable from an appropriate distribution (Sec. 14-2) with a mean of zero and unit variance, σ is the standard deviation of Q, ρ is the lag-1 serial correlation coefficient, and $\bar{Q}$ is the mean of Q. When the model parameters $\bar{Q}$, σ, and ρ have been determined from the data, sequences of length n can be generated using a simple digital computer algorithm. The Q_i are obtained by Monte Carlo sampling from the probability distribution t. The computation can be made by hand using a table of random numbers, but the process would be time-consuming. Because the model contains no structure for generating very long period synthetic flows, n should generally be less than 100.

Numerous investigators have found that generation using the normal distribution is the most effective method. This is particularly true when generating flows at multiple sites where the properties of the multivariate normal distribution are used to advantage. Transformations of data are discussed in Sec. 14-5.

Equation (14-2) states that the flow in period i is the average value from a linear regression of Q_i on Q_{i-1} plus a random element to preserve the variance σ^2. Programming of Eq. (14-2) into convenient computer code is made simpler (particularly when using multiple sites) by making the Q_i into a zero mean, unit variance process X,

$$X_i = \frac{Q_i - \bar{Q}}{\sigma} \tag{14-3}$$

The generation algorithm becomes

$$X_i = \rho X_{i-1} + \epsilon_i \tag{14-4}$$

and

$$Q_i = \sigma X_i + \bar{Q} \tag{14-5}$$

where values of ϵ_i are drawn from the appropriate population having zero mean and variance $\sigma_\epsilon^2 = 1 - \rho^2$. Inspection of Eq. (14-4) indicates that the variance and lag-1 serial correlation of the process are preserved. This can be verified by showing that $E[X_iX_i] = 1.0$ and $E[X_iX_{i-1}] = \rho$, where $E[\cdot]$ is the mathematical expectation and $E[X_iX_i]$ is the variance of X. Since Eq. (14-4) is linear, σ^2 and $\bar{Q}$ are preserved.

Other annual models. Models capable of preserving the Hurst phenomenon have been described by Lawrance and Kottegoda [17] and Lettenmaier and

---

† The notation of Box and Jenkins [6] is used for convenience. An autoregressive process of order p is referred to as an AR(p) process. A moving average process of order q is referred to as an MA(q) process. A combined AR(p), MA(q) process is referred to as an autoregressive moving average process ARMA(p, q).

Burges [14]. The basic models are the broken line [8], autoregressive moving average [18], fractional gaussian noise (FGN) [7] and its computationally convenient form, fast fractional gaussian noise (FFGN) [19], and ARMA-Markov [14]. The ARMA-Markov has many appealing features, particularly computational economy and its ability to approximate FGN while preserving the lag-1 correlation coefficient. Its disadvantage is that it requires five parameters. Details on the models that preserve low-frequency effects should be obtained from the source references.

Some hydrologic modelers feel that the best modeling approach is to fit a Box-Jenkins [6, 20, 21] model to the annual time series. These modelers make extensive use of residuals diagnostic checks to select their models. It is possible to obtain physically implausible models if these procedures are followed blindly.

Seasonal flow modeling. The earliest seasonal flow modeling was done through use of month-to-month autoregressive models. This method has been satisfactory in some cases [9], but not in others [22]. It is usually satisfactory if the annual correlation is between 0 and 0.1 but is usually unsatisfactory if the correlation exceeds 0.1. The equation most commonly used is

$$Q_{i,j} = \bar{Q}_j + \rho_j \frac{\sigma_j}{\sigma_{j-1}} (Q_{i-1,j-1} - \bar{Q}_{j-1}) + t_{i,j}\sigma_j \sqrt{1 - \rho_j^2} \qquad (14\text{-}6)$$

where the subscript j refers to seasons or month. For monthly synthesis j varies from 1 to 12 throughout the year. The subscript i is a serial designation from year 1 to year n as in Eq. (14-2), and ρ_j is the serial correlation coefficient between Q_j and Q_{j-1}. Other symbols are as in Eq. (14-2). Equation (14-6) is used by first determining $\bar{Q}$, σ, and ρ for each month or season. A starting value of $Q_{i-1,j-1}$ is assumed (usually drawn at random from the marginal distribution). It is best to start at the beginning of the water-year.

The appeal of the zero-mean, unit variance approach is clear after examining Eq. (14-6). Let $Y_{i,j} = (Q_{i,j} - \bar{Q}_j)/\sigma_j$ be the residual flow in month j, year i (X_i was used for annual residuals). Then

$$Y_{i,j} = \rho_j Y_{i-1,j-1} + (1 - \rho^2)^{1/2} t_{i,j} \qquad (14\text{-}7)$$

and

$$Q_{i,j} = Y_{i,j}\sigma_j + \bar{Q}_j \qquad (14\text{-}8)$$

A more satisfactory approach to seasonal flow modeling is to disaggregate generated annual flows. This ensures that important annual characteristics are preserved (see Sec. 14-5). A single season generator is appropriate for large reservoirs where seasonal variation in flow cannot materially affect the storage requirement or in cases when the flow is highly seasonal, with the high-flow season preceding the high-use season. In other cases monthly or seasonal flows should be used.

14-2 The Random Variate

Unless the random variate t is drawn from an appropriate distribution, Eq. (14-2) or (14-6) cannot reproduce the historic distribution even though it may satisfactorily duplicate the mean and variance. If the historic distribution is normal, the problem is simple; one draws from a normal random-number generator. If streamflow volumes are distributed log-normally, generation is effected by using transformed variables that are normally distributed. Exponentiation yields log-normal sequences.

There is no a priori basis for choosing a distribution, and no single distribution can be applicable to all streams. In addition, the relatively short historic flow records usually available do not clearly define the properties (parameters) of the distribution. One therefore selects a distribution that fits the observed data within the bounds of acceptance criteria. This selection is influenced by the fact that certain distributions are amenable to generation techniques while others are extremely difficult to apply. For example, a three-parameter, log-normal distribution might fit the observed data as satisfactorily as a general

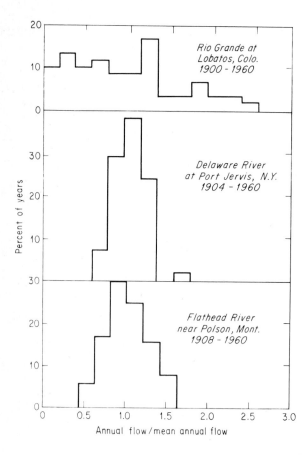

Figure 14-1 Distribution of annual streamflows for three different catchments.

gamma distribution. One would elect the former because it can be easily programmed. Probability distributions can be fitted using bayesian or classical
methods.

Figure 14-1 shows the distributions of annual flows for three streams representing widely varied climatic regimes. The data suggest distributions ranging
from nearly normal (Delaware and Flathead Rivers) to approximately exponential (Rio Grande). Figure 14-2 shows a similar comparison of distributions for
selected months at a single station. Distribution selection is most difficult for
streams where zero flows are frequent. There is little alternative to selecting the
distribution to which the historic data seem to conform. If there is serious doubt
as to the best distribution, it is wise to generate two (or more) synthetic sequences from different distributions to determine the effect on the project plan
resulting from differences in the sequences.

Statistical goodness-of-fit tests have exceedingly weak discrimination
power. Usually many distributions will "fit" the empirical cumulative distribution of the data under study. It is sensible to use the distribution of greatest
convenience. If there are more than about 30 observations (this should not be

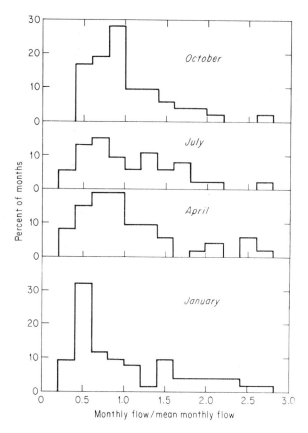

Figure 14-2 Comparison of selected streamflow distributions for the Flathead River at Polson, Mont. [7096 mi^2 (18,379
km^2); data from 1908–1960].

generalized into a rule of thumb!), it is productive to fit a three-parameter distribution to the data. Otherwise a two-parameter distribution is appropriate. If the low flow range is most important to the model application, the distribution parameters can be adjusted to achieve this [22]. A convenient technique for judging the goodness of fit in the region of the lower tail of the distribution is via quantile-quantile plots where theoretical distribution quantiles are plotted against empirical quantiles [23].

14-3 Definition of Parameters

Most probability distributions of importance in hydrology require two or more parameters to define them. These parameters can be directly related to such quantities as the arithmetic mean, variance, and skew of the observed data. At best, however, it is only possible to estimate the true value of the parameters, because the observed time series is only a short sample of the total time series. If the true arithmetic mean of the population is μ, we can only approximate μ by the sample mean $\bar{X}$. Thus for a normal distribution $N(\mu, \sigma_p^2)$ having population mean μ and population variance σ_p^2 we approximate the distribution by $N(\bar{X}, \sigma^2)$.

The most commonly used method for computation of parameters is the method of moments [24]. The method of maximum likelihood has been receiving increasing attention with the wide availability of digital computers. For many skewed distributions, maximum-likelihood parameter estimates are the best. For many stochastic modeling applications, a three-parameter log-normal distribution (LN3) is convenient for transforming data into the normal domain. Moment estimators are usually suitable for this general distribution, although in some applications a quantile method [25] is preferred.

For any probability distribution the arithmetic mean of n observations $\{X_i\}$ is approximated by

$$\bar{X} = \sum_{i=1}^{n} \frac{X_i}{n} \tag{14-9}$$

The standard deviation computed for a sample from any distribution is biased by both persistence and the method of computation [26]. For samples of size 30 from LN3 populations with skew coefficients between 1 and 2, the sample standard deviation is underestimated on the average by 3 to 5 percent when computed from

$$\sigma^2 = \frac{1}{n} \sum_{i=1}^{n} (X_i - \bar{X})^2 \tag{14-10}$$

Some of this bias could be removed by using $n - 1$ instead of n in the denominator, but some bias is also introduced by the persistence structure [22], i.e.,

$$E[\sigma^2] = \sigma_p^2 \left[1 - \frac{2}{n(n-1)} \sum_{k=1}^{n-1} (n-k)\rho_k \right] \tag{14-11}$$

where ρ_k is the population autocorrelation coefficient at lag k defined as $E[X_i X_{i-k}]/E[X_i X_i]$ and $E[\cdot]$ is the mathematical expectation. For a lag-1 Markov situation $\rho_k = \rho_1^k$ so that Eq. (14-11) becomes

$$E[\sigma^2] = \sigma_p^2 \left[1 - \frac{2\rho_1}{n(n-1)} \frac{n(1-\rho_1) - (1-\rho_1^n)}{(1-\rho_1)^2} \right] \qquad (14\text{-}12)$$

The estimation of ρ_1 contains bias and is corrected according to [27]

$$E[r_1] = \rho_1 - \frac{1}{n}(1 + 4\rho_1) \qquad (14\text{-}13)$$

with r_1, the sample estimate of ρ_1, defined as

$$r_1 = \frac{\sum_{j=1}^{n-1}(X_j - \overline{X})(X_{j+1} - \overline{X})}{\sigma^2} \qquad (14\text{-}14)$$

The calculated value of r should not be accepted without some thought. Anderson [28] suggested a test for the significance of the correlation coefficient which is displayed in Fig. 14-3. With short records, sample values of $r_1 \leq 0.3$ are not statistically different from zero. However, even a small correlation reduces the effect of the random term and increases the persistence term in the generating equation.

There is no known natural explanation of negative values of ρ. If negative values are statistically not distinguishable from zero, they should be set to zero. If negative correlations are thought to be significant, the analyst should immediately check the flow record and catchment for deterministic causes. If a cause is found, it must be incorporated into the generation model. A scattergram is helpful for examining correlation coefficients. A plot of Q_i versus Q_{i-1} may indicate (Fig. 14-4) that only a few chance flow sequences materially influence the correlation. For Yegua Creek, 26 years of data yield $r_1 = -0.29$, which is not significantly different from zero by Anderson's test. If point A is omitted, $r_1 = -0.10$, while with 35 years (omitting point A) $r_1 = +0.007$. Similarly, a

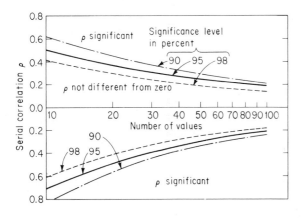

Figure 14-3 Significance of serial correlation coefficient as a function of sample size. (*Adapted from* [28].)

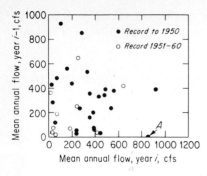

Figure 14-4 Serial plot of annual flows for Yegua Creek, near Somerville, Tex. (*Data from U.S. Geological Survey.*)

chance occurrence of two high years in sequence may lead to a large positive value of r_1. Test calculations can be made omitting outliers from the main body of data to define the probable values of ρ more accurately. Fairly high values of intermonthly correlation may often occur, but interannual correlations in excess of 0.6 are most unlikely, and if the last months of the year are poorly correlated, the interannual correlation must be near zero (under the assumption of a Markov relationship).

In practice ρ_1 is estimated from Eq. (14-13) using r_1 estimated from the data. This value of ρ_1 is used in Eq. (14-12) to estimate σ_p^2 if a lag-1 Markov model is to be used. Generation at the annual scale is accomplished using $\bar{X}$, and estimates of σ_p and ρ_1. Adequate procedures for correcting estimates of model parameters are not fully developed. The corrections are strongly influenced by n and ρ_k. The best approach is to use the longest possible record for deriving parameters including possibly the extension of the available record by deterministic methods.

14-4 Multisite Models—Annual

It is frequently necessary to generate synthetic flows at several sites. Models which preserve annual scale high-frequency phenomena are available [16], but only the multivariate broken line model [29] can deal with Hurst-type persistence. Multivariate models will be illustrated by the first-order autoregressive model introduced by Matalas [16] and extended to second order by Clarke [30].

Using zero-mean, unit-variance notation with variables at each of N sites that have been transformed to multivariate normal,†

$$\mathbf{X}_i = \mathbf{A}\mathbf{X}_{i-1} + \mathbf{B}e_i \qquad (14\text{-}15)$$

where $\mathbf{X}_i$ is a vector of flows in time period i (dimension $N \times 1$), $\mathbf{X}_{i-1}$ is a similar vector at time $i - 1$, $\mathbf{A}$ and $\mathbf{B}$ are $N \times N$ coefficient matrices, and e_i is an N-dimensional vector of random deviates drawn from a standard normal dis-

---

† Equation (14-15) is the multivariate extension of Eq. (14-4); hence scalars become vectors and coefficients become matrices. Because zero-mean, unit-variance relations are used, $E[\bar{\mathbf{X}}_i\bar{\mathbf{X}}_i^T] =$ covariance of X = lag-zero correlation matrix or the matrix of cross correlation between sites.

tribution $N(0,1)$. Postmultiplying Eq. (14-15) by X_{i-1}^T and taking expectations yields

$$A = C_1 C_0^{-1} \qquad (14\text{-}16)$$

where $C_1 = E[X_i X_{i-1}^T]$ and $C_0 = E[X_i X_i^T]$ are the lag-1 and lag-zero correlation matrices, respectively, and X_i^T is the transpose of X_i. Similar manipulations yield

$$BB^T = C_0 - C_1 C_0^{-1} C_1^T \qquad (14\text{-}17)$$

There are many possible solutions for B. One solution takes advantage of the fact that BB^T is symmetric, with a lower triangular matrix B that will satisfy Eq. (14-17)

$$B = \begin{matrix} b_{11} & 0 & \dots & 0 \\ b_{21} & b_{22} & \dots & 0 \\ . & . & \dots & . \\ b_{n1} & b_{n2} & \dots & b_{nn} \end{matrix} \qquad (14\text{-}18)$$

Carrying out the relevant operations (BB^T) indicates that b_{11}^2 is equal to element $(1,1)$ in the right-hand side of Eq. (14-17). The solution for the elements of B then proceeds sequentially by equating elements from the left- and right-hand sides of Eq. (14-17).

There is no guarantee that B has a solution. The assumed lag-1 Markov structure may be incorrect, and sampling errors in the observed data may cause errors in computed parameters which cause B to be singular or not positive semidefinite [31].

A vector of streamflow for each site at time i is obtained by scaling,

$$Q_i = X_i S^T + \overline{Q} \qquad (14\text{-}19)$$

where S is a vector of standard deviations and $\overline{Q}$ a vector of mean annual flows at the N sites. A time series of annual flow at each of the sites is obtained with Eqs. (14-15) and (14-19) for $i = 1, \dots, n$. If data have been transformed to make them normal, Q_i are not actual flows but must be inverse transformed to obtain flow in the original units.

14-5 Disaggregation Models

The monthly model of Eq. (14-6) requires 12 sets of parameters. If an attempt were made to convert it to multisite (Sec. 14-4), the matrices would be very large, and the likelihood of B becoming singular increased. A practicable alternative is to generate sequences of annual flows that preserve the properties of the annual data and to disaggregate the annual values to monthly or seasonal ones [32]. Lane [33] has developed a simplified disaggregation model which preserves the most important features of the correlations between seasonal flows at multiple sites. The disaggregation approach can be applied for yearly to monthly, monthly to daily, or even shorter intervals, provided the process is adequately represented in the model.

Lane generates flows at key stations using an appropriate multivariate annual model. These flows are the basis for estimates of annual flows at other stations of interest. Finally the annual flows are disaggregated to seasonal values. Only the annual to seasonal disaggregation is discussed here. Let the transformed flow in season j, year i, $M_{i,j}$, be defined by

$$\mathbf{M}_{i,j} = \mathbf{F}_j \mathbf{X}_i + \mathbf{H}_j \mathbf{R}_l + \mathbf{G}_j g_{i,j} \tag{14-20}$$

where

$$\mathbf{R}_l = \mathbf{M}_{i-1,k} \quad j = 1$$

$$\mathbf{R}_l = \mathbf{M}_{i,j-1} \quad j = 2, 3, \ldots, k$$

and $\mathbf{X}$ is a column vector of annual transformed standardized flows, $M_{i,j}$ is a vector of flows in season j at sites 1 to N in year i, $\mathbf{F}$, $\mathbf{H}$, and $\mathbf{G}$ are coefficient matrices of dimension $N \times N$, g is a $N \times 1$ vector of random numbers drawn from $N(0,1)$, and k is the number of seasons. Equation (14-20) yields a distinct model for each season.

Model parameters are estimated by postmultiplying Eq. (14-20) by $\mathbf{X}_i^T$ and taking expected values. The equation is then postmultiplied by $\mathbf{R}_l^T$ and expected values are taken again. $\mathbf{F}_j$ and $\mathbf{H}_j$ are obtained from these two equations:

$$\mathbf{F}_j = [\mathbf{C}_{MX}(j, 1) - \mathbf{C}_{MM}(j, j-1)\mathbf{C}_{MM}^{-1}(j-1, j-1)\mathbf{C}_{MX}(j-1, i)]$$
$$\times [\mathbf{C}_{XX}(i, i) - \mathbf{C}_{XM}(i, j-1)\mathbf{C}_{MM}^{-1}(j-1, j-1)\mathbf{C}_{MX}(j-1, i)]^{-1} \tag{14-21}$$

$$\mathbf{H}_j = [\mathbf{C}_{MM}(j, j-1) - \mathbf{F}_j \mathbf{C}_{XM}(i, j-1)]\mathbf{C}_{MM}^{-1}(j-1, j-1) \tag{14-22}$$

where $\mathbf{C}_{\alpha\beta}(\gamma, \delta)$ is a correlation matrix defined by $E[\alpha_\gamma \beta_\delta^T]$. Postmultiplication of Eq. (14-20) by $\mathbf{M}_{i,j}^T$ yields

$$\mathbf{G}_j \mathbf{G}_j^T = \mathbf{C}_{MM}(j, j) - \mathbf{F}_j \mathbf{C}_{XM}(i, j) - \mathbf{H}_j \mathbf{C}_{MM}(j-1, j) \tag{14-23}$$

The actual synthetic flows are computed from

$$\mathbf{Z}_{i,j} = \mathbf{M}_{i,j} \mathbf{S}_j^T + \mathbf{U}_j \tag{14-24}$$

where $\mathbf{U}_j$ is a vector of mean values and $\mathbf{S}_j$ a vector of standard deviations of transformed data at each of N sites for season j. Then $\mathbf{Q}_{i,j}$ is the inverse transform of $\mathbf{Z}_{i,j}$.

Skewed data may be transformed in several ways. The general gamma or three-parameter log-normal distributions are frequently used, as are general Box-Cox transforms [20]. Since LN3 and general gamma are very similar for hydrologic data, the simplest approach is to fit LN3 distributions to seasonal and annual data [34] using moments or quantiles [25]. The theoretical fit that is most suitable to the task should be used. For example, quantile fitting methods usually give excellent representation of the lower tail of an empirical cumulative distribution. Hirsch [35] describes a useful scheme to provide a two-part logarithmic transformation. Maximum-likelihood techniques are generally unsatisfactory for estimating parameters of the LN3 distribution. Hence, only details of moment relationships are discussed below.

The relationships between the location parameter a, the distribution parameters in log space (μ_y and σ_y), and the parameters in flow space $\bar{x}$, σ, and G (mean, standard deviation, and skew coefficient) are [34]

$$\mu\bar{x} = a + \exp\left(\mu_y + \frac{\sigma_y^2}{2}\right) \qquad (14\text{-}25)$$

$$\sigma^2 = [\exp(\sigma_y^2) - 1]\,\exp(2\mu_y + \sigma_y^2) \qquad (14\text{-}26)$$

$$G = \frac{\exp(3\sigma_y^2) - 3\exp(\sigma_y^2) + 2}{[\exp(\sigma_y^2) - 1]^{3/2}} \qquad (14\text{-}27)$$

For positive skew $(G > 0)$, $\ln(x - a)$ is normally distributed. For negative skew, $\ln(a - x)$ is normally distributed. Equation (14-26) gives only one real root of σ_y for a given positive skew coefficient G. The root of interest is determined from

$$\phi = [\beta + (\beta^2 - 1)^{1/2}]^{1/3} + [\beta - (\beta^2 - 1)^{1/2}]^{1/3} - 1 \qquad (14\text{-}28)$$

where $$\beta = 1 + \frac{G_x^2}{2} \qquad (14\text{-}29)$$

$$\exp(\sigma_y^2) = \phi \qquad (14\text{-}30)$$

Solutions for μ_y and a are straightforward. If skew equals zero for some sites or seasons (i.e., the untransformed data are normally distributed), LN3 can still be used as an approximation [36]. When a has been determined for each site and season, the transformation

$$Y_{i,j}^* = \ln(X_{i,j}^* - a) \qquad (14\text{-}31)$$

must be made for each site, using the seasonal flow data $X_{i,j}^*$. Model parameters are calculated from these transformed data. When LN3 transformations have been used, Q is determined from

$$\mathbf{Q}_{i,j} = \exp(Z_{i,j}) + aj \qquad (14\text{-}32)$$

The disaggregation model discussed above does not conserve mass properly because of the limited portions of the correlation matrices that are built into the model structure. It is therefore necessary to apply an adjustment to ensure that the sum of the seasonal flow volumes equals the annual total. A simple scheme for doing this is

$$\mathbf{Q}'_{i,j} = \frac{\sigma_j}{\displaystyle\sum_{j=1}^{k} \sigma_j} \cdot \Delta_i + \mathbf{Q}_{i,j} \qquad (14\text{-}33)$$

where $\mathbf{Q}'_{i,j}$ is the corrected flow, σ the standard deviation of the flow in season j, and Δ_i is the difference between the annual flow and the sum of the seasonal flows. This is repeated for each of the N sites. More detail on the disaggregation process can be found in [30] and [37].

14-6 Flow-Generation Strategies

The physical basis of stochastic modeling is not well established. The models that have been developed are largely descriptive. Short-memory models [such as AR(1)] have some physical basis in that the correlation reflects release of water to the stream from moisture stored in a previous time period. No clear relationships have been found in attempts to relate model properties and catchment physics [38, 39].

Since no one model can describe all relevant interactions for generating synthetic streamflow (or precipitation), it may be advisable to use more than one model for a design study. When the Hurst coefficient is larger than about 0.6, a simple AR(1) model is unable to replicate droughts satisfactorily. It is appropriate to use both the AR(1) and a model which reflects low-frequency phenomena to examine model-system interactions. In some cases there will be no noticeable difference in results, while in others clear differences will be evident [22].

Synthetic sequences should be generated independently, by generating a sequence (or multisite set of sequences) of length n and using it in a model of the system under study to calculate summary information. A second sequence is then generated independently. The process is repeated until a sufficiently large number of sequences has been analyzed to explore all interactions adequately. Burges and Linsley [40] found that 1000 sequences were adequate when exploring tail probabilities in reservoir reliability studies. If the central portion of the probability distribution is more important to decision making, fewer sequences are needed.

It is important to generate sequences equal to the period of interest (usually much fewer than 100 years). One long sequence broken into sub-sequences yields interrelated sequences unsuitable for making probabilistic statements about expected system performance. A random or conditional start [2] can be used. Considerable care must be given to initialization when using models of long-term persistance [14].

14-7 Storage Requirement from Stochastic Data

If a large number of sequences of streamflow are routed through a reservoir algorithm and the maximum storage requirement for each trace is determined, data for a probability analysis of storage requirements are available [41–43]. Burges [44] showed that when approximately 1000 sequences are used the probability distribution conforms to the extreme-value Type I distribution (Gumbel) and that a replication with a second group of 1000 sequences reproduces a virtually identical probability curve. When annual flows are used, reservoir analysis should normally involve at least 1000 sequences, each as long as the expected useful life of the reservoir. If monthly flows are used, it is likely that fewer sequences will be adequate. Because many trials are likely in a planning study, it is wise to determine by trial the number of sequences that should be used. This number may vary from about 300 for streams with low variability to 1000 or more for those streams which have high variability.

The initial condition of the reservoir plays an important role in determining reliability. The reservoir can be assumed to be initially empty or full or anywhere in between. If it is known that a specific condition will exist, it should be used as the starting point for all sequences. Otherwise a random distribution may be appropriate. Such a distribution is likely to result if the sequent peak algorithm is used (Sec. 16-3).

14-8 Reservoir Reliability

It has been customary to refer to the safe yield of a reservoir as if it were a guaranteed minimum yield. Determinations based on analysis of the historic record provide no evidence regarding the reliability of a reservoir. When a stochastic analysis is accepted as a realistic example of what may happen in the future, storage-probability curves like Figs. 14-5 and 14-6 provide useful information. Such curves indicate the probability that flows during any future period equal to the sequence length will prove adequate to sustain the desired demand without deficiency. A reliability of 0.99 indicates that only 1 out of every 100 sequences showed a deficiency; i.e., a reservoir with the indicated capacity offers a 99 percent assurance of successful operation during the project life.

Storage reliability depends on the operating period, the statistical properties of the streamflow, and the withdrawal required. In Figs. 14-5 and 14-6 storage computed with a sequent peak algorithm is expressed as S^*, the storage divided by the mean annual flow. A constant demand D^*, the demand divided by mean annual flow, is imposed for the operating period T. Figure 14-5 shows the effects of coefficient of variation CV, skew G, lag-1 correlation ρ_1, and model memory. Hurst coefficient $h = 0.5$ is a Markov process, while $h = 0.7$ is an average long-term persistence coefficient. The impact of demand level and CV on storage is evident. Figure 14-6 shows the effect of parameter values for long-term persistence models. The impact of skew and demand is evident, especially for $D^* = 0.5$ and 0.7. These plots emphasize the importance of proper modeling of the characteristics of the marginal distribution.

14-9 Time Trends

Stochastic analysis usually assumes a stationary series; i.e., there are no time trends. In the light of our knowledge of secular trends in weather, this assumption is probably the wisest one to use. We are concerned with a fairly short period in the future—rarely more than 100 years, commonly less. No established secular trends will cause a significant change in such a short span of time. Changes anticipated as a result of human activities must be incorporated in the parameters used for stochastic generation. That is, if runoff is expected to increase or decrease as a result of other activities in the catchment, some

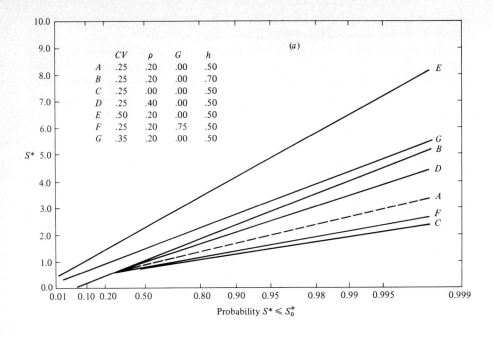

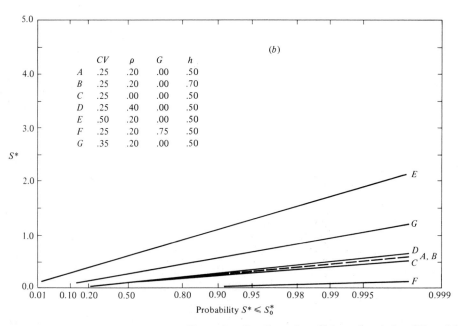

Figure 14-5 Storage reliability curves illustrating the effect of coefficient of variation CV, serial correlation ρ, skew G, and memory h. (*a*) $D^* = 0.7$, (*b*) $D^* = 0.5$.

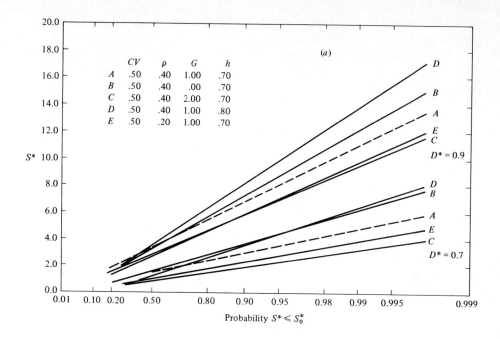

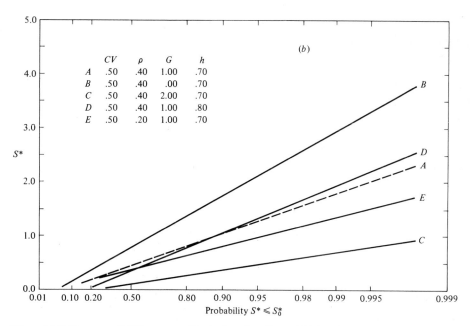

Figure 14-6 Storage reliability curves illustrating the effect of high values of coefficient of variation, serial correlation, skew, and memory. (*a*) $D^* = 0.7$ and 0.9, (*b*) $D^* = 0.5$.

adjustment must be made. Increased evaporation loss from a reservoir should be included in the storage simulation.

Time trends in the historic data on which the stochastic generation is based are another problem. Many streamflow records display obvious trends as a result of changes in land use in the catchment, diversions for irrigation or other purposes, construction of reservoirs, etc. Adjustment for these trends should be made before parameters are calculated. It is probably best to adjust to natural flow to define parameters for the stochastic generation and then to incorporate reservoirs or diversions into the storage simulation. If the change is permanent and uncontrolled, such as the effects of urbanization, it may be wiser to adjust to current or future conditions. The adjustment to natural flow is usually accomplished by adding known diversions and reservoir-storage changes to the observed flows (Sec. 4-16). Adjustment to current or future conditions is probably best accomplished by simulation from precipitation data. An alternative is to fit a trend line to the time series of historic data. This approach achieves only an approximate correction and should be resorted to only if other alternatives are impractical because of data deficiencies.

14-10 Stochastic Analysis of Precipitation

It is the stochastic variation of precipitation that is largely responsible for the stochastic characteristics of streamflow. Monthly or annual precipitation may be treated by stochastic methods similar to those applied to streamflow. Daily and hourly precipitation can also be treated stochastically, but the methods differ somewhat from those used for streamflow. Stochastically generated precipitation may be converted to streamflow by use of an appropriate deterministic model. This approach may provide a basis for dealing with an engineering problem when adequate streamflow data are not available. It also offers the possibility of dealing with short-interval data such as flood peaks.

Monthly flow volumes are usually adequate for reservoir studies, but greater detail is required for small reservoirs or flood problems. The techniques described earlier in this chapter for analysis of streamflow are not usually applicable to daily data because the sequential relations between daily flows are not wholly random and cannot be described by a simple function. From point A to point B (Fig. 14-7), flow on any day is related to that of the previous day by the recession equation [Eq. (7-1)]. Only a small random component caused by the variation of evapotranspiration from day to day may be present. Between points B and C the relation between successive days is poor, since the controlling factor is a separate stochastic factor—precipitation as filtered by the catchment. Flow on day $B + 1$ is only partially determined by flow on day B. A stochastic daily-flow generating scheme must recognize these differences and include an algorithm to determine the occurrence of streamflow rises. During a rise, the flows must follow a pattern consistent with the catchment characteristics [45].

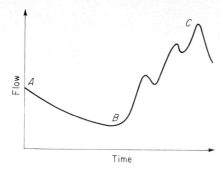

Figure 14-7 Mean daily flow hydrograph showing the differing regimes.

The principal problem in stochastic analysis of short-interval precipitation is that of coping with the spatial variation as well as the time sequences [48]. Thus, for small basins where a single station is adequate, or for moderate areas where precipitation is from frontal systems, a combined stochastic-deterministic approach may be feasible. The key constraint is that the rainfall data must be adequate for a good calibration of the deterministic model. For large catchments subject mainly to air-mass thunderstorms, difficulty will be experienced in the stochastic generation (and in the deterministic simulation if the available data are inadequate).

For single-station rainfall a matrix of precipitation in interval t versus that in interval $t - 1$ from which values may be chosen by a random procedure may be adequate [46] if combined with an empirical distribution of interstorm lengths. Franz [47] developed a method for multistation analysis using multivariate normal statistics similar to the procedure described in Sec. 14-4. Whatever procedure is employed should be subjected to careful testing before it is accepted for use. In addition, calibration of the deterministic model must be done with care, since any bias it contains will be included in the streamflow data developed.

PROBLEMS

14-1 Using a historic data series of 25 years or more of annual flow volumes and the sequent peak method, determine the storage required to satisfy a demand of 30, 50, 70, and 90 percent of mean flow. If your classmates analyze other streams, express the storage as a percent of mean flow and compare. Are the variations between streams as you might expect?

14-2 For the stream of Prob. 14-1, calculate mean, variance, serial correlation, and Hurst coefficient h. Plot the flow data and inspect the plot for occurrences which might bias the computed serial correlation. Can you interpret h in the light of Klemes [13]?

14-3 Use the parameters of Prob. 14-2 to generate a sequence of annual flows. Develop a reliability curve like Fig. 14-5. What probability is applicable to the results of Prob. 14-1? How many synthetic sequences are needed to obtain a stable storage reliability curve?

14-4 Analyze the parameters for the monthly data from the stream of Prob. 14-1. Do all the months exhibit the same distribution? Can you explain the variation in the serial correlation throughout the year? In the variance?

14-5 Build a seasonal flow generation model and examine storage needs for the demand levels of Prob. 14-1 when the demand is distributed approximately as a sinusoid, where (a) peak demand is in phase with peak inflows and (b) peak demand occurs at the time of lowest flow. Compare your results with those of Prob. 14-3.

REFERENCES

1. Bernouilli, J.: "Ars Conjectandi," p. 213, Basel, 1713.
2. Lettenmaier, D. P., and S. J. Burges: Climate Change: Detection and Its Impact on Hydrologic Design, *Water Resour. Res.,* vol. 14, no. 4, pp. 679–687, 1978.
3. Hazen, A.: Storage to Be Provided in Impounding Reservoirs for Municipal Water Supply, *Trans. ASCE,* vol. 77, pp. 1539–1669, 1914.
4. Sudler, C. E.: Storage Required for the Regulation of Streamflow, *Trans. ASCE,* vol. 91, pp. 622–704, 1927.
5. Thomas, H. A., Jr., and M. B. Fiering: Mathematical Synthesis of Stream-Flow Sequences for the Analysis of River Basins by Simulation, Chap. 12 in A. Maass et al.: "Design of Water Resources Systems," Harvard University Press, Cambridge, Mass., 1962.
6. Box, G. E. P., and G. M. Jenkins: "Time Series Analysis: Forecasting and Control," Holden-Day, San Francisco, 1970.
7. Mandelbrot, B. B., and J. P. Wallis: Computer Experiments with Fractional Gaussian Noise, *Water Resour. Res.,* vol. 5, pp. 228–267, February 1969.
8. Mejia, J. M., I. Rodriguez-Iturbe, and D. R. Dawdy: Streamflow Simulation: The Broken Line Process as a Potential Model for Hydrologic Simulation, *Water Resour. Res.,* vol. 8, pp. 931–941, August 1972.
9. Fiering, M.: "Streamflow Synthesis," Harvard University Press, Cambridge, Mass., 1967.
10. Hurst, H. E.: Long-Term Storage Capacity of Reservoirs, *Trans. ASCE,* vol. 116, pp. 770–808, 1951.
11. Hurst, H. E.: A Suggested Statistical Model of Some Time Series Which Occur in Nature, *Nature,* vol. 180, p. 494, 1957.
12. Hurst, H. E., R. P. Black, and Y. M. Simaika: "Long-Term Shortage, An Experimental Study," Constable, London, 1965.
13. Klemes, V.: The Hurst Phenomenon: A Puzzle? *Water Resour. Res.,* vol. 10, no. 4, pp. 675–688, 1974.
14. Lettenmaier, D. P., and S. J. Burges: Operational Assessment of Hydrologic Models of Long-Term Persistence, *Water Resour. Res.,* vol. 13, no. 1, pp. 113–124, 1977.
15. McLeod, A. I., and K. W. Hipel: Preservation of Rescaled Adjusted Range, 1, A Reassessment of the Hurst Phenomenon, *Water Resour. Res.,* vol. 14, no. 3, pp. 491–508, 1978.
16. Matalas, N. C.: Mathematical Assessment of Synthetic Hydrology, *Water Resour. Res.,* vol. 3, no. 4, pp. 937–945, 1967.
17. Lawrance, A. J., and N. T. Kottegoda: Stochastic Modelling of Riverflow Time Series, *J. Roy. Stat. Soc.,* A, Pt. 1, pp. 1–31, 1977.
18. O'Connell, P. E.: Stochastic Modeling of Long-Term Persistence in Stream-Flow Sequences, *Imp. Coll. Dept. Civ. Eng. Hydrol. Sec. Rep.* 1974-2, London, 1974.
19. Mandelbrot, B. B.: A Fast Fractional Gaussian Noise Generator, *Water Resour. Res.,* vol. 7, no. 3, pp. 543–553, 1971.
20. Hipel, K. W., A. I. McLeod, and W. C. Lennox: Advances in Box Jenkins Modeling, 1, Model Construction, *Water Resour. Res.,* vol. 13, no. 3, pp. 567–576, 1977.
21. McLeod, A. I., K. W. Hipel, and W. C. Lennox: Advances in Box Jenkins Modeling, 2, Applications, *Water Resour. Res.,* vol. 13, no. 3, pp. 577–586, 1977.

22. Hoshi, K., S. J. Burges, and I. Yamaoka: Reservoir Design Capacities for Various Seasonal Operational Hydrology Models, *Japan Soc. Civ. Eng.,* no. 273, pp. 121–134, May 1978.
23. Wilk, M. B., and R. Ganadesikan: Probability Plotting Methods for the Analysis of Data, *Biometrica,* vol. 55, no. 1, pp. 1–17, 1968.
24. Benjamin, J. R., and C. A. Cornell: "Probability, Statistics, and Decision for Civil Engineers," McGraw-Hill, New York, 1970.
25. Stedinger, J. R.: Fitting Log Normal Distributions to Hydrologic Data, *Water Resour. Res.,* vol. 16, no. 3, pp. 481–490, 1980.
26. Wallis, J. R., N. C. Matalas, and J. R. Slack: Just a Moment! *Water Resour. Res.,* vol. 10, no. 2, pp. 211–219, 1974.
27. Kendall, M. B.: Note on the Bias in the Estimation of Autocorrelation, *Biometrica,* vol. 42, pp. 403–404, 1954.
28. Anderson, R. L.: Distribution of the Serial Correlation Coefficient, *Ann. Math. Stat.,* vol. 13, pp. 1–13, 1962.
29. Curry, K., and R. L. Bras: Theory and Application of the Multivariate Broken Line, Disaggregation and Monthly Autoregressive Streamflow Generators to the Nile River, *Technology Adaptation Program Rep.* 78-5, Mass. Inst. Tech., Cambridge, Mass., 1978.
30. Clarke, R. T.: Mathematical Models in Hydrology, *Irrigation and Drainage Pap.* 19, Food and Agriculture Organization in the United Nations, Rome, 1973.
31. Matalas, N. C., and J. R. Wallis: Generation of Synthetic Flow Sequences, chap. 3 in A. K. Biswas (ed.): "Systems Approach to Water Management," McGraw-Hill, New York, 1976.
32. Valencia, D. R., and J. C. Schaake: Disaggregation Processes in Stochastic Hydrology, *Water Resour. Res.,* vol. 9, no. 3, pp. 580–585, 1973.
33. Lane, W.: Applied Stochastic Techniques, User's Manual, U.S. Bureau of Reclamation, Denver, Colo., 1979.
34. Lettenmaier, D. P., and S. J. Burges: An Operational Approach to Preserving Skew in Hydrologic Models of Long-Term Persistence, *Water Resour. Res.,* vol. 13, no. 2, pp. 281–290, 1977.
35. Hirsch, R. M.: Synthetic Hydrology and Water Supply Reliability, *Water Resour. Res.,* vol. 15, no. 6, pp. 1603–1615, 1979.
36. Burges, S. J., and K. Hoshi: Approximation of a Normal Distribution by a Three Parameter Log Normal Distribution, *Water Resour. Res.,* vol. 14, no. 4, pp. 620–622, 1978.
37. Salas, J. D., J. W. Delleur, V. Yevjevich, and W. Lane: "Applied Modeling of Hydrologic Series," Water Resources Publications, Fort Collins, Colo., 1980.
38. Moss, M.: Reduction of Uncertainties in Autocorrelation by the Use of Physical Models, *Proc. International Symposium on Uncertainties in Hydrologic and Water Resource Systems,* vol. 1, University of Arizona, Tucson, pp. 203–229, 1972.
39. Klemes, V.: Physically Based Stochastic Hydrologic Analysis, in V. T. Chow (ed.): "Advances in Hydroscience," vol. 11, pp. 285–357, Academic Press, New York, 1978.
40. Burges, S. J., and R. K. Linsley: Some Factors Influencing Required Reservoir Storage, *J. Hydraul. Div. ASCE,* vol. 97, no. HY7, pp. 977–991, 1971.
41. Thomas, H. A., Jr., and M. B. Fiering: The Nature of the Storage Yield Function, in "Operations Research in Water Quality Management," Harvard University Water Program, 1963.
42. Klemes, V.: Storage Mass-Curve Analysis in a Systems-Analytic Perspective, *Water Resour. Res.,* vol. 15, no. 2, pp. 359–370, 1979.
43. Hufschmidt, M., and M. Fiering: "Simulation Techniques for the Design of Water-Resource Systems," Harvard University Press, Cambridge, Mass., 1962.
44. Burges, S. J.: Use of Stochastic Hydrology to Determine Storage Requirements for Reservoirs: A Critical Analysis, *Stanford Univ. Progr. Eng. Econ. Plann. Rep. EEP-34,* 1970.
45. Kelman, J.: Stochastic Modeling of Hydrologic Intermittent Daily Processes, *Hydrol. Pap.* 89, Colorado State University, Fort Collins, Colo., 1977.
46. Pattison, A.: Synthesis of Rainfall Data, *Stanford Univ. Dept. Civ. Eng. Tech. Rep.* 40, 1964.
47. Franz, D. D.: Hourly Rainfall Synthesis for a Network of Stations, *Stanford Univ. Dept. Civ. Eng. Tech. Rep.* 126, 1969.

48. Bras, R. L., and I. Rodriguez-Iturbe: Rainfall-Runoff as Spatial Stochastic Processes: Data Collection and Synthesis, *Mass. Inst. Tech. Ralph M. Parsons Lab. for Water Resour. and Hydrodynamics Rep.* 196, 1975.

BIBLIOGRAPHY

Bayley, G. U., and J. M. Hammersley: The Effective Number of Independent Observations in an Autocorrelated Time Series. *J. R. Statist. Soc.,* vol. 8, pp. 184–197, 1946.

Benson, M. A., and N. C. Matalas: Synthetic Hydrology Based on Regional Statistical Parameters, *Water Resour. Res.,* vol. 3, pp. 931–935, 1967.

Bobee, B., and R. Robitaille: Correction of Bias in the Estimation of the Coefficient of Skewness, *Water Resour. Res.,* vol. 11, pp. 851–854, 1975.

Burges, S. J., D. P. Lettenmaier, and C. L. Bates: Properties of the Three-Parameter Log Normal Probability Distribution, *Water Resour. Res.,* vol. 11, pp. 229–235, 1975.

—— and ——: "Operational Comparison of Stochastic Stream-Flow Generation Procedures," *Tech. Rep.* 45, Harris Hydraul. Lab., Dept. of Civil Eng., Univ. Washington, 1975.

Crosby, D. S., and T. Maddock III: Estimating Coefficients of a Flow Generator for Monotone Samples of Data, *Water Resour. Res.,* vol. 6, pp. 1079–1086, 1970.

Fiering, M. B.: Schemes for Handling Inconsistent Matrices, *Water Resour. Res.,* vol. 4, pp. 291–297, 1968.

—— and B. B. Jackson: Synthetic Streamflows, *Water Resour. Monogr.* 1, American Geophysical Union, 1971.

Finzi, G., E. Todini, and J. R. Wallis: Comment on Multivariate Synthetic Hydrology, *Water Resour. Res.,* vol. 11, pp. 844–850, 1975.

Frame, J. W.: A New BMD Program for the Description and Estimating of Missing Data, *Proc. Statist. Comp. Sec., Am. Statist. Assoc.,* pp. 110–113, 1975.

Gilroy, E. J.: Linear Least Squares Prediction for Multivariate Time Series with Missing Observations, *Proc. Symp. on Statist. Hydrol., U.S. Dept. Agr. Misc. Pub.* 1275, pp. 78–83, 1974.

Hardison, C. H.: Accuracy of Streamflow Characteristics, *U.S. Geol. Surv. Prof. Pap.* 650-D, pp. D210–D214, 1967.

——: Prediction Error of Regression Estimates of Streamflow Characteristics at Ungaged Sites, *U.S. Geol. Surv. Prof. Pap.* 750-C, pp. C228–C236, 1971.

Hufschmidt, M. M., and M. B. Fiering: "Simulation Techniques for the Design of Water-Resource Systems," Harvard University Press, Cambridge, 1966.

Jackson, B. B.: The Use of Streamflow Models in Planning, *Water Resour. Res.,* vol. 11, pp. 54–63, 1975.

——: Markov Mixture Models for Drought Lengths, *Water Resour. Res.,* vol. 11, pp. 64–74, 1975.

——: Birth Death Models for Differential Persistence, *Water Resour. Res.,* vol. 11, pp. 75–95, 1975.

Mandelbrot, B. B., and J. W. Van Ness: Fractional Brownian Motions, Fractional Noises, and Applications, *Soc. Ind. Appl. Math. Rev.,* vol. 10, p. 422, 1968.

Matalas, N. C., and E. J. Gilroy: Some Comments on Regionalization in Hydrologic Studies, *Water Resour. Res.,* vol. 4, pp. 1361–1369, 1968.

——: Developments in Stochastic Hydrology, *Rev. Geophys, and Space Physics,* vol. 13, pp. 67–73, 1975.

Mejia, J. M., and I. Rodriguez-Iturbe: Correlation Links between Normal and Log Normal Processes, *Water Res. Resour.,* vol. 10, pp. 689–690, 1974.

——, ——, and J. R. Cordova: Multivariate Generation of Mixtures of Normal and Log Normal Variables, *Water. Resour. Res.,* vol. 10, pp. 691–693, 1974.

—— and ——: On the Synthesis of Random Field Sampling from the Spectrum: An Application to the Generation of Hydrologic Spatial Processes, *Water Resour. Res.,* vol. 10, pp. 705–712, 1974.

———— and J. M. Rousselle: Disaggregation Models in Hydrology Revisited, *Water Resour. Res.,* vol. 12, pp. 185–186, 1976.

O'Connell, P. E.: A Simple Stochastic Modelling of Hurst's Law, *Int. Symp. on Math. Models in Hydrol., Int. Assoc. Sci. Hydrol.,* Warsaw, 1971.

————: Multivariate Synthetic Hydrology: A Correction, *J. Hydraul. Div. ASCE,* vol. 99, pp. 2392–2396, 1973.

Rodriguez-Iturbe, I., J. M. Mejia, and D. R. Dawdy: Streamflow Simulation. 1—A New Look at Markovian Models, Fractional Gaussian Noise and Crossing Theory, *Water Resour. Res.,* vol. 8, pp. 921–930, 1972.

Tao, P. C., and J. W. Delleur: Multistation, Multiyear Synthesis of Hydrologic Time Series by Disaggregation, *Water Resour. Res.,* vol. 12, pp. 1303–1311, 1976.

Texas Water Development Board: "Stochastic Optimization and Simulation Techniques for Management of Regional Water Resources Systems, vol. II-B, FILLIN-1 Program Description," Austin, Tex., 1970 (NTIS:PB 202-938).

————: "Economic Optimization and Simulation Techniques for Management of Regional Water Resources Systems—Multi-Site Data Fill-in and Sequence Generation Program; MOSS III, Program Description," Austin, Tex., 1971.

Thomas, D. M., and M. A. Benson: Generalization of Streamflow Characteristics from Drainage-Basin Characteristics, *U.S. Geol. Surv. Water-Supply Pap.* 1975, pp. 26–31, 1970.

Vicens, G. J., I. Rodriguez-Iturbe, and J. C. Schaake, Jr.: A Bayesian Approach to Hydrologic Time-Series Modelling, *Rep. 181, Mass. Inst. Tech. Ralph M. Parsons Lab. for Water Resour. and Hydrodynamics,* 1974.

————, ————, and ————: A Bayesian Framework for the Use of Regional Information in Hydrology, *Water Resour. Res.,* vol. 11, pp. 405–414, 1975.

Wallis, J. R., and N. C. Matalas: Sensitivity of Reservoir Design to the Generating Mechanism of Inflows, *Water Resour. Res.,* vol. 8, pp. 634–641, 1972.

———— and ————: Small Sample Properties of H and K—Estimators of the Hurst Coefficient, h, *Water Resour. Res.,* vol. 6, pp. 1583–1594, 1970.

Wilk, M. B., and R. Ganadesikan: Probability Plotting Methods for the Analysis of Data, *Biometrica,* vol. 55, pp. 1–17, 1968.

FIFTEEN

HYDROLOGY AND WATER QUALITY†

Falling precipitation carries gases and particulate matter from the atmosphere. As it strikes the ground it may dislodge sediment which overland flow transports to a stream (Chap. 11) together with material dissolved from the land surface. Infiltrating water undergoes chemical exchange with the soil, giving up some materials and dissolving others. Thus surface runoff, interflow, and groundwater have chemistries characteristic of the rocks and soils encountered along their paths of flow.

It has long been known that waste can be discarded into streams and lakes, where it sinks to the bottom or is carried away. We now know that some wastes discarded on the land are washed into streams, together with fertilizers, pesticides, lead and oil from automobiles, and many other industrial chemicals. These materials can be a serious threat to the aquatic environment and to those who drink the water. Hydrologic factors play a major role in determining the concentration, rate of movement, and final deposition of pollutants. Prediction of water quality and of the changes which might result from control measures requires an understanding of the hydrologic, physical, chemical, and biological processes in water bodies.

15-1 Physical and Chemical Properties of Water

Water has unique properties responsible for the interactions with the material which water contacts throughout the hydrologic cycle. These properties include solvent action, dissociation, and transparency.

---

† The original draft of this chapter was written by John Imhoff, Environmental Engineer, Anderson-Nichols Inc., Palo Alto, Calif.

Solvent action. Of all naturally occurring liquids, water is the most effective solvent. The polarity of charge in the water molecule favors the disruption of ionic crystals by reducing the interionic attractive forces. Sodium chloride and salts of potassium are readily dissolved in water. Water can also solvate separated ions. *Solvation* entails the surrounding of charged solute particles by solvent molecules in response to attractive forces. *Cations,* positively charged ions, are effectively solvated by compounds of elements in the first row of the periodic table that have unshared electron pairs. *Anions,* negatively charged ions, are easily solvated by a solvent in which a strongly electronegative element such as oxygen is bonded to hydrogen. Water satisfies both requirements and consequently is effective in solvating both cations and anions.

Water molecules are capable of forming hydrogen bonds to molecules containing oxygen atoms, hydroxyl groups OH^-, or nitrogen bound to hydrogen. Examples of such molecules are ammonia, NH_3; nitrate, NO_3^-; phosphate, PO_4^{2-}; carbonate, CO_3^{2-}; sulfate, SO_3^{2-}; ethyl alcohol, C_2H_5OH; and various sugars and organic acids. Such bonds are an extreme form of solvation, and result in a dissolved and readily accessible supply of vital compounds for the organisms associated with the aquatic environment.

Dissociation. Water is not only a solvent for other substances but is itself capable of dissociating into two charged ions. Hence water acts as both a base and an acid. An example of this process is the solution of carbon dioxide in water:

$$O{=}C{=}O + H{-}O{-}H \rightarrow O{=}C\overset{\displaystyle OH}{\underset{\displaystyle OH}{\big\langle}} \rightarrow O{=}C\overset{\displaystyle OH}{\underset{\displaystyle O^-}{\big\langle}} + H^+ \qquad (15\text{-}1)$$

Transparency. Solar radiation arriving at the earth's surface is characterized by wavelengths between 0.3 and 1.3 μm. Nearly all radiation outside the range of visible wavelengths (0.4 to 0.8 μm) is absorbed in the first meter of water. Radiation with wavelengths greater than 0.8 μm (infrared) results in significant heat transfer to water. The various spectral bands of visible radiation are absorbed differentially. The red component penetrates no more than 4 m (13 ft) of water while some of the blue component penetrates 70 m (230 ft) or more. The intensity of any component of light at depth z within a column of pure water is given by *Beer's law:*

$$q_z = q_0 e^{-\eta z} \qquad (15\text{-}2)$$

where q_z is the light intensity at depth z, q_0 is the intensity of light at the water surface, and η is an extinction coefficient for a specific wavelength of light. Table 15-1 lists values of η for various wavelengths. Approximately 50 percent of the total incident light is scattered or transformed into heat within the first meter of water.

Table 15-1 Values of light-extinction coefficient η in pure water

Wave-length, μm	Extinction coefficient η, cm^{-1}	Wave-length, μm	Extinction coefficient η, cm^{-1}
0.31	0.0081	0.90	0.0655
0.35	0.00333	0.95	0.288
0.40	0.00072	1.00	0.397
0.45	0.00018	1.05	0.177
0.50	0.00016	1.10	0.203
0.55	0.00027	1.15	0.848
0.60	0.00125	1.20	1.232
0.65	0.0021	1.25	1.153
0.70	0.0084	1.30	1.50
0.75	0.0272		
0.80	0.0240		
0.85	0.0442		

Source: Adapted from table, "Heat and Mass Transfer between a Water Surface and the Atmosphere," Tennessee Valley Authority, Report no. 0-6803, 1972.

The ability of water to transmit light is critical to the organisms living in the aquatic environment. The primary producers of the aquatic community, *phytoplankton,* have molecular pigments capable of capturing energy from solar radiation. This energy is transformed by photosynthesis into energy stored in chemical bonds of organic molecules.

15-2 Quality of Precipitation

Precipitation contains dissolved substances largely determined by the air quality and wind patterns of the region. In areas where there is heavy air pollution the atmosphere is a complex chemical system controlled by dozens of chemical and photochemical transformations [1]. Atmospheric water in such areas accumulates carbon dioxide, nitrates, and inorganic forms of phosphorus and sulfur. This moisture becomes chemical-laden precipitation hundreds of miles from the original pollution source [2].

Reported annual deposition rates of nitrogen and phosphorus from precipitation range [3] from 5 to 10 kg/hectare (4.5 to 8.9 lb/acre) and 0.5 to 0.6 kg/hectare (0.45 to 0.54 lb/acre), respectively (Table 15-2). Data from areas relatively unaffected by human activity indicate that much of the instream burden of nitrogen comes from precipitation, while only a small portion of the instream burden of phosphorus is precipitation-borne [4].

The majority of the chemicals in precipitation are absorbed by soil and become part of the nonpoint-source pollution which may eventually be washed into streams or lakes. Precipitation which falls directly on surface waters can

Table 15-2 Nitrate and phosphorus input from precipitation to six lakes in New York [4]

Lake	Annual precipitation mm or l/m²	Measured nutrient content in melted snow[a]		Nutrient input directly onto lake surface per year[e]		Nutrient input from entire watershed into lake per year[e]	
		NO_3, mg/l	P^f, µg/l	NO_3, g/m³	P^f, g/m³	NO_3, g/m³	P^f, g/m³
Canadice	780[b]	2.72	16.2	2.14	0.013	25.2	0.153
Conesus	788[b]	2.92	16.2	2.30	0.013	32.2	0.182
Honeoye	788[b]	2.98	7.7	2.35	0.006	34.3	0.088
Lime	1045[c]	4.24	31.0	4.43	0.032	22.6	0.163
Java	1045[c]	4.54	34.0	4.74	0.036	71.1	0.540
Devil's Bathtub	804[d]	3.13	16.3	2.52	0.013	15.1	0.078

[a] The higher values for Lime and Java may reflect the industrialized Buffalo area.
[b] Climatic data (1964), Hemlock station, a mean value from 61 to 63 years of records.
[c] Climatic data (1964), Arcade station, a mean value from 29 to 36 years of records.
[d] Climatic data (1964), Honeoye Falls station, a mean value from 5 to 7 years of records.
[e] Calculated values, assuming nutrient content of snow and rain are generally similar.
[f] Total phosphorus, dissolved and suspended.

have varied effects. Dilution by relatively clean rainfall can improve water quality in lakes. Precipitation containing large amounts of nitrogen and sulfur is highly acidic and can disrupt the entire biochemical environment of a water body. Destruction of the fish population in some lakes has been attributed to "acid rain" [5].

15-3 Surface Runoff

As water moves across the land surface during or after a storm, it transports dissolved and suspended materials which have been picked up along the path of flow (Chap. 11). In many cases the pollutants carried to streams and lakes by surface runoff are a major contribution to water pollution. Washoff materials of most concern include sediment, mineral salts, heavy metals, nutrients, pesticides, biodegradable organics, and microbial pollution.

The chemical characteristics of surface runoff are affected by two phenomena—the accumulation of pollutants on the land surface and the transport mechanisms that move the pollutants from the land into the water. Pollutant accumulation is site-specific and depends on land use, season, and climate. The amount of material available for washoff is determined by the total accumulation less removal by wind, biochemical decay, and activities such as street cleaning.

Pollutants in runoff may be in solution or suspension, or attached to particles of sediment. The amount of pollutant washed off is proportional to the amount available on the land surface and to the intensity of runoff. A common expression for surface washoff ΔP of a pollutant during the interval t is

$$\Delta P = P_0 - P = P_0(1 - e^{-kt}) \tag{15-3}$$

where P_0 is the initial amount of pollutant on the land surface, P is the quantity at the end of the interval, and k is a function of overland flow rate. Pollutants accumulated on an impervious area are more easily washed off by overland flow than are those on pervious soil. As much as 90 percent of accumulated pollutant may be washed off an impervious area during 1 hr of overland flow at a rate of 13 mm/hr (0.5 in/hr). The wash-off from a pervious area is typically one-half or less that from an impervious area with the same runoff intensity [6].

Sources of nonpoint pollution are usually described in terms of the land uses that produce the pollutants. The mix of land uses in urban areas determines the relative quantities of individual pollutants. Mining produces pollutants such as acids, heavy metals, and salts dissolved in the runoff water. Agriculture, construction, and silviculture result in wash-off of sediment, nutrients, and pesticides. About 60 percent of the nitrogen and 40 percent of the phosphorus input to water supplies is attributed to agricultural wash-off [7].

Table 15-3 summarizes the water quality characteristics of runoff with various land uses [6]. Pollutant concentrations vary by orders of magnitude from one basin to another, from one storm to the next, and within a single storm [8]. The overall effect of nonpoint-source pollution on water quality is best expressed as *total pollutant mass loading*, the product of concentration and flow. Values of total pollutant mass loading associated with various land uses are available [9–11].

15-4 Transport of Materials through the Soil Profile

To a large extent soil type and texture determine the interactions in which water and its chemical burden are involved. Fine-textured soils with high clay content do not drain well and retain large amounts of water for long periods. Aeration in these soils is limited, and processes such as organic decomposition, ammonia volatilization, and nitrification are retarded. Although clay particles are active sites for ion exchange and adsorption, the effective adsorption of dissolved materials is reduced since the water molecules are polar and compete for adsorption sites. Coarse-textured soils conduct large quantities of air and water, and oxidative processes are encouraged. At the same time the rates of evaporation, lateral transmissibility, and percolation are higher for coarse-textured soils.

As a rule, passage through the soil profile results in purification of water because of adsorption, volatilization, decomposition or degradation, nitrification, denitrification, and plant uptake.

Adsorption-desorption. The interaction between the chemicals in water and the soil through which the water passes largely determines water quality changes within the soil profile. The adsorption process entails the removal of chemicals from solution and retention on the surface of soil particles by chemical or physical bonding. If the bonds formed between the adsorbate and soil are chemical, the process is almost always irreversible. If the bonds are physical, e.g., van der Waals forces [12], the chemicals are easily removed or desorbed by a change in solution concentration of the adsorbate.

The quantity of a chemical that can be adsorbed by soil depends on concentration of adsorbate and soil temperature. In most research on the adsorption-desorption process, the amount of a chemical adsorbed is determined as a function of concentration at a constant temperature, and the resulting function is called an adsorption isotherm [13]. The most common representation of the adsorption-desorption process in soil is the *Freundlich isotherm:*

$$\frac{X}{M} = kc^{1/n} \tag{15-4}$$

where X is the amount of chemical adsorbed, M is the amount of soil adsorbent, c is the equilibrium concentration of adsorbate in solution after adsorption, and k and n are empirical constants.

Organic wastes, pesticides, ammonia, and phosphorus are adsorbed to an extent dependent on the clay content, organic matter content, and cation-exchange capacity of the soil. Adsorption usually assures that chemicals remain in the soil long enough for processes such as decomposition and plant uptake to occur.

Volatilization. The loss of a chemical from the soil-water system by vaporization into the atmosphere is termed *volatilization.* Certain chemicals move to the soil surface by diffusion or mass flow. At the surface the rate of volatilization q of a chemical may be approximated by the equation

$$q = -k_v \lambda m^{-1/2} \tag{15-5}$$

where k_v is a wind- and temperature-dependent vaporization constant for the chemical, λ is the saturation vapor concentration, and m is the molecular weight of the chemical. A high volatilization rate requires that vaporized chemicals move away from the soil surface so that additional chemicals can move into the vapor phase. Consequently, wind speed is a critical determinant of volatilization rate. Volatilization can remove large quantities of chemicals such as ammonia and pesticides from the soil, particularly during the initial period after application [14].

Decomposition or degradation. Organic materials in the soil break down to form carbon dioxide, water, and inorganic elements such as nitrogen and chloride. Degradation rates depend on soil temperature, moisture, strength of binding by soils, soil type, and soil microorganisms. Oxidation, hydrolysis, and microbial enzyme action are the most common methods of degradation. In many soils the

Table 15-3 Water quality characteristics of runoff resulting from various land uses[a] [6]

	Total solids		Suspended solids		BOD		NO₃-N		Total N		Total P	
	Mean[b]	Range	Mean[b]	Range	Mean[b]	Range	Mean[b]	Range	Mean[b]	Range	Mean[b]	Range
General characteristics:												
Precipitation				11–13		12–13		0.14–1.1		1.2–1.3		0.02–0.04
Forested land								0.1–1.3		0.3–1.3		0.01–0.11
Agricultural cropland		194–8620	7				0.4		9			0.02–1.7
Urban land drainage				5–7340		12–160			3			0.2–1.1
Animal feedlot runoff						1000–11,000		10–23		920–2100		290–360
Individual studies:												
Kansas												
Beef cattle feedlot		10,000–25,000				1000–11,000				200–450[c]		
Waynesboro, Va.												
Forested (site 2)				15–311						1.05–1.68[c]		0–0.33
Durham, N.C.												
Urban (Bryan study)	2730	274–13,800			14.5	2–232					0.58[d]	0.15–2.5[d]
Durham, N.C.												
Urban (Colston study)	1440	194–8620	1223	27–7340			0.96[c]			0.1–11.6[c]	0.82	0.2–16
Cincinnati, Ohio												
Urban			227	5–1200	17	1–173					1.1[d]	0.02–7.3[d]
Coshocton, Ohio												
Rural			313	5–2074	7	0.5–23					1.7[d]	0.25–3.3[d]
Seattle, Wash.												
Urban industrial												
SS3 site	140[e]		80		19		0.83			2.91[f]	0.32[a]	
Seattle, Wash.												
Urban commercial												
CBD site	303[e]		190		22		0.72			2.82[f]	0.87[a]	

Table 15-3 (*Continued*)

	Total solids		Suspended solids		BOD		NO$_3$-N		Total N		Total P	
	Mean[b]	Range	Mean[b]	Range	Mean[b]	Range	Mean[b]	Range	Mean[b]	Range	Mean[b]	Range
Individual studies (*cont'd*):												
Tulsa, Okla.[b]												
Urban mixed land uses	545	199–2242	367	84–2052	11.8	8–18			0.85[i]	0.36–1.48[i]	1.15[j]	0.54–3.49[j]
Madison, Wis.												
Urban residential			280				0.60		4.55[k]		0.98	
Eastern South Dakota agriculture runoff												
Cultivated (rain)	1241		1021				1.5		4.1[k]		1.05	
Cultivated (snow)	187		51				1.0		3.1[k]		0.44	
Pasture (snow)	150		18				0.9		4.2[k]		0.67	
Grassland (snow)	134		42				0.8		3.6[k]		0.43	

[a] Data presented here are for comparison purposes only. Different sampling methods, numbers of samples, and analysis techniques were used in individual studies.

[b] Individual values may indicate average or median values.

[c] Total Kjeldahl nitrogen, mg/l N.

[d] Total phosphate, mg/l P.

[e] Suspended plus settleable solids.

[f] Sum of organic, ammonia, nitrite, and nitrate as mg/l N.

[g] Hydrolyzable and ortho as mg/l P.

[h] Values refer to the mean and range of mean values for 15 test areas.

[i] Organic Kjeldahl nitrogen.

[j] Only soluble orthophosphate.

[k] Sum of organic, ammonia, and nitrate as mg/l N.

combined processes of adsorption and degradation can remove 99 percent or more of the organic content of heavily polluted water [15].

Nitrification. The two-step process in which ammonia (NH_3) is oxidized to nitrite (NO_2^-) and then to nitrate (NO_3^-) is termed *nitrification*. This is an important reaction in the soil-water system because a largely immobile form of nitrogen (ammonia) is converted to a highly mobile form (nitrate) which may be absorbed by plants or lost by leaching and denitrification.

Denitrification. The denitrification process involves the conversion of nitrate to gaseous nitrogen species such as elemental nitrogen gas, nitrous oxide, or nitric acid. A relatively large group of bacteria accomplish denitrification by using nitrate as an oxygen source in their respiration. Biological denitrification can cause a 5 to 10 percent reduction in total nitrogen in percolating water [16].

Plant uptake. In soils with heavy vegetal cover, the major mechanism for removal of inorganic nitrogen and phosphorus is uptake by plants. Flow of water toward roots in response to transpiration results in the transport of nonadsorbed nutrients with high solubilities, such as nitrate. Diffusion is the most active mechanism for transporting adsorbed species (e.g., phosphorus, potassium, iron) to plant roots [17].

Table 15-4 demonstrates the overall effectiveness of chemical removal during percolation [15]. The data indicate a high degree of purification for water carrying an extreme chemical burden, i.e., domestic wastewater. Data are provided for the effectiveness of two land application processes for wastewater treatment, i.e., slow rate application and rapid infiltration. Slow rate application is the application of wastewater to vegetated lands with moderate permeability. Rapid infiltration refers to application to highly permeable soils with minimal vegetation. Since natural waters carry a much lighter chemical burden than wastewater, a high degree of purification can be expected for most waters passing through soil during the runoff cycle.

15-5 Groundwater Quality

Groundwater comes from aquifers whose waters have been in contact with the atmosphere as recently as a few hours ago or as long ago as a few centuries [18]. The quality characteristics of groundwater are affected by the downward movement of water in recharge areas (percolation) and the lateral movement through aquifers (underflow).

The percolation of water through the surface soils of recharge areas generally results in significant purification. However, the effectiveness of this purification process is a function of the depth of soil above the water table, the type of soil, and the concentration of pollutants in percolating water. Where the water table is shallow and the soil is porous, dissolved gases, nitrates, sulfates, soluble organic compounds, and dissolved salts may be introduced into the

Table 15-4 Effectiveness of wastewater treatment by percolation through soil [15]

| Constituent | Characteristics of applied wastewater | | Characteristics after treatment by | | | |
| | | | Method 1[†] | | Method 2[†] | |
	Average, mg/l	Maximum, mg/l	Average, mg/l	Maximum, mg/l	Average, mg/l	Maximum, mg/l
BOD	200	300	<2	<5	2	<5
Suspended solids	200	350	<1	<5	2	<5
Ammonia nitrogen as N	25	50	<0.5	<2	0.5	<2
Total nitrogen as N	40	85	3	<8	10	<20
Total phosphorus as P	10	20	<0.1	<0.3	1	<5

† Method 1. Percolation of primary or secondary effluent through 1.5 m (5 ft) of vegetated soil with moderate permeability.
Method 2. Percolation of primary or secondary effluent through 4.5 m (15 ft) of highly permeable soils with limited vegetation.

groundwater system. If land disposal sites for solid wastes are not chosen wisely, high concentrations of chemical and gaseous products of organic decay are leached through the soil and into the phreatic zone (Chap. 6). In areas where the water table is relatively deep or the soil is less porous, the purification process is more thorough, and the aquifer recharge may be void of organic materials. Regardless of the depth of water table or type of soil, the salt content of some percolated waters is high, particularly in areas underlying irrigated land, septic tank systems, and defective industrial wastewater retention ponds [19].

As groundwater moves through an aquifer, contact with soluble rocks may result in a dramatic increase in dissolved minerals, particularly calcium bicarbonate, $Ca(HCO_3)_2$; magnesium bicarbonate, $Mg(HCO_3)_2$; calcium sulfate, $CaSO_4$; and magnesium sulfate, $MgSO_4$. Representative concentrations of these minerals and other important characteristics of groundwater are presented in Table 15-5.

15-6 Stream Transport

The water quality characteristics of streams are determined by the inflows to the stream, the amount of turbulence, interactions between water and the channel rocks and soils, and interactions at the air-water interface. The stream channel serves as the meeting place of water from surface runoff, interflow, groundwater, and municipal and industrial discharges. Thorough mixing of materials within the stream is typical. Dissolved substances, such as inorganic nutrients and gases, are effectively distributed both laterally and longitudinally by turbulent flow. A stream's carrying capacity for suspended materials, such as inorganic sediment and particulate organics, increases with increased velocity and turbulence (Sec. 11-8). Since sediment adsorbs anions and cations, the concentrations of many charged particles are directly related to the ability of the stream to maintain sediment in suspension.

Organisms living in streams have adapted to turbulent flow. Most biota restrict their activities to the shore (*littoral*) and bottom (*benthic*) zones of the stream. For example, benthic algae attach themselves to the channel bottom to avoid being washed downstream. Small invertebrate animals called *zooplankton* live among the rocks and debris of the channel in areas where the velocity of the stream is reduced. On the other hand, many forms of free-floating algae have adapted to turbulent flow areas by taking advantage of shallow, well-lighted swirls for exposure to light as well as spatial dispersion.

The atmosphere can be a significant source or sink of the biologically important gases, carbon dioxide and oxygen. Gas transfer between the atmosphere and the water is most commonly represented by the "stagnant boundary layer" model [20]. In this model, gas exchange is depicted as diffusion of gas through a stagnant water layer of variable thickness at the surface. The net flux F is dependent on *Fick's law* of diffusion:

$$F = \frac{c_0 - p^\alpha}{z} D \qquad (15\text{-}6)$$

Table 15-5 Quality of selected groundwaters in the United States

Substance[a]	Huntsville, Ala.[b]	Willamansett, Mass.[c]	Jefferson City, Tenn.[d]	Elizabeth City, N.C.[e]	Owyhee County, Idaho[f]	Wake County, N.C.[g]
Silica	8.4	15.0	8.4	41.0	99.0	22.0
Iron (total)	0.04	0.01		8.1	0.04	0.20
Manganese		0.01		0.0		
Calcium	46.0	96.0	40.0	45.0	2.4	2.5
Magnesium	4.2	19.0	22.0	17.0	1.4	2.1
Sodium	1.5	18.0	0.4	63.0	100.0	182.0
Potassium	0.8	1.5	1.2		2.9	
Bicarbonate	146.0	133.0	213.0	208.0	111.0	412.0
Sulfate	4.0	208.0	4.0	29.0	130.0	3.5
Chloride	3.5	25.0	2.0	83.0	10.0	9.5
Fluoride	0.0	0.4	0.0	0.2	22.0	1.7
Nitrate	7.3	0.4	4.8	0.6	0.5	0.6
Dissolved solids	130.0	468.0	180.0	405.0		452.0
Hardness (CaCO$_3$)	132.0	318.0	190.0	182.0	12.0	15.0
Specific conductance	250.0	690.0	326.0	638.0	449.0	718.0
pH	7.0	7.8	7.4	6.7	9.2	8.7
Color	5.0	3.0	5.0	23.0		1.0
Aluminum			1.4			

Source: Adapted from J. D. Hem, Study and Interpretation of the Characteristics of Natural Water, *U.S. Geol. Surv. Water-Supply Paper 1473*, 1970.

[a] All expressed in mg/l except conductance (micromhos at 25°C), pH, and color.
[b] Big Spring, limestone, 61°F.
[c] Industrial well, 120 ft deep, Chicopee shale, 54°F.
[d] Spring, Knox dolomite, 58°F.
[e] 30 to 80 ft deep, coastal plain deposits.
[f] 800 ft deep, 122°F.
[g] 184 ft deep, coastal plain sediments.

where c_0 is the concentration of gas in the water below the stagnant layer, p is the partial pressure of gas in the atmosphere, α is the solubility of gas in water, and D is the molecular diffusion rate of gas in water. The thickness of the stagnant boundary layer z is dependent on wind speed and the hydrodynamic properties of flow. High wind or turbulent water reduces the thickness of the boundary layer, and increases the gas exchange. Stream waters are less likely to be deficient in dissolved gases than are lake waters.

15-7 Lake Transport

Wind blowing across the water surface produces currents. These currents move across the lake until they are interrupted by a shoreline where the water moves downward and returns along the lake bottom.

Lakes may undergo an annual stratification cycle. During the summer, heat from solar radiation causes a decrease in the density of water near the surface. This difference in density between the water near the surface and that below results in a stable condition which resists mixing by currents. As a result, heat, dissolved gases, and nutrients present in the water near the surface are not readily distributed to lower layers. Thermal stratification separates the ecological processes that affect water quality into two zones. In the upper zone (*epilimnion*), the photosynthetic activity of phytoplankton, combined with oxygen from the atmosphere, maintains an aerobic environment. As particulate organic material settles out of the upper zone and into the lower zone (*hypolimnion*), bacterial growth is stimulated there. As bacteria break down the particulate organics, oxygen is consumed and nutrients are released. Throughout the summer months the dissolved oxygen content in the hypolimnion approaches zero, and all aerobic forms of life are threatened. In autumn, surface water cools and its density increases until it is denser than the deeper water. At this point the nutrient-rich, oxygen-depleted waters of the hypolimnion may circulate to the surface, causing temporary water quality problems.

Differences in water temperature affect the movement of water and its associated constituents in another manner. The river water entering a stratified lake travels downward through the lake profile until water of a greater density is reached, and then flows horizontally. Such currents are called *density currents*. Since increased suspended sediment load increases the density of water, muddy river waters may also induce a density current, or *turbidity current* (Sec. 11-14).

15-8 Water Temperature

The density and viscosity of water, and the solubility and diffusivity of gas in water are dependent upon water temperature. Both water density and viscosity decrease as temperature increases (Appendix Tables A-9, A-10). Even slight changes in density can result in stratification of lakes and impoundments, resulting in the water quality problems outlined in the preceding section. As temperature increases, the ability of water to carry suspended materials decreases according to Stokes' law (Sec. 11-8). The velocity of water below which particles settle out of suspension, the *settling velocity*, is directly related to water density and inversely proportional to viscosity. The net result of an increase in temperature (above 4°C or 39°F) is an increase in settling velocity and increased sedimentation in sludge deposits.

In general, solubility of gases in water decreases with increasing temperature (Appendix Table A-11). The relationship between oxygen saturation c_s in milligrams per liter and temperature T in degrees Celsius can be expressed by [21]

$$c_s = 14.652 - 0.4102T + 0.0079910T^2 - 0.00077794T^3 \qquad (15\text{-}7)$$

All biochemical reactions are sensitive to variations in temperature. Organisms consist of heat-sensitive proteins and enzymes which control the growth, respiration, reproduction, and death rate of each species. Within the range of tolerance for a species, an increase in temperature increases metabolic reaction rates. A generalized expression for the variation of a reaction-rate constant with temperature has been derived from the *van't Hoff-Arrhenius law* [22]:

$$\frac{k_1}{k_2} = \theta^{(T_2 - T_1)} \qquad (15\text{-}8)$$

where k_2 is the rate constant at temperature T_2, k_1 the rate constant at temperature T_1, and θ the temperature correction constant for the reaction. A value of 1.072 for θ corresponds to a twofold increase in reaction for each 10 Celsius degrees (18 Fahrenheit degrees) increase in temperature (*van't Hoff principle*). Although this principle serves as a good rule of thumb, research has shown that θ values for certain biologically mediated reactions, such as organic decomposition by bacteria, can vary significantly from 1.072 [23].

Exposure of organisms to temperatures outside their range of tolerance impedes or destroys enzymatic systems. Mathur [24] provides a summary of the biologic effects of water temperature fluctuations.

Although water temperature affects many processes important to the aquatic environment, the affected processes do not exert any material feedback on temperature. Consequently, changes in the heat content of a volume of water are due only to the transport of thermal energy across the volume boundaries. Within a body of water, changes in heat content occur by advection from water flowing into and out of the system, and heat transfer at the air-water and water-soil boundaries. Heat transfer between the water surface and the atmosphere is a major factor in the heat balance, but the transfer at the water-soil boundary is usually negligible. Heat discharges from industries and power plants can also be important factors in determining the thermal regime of a stream. Mechanisms which increase the heat content of water include absorption of shortwave and longwave radiation, and conduction-convection. Heat is lost by longwave back radiation, evaporation, conduction-convection, and shortwave and longwave reflection at the water surface. These elements of heat exchange are discussed in a general way in Secs. 2-1 through 2-3. The energy balance of a lake is treated in Sec. 5-3.

15-9 Biochemistry of Natural Waters

The discussion of nonliving materials is divided categorically into inorganic and organic chemicals. Organic chemicals are further divided into biodegradable and nonbiodegradable substances. The section concludes with a discussion of living materials in natural waters.

Inorganic chemicals. The inorganic chemicals of most importance to the aquatic environment include dissolved oxygen, free hydrogen ions, and compounds of carbon, nitrogen, and phosphorus. The survival of many aquatic organisms and the aerobic decomposition of waste materials depend on the maintenance of adequate dissolved oxygen in water. The major sources of oxygen are reaeration at the water surface and production of oxygen as a by-product of photosynthesis. The generalized equation for rate of reaeration of water is

$$\frac{dc}{dt} = k_i(c_s - c) \tag{15-9}$$

where k_i is the atmospheric reaeration coefficient (typically $0 \le k_i \le 10$ days) and c_s is the saturation concentration of dissolved oxygen. The most effective method of determining k_i depends on the hydraulic characteristics of the system [25]. In streams, empirical equations relating reaeration to depth and velocity of water [26, 27, 28] or drop in energy line along the length of the stream [29] have proved reliable. For lakes, wind speed and depth are the critical factors in determination of reaeration.

For every gram of new algal mass synthesized through photosynthesis, almost two grams of oxygen are produced. However, during the night photosynthesis ceases, while algae continue to consume oxygen to metabolize stored food. Thus, large algal blooms can be responsible for wide diurnal fluctuations in dissolved oxygen concentration. Other biochemical processes which deplete the oxygen supply of natural waters include oxidation of carbonaceous organic wastes by bacteria, nitrification, benthal oxygen demand, and respiration by organisms other than algae. Thomann [30] discusses the relative effects of these processes on the oxygen balance of streams and lakes.

The availability of free hydrogen ions in water is measured by the pH, defined in terms of the hydrogen-ion concentration $[H^+]$ as

$$\text{pH} = \log_{10} \frac{1}{[H^+]} = -\log_{10}[H^+] \tag{15-10}$$

Pure water has a pH of 7.0, representing absolute neutrality. Values lower than 7.0 are indicative of acidic waters, and values higher than 7.0 of alkaline waters. Most natural waters have a pH value in the range from 6.0 to 9.0, although accumulation of acids in volcanic lakes can result in a pH as low as 1.7, and basins without outlets may accumulate alkaline substances, with pH values higher than 12.0 [31]. Most organisms are highly sensitive to changes in pH. Even a small change in the pH of a system can result in competitive advantage and eventual dominance of species other than those historically within the system. Major changes in pH, such as those caused by spills from retaining ponds associated with mining activities, can cause total destruction of fish and all bottom-dwelling organisms for miles of river. After such acute stress, restoration of organisms to the stream may require months [32].

The value of pH is controlled by the inorganic carbon system. Consequently, processes which increase or decrease carbon dioxide concentration in water directly affect pH. These processes include carbon dioxide invasion at the water surface and biological activities such as photosynthesis, respiration, denitrification, and bacterial decomposition of organics.

The importance of the various species of inorganic carbon to the aquatic environment results from their ability to buffer water against rapid changes in pH. This buffering ability is brought about by the combination of carbon dioxide with water to form carbonic acid (H_2CO_3), and the further reaction of carbonic acid to produce neutral salts or bases. Depending on the hydrogen-ion concentration of water, carbonic acid dissociates into free hydrogen ions and bicarbonate (HCO_3^-) and carbonate (CO_3^{2-}) ions:

$$CO_2 + H_2O \rightarrow H_2CO_3 \rightarrow H^+ + HCO_3^- \rightarrow 2H^+ + CO_3^{2-} \qquad (15\text{-}11)$$

Acid in natural waters combines with the basic carbonate ion to form bicarbonate. If excess acidity remains after neutralization of all carbonate, bicarbonate is driven back into the carbonic acid form, and finally carbonic acid breaks down into water and free carbon dioxide. When the carbon dioxide concentration in the water exceeds solubility, carbon dioxide is released to the atmosphere. If alkaline conditions arise in natural waters, alkaline compounds combine with carbonic acid to form water and bicarbonate.

Nitrogen, in its various forms, can deplete dissolved oxygen levels, stimulate aquatic growth, or act as a toxic substance to aquatic organisms. Inorganic nitrogen compounds found in water include ammonia (NH_3), nitrite (NO_2^-), and nitrate (NO_3^-). In most unpolluted fresh waters the concentrations of these compounds are low [33], but organic wastes, and/or fertilizers can result in appreciable increases. At the end of the first-stage oxidation of organic wastes, most nitrogen is in the form of free ammonia. Nitrogenous oxygen demand (NBOD) is exerted on natural waters by the oxidation of ammonia to nitrite and nitrate by the chemoautotrophic bacteria *Nitrosomonas* and *Nitrobacter*:

$$NH_3 + \tfrac{3}{2}O_2 \xrightarrow{\text{Nitrosomonas}} NO_2^- + H_2O + H^+ \qquad (15\text{-}12)$$

$$NO_2^- + \tfrac{1}{2}O_2 \xrightarrow{\text{Nitrobacter}} NO_3^- \qquad (15\text{-}13)$$

For each 14 g of ammonia nitrogen oxidized, 48 g of oxygen is required for the conversion to nitrite nitrogen, and an additional 16 g for conversion of nitrite nitrogen to nitrate. Nitrate is the end product of aerobic stabilization of organic nitrogen.

All the inorganic forms of nitrogen can be used as nutrients by most green plants. Excessive concentrations of inorganic nitrogen can lead to algal blooms and overstimulation of growth of plants attached to the sides and bottoms of streams and lakes. In turn, the respiration of these plants while they are living, and their decomposition upon death, can cause severe depletion of oxygen within the water.

Free ammonia concentrations greater than 2.5 mg/l in neutral or alkaline water are toxic to many freshwater species, especially fish [34]. In acid waters ammonia exists in the nontoxic form ammonium hydroxide, NH_4OH. The Environmental Protection Agency Design Manual for Nitrogen Control [35] discusses all aspects of inorganic nitrogen chemistry in natural waters.

Phosphorus is vital to energy transfer systems in biota, and in many cases is the growth-limiting factor for algal communities. Like nitrogen, phosphorus passes through a cycle of uptake and assimilation into organic matter through photosynthesis, followed at death by bacterial decomposition into phosphates PO_4^{2-}. Total phosphorus concentrations in excess of 0.2 mg/l generally indicate that domestic wastes, industrial wastes, or fertilizers have been introduced into the water. High levels of phosphorus are believed to be the critical factor in the rate of eutrophication of lakes [36].

Organic chemicals. Organic materials found in the aquatic environment include natural compounds such as sugar, starch, fat, and oil, and synthetic compounds such as surfactants, phenols, and pesticides. All organic material eventually breaks down into inorganics, but the time span required for degradation varies significantly. Substances which are readily decomposed by the bacteria common to natural waters are termed *biodegradable,* and those which persist in the aquatic system for several weeks or longer are labeled *nonbiodegradable.* The distinction between degradable and nondegradable is not precise.

Forty to 80 percent of natural organics in water can be considered biodegradable [37]. The kinetics of aerobic decomposition can be represented by a first-order reaction rate formula:

$$\frac{dL_t}{dt} = -kL_t \tag{15-14}$$

where L_t is the amount of degradable material in water and k is the rate constant for a given system. Depending on natural population levels for decomposing bacteria and environmental factors such as water temperature and availability of oxygen, k varies for different waters from 0.1 to 0.6 per day [38]. Within a 5-day period, oxidation of degradable materials by aerobic bacteria is 60 to 70 percent complete, and 95 to 99 percent decomposition is accomplished in 20 days [39].

Settling of material to the stream bed or lake bottom may result in loss of organic material from water at a rate one or two orders of magnitude higher than the in-solution decomposition rate [40]. The decomposition of settled organics in aerobic waters still requires oxygen, but the oxygen demand exerted on the system may be spread over greater lengths of time.

In anaerobic waters, such as those sometimes found in the lower zone of lakes and reservoirs, the process of decomposition is carried out by anaerobic bacteria, and end products include methane and hydrogen sulfide in addition to ammonia, phosphate, and carbon dioxide, the end products of both aerobic and anaerobic decomposition. The end products of anaerobic decomposition may be toxic to freshwater organisms [41]. Synthetic organics such as phenols, certain pesticides, and the most common surfactant, linear alkyl sulfonate (LAS), are included in the category of biodegradable organics. Large concentrations of any of these synthetic materials can drastically deplete the oxygen supply of natural waters.

Nonbiodegradable organics include a fraction of the naturally occurring organics from plant and animal biomass, petroleum products, and many synthetic organic compounds. Naturally occurring nonbiodegradable substances do not directly upset the chemical balance of streams or lakes. However, since many portions of these substances are particulate, their settling may effectively smother the bottom of streams and lakes, resulting in the disruption of benthic algae, zooplankton, and fish communities.

Petroleum products can (1) form films on the water surface, resulting in decreased gas exchange across the surface; (2) emulsify in water and coat and destroy fish, algae, and other plankton; or (3) settle onto bottom areas, eliminating benthic algae and interfering with fish spawning.

Most current concern for nonbiodegradable organics in water is related to synthetic organic chemicals. Hundreds of new chemical compounds are introduced annually into commercial use, and each has its own characteristics relating to potential toxicity, persistence, and eventual breakdown products [42]. In some cases, the eventual breakdown products are more toxic than the original chemical. Other substances, such as the pesticide DDT or the family of plasticizers called polychlorinated biphenyls (PCBs), are not only persistent, but they accumulate in the tissues of animals. This process, known as *bioaccumulation*, can result in the concentration of an otherwise highly diluted substance into quantities which become toxic to fish or other animals higher in the food chain.

Biological organisms. The major division of organisms in water is the differentiation between those organisms which obtain the carbon that they need for biosynthesis from carbon dioxide, and those which obtain carbon from organic compounds. The first category are called *autotrophs*. Important autotrophs in the aquatic community include free-floating algae (phytoplankton) and algae attached to the bottom and sides (benthic algae). Because these photosynthetic organisms utilize energy from light in the initial production of organic matter, they are called *primary producers*. The rate of growth of primary producers is determined by the water temperature, pH, and the availability of inorganic carbon, nitrogen, and phosphorus [43–45].

Organisms which use organic compounds as their carbon source are called *heterotrophs* and include zooplankton, macroinvertebrates, fish, and decomposing bacteria. *Zooplankton,* the first level of predators in the aquatic community, ingest water and extract food from the water with simple filtering mechanisms. By feeding on algae, and in turn being consumed by larger animals, zooplankton assure that energy found in the organic matter of primary producers reaches the later stages of the aquatic food chain. In addition to algae eaters (*copepods*), there are certain species of zooplankton known as *cladocerans* which feed on both phytoplankton and suspended particulate organic matter.

Macroinvertebrates are larger organisms which serve as food for most types of fish and amphibians. Included in this category are insect larvae, crayfish, and aquatic snails and worms. As a group, these organisms are omnivorous, although certain species do exhibit a strong preference for a particular type of food. Vertebrates, including fish and amphibians, comprise the final step upward in the aquatic food chain.

The final group of heterotrophs, the decomposing bacteria, are critical to the aquatic environment. They are responsible for the breakdown and recycling of dead organics resulting from the excretion and death of all other life forms within the system, as well as organics released into water from external sources such as wastewater. In the aerobic zone, bacteria convert organic materials such as starch, fat, cellulose, and chitin into carbon dioxide and oxidized inorganic nutrients. Organic materials which settle into aerobic waters are decomposed into organic acids, methane, hydrogen gas, and carbon dioxide [46].

15-10 Water Quality Simulation

The concentration of chemicals and biota in natural waters varies from day to day and even hour to hour depending on weather, flow conditions, and time-varying inputs by people. Additional changes in concentration result from interactions between the various materials present in water. For example, nitrate concentration depends on availability of inorganic nitrogen in the form of ammonia or nitrite, the presence of nitrifying bacteria, water temperature, and the uptake of nitrate as a nutrient by phytoplankton. Figure 15-1 is a flow diagram illustrating additional interactions between significant water quality parameters. The fluctuation of material concentrations and the complexity of interactions within water make it difficult to evaluate the relative health or deterioration of a stream or lake. While data on isolated events or average concentration values may indicate that overall quality of a water body is good, serious problems may occur during several hours of the day or at certain times of the year. In one system critical conditions may occur during low-flow conditions in summer, while another system may suffer its worst deterioration due to non-point-source pollution shortly after storm events.

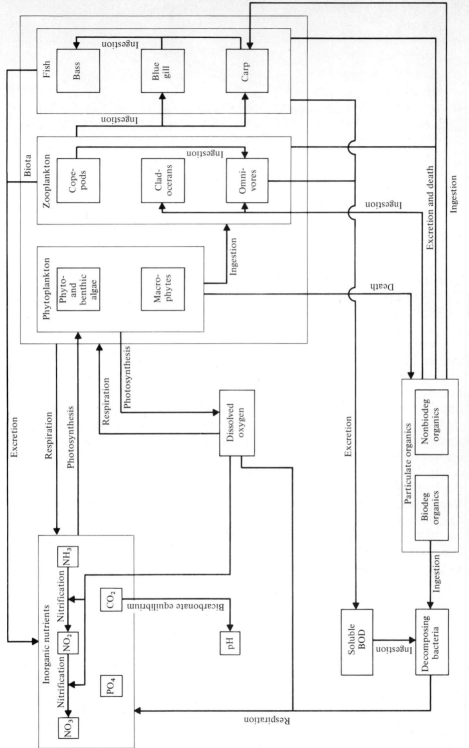

Figure 15-1 Flow diagram of important material interactions between water quality parameters.

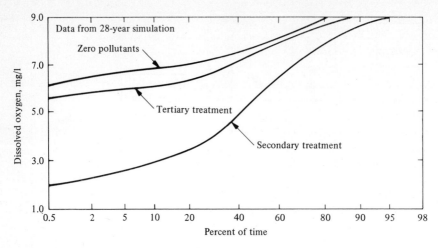

Figure 15-2 Percent of time dissolved oxygen is less than or equal to indicated values, South Platte River, Denver, Colo. (*Black and Veatch.*)

A useful way to describe pollution is in terms of tables or graphs showing the percentage of time that a quality characteristic is above or below critical levels [47]. Figure 15-2 represents the expected dissolved oxygen characteristics of a stream receiving domestic wastewater, assuming three levels of pollution control. In this example the concentration of dissolved oxygen is less than 5 mg/l 40 percent of the time with conventional secondary treatment of the wastewater. Since such low oxygen levels are detrimental to many fish species, the data indicate a need for additional treatment. If tertiary treatment measures are applied to the wastewater, the data indicate that the stream dissolved oxygen level would be above the critical level for fish at all times. If no wastewater were discharged to the stream at all, a small additional increase in dissolved oxygen would result.

Long records of water quality data suitable for development of probability curves are scarce. Mathematical models supported by coordinated data-collection programs can provide the data required for probability curves. Three types of water quality models are available. *Event models* utilize measured or hypothetical hydrographs (runoff versus time) and pollutographs (mass loadings versus time) to simulate water quality for selected critical events. This type of model does not provide probability information but can be used to estimate "worst case" conditions. The *steady-state models* assume a constant stream-flow and estimate the changes in water quality downstream from the source of pollution. The reliability of the output data is restricted by the simplifying assumptions used in the model. *Continuous models* deal with meteorological, hydrologic, hydraulic, and biochemical processes as continuously varying in time. Given the proper input data, such models can reproduce a continuous record of water quality characteristics over a period of years. Five to 10 years of data are required to produce a reasonably reliable probability curve. The effective use of a continuous simulation model requires six steps:

1. *Data assembly*. Required input data include meteorological time series; flow-rate and chemical composition of point-source inflows; instream water quality data; land use, soil types, and vegetal cover information for the study basin; and information on storm sewer discharges or other significant pollution sources.
2. *Hydrologic calibration*. A calibration procedure similar to that outlined in Chap. 12 is used.
3. *Quality sampling*. In most regions a special data-collection program is required to obtain an adequate spatial and temporal characterization of certain point sources, storm-water runoff events, and water quality characteristics at critical instream locations.
4. *Quality calibration*. Model parameters are adjusted to produce close agreement between simulated and real values. The calibration effort should include comparison of simulated and recorded values which represent both diurnal and seasonal variations in quality parameters.
5. *Establish base condition*. Simulation of flow and quality constituents for a period of 5 to 10 years using "as is" conditions with respect to catchment practices. This provides information on current problems within study area.
6. *Test alternative management plans*. Guided by base conditions, a list of possible alternative water quality control measures can be developed. The effectiveness of the measures which appear most promising is simulated by modifying the input data to the model to represent implementation of the measure. For example, the effect on stream quality of upgrading an existing secondary treatment plant to tertiary treatment may be modeled by modifying the chemical characteristics of the point source used to represent the existing plant.

Numerous water quality models are available. Models vary in number and type of water quality parameters considered, mathematical solution techniques, and number of dimensions modeled. Some models are specific to only one phase of the hydrologic cycle (i.e., surface runoff, receiving water, or groundwater), and others simulate a combination of phases. Table 15-6 summarizes the basic capabilities of a number of currently available water quality models. The ability of each model to simulate critical hydrologic, hydraulic, and water quality processes is indicated. Some of the models are structured so as to allow their use in more than one mode (i.e., event, steady-state, or continuous). Possible run modes for each model are also included in the summary table. Choice of the suitable computer model for an application depends on the questions which must be answered, the required accuracy of the answers, and the complexity of the system which is to be modeled.

Table 15-6 Comparative capabilities of simulation models†

| | Land surface simulation capabilities | | | | |
| | Hydrology | | | Water quality | |
Models	Pervious lands	Impervious lands	Snowmelt	Physical/ chemical constituents	Agricultural chemicals
HSPF [48]	●	●	●	●	●
ARM [49]	●		●	●	●
NPS [6]	●	●	●	●	
HSPX [50]	●	●	●	●	
SWMM [51]		●		●	
QUAL-II [52]					
SSARR [53]	●	●	●		
WQRRS [54]					
MIT model [55]	●	●	●	●	
STORM [56]	●	●	●	●	

† Numbers in brackets correspond to references at end of chapter.

15-11 Water Quality Sampling

Water quality sampling and analysis programs are necessary to determine the suitability of water for domestic, industrial, and agricultural uses. Data from carefully designed programs can provide understanding of the geochemical and hydrologic relationships in natural systems and the influence of human activities on these systems. Samples must be truly representative of the water body, or portion of the water body, which they are intended to portray. Successful representation of the dissolved and suspended matter present in natural waters depends on prudent selection of sample sites, type and frequency of observations, and sampling equipment and procedures.

Critical factors in site selection vary with the type of water (i.e., surface runoff, receiving water, or groundwater) being sampled, and the questions which the sampling program is intended to answer. For example, the purpose of most surface runoff studies is to characterize the runoff quality resulting from specific land uses within a given geological and hydrologic regime. For such

Table 15-6 (*Continued*)

| | Receiving water simulation capabilities | | | | | | | | |
| | Hydraulics | | | | Water quality | | | | |
Models	One-dimensional streams	Unstratified reservoirs	Stratified reservoirs	Estuaries	Basic parameters	Sediments	Nutrients	Algae	pH
HSPF [48]	●	●			●	●	●	●	●
ARM [49]									
NPS [6]									
HSPX [50]	●	●	●		●		●	●	●
SWMM [51]	●	●	●	●	●		●		
QUAL-II [52]	●				●		●	●	
SSARR [53]	●	●							
WQRRS [54]	●	●	●		●	●	●	●	●
MIT model [55]	●			●	●		●		
STORM [56]					●		●		

studies it is essential to locate each sampling site so that the runoff collected is representative of a well-defined and homogeneous land segment. In the case of receiving water, the usual goal is to determine the impact of specific point-source loadings on stream water quality. To accomplish this, sites must be situated directly upstream and downstream of each major point source. If the purpose of a groundwater analysis is to evaluate the general quality of an aquifer, a large number of areally distributed samples must be collected to determine direction and rate of flow, as well as chemical variability.

Sample type and frequency are also dictated by the type of water being sampled. For characterization of storm-water runoff, sequential discrete samples are recommended [57]. A *sequential discrete sample* is a series of "grab samples" taken at regular or irregular (but known) time intervals. The sampling should continue through a runoff event and each sample should have a concurrent flow measurement so that mass loading can be calculated. If average pollutant concentrations for storm events are adequate, a *flow-proportional composite sample* may be used. This consists of a number of samples at regular

Table 15-6 (***Continued***)

	Possible run modes					
	Event		Steady state		Continuous	
Models	Hydrology	Water quality	Hydrology	Water quality	Hydrology	Water quality
HSPF [48]	●	●	●	●	●	●
ARM [49]	●	●	●	●	●	●
NPS [6]	●	●	●	●	●	●
HSPX [50]	●	●	●	●	●	●
SWMM [51]	●	●	●	●		
QUAL-II [52]			●	●		
SSARR [53]	●		●		●	
WQRRS [54]	●	●	●	●	●	●
MIT model [55]	●	●	●	●		
STORM [56]	●	●	●	●		

intervals with their volumes proportional to the instantaneous flowrate at the time of sampling. For stream sampling, sequential composites are commonly collected manually or with automatic equipment (Fig. 15-3). With this equipment samples can be taken at regular intervals (e.g., hourly) and retained separately for analysis. Such samples make it possible to study diurnal fluctuations such as those produced by algal photosynthesis. Sequential composites are collected periodically throughout the year to monitor seasonal variability. Since groundwater quality does not vary rapidly, most programs call for monthly or even quarterly samples.

Equipment and sampling procedures for water quality are highly specialized. The values of some parameters change rapidly in the sample bottle, and field analysis is necessary. Dissolved oxygen, temperature, conductivity, and pH are usually determined on-site. Special procedures are required for samples to be analyzed for heavy metals or trace organics to avoid contamination, oxidation, or adsorption to the sampling equipment [58]. The U.S. Geological Survey has published a series of manuals dealing with sampling and analysis procedures for storm-water runoff [57], receiving water [59, 60], and groundwater [58].

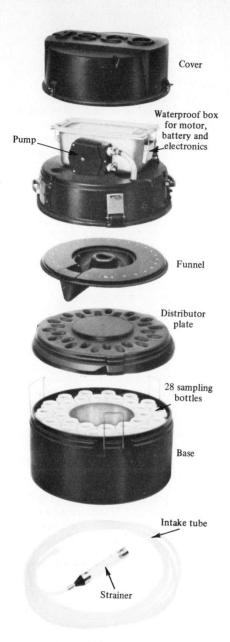

Cover

Waterproof box
for motor,
battery and
electronics

Pump

Funnel

Distributor
plate

28 sampling
bottles

Base

Intake tube

Strainer

Figure 15-3 Exploded view of an automatic water quality sampler. Water from the source is pumped at timed intervals into successive sampling bottles for latter analysis. Ice can be used to cool the unit if necessary. (*Instrumentation Specialties Co.*)

PROBLEMS

15-1 Obtain a copy of the water quality standards for your state. Are these standards written in terms of the frequency or duration of events in violation of the specified concentration levels? Do the standards for surface water and groundwater differ? Which seem to be more strict?

15-2 Chlorophyll strongly absorbs light in the regions 0.40 to 0.45 μm (violet) and 0.64 to 0.66 μm (red). Using the information of Table 15-1, determine the percentage of each of these two colors which is available for photosynthesis at depths of 10 cm, 1 m, and 10 m below the water surface.

15-3 Observations on a number of small Canadian lakes show the stagnant boundary layer to be about 310 μm thick. The oxygen partial pressure remains at or above the atmospheric equilibrium value of 0.21×40.32 mg O_2/l per atm. Calculate the net flux of oxygen at the water surface at 20°C if the water below the stagnant layer is anoxic (dissolved oxygen concentration near zero). Gas diffusivity of oxygen at 20°C is 0.000081 ft²/hr. How does this flux compare with that expected if the water below the stagnant layer has sufficient oxygen for a healthy trout population, about 7.0 mg/l? Note that a negative value indicates the addition of the gas to the water while a positive value connotes a loss of oxygen.

15-4 Two alternative empirical relationships for calculating the reaeration coefficient k_i in Eq. (15-9) are:

(1)
$$k_i = \frac{11.61 v^{0.969}}{D^{1.673}}$$
(Churchill [26])

(2)
$$k_i = \frac{21.73 v^{0.67}}{D^{1.85}}$$
(Owens [28])

where k_i is in days^{-1}, v is the stream velocity in feet per second and D is the average depth of the stream in feet. Calculate k_i from each formula assuming $D = 4$ ft and $v = 2$ ft/s. If the stream contained no oxygen, how long would it take to raise the oxygen level to 98 of the saturation level at 20°C (assume sea-level elevation and zero chloride concentration, Table A-11)? Calculate the time for each value of k_i.

15-5 If you have access to a computer model for quality simulation, familiarize yourself with the model and the required input sequence. Select two or three parameters and make a series of short runs, changing the value of one parameter at a time. Which parameter is most sensitive to change? Why?

REFERENCES

1. MacCracken, M. C., and G. D. Sauter (eds.): Development of an Air Pollution Model for the San Francisco Bay Area, *Lawrence Livermore Lab. Rep. UCRL*-51920, vols. 1, 2, 1975.
2. U.S. Environmental Protection Agency: Acid Rain, *Office of Research and Development Summary Rep.* 600/8-79-028, pp. 3–6, 1979.
3. McElroy, A. D., et al.: Loading Functions for Assessment of Water Pollution from Nonpoint Sources, *U.S. Environmental Protection Agency Rep.* 600/2-76-151, pp. 92–94, 1976.
4. Stewart, K. M., and S. J. Markello: Seasonal Variations in Concentrations of Nitrate and Total Phosphorus, and Calculated Nutrient Loading for Six Lakes in Western New York, *Hydrobiologia*, vol. 44, no. 1, pp. 61–89, 1974.
5. See [2], pp. 1–3.
6. Donigian, A. S., Jr., and N. H. Crawford: Modeling Nonpoint Pollution from the Land Surface, *U.S. Environmental Protection Agency Rep.* 600/2-76-043, pp. 33–40, 1976.
7. Sources of Nitrogen and Phosphorus in Water Supplies, Task Group Rep., *J. Am. Water Works Assoc.*, vol. 59, no. 344, 1967.
8. See [6], pp. 5–9.

9. Beyerlein, D. C., and A. S. Donigian, Jr.: The Effects of Soil and Water Conservation Practices on Runoff and Pollutant Losses from Small Agricultural Watersheds, Hydrocomp Inc., Palo Alto, Calif., 1978.

10. Loehr, R. C.: Characteristics and Comparative Magnitudes of Nonpoint Sources, *J. Water Poll. Cont. Fed.,* vol. 46, pp. 1849–1870, 1974.

11. Singh, R.: Statistical Characterization of Urban Pollutants Loading Rates, *ASCE preprint* 2833, pp. 1–20, 1977.

12. Eastman, R. H.: "General Chemistry: Experiment and Theory," chap. 13, Holt, Rinehart and Winston, New York, 1970.

13. Metcalf and Eddy, Inc.: "Wastewater Engineering; Collection, Treatment, Disposal," chap. 9, McGraw-Hill, New York, 1972.

14. Farmer, W. J., and J. Letey: Volatilization Losses of Pesticides from Soils, *U.S. Environmental Protection Agency Rep.* 660/2-74-054, pp. 24–51, 1974.

15. U.S. Environmental Protection Agency: Process Design Manual for Land Treatment of Municipal Wastewater, *Rep.* 625/1-77-008, chap. 2, 1977.

16. See [15], chap. 2.

17. Davidson, J. M., et al.: Simulation of Nitrogen Movement, Transformation, and Uptake in the Plant Root Zone, *U.S. Environmental Protection Agency Rep.* 600/3-78-029, pp. 21–33, 1978.

18. Gehm, H. W., and J. I. Bregman (eds.): "Handbook of Water Resources and Pollution Control," chap. 1, Van Nostrand Reinhold, New York, 1976.

19. National Water Well Association: Ground Water Pollution Control, pp. 1–16, 1978.

20. Emerson, S., W. Broecker, and D. W. Schindler: Gas-Exchange Rates in a Small Lake as Determined by the Radon Method, *J. Fish. Res. Board Can.,* vol. 30, no. 10, pp. 1475–1484, 1973.

21. Committee on Sanitary Engineering Resources: Solubility of Atmospheric Oxygen in Water, Twenty-ninth Progress Rep., *J. San. Eng. Div. ASCE,* vol. 86, no. SA4, p. 41, 1960.

22. See [18], chap. 18.

23. Thomann, R. V.: "Systems Analysis and Water Quality Management," chap. 5, McGraw-Hill, New York, 1972.

24. See [18], pp. 727–728.

25. Covar, A. P.: Selecting the Proper Reaeration Coefficient for Use in Water Quality Models, *Proc. Conf. Env. Model. and Simulation,* Cincinnati, Ohio, pp. 340–342, 1976.

26. Churchill, M. A., et al.: Prediction of Stream Reaeration Rates, *J. San. Eng. Div. ASCE,* Proc. Paper 3199, p. 1, 1962.

27. O'Connor, D. J., and W. E. Dobbins: Mechanisms of Reaeration in Natural Streams, *Trans. ASCE,* vol. 123, p. 655, 1958.

28. Owens, M., et al.: Some Reaeration Studies in Streams, *Int. J. Air Water Poll.,* vol. 8, pp. 469–486, 1964.

29. Tsivoglou, E., et al.: Tracer Measurement of Stream Reaeration, II. Field Studies, *J. Water Poll. Cont. Fed.,* vol. 40, pp. 285–305, 1968.

30. See [23], chap. 5.

31. Reid, G. K.: "Ecology of Inland Waters and Estuaries," pp. 156–163, Van Nostrand Reinhold, New York, 1961.

32. Crossman, J. S., et al.: The Use of Cluster Analysis in the Assessment of Spills of Hazardous Materials, *American Midland Naturalist,* vol. 92, pp. 94–114, University of Notre Dame Press, South Bend, Ind., 1974.

33. See [31], pp. 183–187.

34. See [31], p. 185.

35. Process Design Manual for Nitrogen Control, U.S. Environmental Protection Agency, Office of Technology Transfer, 1975.

36. McKee, J. E., and H. W. Wolf (eds.): "State of California Water Quality Criteria," 2d ed., State Water Resources Control Board, p. 242, 1971.

37. Foree, E. G., and P. L. McCarty: Anaerobic Decomposition of Algae, *Env. Sci. and Tech.,* vol. 4, pp. 842–849, 1970.

38. See [23], p. 97.
39. See [13], pp. 241–254.
40. See [23], pp. 92–96.
41. Haan, R. W., Jr., and R. G. Willey: "Water Quality Determinations," chap. 2, U.S. Army Corps of Engineers, Hydrologic Engineering Center, Davis, Calif., 1972.
42. U.S. Council on Environmental Quality: Toxic Substances, pp. 1–22, 1971.
43. Sorokin, C., and R. W. Krauss: Effects of Temperatures and Illuminance on Chlorella Growth Uncoupled from Cell Division, *Plant Physiol.*, vol. 37, pp. 37–42, 1962.
44. MacIsaac, J. J., and R. C. Dugsdale: The Kinetics of Nitrate and Ammonia Uptake by Natural Populations of Marine Phytoplankton, *Deep Sea Res.*, vol. 16, pp. 415–422, 1969.
45. Thomas, W. H., and A. N. Dodson: Effects of Phosphate Concentration on Cell Division Rates and Yield of a Tropical Oceanic Diatom, *Biol. Bull.*, vol. 134, no. 1, pp. 199–208, 1968.
46. Brock, T. D.: "Biology of Microorganisms," pp. 557–560, Prentice-Hall, Englewood Cliffs, N.J., 1974.
47. Planning for Water Quality Management, *Tech. Note* 1, pp. 1–12, Hydrocomp Inc., Palo Alto, Calif., 1977.
48. Johanson, R. C., J. C. Imhoff, and H. H. Davis, Jr.: User's Manual for the Hydrologic Simulation Program-Fortran (HSPF), *U.S. Environmental Protection Agency Rep.* EPA-600/80-015, April 1980.
49. Donigian, A. S., Jr., and H. H. Davis, Jr.: User's Manual for Agricultural Runoff Management (ARM) Model, *U.S. Environmental Protection Agency Rep.*, 600/3-78-080, 1978.
50. Hydrocomp Water Quality Operations Manual, 3d ed., Hydrocomp Inc., Palo Alto, Calif., 1977.
51. Metcalf & Eddy, Inc.: University of Florida, and Water Resources Engineers Inc.: Storm Water Management Model, vols. 1–4, *U.S. Environmental Protection Agency Rep.* 110224DOC, 1971.
52. Roesner, L. A., J. R. Monser, and D. E. Evenson: Computer Program Documentation for the Stream Water Quality Model QUAL/II, Water Resources Engineers Inc., Walnut Creek, Calif., 1973.
53. Streamflow Synthesis and Reservoir Regulation: Program Description & User Manual for SSARR Model, U.S. Army Corps of Engineers, North Pacific Division, Portland, Oreg., 1972.
54. Water Quality for River-Reservoir Systems, U.S. Army Corps of Engineers, Hydrologic Engineering Center, Davis, Calif., 1978.
55. Harleman, D. R. F., et al.: User's Manual for the M. I. T. Transient Water Quality Network Model, *U.S. Environmental Protection Agency Rep.* 600/3-77,010, 1977.
56. Urban Storm Water Runoff: STORM User's Manual, U.S. Army Corps of Engineers, Hydrologic Engineering Center, Davis, Calif., 1976.
57. Guide for Collection, Analysis, and Use of Urban Stormwater Data, Conference Report by The Engineering Foundation, U.S. Geological Survey, and ASCE, published by ASCE, pp. 1–92, 1976.
58. "Guidelines for Collection and Field Analysis of Ground-Water Samples for Selected Unstable Constituents," Book 1, chap. D2, U.S. Geological Survey, 1976.
59. "Methods for Collection and Analysis of Water Samples for Dissolved Minerals and Gases," book 5, chap. Al, U.S. Geological Survey, 1970.
60. "Methods for Collection and Analysis of Aquatic Biological and Microbiological Samples," book 5, chap. A4, U.S. Geological Survey, 1977.

BIBLIOGRAPHY

Cairns, J., Jr., and K. L. Dickson (eds.): Biological Methods for the Assessment of Water Quality, *ASTM Tech. Publ.* 528, pp. 3–175, 1973.

——, ——, and A. W. Maki (eds.): Estimating the Hazard of Chemical Substances to Aquatic Life, *ASTM Tech. Publ.* 657, pp. 3–191, 1978.

Debo, T. N., A. M. Lumb, and R. K. Linsley: "Storm Water Management Handbook," Hydrocomp Inc., Palo Alto, Calif., 1977.

Loehr, R. C., et al. (eds.): "Best Management Practices for Agriculture and Silviculture," Ann Arbor Science Publishers, Inc., Ann Arbor, Mich., 1979.

O'Connor, D. J., R. V. Thomann, and D. M. Ditoro: Dynamic Water Quality Forecasting and Management, *U.S. Environmental Protection Agency Rep.* 660/3-73-009, 1973.

Skelly and Loy Engineers—Consultants: Processes, Procedures, and Methods to Control Pollution from Mining Activities, *U.S. Environmental Protection Agency Rep.* 430/9-73-011, pp. 17–359, 1973.

Tennessee Valley Authority: Heat and Mass Transfer between a Water Surface and the Atmosphere, *Lab. Rep.* 14, pp. 1.1–4.35, 1972.

Zison, S. W., et al.: Rates, Constants, and Kinetic Formulations in Surface Water Quality Modeling, *U.S. Environmental Protection Agency Rep.* 600/3-78-105, 1978.

CHAPTER
SIXTEEN

APPLICATIONS OF HYDROLOGY

Most decisions regarding the management of water resources are based, in part at least, on estimates of the quantity of water to be managed or the rate of flow to be regulated. Hydrology is therefore basic to the planning of water resources projects and to the subsequent operation of these projects. The project does not have to be a physical structure—it may be only a management plan.

Traditionally, hydrology has been considered a tool which could provide the design parameters of a project—reservoir capacity, spillway capacity, design flood, etc. The *designer* needs only a single value for each parameter. In contrast, the *planner* requires data in probabilistic form in order to make meaningful socioeconomic evaluations and an effective choice among the possible alternatives. This chapter describes the way in which hydrologic techniques may be used to provide information for a variety of problems.

With minor exceptions there are no "standard" methods in hydrology. This situation is in part justified by the fact that differing climatic and physical characteristics, varying levels of data availability and reliability, and differing end purposes do require differences in the details of analysis. On the other hand different countries tend to use their own methods, and within the United States each government agency concerned with hydrology has its own manual of procedures. Hydrology is highly data-dependent, and the steps required to prepare data for analysis (Secs. 16-1 and 16-2) are an important part of every study. The approaches chosen for this chapter are the simplest ones available consistent with reliable answers. There is, however, no way that the hydrologist can be relieved of the necessity of using judgment in the selection and application of procedures. The many simple empirical procedures which abound in the literature of hydrology cannot be recommended, since the scanty tests available show them to be highly unreliable [1–3].

The planning task should always include consideration of the economics of a project. The details of this evaluation are not within the scope of this text, but there are good discussions elsewhere [4–6]. It is important to emphasize that reliable hydrologic analysis is essential to reliable economic evaluation. Planning may also require four different phases of analysis:

1. Defining preproject conditions as a base for project evaluation.
2. Defining postproject conditions to determine project effects.
3. Development of an operating procedure if required.
4. Definition of specific design events such as project design flood or spillway design flood.

16-1 Data Preparation

Before beginning any hydrologic analysis it is important to be sure that the data are homogeneous. This requires checking the station history and/or double-mass analysis (Sec. 3-10). Streamflow data can be further evaluated by checking the changes which have occurred in the basin during the period of record. In each case the data should be appropriately adjusted to either current conditions or natural conditions (Sec. 4-16). The errors resulting from lack of homogeneity in data are especially serious because they lead to bias in the final answers.

Random errors in the data are somewhat less serious but nevertheless the data should be inspected for large random errors resulting from observer error, instrument failure, and unnatural events. For example, flood flows resulting from a dam failure should generally be removed from streamflow data. Large random errors can often be detected by comparing records at different stations for consistency. A dam failure will, in many cases, be a matter of record.

For many purposes missing records must be estimated. Small segments of missing precipitation data can be estimated by several techniques (Sec. 3-9). If long periods of data are missing from a station record, it may be wise to drop the station from the analysis. Similarly, if records are missing at several stations, it may be better to omit that period from the analysis.

16-2 Record Extension

One of the most serious sources of error in probability estimates is limited data (Sec. 13-1). Streamflow data can be extended in several ways. If there is a record at the site of 15 to 20 years or more, correlation with a nearby streamflow record may be satisfactory for monthly and annual data. The equation should include a random term to maintain the variance of the flow and thus would take the form

$$Q_x = a + bQ_y + tS_x \qquad (16-1)$$

where the subscript x refers to the station being estimated and y refers to the record which is the base for the extension. S_x is the standard error of estimate of Q_x, and t is a random normal deviate with mean of zero and unit standard deviation. Where the correlation is very good, a similar approach might work for daily flow. An alternative is the use of a deterministic simulation model (Chap. 12) if suitably long rainfall records are available. If only flood peaks are required, the two or three largest values each year might be estimated with a rainfall-runoff relation and unit hydrograph (Chaps. 7 and 8). An extended rainfall record might be developed by regression using Eq. (16-1) and time periods of a month or more. In some locations it may be necessary to have a different equation for each month or season.

If there is no streamflow record at or near the site, monthly or annual flow volumes may be transferred from another location with adjustment for size of drainage area and mean annual precipitation. The two catchments should be similar in all respects if the estimate is to be reliable. A large part of the runoff volume in some catchments appears as base flow subject to geological controls, and dissimilarities in this regard may cause serious error. A deterministic simulation model can also be used by calibrating on one or more similar streams and making adjustments to the parameters as judgment indicates.

16-3 Water-Supply Reservoirs

Water-supply reservoirs store water in periods of surplus for use during periods of deficit. The water may be for city water supply, irrigation, hydroelectric power, or other uses. In all cases the hydrologic approach is essentially the same, the only difference being the estimate of required water supply. The analysis may attempt to determine the firm or average yield or the reliability with which a specific demand can be met. *Firm yield* is the minimum yield during the life of the reservoir. *Average yield* is the arithmetic average of the yields available in each year of the project life. A *short-term reservoir* is planned to operate on an annual (or shorter) cycle. A *long-term reservoir* is intended to carry water over two or more years to cope with long droughts.

The first step in the analysis of a reservoir is the determination of the area-elevation and volume-elevation curves for the site (Sec. 9-4). Care should be taken to use maps with scales and contour intervals adequate to define the reservoir area and capacity accurately. It is also necessary to have an estimate of the required yield to meet the purpose of the reservoir [7]. Finally, it is usually necessary to estimate the *usable storage,* commonly taken as equal to the total storage less the volume required to store the sediment accumulation during the life of the project (Chap. 11). Other factors may sometimes control.

The traditional estimate of yield has been based on a *critical period,* usually the driest period in the historic record. No meaningful estimate of probability can be made for such a critical period. Almost certainly a more severe drought will occur, but the traditional analysis provides no estimate of risk.

Short-term reservoirs. Since the short-term reservoir completes a storage cycle annually or more frequently, a reasonably long record of streamflow should prove to be an adequate data base. One needs to use these data to simulate the performance of the reservoir. The time increment for this simulation should be no longer than 1 month, and in some cases, daily increments should be used.

Illustrative example 16-1 An offstream reservoir is to be built adjacent to Antietam Creek near Sharpsburg, Md. It will be filled by diversion in a canal with 200 cfs capacity. Using the data of Fig. 4-15, determine the yield of the reservoir during water year 1971 and the yield which would be assumed if monthly and annual data were used in the analysis. Ignore evaporation and seepage loss.

Summing the daily flows taking only 200 cfs when the flow equals or exceeds that value gives a total yield of 67,789 cfs-days. If monthly mean flows are used, the yield appears to be 69,002 cfs-days (1.7 percent high). The annual data suggest that the yield is $200 \times 365 = 73,000$ cfs-days (7.5 percent high). Note that the error generally tends to overestimate yield. Even the calculation based on mean daily flows may be slightly high.

A storage model is used to compute the yield of the reservoir each year. The model performs computations similar to those illustrated in Table 16-1 but can easily be programmed for solution on a small computer. In the table, precipitation on the reservoir and evaporation from it (columns 6 and 7) are calculated by multiplying the estimated rates (columns 2 and 3) by the average reservoir area for the month taken as the average of the areas at the beginning and end of the month (column 11). A special release to meet downstream water rights is required (column 8) in the amount of 400 cfs or the natural inflow, whichever is less. Column 5 indicates an assumed variable monthly withdrawal for use. For a short-term reservoir without carryover storage as in Table 16-1, the annual yield is given by

$$\text{Yield} = \text{inflow} - \text{evaporation} + \text{precipitation} - \text{releases} \quad (16\text{-}2)$$

or $$\text{Yield} = \text{withdrawals} + \text{change in storage} \quad (16\text{-}3)$$

For the example year of Table 16-1, yield = 464 acre · ft. In simple cases, Eq. (16-2) may be solved with annual totals, but if there is a large area change or the calculation of releases (including spills) requires monthly or daily computations, the pattern of Table 16-1 should be followed.

When a yield value has been determined for each year of record, these values are arranged in order of magnitude and plotted as a flow-duration curve (Fig. 16-1) showing the percent of the years with yields equal to or less than various values. In the example, the firm yield based on the worst years in the record is zero, the average yield is 4815 acre · ft, the median yield is 3700 acre · feet and a yield of 1500 acre · ft is indicated in four years out of five. By

Table 16-1 Operation study for a storage reservoir

Month (1)	Precipitation, ft (2)	Evaporation, ft (3)	Inflow, acre·ft (4)	Withdrawal, acre·ft (5)	Precipitation, acre·ft (6)	Evaporation, acre·ft (7)	Release, acre·ft (8)	Storage change, acre·ft (9)	Storage,† acre·ft (10)	Area,† acres (11)
Sept.									5421	
Oct.	0.034	0.252	0	255	7	54	0	−302	5119	213
Nov.	0.066	0.137	0	132	14	29	0	−147	4972	210
Dec.	0.381	0.074	0	102	80	16	0	−38	4934	210
Jan.	0.745	0.083	1090	104	160	18	400	728	5662	220
Feb.	0.100	0.102	33	136	22	22	33	−136	5526	218
Mar.	0.280	0.210	347	146	61	46	347	−130	5395	216
Apr.	0.021	0.274	12	260	45	59	12	−274	5121	212
May	0.155	0.398	11	458	31	79	11	−506	4615	198
June	0	0.499	3	458	0	95	3	−553	4062	185
July	0	0.547	0	458	0	96	0	−554	3508	168
Aug.	0	0.465	0	416	0	75	0	−491	3017	155
Sept.	0	0.384	0	385	0	57	0	−442	2575	144
Total	1.782	3.425	1496	3310	420	646	806	−2845		

† At end of month.

446

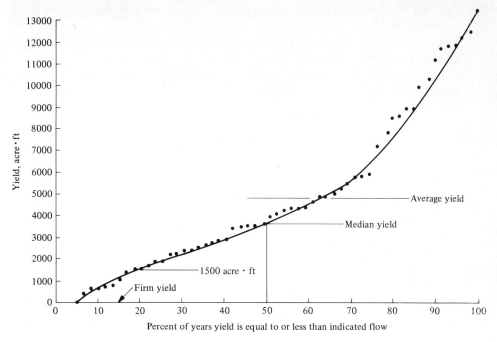

Figure 16-1 A flow-duration curve.

assuming a larger reservoir and thus reducing the spills, these yield values might be increased. Hence, the process should be repeated for several reservoir capacities to find the most economic capacity for the desired yield.

Long-term reservoirs. Long-term reservoirs have sufficient storage capacity to meet demands during periods of two or more dry years. The kind of annual probability suggested for short-term reservoirs is not adequate. Generally this problem is best solved by stochastic methods [8] and the hydrologic question is "What is the probability of a drought leading to deficits during the project life?" or "How large should the reservoir be to make the probability of deficit small enough to be acceptable."

To accomplish this it is suggested that the following steps be followed:

1. Extend the available flow record for the longest possible period (Sec. 16-2).
2. Determine the parameters for a suitable stochastic model (Chap. 14).
3. Generate a large number (say 1000) of streamflow sequences, each sequence having a length equal to the expected useful life of the reservoir (20 to 100 years).
4. Estimate the project water demand including any changes expected during the project life. The demand figure should include estimated evaporation

loss and an allowance for any water which must be released for instream or downstream uses.

5. Use a simple algorithm such as the sequent peak algorithm to determine the storage required to meet the demand for each sequence.
6. Construct a reliability function for the reservoir and select a capacity offering an acceptable level of reliability. Include an allowance for sediment storage if necessary.

The storage model. Many methods have been suggested for determination of the storage required in a reservoir or the yield which can be obtained from a fixed storage. All are based on the storage equation (Chap. 9). For a simple reservoir the *sequent peak algorithm* is quite useful if a computer solution is desired. In this algorithm the cumulative sum of inflow minus demand is calculated (Fig. 16-2). The first peak on the cumulative net inflow diagram is located and the next larger peak (sequent peak) is identified. The storage required is the difference between the initial peak and the lowest trough in the interval preceding the sequent peak. The process is repeated for all sequent peaks in the flow sequence and the largest value of storage so determined governs [9]. Because the sequent peak procedure makes simplifying assumptions as to evaporation losses and possibly downstream releases, it may be desirable to test several of the more critical sequences by a more detailed study such as that illustrated in Table 16-1.

It is important to use an adequate number of flow sequences in the development of reliability curves (Sec. 14-6 and Fig. 16-3). If the demand on the reservoir is expected to grow over its life, the varying demand should be used in the analysis. Figure 16-4 shows the difference between the reliability curves for a reservoir assuming various initial demands and rate of growth of demand. If the reservoir is analyzed at its ultimate demand rate a larger storage is required than if an increasing rate is assumed.

Multiple reservoirs and multiple-purpose reservoirs will usually require a special model constructed to meet the specifics of the particular system. Such a

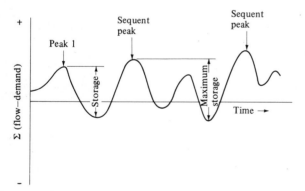

Figure 16-2 Illustration of the sequent peak algorithm.

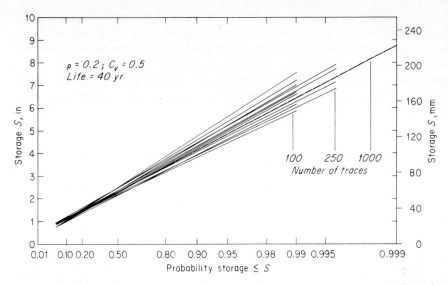

Figure 16-3 Reliability curves showing the influence of the number of sequences used in their development. (*From* [8].)

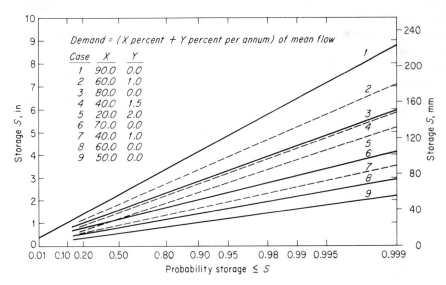

Figure 16-4 Reliability curves showing the effect of linearly increasing demand over time assuming a first-order Markov process with $CV = 0.5$ and $\rho = 0.2$. (*From* [8].)

system can, if necessary, include pipelines or canals as constraints on water transfers. If all outputs can be assigned money values, their combined value can be computed. Such models require that a set of operating rules be devised. This may be done by judgment on the part of the operators, or rules may be determined by formulating the problem in an optimization framework. It is important to consider the initial reservoir condition, and if the water use is expected to grow over time, the simulation should incorporate a time-varying demand.

16-4 Flood Regulation

Flood-mitigation reservoirs reduce downstream flood peaks by retaining a portion of the floodwater until it can be safely released. The traditional basis for design has been the selection of a *project design flood* (PDF). Reservoir capacity is selected such that it would store sufficient water to reduce the peak of the PDF to a safe flow downstream. It is assumed that any lesser flow would be adequately regulated. This assumes that the reservoir is at or below some specified initial level at the onset of the flood. It also assumes that the peak and hydrograph shape of the PDF is truly representative, since a different shape would mean a different volume.

A more thorough study of a flood-mitigation reservoir should include the following steps:

1. Develop a flood-peak probability curve at the point or points to be protected.
2. Develop a curve relating peak flow to damage at each protected point. This requires estimates of flood damage in several historic floods adjusted to current cost levels.
3. Combine the flow-damage curve and flow-probability curve into a damage-probability curve (Fig. 16-5).
4. Route the historic series of floods through an assumed reservoir and downstream to the protected area. Construct a modified flood-probability curve for the protected site and transform it into a damage-probability curve as before.
5. Repeat step 4 for various reservoir capacities and operating policies, and compare the cost of the alternate projects with the corresponding reduction in damages to find the most cost-effective alternative.

The above outline is based on the fact that the average annual flood damage is equal to the area under the damage-probability curve when plotted on natural scales, i.e.,

$$\bar{D} = \int_0^1 Dp \ dp \qquad (16-4)$$

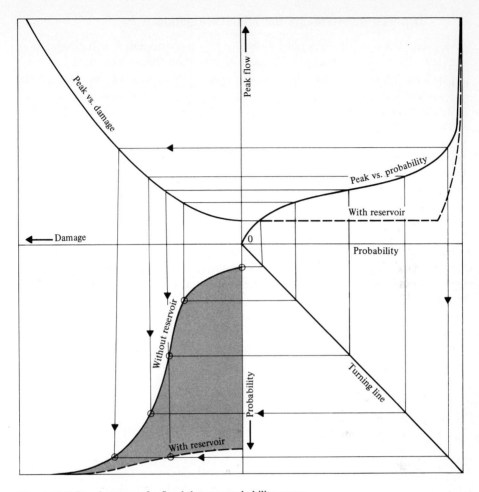

Figure 16-5 Development of a flood-damage probability curve.

where D is damage and p is probability. This permits a comparison of average annual benefits (damage reduction) and costs in the economic analysis. Because probability analysis is involved, this process requires a reasonably long record. If the available record is short, it should be extended by deterministic simulation. The use of the entire flood series is much more reliable as a base for economic analysis than use of an arbitrary PDF.

In the routing of the floods through the reservoir, it should be noted that by using actual data, all the uncertainty in the actual real-time operation is hidden. Few reservoirs are able to approach 100 percent effectiveness in operation. Actual effectiveness may be closer to 50 percent. An investigation simulating the operation of a flood control reservoir, using simulated inflow forecasts which reflect "real world" accuracy, would be informative.

16-5 Channel Improvements for Flood Mitigation

Assuming that a levee does not fail before it is overtopped, it will eliminate all damages from floods whose peak stage is less than the elevation of the levee crest. The hydrologist may be asked "What is the trace of the peak of the design flood as it passes through the leveed reach?" To answer this question, cross sections of the leveed reach must be assumed. The design flood is then determined and routed through the leveed reach by kinematic, zero-inertia, or dynamic routing (Chap. 10). The computed trace of the flood peak as it passes through the reach defines the effective height of levee required (with suitable freeboard added). Trials with differing spacing between levees can be made to determine the most cost-effective combination of channel width and levee height.

If the channel improvement consists of cleaning, lining, or straightening, the effect is to lower the peak stage for a given flood flow. Routing by one of the methods of Chap. 10 through the assumed new channel configuration will determine the new stages. An analysis similar to that illustrated in Fig. 16-5 can be used for determining the damage reduction. Stage would replace flow values in the plot.

The usual approach to analysis of channel improvements is the use of standard backwater computations, assuming a steady flow equal to the peak flow of the design flood. This approach may be satisfactory in some cases where the modified reach is short, but it neglects the change in the flood peak as it moves downstream under projected conditions.

16-6 Floodplain Mapping

Floodplain management is becoming an accepted part of flood-damage mitigation. The hydrologic problem is to define the area which will be flooded during the occurrence of a flood of specified return period. The standard procedure for this determination in the United States is to find the 100-year flood at a key point and use a backwater computation to determine the stages upstream. The 100-year flood is usually estimated from a stream-gage record if available, or by regional frequency methods (Sec. 13-7), or by loss rate and unit hydrograph applied to the 100-year rainfall.

For analysis of a short reach well downstream on a large river where the change in flow through the reach is negligible, the standard procedure is probably satisfactory provided the flood frequency can be determined from an adequate streamflow record. For small streams, it is important that the flood record be extended as far as the data will permit, since an accurate determination of the 100-year flood is basic to the whole process. If no flow record exists, it is probably better to develop a synthetic record by simulation with a deterministic model. The model should employ a routing technique to construct the hydrographs. This permits the calculation of peak flows at the end of each reach.

Thus a frequency curve can be constructed for each such point and the 100-year flood peak determined at that point. This allows for the variation in flow throughout the basin and eliminates the need to assume that all points experience the 100-year flood from the same storm. If the reach lengths are too long to provide needed detail, backwater computations can be used to interpolate within the reaches.

The lateral slope of the floodplain is an important control on the level of precision required in the analysis. A steeply sloping floodplain in which a change in stage of several feet inundates only a small area does not usually require the care in the analysis that is needed on a relatively flat floodplain where a small change in stage extends the inundation over a large area.

16-7 Urban Storm Drainage

Urban storm drainage is flood control in microcosm with the added peculiarity that observed flow data are almost never available. In urban drainage the classical solution is channel improvement, either improved ditches or buried storm drains. In some instances small reservoirs are useful for reducing local flood peaks. However, since the land-use changes associated with urbanization are important, only rarely does a meaningful record of streamflow exist in the area. The traditional method utilizes rainfall-intensity-frequency data and the rational equation ($q = CiA$) and generally leads to a substantial overestimate of

Table 16-2 Hydrograph computations for Redwood Creek

Time, min I	Sta. 1 I	0.06 I_2	0.06 I_1	0.88 0_1	Sta. 1	Sta. 2	(6) + (7)	(8) lagged	(9) routed to sta. 3	Urban I	(11) routed to sta. 3	Total, (10) + (12)
(1)	(2)	(3)	(4)	(5)	(6)	(7)	(8)	(9)	(10)	(11)	(12)	(13)
0	0	0	0	0	0	0	0	0	0	0	0	0
5	67	4	0	0	4	0	4	0	0	86	22	22
10	134	8	4	4	16	1	17	4	1	120	64	65
15	201	12	8	14	34	6	40	17	6	120	93	99
20	268	16	12	30	58	16	74	40	20	120	107	127
25	335	20	16	51	87	32	119	74	41	120	113	154
30	402	24	20	77	121	50	171	119	73	120	116	189
35	470	28	24	106	158	67	225	171	116	120	118	234
40	537	32	28	139	199	86	285	225	164	120	119	283
45	537	32	32	175	239	100	339	285	217	120	119	336
50	470	28	32	210	270	99	369	339	272	86	110	382
55	402	24	28	238	290	86	376	369	319	0	75	394
60	335	20	24	255	299	66	365	376	350	0	36	386
65	268	16	20	263	299	50	349	365	362	0	17	379
70	201	12	16	263	291	38	329	349	359	0	8	367
75	134	8	12	256	276	24	300	329	347	0	4	351
80	67	4	8	243	255	19	274	300	328	0	2	330
85	0	0	4	224	228	15	243	274	303	0	1	304
90	0	0	n0	201	201	11	212	243	276	0	0	276

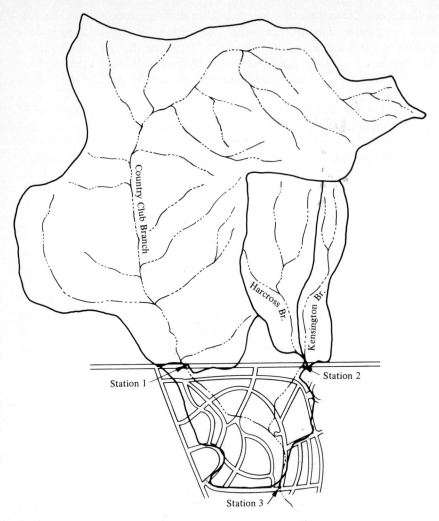

Figure 16-6 Map of Redwood Creek, Redwood City, Calif.

the design flows. Expenditures on urban drainage amount to billions of dollars annually, and minimizing cost by careful hydrologic analysis is well warranted.

Noncomputer approach. A rainfall-intensity pattern can be found by analysis of the local rainfall record or from rainfall-frequency maps (Sec. 13-16). Runoff amounts can then be estimated by applying a runoff coefficient, or subtracting suitable infiltration. Hydrographs from small subareas of the catchment (Fig. 16-6) can be calculated using overland flow computations (Sec. 7-13). Flow times in the trunk sewers or channels are estimated from calculated flow velocities. The subarea hydrographs are then summed using a time delay corresponding to the flow time to the next junction and then routed using the Muskingum method (Sec. 9-8), with K equal to the calculated flow time and $x = 0$ (Table 16-2). If this approach is chosen, the critical duration of rainfall for the smallest subareas can be estimated from

$$t_R = \frac{41 k L_0^{1/3}}{s^{1/3} i_e^{2/3}} \tag{16-5}$$

Equation (16-5) is derived from Eqs. (7-20) to (7-24), ignoring the intensity term in Eq. (7-24). Rainfall excess is represented by i_e in the equation. By estimating mean values of k, L_0, and s for the area, a trial solution can be made assuming a value of i_R appropriate for some t_R and solving to see whether the computed t_R corresponds with the assumed value.

Although this approach includes some approximations, it avoids the main defects of the traditional approach, namely, use of a value of t_R which is much too short and ignoring the effect of channel storage on the hydrograph. It seems to be a good solution for a small job and can be carried out conveniently with a desk calculator.

Illustrative example 16-2 For the catchment above station 1 of Fig. 16-6 the 30-year rainfall intensity is given by $i_R = 7.33/t_R$, where t_R is duration in minutes. Random sampling gives $L_0 = 200$ ft and $s_0 = 0.1$. Assume a runoff coefficient of 0.4 and a retardance coefficient $k = 0.05$. Find the peak flow at station 1 resulting from a rainfall with a 30-year return period.

Try
$$t_R = 45 \text{ min}$$
$$i_R = 7.33/6.71 = 1.09 \text{ in/hr}$$
$$i_e = 0.4 \times 1.09 = 0.44 \text{ in/hr}$$
$$t_R = \frac{41 \times 0.05 \times 5.85}{0.46 \times 0.58} = 45 \text{ min}$$

Hence, 45 min is an acceptable estimate of time to equilibrium. Based on physical data on the channel, the flow time from the headwaters to station 1 is calculated as 40 min. The overland flow hydrograph can be approximated as a trapezoid with a base length $= 40 + 45 = 85$ min and since t_R is greater than the flow time the top width is $45 - 40 = 5$ min. The area of the trapezoid equals the volume of runoff from the catchment (1220 acres) or

$$Q = 0.4 \times 1.09 \times 1220 \times \frac{45}{60} \times \frac{43,560}{12} = 1.448 \times 10^6 \text{ ft}^3$$

The peak height of the trapezoid is then

$$Q = \frac{85 + 5}{2} \times 60 \times q_p = 1.448 \times 10^6$$
$$q_p = 537 \text{ ft}^3/\text{s}$$

The trapezoidal hydrograph is given in column 2, Table 16-2. Routing with $K = 40$ min provides the required hydrograph at station 1 (column 6). If we wish to continue the analysis for the balance of the area, a similar calculation yields a hydrograph at station 2 (column 7). The urban area between stations 1, 2, and 3 is 240 acres. Flow time through the street drains and gutters is calculated at 4 min and travel time in the main channel from stations 1 and 2 to station 3 is 6 min. Using the value of i_e derived above, the hydrograph of inflow to the channels is estimated to have a base length of 52 min ($45 + 4 + (6/2)$, and a maximum flow of 120 ft³/s (column 11). The combined inflows from stations 1 and 2 are lagged 5 min (an approximation to the 6 min flow time) and routed to station 3 (column 10) using $K = 6$ min. The hydrograph from the urban area is routed to station 3 using $K = 7$ min (column 12) and added to the other flows to get the total at station 3 (column 13).

Note that for design of sewers within the urban area a shorter duration of rain would be used, but in this computation to determine total flow at station 3, rainfall duration is assumed to be 45 min for the entire area.

Computer approach. For large complex systems a continuous deterministic simulation model is helpful. The model should employ kinematic routing. The first step is a layout of the system in plan with a first estimate of pipe diameters (or channel cross sections) and slopes. This system is divided into reaches ending at points where a change in pipe size is anticipated. The available rainfall data are then input to the model using a time increment appropriate to the flow times in the reaches, and simulation is performed. The result is flow data at the end of each reach which can be used to derive a flood-frequency curve (Fig. 16-7). The design flow can be determined at each point using appropriate return periods, and the initial assumption as to pipe size can be checked and adjusted if necessary. Some test runs using the larger floods can be made to confirm that the pipe-size changes have not significantly altered the flow frequencies.

This approach should, of course, be preceded by a calibration of the model on several years of data for a catchment in the area. The parameters derived in calibration should be modified to fit the urban scene. This usually requires a change in the parameter representing impervious area. If the drainage system collects primarily surface runoff, the subsurface contribution to streamflow should be shut off during the simulation. If the model includes an overland flow routine, the parameters representing length, slope, and roughness of the overland flow plane will require changing. The simulation should extend over a period at least twice as long as the return period assumed for design. If only one rain gage is available, and if thunderstorms are the primary cause of the design floods, this approach may overestimate flows in the downstream portion of the system since it assumes the unlikely condition of uniform rain over the entire area. For subareas up to several square miles, the computed flows should be close, since we are assuming that the thunderstorms observed at the rain gage

APPLICATIONS OF HYDROLOGY **457**

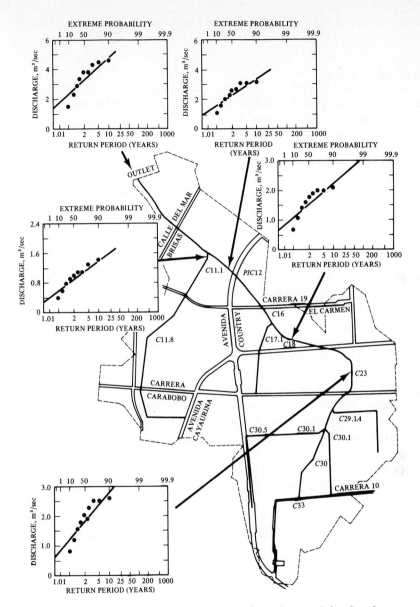

Figure 16-7 Schematic showing how continuous simulation can define flow frequency throughout an area. (*From Hydrocomp, Inc.*)

could have occurred elsewhere in the area and with the same frequencies as at the station.

16-8 Highway Culverts

Highway drainage costs about one-fourth as much as the rest of the highway system. The costs total more than a billion dollars a year, making accurate design of culverts a matter of considerable economic importance. The standard procedure in most states is the use of the rational equation and rainfall from intensity-frequency maps or charts. Since most catchments which contribute to culverts are rural and many are several square miles in area, the rational method is a very poor choice. Hiemstra and Reich [1] found that the rational method overestimated in approximately two-thirds of the basins in a test sample, and the average of the calculated peaks was about twice the observed peaks.

The noncomputer approach suggested for urban drainage could be applied in highway drainage and would probably be a considerable improvement over current methods. Simulation could also be utilized. However, the approach of calibration and then simulation would be costly if it were repeated separately for each culvert. A more effective approach would be to use the simulation model to develop a reasonably long series of hourly runoff increments which can be stored in the direct-access memory of a computer. If these runoff increments can be considered applicable anywhere within a relatively large area, it would possible to use them as inputs to a routing program which could generate the hydrographs, given information on the reaches within the catchment up stream of a culvert location. Thus, in use, the design engineer would merely input information on the length, slope, roughness, and cross section for each reach and the catchment area tributary to the reach. The computer could simulate floods throughout the entire data series, summarize the annual peaks, and construct a flood-frequency curve from the summarized data. This type of approach would be much more reliable than the present methods, would take very little of the design engineer's time, and would be inexpensive in the long run. With a complete frequency curve available, it would allow an approach in which culvert size is selected on the basis of an analysis of the consequences of overtopping at each location in contrast to the system of designing all culverts in a given class of roads to the same frequency standards.

16-9 Spillway Design

Most reservoirs of moderate to large capacity are provided with a spillway capable of discharging the probable maximum flood (Secs. 13-19 and 13-20). Lesser dams may have spillways able to cope with a lesser flood estimated on a frequency basis, but it might be wise to calculate the PMF and determine the consequences of its occurrence in the catchment tributary to the reservoir. Estimates based on enveloping curves, empirical formulas, or criteria such as

Table 16-3 Probable maximum precipitation for the Neosho River basin above Council Grove, Kan.

Duration, hr	3	6	9	12	24
PMP, in	14.6	19.2	22.0	23.1	24.8
PMP, mm	371	488	559	587	630

one-half PMF should not be used. The empirical procedures do not give reliable estimates of a maximum or of an event of known frequency. Use of a criterion such as one-half PMF does not indicate the frequency of its occurrence, and in some areas of the world one-half PMF would be less than the maximum observed flood.

Calculation of the PMF begins by obtaining an estimate of the probable maximum precipitation (Sec. 13-19). For the example in this section, PMP rain (Table 16-3) is determined from *Hydrometeorological Report* 51 [10].

To illustrate the hydrologic aspects of spillway design, let us assume that a major flood-control reservoir is to be constructed on the Neosho River just upstream of Council Grove, Kan. For simplicity it is assumed that the discharge record at Council Grove (250 mi^2, or 650 km^2) is applicable to the dam site. Possible loss of life in case of a dam failure dictates a spillway design flood well beyond that obtainable through frequency analysis.

The discharge record at Council Grove began in 1939, and the maximum observed discharge before July 1951 was 69,500 ft^3/s in October 1941 (1970 m^3/s). Flood marks indicate that the 1903 flood reached a stage of 37.3 ft (11.4 m), or about 0.2 ft (6 cm) higher than the 1941 flood. A peak of 121,000 ft^3/s (3420 m^3/s) occurred at 8:30 a.m., July 11, 1951, and just 24 hr later a second distinct peak of 71,100 ft^3/s (2010 m^3/s) occurred [11]. The first of these two floods and the 3-hr unit hydrograph derived therefrom are shown in Fig. 16-8a. The derived unit hydrograph fits the July 11 flood reasonably well, but it is more peaked than one derived from lesser storms of record (Sec. 7-7).

The July 1951 storm which produced the maximum observed discharge at Council Grove was not nearly as severe as the maximum observed storm in that region. On Sept. 12, 1926, a storm centered at Neosho Falls, 60 mi (100 km) to the southeast, deposited an average of 11.4 in (290 mm) of rain over 200 mi^2 (518 km^2) in 6 hr. The depth-area-duration data for this storm and the one at Council Grove in July 1951 are given [12] in Table 16-4. The 9-hr, 250-mi^2 rain for the 1926 storm is about 11.2 in (284 mm), and the time distribution of the storm was such that 3-hr increments of 1.5, 9.5, and 0.2 in (38, 241, and 5 mm) appear reasonable. From a runoff relation (similar to Fig. 8-6), this rainfall would yield runoff increments of 0.4, 8.5, and 0.2 in (10, 216, and 5 mm) if the basin were initially very wet and 0.2, 6.0, and 0 in (5, 152, and 0 mm) if the basin were initially very dry. Application of the unit hydrograph derived from the 1951 storm to these runoff increments results in the synthesized hydrographs shown in Fig. 16-8b.

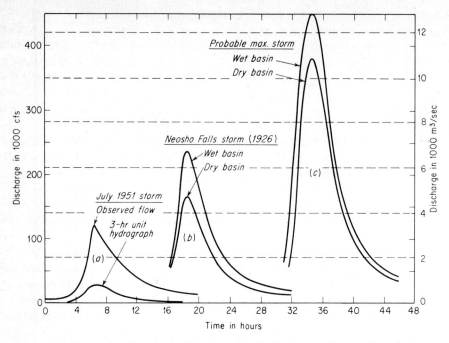

Figure 16-8 Observed and synthesized hydrographs for the Neosho River at Council Grove, Kan.

From a meteorological viewpoint, there is every reason to expect that the 1926 storm is far short of the maximum which can occur over the basin (Table 16-2). Storms approaching this magnitude can occur from June through September [10]. From an applicable runoff relation it is found that such a storm occurring on an initially wet basin in June or July would generate runoff increments of 1.5, 16.0, and 1.2 in (38, 406, and 30 mm). Should it occur in August or

Table 16-4 Depth-area-duration data for major storms centered at Neosho Falls (September 1926) and Council Grove, Kan. (July 1951) [12]

Area, mi²	Rainfall depth, in					
	Neosho Falls			Council Grove		
	6 hr	12 hr	18 hr	6 hr	12 hr	18 hr
Point	13.6	13.8	14.0	5.8	7.5	8.2
10	13.4	13.7	13.9	5.3	7.0	7.9
100	12.2	12.5	12.7	4.7	6.4	7.4
200	11.4	11.7	11.9	4.6	6.2	7.2
500	9.5	10.0	10.2	4.3	5.8	6.7

September with the basin initially dry, the corresponding runoff increments would be 0.2, 14.0, and 1.1 in (15, 356, and 28 mm). Applying the unit hydrograph provides the two flood hydrographs shown in Fig. 16-8c.

The observed and synthesized hydrographs are indicative of flow potentialities under natural conditions. If the reservoir capacity and plan of operation are such that some water could be stored at time of peak discharge, the hydrograph should be modified by routing (Sec. 9-6). Any anticipated flow through sluiceways should also be subtracted from the spillway flow. If a simulation model has been used in other aspects of the hydrologic study, it can also be used in lieu of a rainfall-runoff relation and unit hydrograph to estimate the PMF hydrograph. If the model incorporates hydraulic routing methods, it may be a more reliable basis for extrapolating to an extreme flood than the unit hydrograph.

16-10 Cooling-Pond Design

Large thermal electric power plants require vast quantities of water for cooling the condensers. Direct return of heated water to a lake or river usually mini-

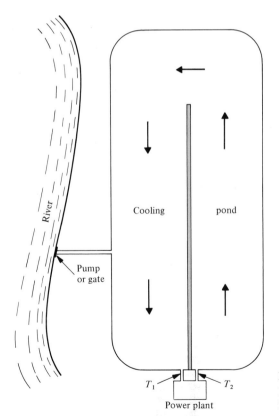

Figure 16-9 Sketch of a thermal electric power plant and recirculating cooling pond.

mizes water consumption and project costs, but is rarely environmentally acceptable. Sometimes the heated water is discharged into a *cooling pond* specially constructed for the purpose, where heat is dissipated to the atmosphere through radiation, evaporation, and sensible heat exchange (Chap. 5). If environmentally acceptable, warm water from the pond can be returned to the river, lake, or ocean after specified precooling. In some cases, the cooling pond must be sufficiently large to dissipate all waste heat from the plant. The design problem discussed here pertains to a closed system, with the natural water body serving only for water supply (Fig. 16-9), but the design of a precooling pond follows similar lines in all important respects.

Illustrative example 16-3 Assume that a cooling pond is to be designed for a thermal electric power plant of 1000 MW capacity, that the flowrate through the plant is to be 30 m³/s (1060 cfs), and that the waste heat to be dissipated in the cooling pond Q_v is 3.10×10^{13} cal/day (1500 MW or 1.23×10^{11} Btu/day). To maintain plant efficiency, it has been concluded that the plant inflow temperature should be no more than 10 Celsius degrees (18 Fahrenheit degrees) above the air temperature T_a over critical periods of several days. Representative normal monthly values of dewpoint, precipitation, and wind movement are available as are surface water temperature observations for a nearby shallow lake. Independent studies have provided the following design criteria for a critical 3-day period:

Air temperature $T_a = 30°C$ (86°F)
Natural water temperature $T_0 = 26°C$ (79°F)
Dewpoint $T_d = 18°C$ (64°F)
Atmospheric pressure $p = 1000$ mb (29.53 in Hg)
4-m wind movement $v_4 = 4.4$ m/s = 380 km/day (236 mi/day)

To be determined are the required dimensions of the cooling pond and the projected mean annual withdrawals from the river.

The daily advection of heat into the pond (3.10×10^{13} cal) is imparted to the daily flow; therefore, the inflow-outflow temperature difference at the plant is

$$T_2 - T_1 = \frac{3.10 \times 10^{13}}{(30 \times 10^6)(86,400)} = 12°C \ (21.5°F)$$

To meet the design criteria, the pond area must be sufficient to assure that $T_1 \le (T_a + 10)$ and, under these limiting circumstances, the mean temperature of the pond T_0' would be $30 + 10 + (T_2 - T_1)/2$, or $46°C$ (115°F). It is assumed that mixing maintains an isothermal profile within the pond, so that the mean surface temperature is also $46°C$.

From Eq. (5-12), the volume of evaporation from a pond is

$$E A = 0.122 \ (e_0 - e_a) \ v_4 A \times 10^{-3} \ \text{m}^3/\text{day}$$

where vapor pressures are in millibars, wind is in meters per second, and the water surface area of the pond A is in square meters. The difference in evaporation under modified and natural conditions (equivalent to the induced evaporation) is

$$(E' - E)A = 0.122 (e'_0 - e_0) v_4 A \times 10^{-3} \text{ m}^3/\text{day}$$

where e'_0 and e_0 correspond to the modified and natural water temperatures, respectively. The induced evaporation is also given by

$$(E' - E)A = \frac{\alpha Q_v}{H_v} \times 10^{-6} \text{ m}^3/\text{day}$$

where α is the portion of advected energy contributing to evaporation [Eq. (5-19)] and H_v is the heat of vaporization in calories per gram ($\approx$cm^3). Combining the above two equations,

$$A = \frac{\alpha Q_v}{0.122(e'_0 - e_0)v_4 H_v} \times 10^{-3} \text{ m}^2$$

Solving Eq. (5-31a) with $v_4 = 380$ km/day and $T_0 = 36°C$ (average of natural and modified water surface temperatures) gives $\alpha = 0.75$; solving Eq. (5-28a) with temperatures of 46°C and 26°C gives $(e'_0 - e_0) = 68.3$ mb; and solving Eq. (2-1) with temperature of 46°C gives $H_v = 571.4$ cal/cm^3. Thus

$$A = \frac{0.75(3.10 \times 10^{13})}{0.122(68.3)(4.4)(571.4)} \times 10^{-3} \text{ m}^2$$
$$= 1.11 \times 10^6 \text{ m}^2 = 1.11 \text{ km}^2 \text{ (274 acres)}$$

Solving Eq. (5-28a) with $T_0 = 26°C$ and $T_d = 18°C$ gives $(e_0 - e_a) = 13.08$ mb. Thus the natural evaporation for the critical design period would be

$$E A = 0.122 (13.08)(4.4)(1.11)(10^3) = 7.79 \times 10^3 \text{ m}^3/\text{day (6.32 acre} \cdot \text{ft)}$$

Similarly, with $T'_0 = 46°C$, the modified evaporation would be

$$E'A = 0.122 (81.4)(4.4)(1.11)(10^3) = 4.85 \times 10^4 \text{ m}^3/\text{day (39.3 acre} \cdot \text{ft)}$$

The approach to estimating required pond area naturally depends on the availability and reliability of data. If incoming solar Q_s and atmospheric Q_a radiation data were available in addition to the data used above, a second estimate could be made. Say, for example, that $Q_s = 600$ cal/cm$^2 \cdot$ day and $Q_a = 810$ cal/cm$^2 \cdot$ day. From Eqs. (5-5) and (5-7), and assuming reflectivities of 6 and 3 percent for shortwave and longwave radiation, respectively, the net radiation would be

$$Q_n = 0.94(600) + 0.97(810) - 0.97(11.71)(10^{-8})(46 + 273)^4$$
$$= 174 \text{ cal/cm}^2 \cdot \text{day}$$

From Eqs. (5-2), (5-4), and (5-12),

$$R = 0.66 \frac{T'_0 - T_a}{e'_0 - e_a} = 0.66 \frac{46 - 30}{102.1 - 20.8} = 0.130$$

$$E' = 0.122 \, (81.4)(4.4) = 43.7 \text{ mm}$$

$$Q_e = 43.7 \, (57.14) = 2497 \text{ cal/cm}^2 \cdot \text{day}$$

$$Q_h = R \, Q_e = 0.13 \, (2497) = 324.6 \text{ cal/cm}^2 \cdot \text{day}$$

$$Q_v = Q_e + Q_h - Q_n = \frac{3.10 \times 10^{13}}{A}$$

$$A = \frac{3.10 \times 10^{13}}{2497 + 325 - 174} = 1.17 \times 10^{10} \text{ cm}^2 = 1.17 \times 10^6 \text{ m}^2 \text{ (289 acres)}$$

It is seen from the above computations that the required pond area is independent of the travel time from the point of discharge around to the intake pumps. Even so, the solutions are valid only if the pond depth and channel configuration provide full heat exchange and active circulation over the entire lake area, thus maintaining the specified water surface temperatures. Both approaches further assume that Eq. (5-12) is valid under extreme conditions when the water is much warmer than the overlying air [13].

The normal water requirement for each month of the year is estimated as follows:

1. Determine α from Fig. 5-2 or Eq. (5-31a) using given values of wind and water temperature (adjusted for advection effects).
2. Compute the induced evaporation [($\alpha \, Q_v/H_v$) $\times$ 10^{-6} m^3, where Q_v is the total monthly waste heat from the plant].
3. Compute the natural evaporation from Eq. (5-12) using given values of wind, dewpoint, and water temperature, and convert to cubic meters per month.
4. The monthly water requirement is the difference between the total evaporation (natural plus induced) and precipitation, all expressed in cubic meters. If the seepage is significant, the monthly requirement should be increased accordingly.

The annual requirement for cooling water from the river is determined by summing the monthly values as computed by steps 1 through 4.

16-11 River Forecasting

The previous sections of this chapter treat problems of hydrologic design. Equally important, and in many respects more tangible, is the role of hydrology in the operation of water-control projects. In design, the hydrologist is most often required to estimate the magnitude of extreme events, whereas operation is often dependent on reliable estimates of flow for an ensuing period of hours, days, or possibly longer. The time sequence of flows during critical periods can be of considerable importance. The operation of a flood-control reservoir is necessarily based on anticipated flows into the reservoir and at key points downstream. The extent to which the reservoir yields the anticipated benefits is highly dependent on the accuracy of flow projections. Reliable flow forecasts are particularly important in the case of multipurpose reservoirs, and they are indispensable to the operation of flood-mitigation reservoir systems.

In addition to the role of forecasts in planning water-control operations, flood forecasts constitute a direct means for the reduction of flood damage and loss of life. Advance warning of an approaching flood permits evacuation of people, livestock, and equipment with no loss except for the cost of removal. The relatively low ratio of cost to benefit for a flood forecast and warning service makes it an ideal flood-protection measure in many areas where physical means cannot be economically justified. The soundest approach to the flood problem lies in a planned combination of water-control structures, floodplain zoning, adequate forecasting, and insurance.

The formulation of a river forecast requires reliable information on current hydrologic conditions over the drainage basin, augmented by weather reports and forecasts. The necessary data—precipitation, river stage, water equivalent of snowpack, reservoir storage, etc.—are assembled by telephone, telegraph, and/or radio from an organized network of stations. Station observers report to the forecast office in accordance with specific instructions, and automated stations are queried or programmed to transmit data as dictated by circumstances.

The tools of the river forecaster are the same as those used in design. An increasing number of national river forecast services are gaining access to digital computers, automated networks, and advanced communications. With such facilities at hand, methods based on the application of a deterministic, conceptual model (Chap. 12) will provide the most reliable and beneficial forecasts and warnings. Traditional forecast methods based on rainfall-runoff relations, unit hydrographs, routing procedures, gage relations, etc., are still widely used, however, and the example given in this section treats this approach to the problem.

The specific techniques used in a particular case depend to a large extent on the forecasts needed, the facilities and data available, and the hydrologic characteristics of the basin. A forecast of crest height and the time of occurrence will usually suffice as a warning in small, headwater areas. The rapid rate of rise and fall makes the duration above flood stage so short that there is little value in forecasting the entire hydrograph. In the lower reaches of large rivers where rates of rise are slow, it is important to forecast the time when various critical stages will be reached during the rise and fall. Efficient operation of flood-control reservoirs requires forecasts of the complete hydrograph.

Illustrative example 16-4 To show the manner in which forecasts of the hydrograph are formulated for a series of points over a river basin, let us assume that the area mapped in Fig. 16-10 represents the upper reaches of one of several tributary areas. The forecasts are to be prepared as of 7:00 a.m., April 13, following a 24-hr storm with rainfall as shown. The 12-hr increments of storm runoff are first obtained by entering a relation of the type shown in Fig. 8-6 with antecedent precipitation, week number, retention index, and 12-hr rainfall for each reporting station. The point values are then averaged to determine the mean areal values for the catchment above Riverton and the local area below (1.05 and 0.55 in, respectively). The distributed flow contributions from these increments (labeled "12-hr storm runoff" in Table 16-5) are obtained by applying the indicated unit hydrograph ordinates. The computations have been based on the assumption that the groundwater flow peaks at the cessation of storm runoff. The magnitude of the groundwater peak is estimated subjectively. Adding the groundwater flow and storm runoff to the previous forecast provides the required forecast for Riverton.

Local inflow from the area between Riverton and Greenriver is computed in similar fashion. The Greenriver forecast hydrograph is obtained by routing the Riverton hydrograph (Muskingum method, $K = 10$ hr, $x = 0.2$) and adding the expected flow from the local area. In practice, stages of 7:00 a.m. of April 13 would be available, and the forecast hydrograph would be smoothed into the observed data. For flood-warning purposes, the com-

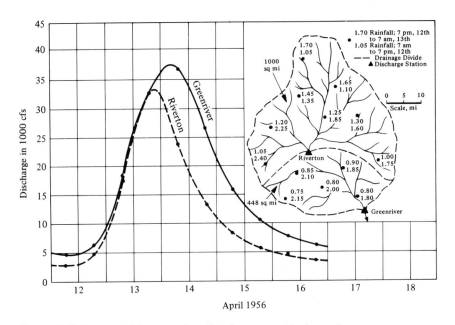

Figure 16-10 Storm rainfall map and predicted consequent hydrographs.

Table 16-5 Runoff distribution and routing computations

	Discharge in 1000 cfs									
	Apr. 12		Apr. 13		Apr. 14		Apr. 15		Apr. 16	
Tabular entry	7 a.m.	7 p.m.	7 a.m.	7 p.m.	7 a.m.	7 p.m.	7 a.m.	7 p.m.	7 a.m.	7 p.m.
Riverton:										
12-hr unit hydrograph		0	4.8	20.5	16.0	7.5	3.5	1.6	0.5	0
12-hr storm runoff (1.05 in)		0	5.0	21.5	16.8	7.9	3.7	1.7	0.5	0
12-hr groundwater runoff				0	0.2	0.4	0.6	0.8	1.0	1.2
Previous forecast (7 p.m., Apr. 12)	2.8	5.0	12.3	11.3	7.1	5.0	4.2	3.6	3.2	2.6
Total	2.8	5.0	17.3	32.8	24.1	13.3	8.5	6.1	4.7	3.8
Local to Greenriver:										
12-hr unit hydrograph		0	2.0	9.8	6.8	3.6	1.3	0.3	0	0
12-hr storm runoff (0.55 in)		0	1.1	5.4	3.7	2.0	0.7	0.2	0	0
12-hr groundwater runoff				0	0.1	0.2	0.3	0.4	0.5	0.4
Previous forecast (7 p.m., Apr. 12)	1.5	3.0	9.0	7.0	4.5	2.8	2.0	1.7	1.5	1.3
Total	1.5	3.0	10.1	12.4	8.3	5.0	3.0	2.3	2.0	1.7
Greenriver, Riverton routing,										
$K = 10$ hr, $x = 0.2$:										
$0.285 I_2$		1.4	4.9	9.3	6.9	3.8	2.4	1.7	1.3	1.1
$0.572 I_1$		1.6	2.9	9.9	18.8	13.8	7.6	4.9	3.5	2.7
$0.143 O_1$		0.5	0.5	1.2	2.9	4.1	3.1	1.9	1.2	0.9
O_2	3.3	3.5	8.3	20.4	28.6	21.7	13.1	8.5	6.0	4.7
Local runoff	1.5	3.0	10.1	12.4	8.3	5.0	3.0	2.3	2.0	1.7
Total	4.8	6.5	18.4	32.8	36.9	26.7	16.1	10.8	8.0	6.4

puted flows would be converted to stages by application of stage-discharge relations.

Although simplified, the example shows how forecasts can be prepared from rainfall reports for numerous points over a river basin. The forecasts are made in sequence proceeding downstream, point by point, routing upstream flows and adding predicted local inflow. While this compounding of forecasts might seem insecure, the damping effect of storage, error compensation, and the curvature of the stage-discharge relation all combine to produce downstream stage forecasts that are generally more accurate than the headwater forecasts from which they are derived.

The approach of routing and combining flows from headwater and local areas to obtain forecasts for required points in downstream order is also used when applying deterministic catchment models [14]. In other respects, the application of models requires a change in methodology. With traditional methods, the forecaster has been in a position to adjust forecasts subjectively when reported flows were not in agreement with the predicted. It is usually found that the flow simulated by a continuous-flow model is in error at the time of forecast preparation, and computerized methods for taking advantage of reported flows can become complex [15].

PROBLEMS

16-1 Under what conditions would the error resulting from the use of monthly or annual data in Illustrative example 16-1 disappear?

16-2 Referring to Illustrative example 16-1, what increase in yield would result if the canal capacity is 300 cfs?

16-3 Obtain a report describing the studies leading up to some water-related action, i.e., reservoir design, floodplain mapping, etc. Does the hydrologic analysis appear to be adequate with respect to general approach, adequacy of data, length of record, etc. Make such check computations as the data permit to estimate the accuracy of the analysis.

16-4 Select a streamflow record and use it (together with other relevant information) to design a reservoir at the gaging station.

16-5 Assume a cooling pond located in your immediate area. What reservoir area is required for cooling a 500-MW plant with a flowrate of 17 m^3/s and a thermal efficiency of 45 percent? Assume plant inflow temperature to be limited to 8 Celsius degrees above the air temperature during the hottest 3-day period of record. Take other necessary data at values appropriate for the location.

REFERENCES

1. Hiemstra, L. A. V. and B. M. Reich: Engineering Judgement and Small Area Flood Peaks, *Hydrology Papers* No. 19, Colorado State University, April 1967.
2. Fleming, G., and D. D. Franz: Flood Frequency Estimating Techniques for Small Watersheds, *J. Hydr. Div. ASCE,* vol. 97, pp. 1441–1460, September 1971.
3. U.S. Water Resources Council: "Estimating Peak Flow Frequencies for Natural Ungaged Watersheds: A Proposed Nationwide Test," Report of the Hydrology Committee, 1981.

4. Grant, E. L., W. G. Ireson, and R. S. Leavenworth: "Principles of Engineering Economy, 6th ed., Ronald, New York, 1976.
5. James, L. D., and R. R. Lee: "Economics of Water Resources Planning," McGraw-Hill, New York, 1971.
6. Kuiper, E.: "Water Resources Development," Butterworth's, London.
7. Linsley, R. K., and J. B. Franzini: "Water Resources Engineering," 3d ed., New York, 1979.
8. Burges, S. J.: Use of Stochastic Hydrology to Determine Storage Requirements for Reservoirs: A Critical Analysis, *Stanford Univ. Progr. Eng. Econ. Plann. Rep.* EEP-34, September 1970.
9. Thomas, H. A., Jr., and M. B. Fiering: The Nature of the Storage-Yield Function in "Operations Research in Water Quality Management," Harvard University Water Program, 1963.
10. National Weather Service: Probable Maximum Precipitation Estimates, United States East of the 105th Merididan, *Hydrometeorol. Rep.* 51, June 1978.
11. Kansas-Missouri Floods of July, 1951, *U.S. Geol. Surv. Water-Supply Paper* 1139, p. 180, 1952.
12. Storm Rainfall in the United States, U.S. Corps of Engineers.
13. Carson, J. E.: Meteorological Consequences of Thermal Discharge from Nuclear Power Plants—Research Needs, "Cooling Tower Environment—1974," ERDA Symposium Series 35, pp. 221–238, 1976.
14. National Weather Service River Forecast System Forecast Procedures, *NOAA Tech. Memo. NWS-Hydro*-14, December 1972.
15. Sittner, W. T., and K. M. Krouse: Improvement of Hydrologic Simulation by Utilizing Observed Discharge as an Indirect Input (Computed Hydrograph Adjustment Technique—CHAT), *NOAA Tech. Memo. NWS Hydro*-38, February 1979.

APPENDIX

TABLES OF PHYSICAL CONSTANTS AND EQUIVALENTS AND PSYCHROMETRIC TABLES

Table A-1 Volume equivalents

Unit	Equivalent						
	in³	gal	Imperial gal	ft³	m³	acre · ft	cfs-day
Cubic inch	1	0.00433	0.00361	5.79×10^{-4}	1.64×10^{-5}	1.33×10^{-8}	6.70×10^{-9}
U.S. gallon	231	1	0.833	0.134	0.00379	3.07×10^{-6}	1.55×10^{-6}
Imperial gallon	277	1.20	1	0.161	0.00455	3.68×10^{-6}	1.86×10^{-6}
Cubic foot	1728	7.48	6.23	1	0.0283	2.30×10^{-5}	1.16×10^{-5}
Cubic meter	61,000	264	220	35.3	1	8.11×10^{-4}	4.09×10^{-4}
Acre · foot	7.53×10^{7}	3.26×10^{5}	2.71×10^{5}	43,560	1233	1	0.504
Cubic foot per second-day	1.49×10^{8}	6.46×10^{5}	5.38×10^{5}	86,400	2447	1.98	1

Table A-2 Discharge equivalents

Unit	Equivalent						
	gal/day	ft³/day	gal/min	Imperial gal/min	acre · ft/day	cfs	m³/s
U.S. gallon per day	1	0.134	6.94×10^{-4}	5.78×10^{-4}	3.07×10^{-6}	1.55×10^{-6}	4.38×10^{-8}
Cubic foot per day	7.48	1	5.19×10^{-3}	4.33×10^{-3}	2.30×10^{-5}	1.16×10^{-5}	3.28×10^{-7}
U.S. gallon per minute	1440	193	1	0.833	4.42×10^{-3}	2.23×10^{-3}	6.31×10^{-5}
Imperial gallon per minute	1728	231	1.20	1	5.31×10^{-3}	2.67×10^{-3}	7.57×10^{-5}
Acre · foot per day	3.26×10^{5}	43,560	226	188	1	0.504	0.0143
Cubic foot per second	6.46×10^{5}	86,400	449	374	1.98	1	0.0283
Cubic meter per second	2.28×10^{7}	3.05×10^{6}	15,800	13,200	70.0	35.3	1

Table A-3 Energy equivalents

Unit	Btu	cal	J	kW · hr	ft · lb	hp · hr
British thermal unit (60°F)	1	252.0	1055	0.0002930	777.9	0.0003929
Calorie (15°C)	0.003969	1	4.186	1.163×10^{-6}	3.087	1.559×10^{-6}
Joule	0.0009482	0.2389	1	2.778×10^{-7}	0.7376	3.725×10^{-7}
Kilowatt-hour	3413	860,100	3.600×10^{6}	1	2.655×10^{6}	1.341
Foot-pound	0.001286	0.3239	1.356	3.766×10^{-7}	1	5.051×10^{-7}
Horsepower-hour	2545	641,300	2.685×10^{6}	0.7457	1.980×10^{6}	1

Equivalent

Table A-4 Energy/area time equivalents

	Equivalent		
Unit	Btu/ft² · min	cal/cm² · min	W/m²
British thermal unit (60°F) per square foot per minute	1	0.2712	189.2
Calorie (15°C) per square centimeter per minute	3.687	1	697.6
Watt per square meter	0.005285	0.001433	1

Table A-5 Pressure equivalents

Unit	Equivalent					
	in Hg	mm Hg	mb	kPa	psi	kg/m²
Inch of mercury (32°F)	1	25.40	33.86	3.386	0.4912	345.3
Millimeter of mercury (0°C)	0.03937	1	1.333	0.1333	0.01934	13.60
Millibar	0.02953	0.7501	1	0.1000	0.01450	10.20
Kilopascal	0.2953	7.501	10.00	1	0.1450	102.0
Pounds per square inch	2.036	51.71	68.95	6.895	1	703.1
Kilograms per square meter	0.002896	0.07356	0.09807	0.009807	0.001422	1

Table A-6 Velocity equivalents

Unit	ft/s	mi/hr	m/s	km/hr	kn
			Equivalent		
Feet per second	1	0.6818	0.3048	1.097	0.5925
Miles per hour	1.467	1	0.4470	1.609	0.8690
Meters per second	3.281	2.237	1	3.600	1.944
Kilometers per hour	0.9113	0.6214	0.2778	1	0.5400
Knots	1.688	1.151	0.5144	1.852	1

Table A-7 Miscellaneous English equivalents and physical constants

$$
\begin{aligned}
1 \text{ cubic foot per second-day per square mile} &= 0.03719 \text{ inch} \\
1 \text{ inch of runoff per square mile} &= 26.89 \text{ cubic feet per second-days} \\
&= 53.33 \text{ acre} \cdot \text{feet} \\
&= 2{,}323{,}200 \text{ cubic feet} \\
1 \text{ miner's inch} &= 0.025 \text{ cubic foot per second in Arizona, California,} \\
&\quad \text{Montana, and Oregon} \\
&= 0.020 \text{ cubic foot per second in Idaho, Kansas,} \\
&\quad \text{Nebraska, New Mexico, North and South Dakota,} \\
&\quad \text{and Utah} \\
&= 0.026 \text{ cubic foot per second in Colorado} \\
&= 0.028 \text{ cubic foot per second in British Columbia} \\
1 \text{ cubic foot per second} &= 0.0002142 \text{ cubic mile per year} \\
&= 0.9917 \text{ acre} \cdot \text{inch per hour} \\
1 \text{ pound of water} &= 0.5507 \text{ inch over 8-inch circle} \\
&= 0.3524 \text{ inch over 10-inch circle} \\
&= 0.2448 \text{ inch over 12-inch circle} \\
1 \text{ horsepower} &= 0.7457 \text{ kilowatt} \\
&= 550 \text{ foot-pounds per second} \\
e &= 2.71828 \\
\log e &= 0.43429 \\
\ln 10 &= 2.30259
\end{aligned}
$$

Table A-8 Map-scale equivalents

Ratio	in/mi	mi/in	mi²/in²	cm/km	km/cm	km²/cm²
1 : 1,000,000	0.0634	15.7828	249.097	0.1000	10.0000	100.000
1 : 500,000	0.1267	7.8914	62.274	0.2000	5.0000	25.000
1 : 250,000	0.2534	3.9457	15.569	0.4000	2.5000	6.250
1 : 126,720	0.5000	2.0000	4.000	0.7891	1.2672	1.606
1 : 125,000	0.5069	1.9728	3.892	0.8000	1.2500	1.562
1 : 90,000	0.7040	1.4205	2.018	1.1111	0.9000	0.8100
1 : 63,360	1.0000	1.0000	1.000	1.5783	0.6336	0.4014
1 : 62,500	1.0138	0.9864	0.9730	1.6000	0.6250	0.3906
1 : 45,000	1.4080	0.7102	0.5044	2.2222	0.4500	0.2025
1 : 31,680	2.0000	0.5000	0.2500	3.1566	0.3168	0.1004
1 : 30,000	2.1120	0.4735	0.2242	3.3333	0.3000	0.0900
1 : 24,000	2.6400	0.3788	0.1435	4.1667	0.2400	0.0576
1 : 12,000	5.2800	0.1894	0.0358	8.3333	0.1200	0.0144
1 : 2,300	26.4000	0.0379	0.001435	41.6667	0.0240	0.00576
1 : 1,200	52.8000	0.0189	0.000358	83.3333	0.0120	0.000144

Table A-9 Properties of water in English units

Temp., °F	Specific gravity	Specific weight, lb/ft³	Heat of vaporization, Btu/lb	Viscosity Dynamic, lb · s/ft²	Viscosity Kinematic, ft²/s	Vapor pressure in Hg	Vapor pressure Millibar	Vapor pressure lb/in²
32	0.99986	62.418	1075.5	3.746×10^{-5}	1.931×10^{-5}	0.180	6.11	0.089
40	0.99998	62.426†	1071.0	3.229	1.664	0.248	8.39	0.122
50	0.99971	62.409	1065.3	2.735	1.410	0.362	12.27	0.178
60	0.99902	62.366	1059.7	2.359	1.217	0.522	17.66	0.256
70	0.99798	62.301	1054.0	2.050	1.058	0.739	25.03	0.363
80	0.99662	62.216	1048.4	1.799	0.930	1.032	34.96	0.507
90	0.99497	62.113	1042.7	1.595	0.826	1.422	48.15	0.698
100	0.99306	61.994	1037.1	1.424	0.739	1.933	65.47	0.950
120	0.98856	61.713	1025.6	1.168	0.609	3.448	116.75	1.693
140	0.98321	61.379	1014.0	0.981	0.514	5.884	199.26	2.890
160	0.97714	61.000	1002.2	0.838	0.442	9.656	326.98	4.742
180	0.97041	60.580	990.2	0.726	0.386	15.295	517.95	7.512
200	0.96306	60.121	977.9	0.637	0.341	23.468	794.72	11.526
212	0.95837	59.828	970.3	0.593	0.319	29.921	1013.25	14.696

† Maximum specific weight is 62.427 lb/ft³ at 39.2°F.

Table A-10 Properties of water in metric units

Temp., °C	Specific gravity	Density, g/cm³	Heat of vaporization, cal/g	Viscosity		Vapor pressure		
				Dynamic, centipoise‡	Kinematic, centistokes§	mm Hg	Millibar	g/cm²
0	0.99987	0.99984	597.3	1.79	1.79	4.58	6.11	6.23
5	0.99999	0.99996†	594.5	1.52	1.52	6.54	8.72	8.89
10	0.99973	0.99970	591.7	1.31	1.31	9.20	12.27	12.51
15	0.99913	0.99910	588.9	1.14	1.14	12.78†	17.04	17.38
20	0.99824	0.99821	586.0	1.00	1.00	17.53	23.37	23.83
25	0.99708	0.99705	583.2	0.890	0.893	23.76	31.67	32.30
30	0.99568	0.99565	580.4	0.798	0.801	31.83	42.43	43.27
35	0.99407	0.99404	577.6	0.719	0.723	42.18	56.24	57.34
40	0.99225	0.99222	574.7	0.653	0.658	55.34	73.78	75.23
50	0.98807	0.98804	569.0	0.547	0.554	92.56	123.40	125.83
60	0.98323	0.98320	563.2	0.466	0.474	149.46	199.26	203.19
70	0.97780	0.97777	557.4	0.404	0.413	233.79	311.69	317.84
80	0.97182	0.97179	551.4	0.355	0.365	355.28	473.67	483.01
90	0.96534	0.96531	545.3	0.315	0.326	525.89	701.13	714.95
100	0.95839	0.95836	539.1	0.282	0.294	760.00	1013.25	1033.23

† Maximum density is 0.999973 g/cm³ at 3.98°C.
‡ centipoise = (g/cm · s) × 10² = (Pa · s) × 10³
§ centistokes = (cm²/s) × 10² = (m²/s) × 10⁶

478

Table A-11 Solubility of oxygen in water exposed to water-saturated air (milligrams per liter) for selected temperatures and chloride concentrations†

Water temp., °C	Chloride concentrations in water, mg/l				
	0	5000	10,000	15,000	20,000
0	14.5	13.8	13.0	12.1	11.3
5	12.8	12.1	11.4	10.7	10.0
10	11.3	10.7	10.1	9.6	9.0
15	10.2	9.7	9.1	8.6	8.1
20	9.2	8.7	8.3	7.9	7.4
25	8.4	8.0	7.6	7.2	6.7
30	7.6	7.3	6.9	6.5	6.1

Source: "Standard Methods for the Examination of Water and Wastewater," 11th ed., p. 1312, American Public Health Association, New York, 1960.

† At standard sea-level pressure $p_0 = 1013.2$ mb $= 760$ mm Hg $= 29.92$ in Hg. For any other pressure p, tabular value should be multiplied by $(p - e_s)/(p_0 - e_s)$, where e_s is the saturation vapor pressure for the existing water temperature.

Table A-12 Ratio of saturation vapor pressure over ice to that over water at the same temperature

Deg	0	1	2	3	4	5	6	7	8	9
					Celsius					
−0	1.000	0.990	0.981	0.971	0.962	0.953	0.943	0.934	0.925	0.916
−10	0.907	0.899	0.890	0.881	0.873	0.864	0.856	0.847	0.839	0.831
−20	0.823	0.815	0.807	0.799	0.791	0.784	0.776	0.769	0.761	0.754
−30	0.746	0.739	0.732	0.725	0.718	0.711	0.704	0.698	0.691	0.685
−40	0.678	0.672	0.666	0.660	0.654	0.648	0.642	0.636	0.630	0.625
					Fahrenheit					
30	0.989	0.994	1.000							
20	0.937	0.942	0.947	0.953	0.958	0.963	0.968	0.973	0.979	0.984
10	0.888	0.893	0.897	0.902	0.907	0.912	0.917	0.922	0.927	0.932
0	0.841	0.845	0.850	0.855	0.859	0.864	0.868	0.873	0.878	0.883
−0	0.841	0.836	0.832	0.827	0.823	0.818	0.814	0.809	0.805	0.801
−10	0.796	0.792	0.788	0.784	0.779	0.775	0.771	0.767	0.763	0.759
−20	0.755	0.750	0.746	0.742	0.739	0.734	0.731	0.727	0.723	0.719
−30	0.715	0.711	0.708	0.704	0.700	0.696	0.693	0.690	0.686	0.682
−40	0.678	0.675	0.672	0.668	0.665	0.661	0.658	0.654	0.651	0.648
−50	0.644	0.641	0.638	0.635	0.631	0.628	0.625	0.622	0.619	0.616

Source: From "Smithsonian Meteorological Tables," 6th ed., p. 370, Smithsonian Institution, Washington, 1966.

Table A-13 Variation of pressure, temperature, density, and boiling point with elevation

U.S. standard atmosphere

Elevation from mean sea level, m	Pressure			Air temp., °C	Air density, kg/m³	Boiling point, °C
	mm Hg	Millibars	cm H$_2$O			
-500	806.15	1074.78	1096.0	18.2	1.285	101.7
0	760.00	1013.25	1033.2	15.0	1.225	100.0
500	716.02	954.61	973.4	11.8	1.167	98.3
1000	674.13	898.76	916.5	8.5	1.112	96.7
1500	634.25	845.60	862.3	5.3	1.058	95.0
2000	596.31	795.01	810.7	2.0	1.007	93.4
2500	560.23	746.92	761.6	-1.2	0.957	91.7
3000	525.95	701.21	715.0	-4.5	0.909	90.0
3500	493.39	657.80	670.8	-7.7	0.863	88.3
4000	462.49	616.60	628.8	-11.0	0.819	86.7
4500	433.18	577.52	588.9	-14.2	0.777	85.0
5000	405.40	540.48	551.1	-17.5	0.736	83.3

Source: "U.S. Standard Atmosphere, 1962," National Aeronautics and Space Administration, U.S. Air Force, and U.S. Weather Bureau.

Table A-14 Variation of relative humidity in percent with temperature and wet-bulb depression on the Fahrenheit scale

Pressure = 30.00 in = 1015.9 millibars = 101.59 kilopascals

Air temp., °F	Wet-bulb depression, deg														
	0	1	2	3	4	6	8	10	12	14	16	18	20	25	30
0	84	56	27												
5	86	63	40	16											
10	89	69	50	30	11										
15	91	74	58	42	26										
20	94	79	65	51	37	10									
25	96	84	71	59	47	24	1								
30	99	88	77	66	56	35	15								
35	100	91	81	72	63	45	27	10							
40	100	92	84	76	68	52	37	22	7						
45	100	93	85	78	71	57	44	31	19	6					
50	100	93	87	80	74	61	49	38	27	16	5				
55	100	94	88	82	76	65	54	43	33	24	14	5			
60	100	94	89	83	78	68	58	48	39	30	21	13	5		
65	100	95	90	85	80	70	61	52	44	35	28	20	13		
70	100	95	90	86	81	72	64	55	48	40	33	26	19	3	
75	100	95	91	87	82	74	66	58	51	44	37	31	24	10	
80	100	96	91	87	83	75	68	61	54	47	41	35	29	15	3
85	100	96	92	88	84	77	70	63	56	50	44	38	33	20	8
90	100	96	92	89	85	78	71	65	58	53	47	41	36	24	13
95	100	96	93	89	86	79	72	66	60	55	49	44	39	28	17
100	100	96	93	89	86	80	74	68	62	57	51	46	42	31	21

Source: U.S. Weather Bureau, Relative Humidity and Dew Point Table, TA 454-0-3E, September 1965.

Table A-15 Variation of relative humidity in percent with temperature and wet-bulb depression on the Celsius scale

Pressure = 990 millibars = 29.24 in = 99.0 kilopascals

Air temp., °C	\multicolumn{16}{c}{Wet-bulb depression, deg}
	0	1	2	3	4	5	6	7	8	9	10	11	12	13	14	15
−10	91	60	31	2												
−8	93	65	39	13												
−6	94	70	46	23	0											
−4	96	74	53	32	11											
−2	98	78	58	39	21	3										
0	100	81	63	46	29	13										
2	100	84	68	52	37	22	7									
4	100	85	71	57	43	29	16									
6	100	86	73	60	48	35	24	11								
8	100	87	75	63	51	40	29	19	8							
10	100	88	77	66	55	44	34	24	15	6						
12	100	89	78	68	58	48	39	29	21	12	4					
14	100	90	79	70	60	51	42	34	26	18	10	3				
16	100	90	81	71	63	54	46	38	30	23	15	8				
18	100	91	82	73	65	57	49	41	34	27	20	14	7			
20	100	91	83	74	66	59	51	44	37	31	24	18	12	6		
22	100	92	83	76	68	61	54	47	40	34	28	22	17	11	6	
24	100	92	84	77	69	62	56	49	43	37	31	26	20	15	10	5
26	100	92	85	78	71	64	58	51	46	40	34	29	24	19	14	10
28	100	93	85	78	72	65	59	53	48	42	37	32	27	22	18	13
30	100	93	86	79	73	67	61	55	50	44	39	35	30	25	21	17
32	100	93	86	80	74	68	62	57	51	46	41	37	32	28	24	20
34	100	93	87	81	75	69	63	58	53	48	43	39	35	30	26	23
36	100	94	87	81	75	70	64	59	54	50	45	41	37	33	29	25
38	100	94	88	82	76	71	66	61	56	51	47	43	39	35	31	27
40	100	94	88	82	77	72	67	62	57	53	48	44	40	36	33	29

Source: From "Radiosonde Observation Computation Tables," Dept. of Commerce–Dept. of Defense, Washington, June 1972.

Table A-16 Variation of dewpoint with temperature and wet-bulb depression and of saturation vapor pressure over water with temperature on the Fahrenheit scale

Pressure = 30.00 in = 1015.9 millibars = 101.59 kilopascals

Air temp., °F	Saturation vapor pressure		Wet-bulb depression, deg															
	Milli-bars	in Hg	0	1	2	3	4	6	8	10	12	14	16	18	20	25	30
0	1.52	0.045	-4	-12	-26												
5	1.91	0.056	2	-5	-14	-31											
10	2.40	0.071	7	2	-5	-15	-34										
15	2.99	0.088	13	8	3	-4	-14										
20	3.71	0.110	18	15	10	5	-2	-27									
25	4.58	0.135	24	21	17	13	8	-7									
30	5.63	0.166	30	27	24	20	16	6	-12								
35	6.89	0.203	35	33	30	27	24	16	5	-16							
40	8.39	0.248	40	38	35	33	30	24	16	4	-18						
45	10.17	0.300	45	43	41	39	36	31	24	16	5	-18					
50	12.27	0.362	50	48	46	44	42	37	32	25	17	5	-17				
55	14.75	0.436	55	53	51	50	48	43	39	33	27	18	7	-15			
60	17.66	0.522	60	58	57	55	53	49	45	40	35	29	20	9	-11		
65	21.07	0.622	65	63	62	60	59	55	51	47	42	37	31	23	12		
70	25.03	0.739	70	69	67	66	64	61	57	53	49	45	39	33	26	-14	
75	29.63	0.875	75	74	72	71	69	66	63	59	56	52	47	42	36	14	
80	34.96	1.032	80	79	77	76	74	72	68	65	62	58	54	50	45	29	-9
85	41.10	1.214	85	84	82	81	80	77	74	71	68	64	61	57	53	39	18
90	48.15	1.422	90	89	87	86	85	82	79	76	73	70	67	63	60	48	33
95	56.24	1.661	95	94	93	91	90	87	85	82	79	76	73	70	66	56	44
100	65.47	1.933	100	99	98	96	95	93	90	87	85	82	79	76	73	64	53

Source: U.S. Weather Bureau, Relative Humidity and Dew Point Table, TA 454-0-3E, September 1965.

Table A-17 Variation of dewpoint with temperature and wet-bulb depression and of saturation vapor pressure over water with temperature on the Celsius scale

Pressure = 1013.2 millibars = 29.92 in = 101.32 kilopascals

Air temp., °C	Saturation vapor pressure		Wet-bulb depression, deg															
	Milli-bars	in Hg	0	1	2	3	4	5	6	7	8	9	10	11	12	13	14	15
-10	2.86	0.085	-11	-16	-24													
-8	3.35	0.099	-9	-13	-20	-33												
-6	3.91	0.115	-7	-11	-16	-24	-32											
-4	4.55	0.134	-5	-8	-12	-19	-22											
-2	5.28	0.156	-2	-5	-9	-14	-16											
0	6.11	0.180	0	-3	-6	-11	-12	-27										
2	7.05	0.208	2	-1	-3	-7	-8	-19	-33									
4	8.13	0.240	4	2	-1	-4	-5	-13	-21	-47								
6	9.35	0.276	6	4	2	-1	-2	-9	-14	-23								
8	10.72	0.317	8	6	4	1	1	-5	-9	-15	-26							
10	12.27	0.362	10	8	6	4	4	-2	-5	-10	-17	-29						
12	14.02	0.414	12	10	8	6	6	1	-2	-6	-11	-18	-34					
14	15.98	0.472	14	12	11	9	9	4	1	-2	-6	-11	-19					
16	18.17	0.532	16	14	13	11	11	7	4	1	-2	-6	-11					
18	20.63	0.609	18	16	15	13	14	9	7	4	2	-2	-6	-10				
20	23.37	0.690	20	19	17	15	16	12	10	7	5	2	-1	-4				
22	26.43	0.780	22	21	19	17	18	14	12	10	8	5	2	-1	-5			
24	29.83	0.881	24	23	21	20	20	16	15	13	11	8	6	3	-1	-5	-10	
26	33.61	0.992	26	25	23	22	22	19	18	15	13	11	9	6	4	0	-4	-9
28	37.80	1.116	28	27	25	24	24	21	19	18	16	14	12	10	7	4	1	-3
30	42.43	1.253	30	29	27	26	26	23	22	20	18	17	15	13	10	8	5	2
32	47.55	1.404	32	31	29	28	27	25	24	22	21	19	17	15	13	11	9	6
34	53.20	1.571	34	33	32	30	29	28	26	25	23	21	20	17	16	14	12	10
36	59.42	1.755	36	35	34	32	31	30	28	27	25	24	22	21	19	17	15	13
38	66.26	1.957	38	37	36	34	33	32	30	29	28	26	25	23	21	20	18	16
40	73.78	2.179	40	39	38	36	35	34	33	31	30	28	27	25	24	22	20	19

Source: U.S. National Weather Service, Marine Surface Observations, Weather Bur. Handb. 1, 1969.

Table A-18 Values of *n* for the Manning Formula [Eq. (4-7)]

Channel condition	*n*†
Plastic, glass, drawn tubing	0.009
Neat cement, smooth metal	0.010
Planed timber, asbestos pipe	0.011
Wrought iron, welded steel, canvas	0.012
Ordinary concrete, asphalted cast iron	0.013
Unplaned timber, vitrified clay, glazed brick	0.014
Cast-iron pipe, concrete pipe	0.015
Riveted steel, brick, dressed stone	0.016
Rubble masonry	0.017
Smooth earth	0.018
Firm gravel	0.020
Corrugated metal pipe and flumes	0.023
Natural channels:	
Clean, straight, full stage, no pools	0.029
As above with weeds and stones	0.035
Winding, pools and shallows, clean	0.039
As above at low stages	0.047
Winding, pools and shallows, weeds and stones	0.042
As above, shallow stages, large stones	0.052
Sluggish, weedy, with deep pools	0.065
Very weedy and sluggish	0.112

† Values quoted are averages of many determinations; variations of as much as 20 percent must be expected, especially in natural channels.

Table A-19 Values of k^t for various values of k and t [Eq. (8-7)]

t	k 0.80	0.82	0.84	0.86	0.88	0.90	0.92	0.94	0.96	0.98
1	0.800	0.820	0.840	0.860	0.880	0.900	0.920	0.940	0.960	0.980
2	0.640	0.672	0.706	0.740	0.774	0.810	0.846	0.884	0.922	0.960
3	0.512	0.551	0.593	0.636	0.681	0.729	0.779	0.831	0.885	0.941
4	0.410	0.452	0.498	0.547	0.600	0.656	0.716	0.781	0.849	0.922
5	0.328	0.371	0.418	0.470	0.528	0.590	0.659	0.734	0.815	0.904
6	0.262	0.304	0.351	0.405	0.464	0.531	0.606	0.690	0.783	0.886
7	0.210	0.249	0.295	0.348	0.409	0.478	0.558	0.648	0.751	0.868
8	0.168	0.204	0.248	0.299	0.360	0.430	0.513	0.610	0.721	0.851
9	0.134	0.168	0.208	0.257	0.316	0.387	0.472	0.573	0.693	0.834
10	0.107	0.137	0.175	0.221	0.279	0.349	0.434	0.539	0.665	0.817
11	0.086	0.113	0.147	0.190	0.245	0.314	0.400	0.506	0.638	0.801
12	0.069	0.092	0.123	0.164	0.216	0.282	0.368	0.476	0.613	0.785
13	0.055	0.076	0.104	0.141	0.190	0.254	0.338	0.447	0.588	0.769
14	0.044	0.062	0.087	0.121	0.167	0.229	0.311	0.421	0.565	0.754
15	0.035	0.051	0.073	0.104	0.147	0.206	0.286	0.395	0.542	0.739
16	0.028	0.042	0.061	0.090	0.129	0.185	0.263	0.372	0.520	0.724
17	0.023	0.034	0.052	0.077	0.114	0.167	0.242	0.349	0.500	0.709
18	0.018	0.028	0.043	0.066	0.100	0.150	0.223	0.328	0.480	0.695
19	0.014	0.023	0.036	0.057	0.088	0.135	0.205	0.309	0.460	0.681
20	0.012	0.019	0.031	0.049	0.078	0.122	0.189	0.290	0.442	0.668
21	0.009	0.015	0.026	0.042	0.068	0.109	0.174	0.273	0.424	0.654
22	0.007	0.013	0.022	0.036	0.060	0.098	0.160	0.256	0.407	0.641
23	0.006	0.010	0.018	0.031	0.053	0.089	0.147	0.241	0.391	0.628
24	0.005	0.009	0.015	0.027	0.047	0.080	0.135	0.227	0.375	0.616
25	0.004	0.007	0.013	0.023	0.041	0.072	0.124	0.213	0.360	0.603
26	0.003	0.006	0.011	0.020	0.036	0.065	0.114	0.200	0.346	0.591
27	0.002	0.005	0.009	0.017	0.032	0.058	0.105	0.188	0.332	0.579
28	0.002	0.004	0.008	0.015	0.028	0.052	0.097	0.177	0.319	0.568
29	0.002	0.003	0.006	0.013	0.025	0.047	0.089	0.166	0.306	0.557
30	0.001	0.003	0.005	0.011	0.022	0.042	0.082	0.156	0.294	0.545
40				0.002	0.006	0.015	0.036	0.084	0.195	0.446
50						0.005	0.015	0.045	0.130	0.364
60							0.007	0.024	0.086	0.298

NAME INDEX

Abbott, M. B., 309, 310
Abdo, M., 90
Ackermann, W. C., 232, 242, 261
Adams, F. R., 173
Addison, H., 132
Aitken, A. D., 355
Alekseev, G. A., 386
Allen, C. D., 172
Allen, C. M., 132
Alley, W. M., 310
Alter, J. C., 59–61
Alvarez, F., 93
Amein, M., 309
Ames, W. F., 309
Anderson, E. A., 43, 169, 261, 350, 354, 355
Anderson, E. R., 43, 169, 262
Anderson, H. W., 173
Anderson, L. J., 169
Anderson, N., 309
Anderson, R. L., 397, 409
Angus, D. E., 172
Appleman, H. S., 44
Arkell, R. E., 91
Armacost, L. V., 310
Armstrong, B. L., 355
Arnold, T. G., 170
Aubenok, L. I., 11, 12
Auciello, E. P., 91, 95

Badon Ghyben, W., 197, 202
Baker, D. R., 43, 92, 169, 170
Balls, B. W., 132
Barclay, P. A., 93
Barnes, A. H., 310
Barnes, B. S., 208, 232
Barron, E. G., 131
Bartholic, J. F., 171
Basco, D. R., 309
Bates, C. L., 410

Battan, L. J., 44, 93
Bayley, G. U., 410
Bear, J., 202
Beard, L. J., 170
Beaumont, R. T., 92
Beausoleil, R. W., 90
Bell, F. C., 386
Belt, G. H., 173
Benjamin, J. R., 409
Benson, M. A., 132, 133, 384, 385, 410, 411
Bergmann, J. M., 355
Berkofsky, L., 43
Bernard, M., 92
Bernoulli, J., 408
Bernstein, A. B., 45
Best, A. C., 44
Betson, R. P., 239, 240, 244, 261
Beyerlein, D. C., 439
Birot, P., 337
Bissell, V. C., 92
Biswas, A. K., 3, 5, 409
Biswas, K. R., 89
Björck, A., 309
Black, R. D., 261
Black, R. P., 408
Black, T. A., 171, 172
Blaney, H. F., 171, 172
Blench, T., 337
Bobee, B., 410
Bodhaine, C. L., 132
Bogardi, J., 337
Bolton, G. C., 337
Boltzman, L., 140
Bornstein, R. D., 45
Bosen, J. F., 43
Boughton, W. C., 354
Bouwer, H., 202
Bowen, I. S., 138, 139, 142, 144, 152, 161, 169
Box, G. E. P., 392*n.*, 393, 400, 408

SUBJECT INDEX

Abrupt waves, 289–290
Absolute (dynamic) viscosity, 183, 477, 478
Acre-foot, 116
Actinometers, 10–13
Adiabatic process, 26
Adjustment of data:
  precipitation, 70–71, 376–377
  streamflow, 117–120
Adsorption, 416–417
Advected energy:
  computation of, 138–140
  definition of, 138
  effect of, on evaporation, 138–140,
    144–146, 150–153
Advection, net, definition of, 145–146
Aeration, zone of, 157, 176
Aerodynamic determination:
  of evaporation, 140–142, 153, 155
  of evapotranspiration, 160–161
  of snowmelt, 251–253
Air density, 30–31
  in U.S. standard atmosphere, 480
Air temperature, 23–28
  Celsius vs. Fahrenheit, $9n$.
  definition of terms, 23–24
  distribution of: geographic, 27
    vertical, 25–26
  effect of, on evaporation, 135, 138–139,
    141–144, 150
  effect on: of cities, 27
    of topography, 27
  inversions, 26
  lapse rates, 25–26
  mean daily, 24
  measurement of, 24–25
  minimum, for growing plants, 156
  normals, 23–24
  in U.S. standard atmosphere, 480
  variation of: with elevation, 25–26, 350, 480
    with time, 27

Albedo, 8–9, 137, 140
Anabranches, 318–320
Analog computers, 194–195
Anchor ice, 112
Anemometers, 37
Anions, 413
Annual autoregressive stochastic model, 392
Annual floods, 359
Annual series, 359, 373–374
Antecedent precipitation index, 242–244
  table of k^t, 485
Anticyclones, 19
Applications of hydrology, 442–468
  channel improvements for flood mitigation,
    452
  cooling ponds, 461–464
  extension of streamflow records, 443–444
  flood regulation, 450–451
    project-design flood, 450
  floodplain mapping, 452–453
  highway culverts, 458
  river forecasting, 464–468
  spillways, 458–461
  urban storm drainage, 242, 453–458
    computer approach, 456–458
    noncomputer approach, 454–456
  water-supply reservoirs (see Reservoirs,
    water-supply)
Aquiclude, 182
Aquifers, 182–195, 198–199
  analysis of, 194–195
  artesian, 198–199
  artificial recharge of, 198
  storage constant for, 189, 191
Aquifuge, 182
Area-elevation data, 315
Area relations of catchments, 313
ARMA-Markov model, 393
Artesian aquifers, 198–199
Artesian water, 176

495

Conversion table (English to metric units)

	To convert from	Multiply by	To obtain metric/SI unit
Area	in²	645.2	mm²
	ft²	0.09290	m²
	mi²	2.590	km²
	acre	0.4047	hectare (ha) = 10⁴ m²
Energy (work or quantity of heat)	Btu (60°F)	1055	joule (J) = N · m
	cal (15°C)	4.186	J
	langley	4.186	J/cm²
	langley/min	697.6	W/m²
Flowrate	cfs	0.02832	m³/s = 10³ l/s
	mgd	0.04381	m³/s = 10³ l/s
	1000 gal/min	0.06309	m³/s = 10³ l/s
Force	lb	4.448	newton (N)
Length	in	25.40	mm
	ft	0.3048	m
	mi	1.609	km
	nmi	1.852	km
Power	ft · lb/s	1.356	W = J/s = N · m/s
	hp	745.7	W
Pressure	lb/in²	6895	N/m² = Pa
	lb/ft²	47.88	N/m²
	in Hg (32°F)	3.386	kPa = 10 mb
Specific heat	ft · lb/slug · °R	0.1672	J/kg · K
Specific weight	lb/ft³	157.1	N/m³
Velocity	ft/s	0.3048	m/s
	mi/hr	1.609	km/hr
Viscosity, dynamic	lb · s/ft²	47.88	N · s/m² = Pa · s
Viscosity, kinematic	ft²/s	0.09290	m²/s = 10⁴ stokes
Volume	ft³	0.02832	m³
	U.S. gal	3.785	l = 10⁻³ m³
Weight (see Force)			

Important quantities

	English unit	Metric/SI unit
Acceleration of gravity	32.17 ft/s²	9.806 m/s²
Density of water (39.4°F, 4°C)	1.940 slug/ft³	1000 kg/m³ = g/cm³
Sea-level pressure (standard atmosphere)	14.70 lb/in²	101.32 kN/m² (kPa)
	29.92 in Hg	760 mm Hg
	33.90 ft H₂O	1013.2 mb
Specific weight of water (59°F, 15°C)	62.4 lb/ft³	9800 N/m³, or 9.8 kN/m³
Stefan-Boltzmann constant	4.112×10^{-8} Btu/ft² · °R · day	11.71×10^{-8} cal/cm² · K · day